Y0-AAG-275

Macromolecular Symposia 210

Reactive Polymers 2003

Dresden, Germany
September 28–October 1, 2003

Symposium Editor:
H.-J. P. Adler, Dresden, Germany

pp. 1–521 · March 2004
ISBN 3-527-31043-6

WILEY-VCH

Macromolecular Symposia publishes lectures given at international symposia and is issued irregularly, with normally 14 volumes published per year. For each symposium volume, an Editor is appointed. The articles are peer-reviewed. The journal is produced by photo-offset lithography directly from the authors' typescripts.
Further information for authors can be found at http://www.ms-journal.de.
Suggestions or proposals for conferences or symposia to be covered in this series should also be sent to the Editorial office (E-mail: macro-symp@wiley-vch.de).

Macromolecular Symposia:
Annual subscription rates 2004
Macromolecular Full Package: including Macromolecular Chemistry & Physics (18 issues), Macromolecular Rapid Communications (24), Macromolecular Bioscience (12), Macromolecular Theory & Simulations (9), Macromolecular Materials and Engineering (12), Macromolecular Symposia (14):

Europe	Euro	6.424 / 7.067
Switzerland	Sfr	11.534 / 12.688
All other areas	US$	7.948 / 8.743

print only **or** electronic only / print **and** electronic

Postage and handling charges included. All Wiley-VCH prices are exclusive of VAT.
Prices are subject to change.

Single issues and back copies are available. Please ask for details at: service@wiley-vch.de.

Orders may be placed through your bookseller or directly at the publishers:
WILEY-VCH Verlag GmbH & Co. KGaA, P. O. Box 10 11 61, 69451 Weinheim, Germany, Tel. +49 (0) 62 01/6 06-400, Fax +49 (0) 62 01/60 61 84. E-mail: service@wiley-vch.de

For USA and Canada: Macromolecular Symposia (ISSN 1022-1360) is published with 14 volumes per year by WILEY-VCH Verlag GmbH Co. KGaA, Boschstr. 12, 69451 Weinheim, Germany. Air freight and mailing in the USA by Publications Expediting Inc., 200 Meacham Ave., Elmont, NY 11003, USA. Application to mail at Periodicals Postage rate is pending at Jamaica, NY 11431, USA. POSTMASTER please send address changes to: Macromolecular Symposia, c/o Wiley-VCH, III River Street, Hoboken, NJ 07030, USA.

Printing: Strauss Offsetdruck, Mörlenbach. Binding: J. Schäffer, Grünstadt

Macromolecular Symposia

Articles published on the web will appear several weeks before the print edition. They are available through:

www.ms-journal.de

www.interscience.wiley.com

Reactive Polymers 2003

Dresden (Germany), 2003

2. Reactive Polymers at Interfaces

Author Index

Preface

The 2nd International Symposium "React 2003", was held in Dresden from September 28 to October 1, 2003.
It was organised in the context of the Collaborative Research Centre 287 (SFB 287) "**Reactive Polymers in Inhomogeous Systems, in Melts and at Interfaces"** of the German Research Foundation (DFG) and the European Graduate College (DFG, EGK 720-1) "Advanced Polymeric Materials" (Dresden, Prague, Gliwice).
After the success of the 1st Symposium "React 2000", which was held from July 16th to 19th, 2000, (see ISBN 3-527-30326-X) the organisers, the Institute for Macromolecular Chemistry and Textile Chemistry, the Institute for Physical Chemistry and Electrochemistry of the Technische Universitaet Dresden, and the Institute of Polymer Research Dresden, again invited well-known scientists in this field from all over the world.
The topic of the symposium is also the name and topic of the SFB 287 which is focused on the development of new reactive materials for novel technologies. This interdisciplinary collaborative research centre combines ideally modern polymer material science with engineering aspects.
Lectures given covered a broad range of areas from the synthesis of new functionalised and reactive polymers, their characterisation, surface and material properties, as well as their application, to new morphologies and supramolecular structures providing again an excellent overview of actual research areas.
Contributions are divided into different sections:

Synthesis and structures of reactive polymers
Reactive polymers at interfaces
Reactive polymers for sensors and actuators

The first part contained lectures about synthesis of new branched polymers, polymer brushes, functional materials and gels, block copolymers as well as modification of polymers, preparation of special resins, use of small molecules for self-assembling processes and even modelling aspects.
A large part was focussed on interfacial phenomena such as surface patterning, adhesion, colloid stability, polyelectrolyte layers, biocompatibility of materials, and polymers in blends and composites, involving modern surface characterisation techniques.
New fields of future applications of reactive polymers in sensors and actuators are described in the third part. A main objective of the SFB is also to study and develop materials and devices based upon stimuli responding micro- and nanostructured hydrogels and thin functional polymer layers. Due to the co-operation of different faculties we were able to develop working micro- and nano-devices.
188 guests were at "React 2003" , 50 % foreign scientists from 22 countries including USA, China, India, Australia and Saudi Arabia together with participants from the whole of Europe.

Renowned as well as younger scientists presented 62 oral lectures and 124 posters and enjoyed vivid and interesting discussions as well as the beauty of Dresden. This significant increase in contributions compared to the "React 2000" shows the interest in and the increasing importance of this research area.
The symposium thus offered an excellent platform for information and discussion and provided the chance to initiate collaborations. The publication of this *Macromolecular Symposia* volume is meant to address those not having been able to participate but hopefully will become interested in the field, and most importantly to stimulate further progress in the field.
The 3rd International Symposium on "Reactive Polymers" will be held in 2006, again in Dresden.

Acknowledgements

We thank all those who contributed to the success of the meeting and to its organisation.
Our special thanks to those who submitted manuscripts for publication in this volume.
The organising committee of the 2nd International Symposium "React 2003" would like to thank the Deutsche Forschungsgemeinschaft (DFG), the Sächsisches Staatsministerium für Wissenschaft und Kunst (SMWK), the Deutscher Akademischer Austauschdienst (DAAD), the Technische Universität Dresden, the companies Parr Instrument Deutschland GmbH, DSM, Geleen and the Büchi Autoklaven & Reaktorsysteme GmbH, Eislingen, for financial support.

H.-J. P. Adler
K.-F. Arndt
B. Voit
T. Wolff
D. Kuckling
J. Hunger
Board of the SFB

Macromol. Symp. **2004**, *210*, 1-9

Monolithic Systems: From Separation Science to Heterogeneous Catalysis

*Said Lubbad, Betina Mayr, Monika Mayr, M. R. Buchmeiser**

Institute of Analytical Chemistry and Radiochemistry, University of Innsbruck, Innrain 52 a, 6020 Innsbruck, Austria

Summary: Recent results that have been obtained in the ring-opening metathesis polymerization (ROMP)-based synthesis of monolithic supports are summarized. We have elaborated a synthetic concept that allows modifying monolithic supports in a way that they can be used both for applications in separation science, for SEC and as supports for catalytically active systems. In all cases, a tailor-made microstructure was accessible due to the controlled character of the transition-metal catalyzed polymerization. Taking advantage of the "living" catalytic sites, an" *in situ*" functionalization was accomplished by subsequently grafting a variety of functional monomers and catalyst precursors onto the rod. Their design and use as supports for high-performance separation devices (e.g. for *ds*-DNA) and catalytic supports (*e. g.* supported Grubbs-type catalysts) is summarized.

Keywords: catalysis; flow reactors; metathesis; ROMP; supports

Introduction

Monolithic separation media evolved from the idea to produce a support with a high degree of continuity that should meet the requirements for fast, yet highly efficient separations.[1] Standard monolithic supports are usually prepared from poly(styrene-divinylbenzene) or poly(acrylate)s and have been used mainly in liquid chromatography including micro-separation techniques.[2] Starting in 1999, our group developed an entirely new concept for the manufacture of

 DOI: 10.1002/masy.200450601

functionalized monolithic supports. It entails the ring-opening metathesis polymerization (ROMP)-based synthesis for these types of materials[3-12] and has lately been extended to the use of these supports in heterogeneous catalysis.[11, 13]

Results and Discussion

Basics and Concepts

Generally speaking, the term "monolith" applies to any single-body structure containing interconnected repeating cells or channels. In this contribution, the term "monolith" shall comprise crosslinked, organic materials which are characterized by a defined porosity and which support interactions/reactions between this solid and the surrounding liquid phase. Besides advantages such as lower backpressure and enhanced mass transfer[14, 15], the ease of fabrication as well as the many possibilities in structural alteration need to be mentioned.

Until now, a considerable variety of functionalized and non-functionalized monolithic materials based on either organic or inorganic polymers are available. Organic monoliths have mostly been prepared from methacrylates or poly(styrene-*co*-divinylbenzene)[2, 16-19] applying almost exclusively free radical polymerization.[20] Despite the comparably poor control over polymerization kinetics in free radical polymerization-based systems, the porosity and microstructure of monolithic materials has successfully been varied.[2] Due to the broad applicability of ROMP and the good definition of the resulting materials, we investigated to which extent this transition metal-catalyzed polymerization could be used for the synthesis of monolithic polymers.[4] We found that this may be accomplished by generating a continuous matrix by ring-opening metathesis copolymerization of suitable monomers with a crosslinker in the presence of porogenic solvents within a device (column).

Manufacture of Metathesis-Based Monolithic Supports

The choice of the suitable initiator represents an important step in creating a well-defined polymerization system in terms of initiation efficiency and control over propagation. Only in the case where a quantitative and fast initiation occurs, the entire system can be designed on a *stoichiometric base*. This is of enormous importance, since for control of microstructure, the composition of the entire polymerization mixture needs to be varied within very small increments, being sometimes less than 1%. In terms of monomers, the copolymerization of norborn-2-ene (NBE) with 1,4,4a,5,8,8a-hexahydro-1,4,5,8-*exo-endo*-dimethanonaphthalene (DMN-H6) or tris(norborn-2-en-5-ylmethylenoxy)methylsilane ($NBE-CH_2O)_3SiCH_3$) in the presence of two porogenic solvents, e. g. 2-propanol and toluene worked best (Scheme 1). In all cases, the less oxygen-sensitive ruthenium-based Grubbs-type initiator $RuCl_2(=CHPh)(PCy_3)_2$ was used as catalyst. In contrast to the highly reactive second-generation Grubbs-type catalysts $RuCl_2(=CHPh)(NHC)(PCy_3)$ (NHC=N-heterocyclic carbene), it possesses a balanced reactivity that avoids highly exothermic reactions.

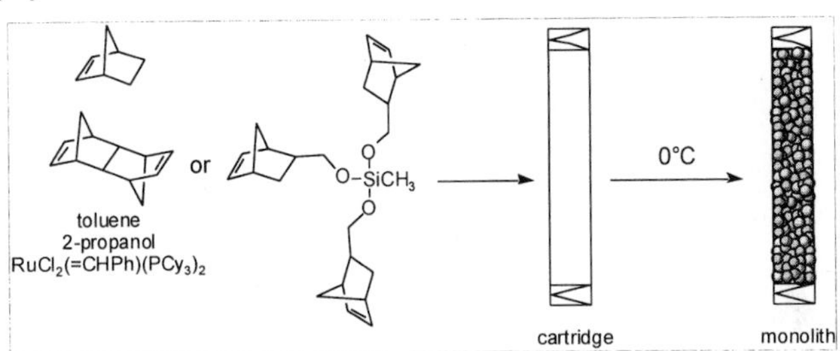

Scheme 1. Synthesis of monolithic supports.

Microstructure of Metathesis-Based Rigid Rods

In order to understand monolithic supports and the effects of polymerization parameters, a brief description of the general construction of a monolith in terms of microstructure, backbone and relevant abbreviations is given in Figure 1.[3, 4] As can be deduced therefrom, monoliths consist of interconnected microstructure-forming microglobules, which are characterized by a certain diameter (d_p) and microporosity (ε_p). In addition, the monolith is characterized by an inter-microglobule void volume (ε_z), which is mainly responsible for the backpressure at a certain flow rate. The volume fractions of both the micropores (ε_p) and voids (intermicroglobule porosity, ε_z) represent the total porosity (ε_t). This value indicates a percentage of pores in the monolith. Together with the pore size distribution, which can be calculated from inverse size exclusion chromatography (ISEC)[21] or mercury intrusion data,[22] it directly translates into a total pore volume, V_p, usually expressed in mL/g. Furthermore, it allows calculation of the specific surface area σ, expressed in m^2/g. For the design of monolithic supports for different tasks, the influence of all components of the polymerization mixture (NBE, DMN-H6 or $NBE-CH_2O)_3SiCH_3$, solvents, free phosphine and initiator as well as temperature) on microstructure formation was investigated. The relative ratios of all components, i. e. NBE, DMN-H6, porogens and catalyst, allowed broad variations in the microstructure of the monolithic material including structures ideal for heterogeneous catalysis.

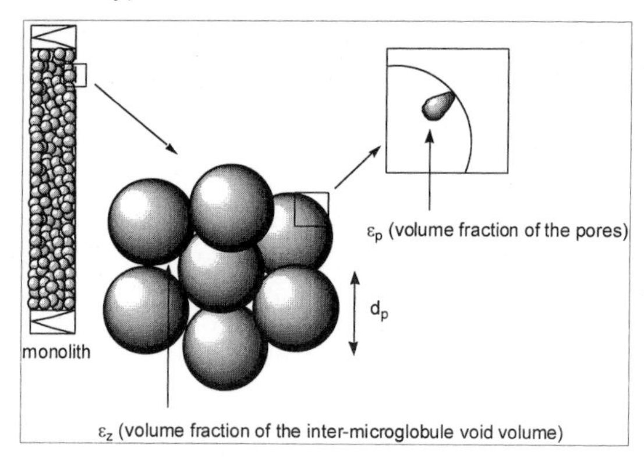

Figure 1. Physical meaning of relevant parameters of monoliths.

In summary, the volume fraction of the interglobular void volume (ε_z) and total porosity (ε_t) were varied within a range of 0 – 50 % and 50 – 80 %, respectively. Figure 2 illustrates some of the microstructures that were generated.

Figure 2. Microstructures of monoliths.

Functionalization, Metal Removal and Metal Content

Using the ROMP-based protocol, the living[23, 24] ruthenium-sites could be used for derivatization after rod-formation was complete. Grafting experiments and ICP-OES-based investigations revealed that more than 98 % of the initial amount of initiator were located at the microglobule surface after microstructure formation.[11] Using the initiator covalently bound to the surface, functional monomers were grafted onto the monolith surface by simply passing solutions thereof through the mold (Scheme 2).[3, 4, 25]

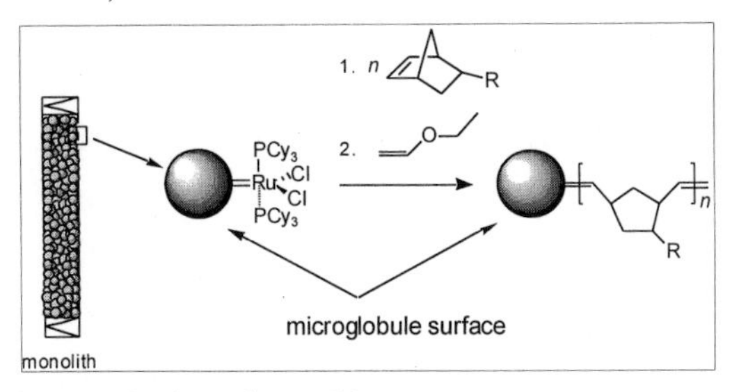

Scheme 2. Surface-functionalization of monoliths.

Since no cross-linking can take place, tentacle-like polymer chains attached to the surface were formed. In addition, microglobules were designed in a way that their pore size was < 1.2 nm, which basically restricted functionalization to their surface.[7] The degree of this graft polymerization of functional monomers varies within almost two orders of magnitude, depending on their ROMP activity. Important enough, the structure of the parent monolith was not affected by the functional monomer and could be optimized regardless of the functional monomer used later. Since the initiator was almost quantitatively located at the surface of the microglobules, the efficiency of *metal removal* from the monolith after polymerization was high. ICP-OES investigations revealed that the remaining ruthenium-concentrations after capping with ethyl vinyl ether (EVE) were below 10 μg/g, corresponding to a metal removal of more than 99.8 %.

Applications of Functionalized Metathesis-Based Monoliths in Catalysis

Grafted Supports for Ring-Closing Metathesis (RCM) and Related Reactions

In heterogeneous catalysis, one wants to combine the general advantages of homogeneous systems such as high definition, activity, etc. with the advantages of heterogeneous catalysis such as increased stability, ease of separation, and recycling. The first successful use of metathesis-based monolithic media for heterogeneous catalysis was accomplished by using these supports as carriers for Grubbs-type initiators based on N-heterocyclic carbenes (NHC-ligands).[26] For this purpose, monoliths with a suitable microporosity (40 %) and microglobule diameter (1.5 ± 0.5 μm) were synthesized. Consecutive „in-situ" derivatization was successfully accomplished using a mixture of NBE and a polymerizable NHC-precursor (Scheme 3).[11, 27-29]

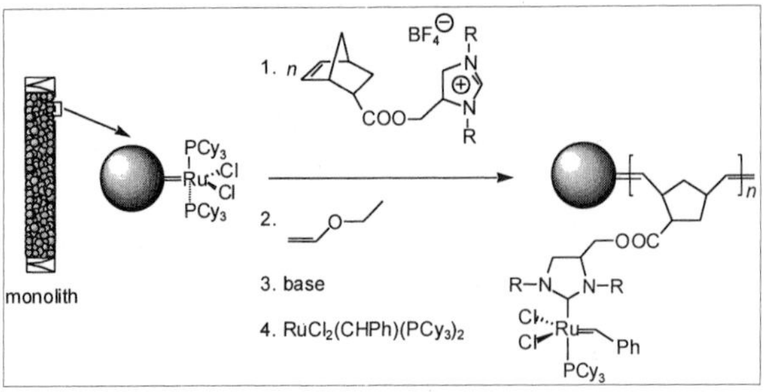

Scheme 3. Immobilization of a second-generation Grubbs catalyst on a monolithic support.

The use of NBE drastically enhanced grafting yields for the functional monomer. Using this setup, tentacles of copolymer with a degree of oligomerization of the functional monomer of 2 – 5 were generated. The free NHC necessary for catalyst formation was simply generated using a strong base such as 4-dimethylaminopyridine (DMAP). In a last step, excess base was removed by extensive washing and the catalyst was immobilized/formed by passing a solution of $Cl_2Ru(CHPh)(PCy_3)_2$ over the rigid rod. Loadings of up to 1.4 % of Grubbs-catalyst on NHC base were achieved. Monolith-immobilized metathesis catalysts prepared by this approach showed high activity in various metathesis-based reactions such as ROMP and RCM. In a benchmark reaction with diethyl diallylmalonate (DEDAM), these properties directly translated into high average turn-over frequences (TOFs) of up to 0.5 s^{-1}.

In an alternative approach, monolith-supported second generation Grubbs catalysts containing unsaturated (e.g. IMes) or saturated (e. g. SIMes) NHCs[30] can be prepared by a synthetic protocol summarized in Scheme 4. Surface-derivatization of a monolith was carried out with 7-oxanorborn-5-ene-2,3-dicarboxylic anhydride followed by conversion of the grafted poly(anhydride) into the corresponding poly-silver salt. This silver salt was used for the halogen exchange with a broad variety of second generation Grubbs catalysts, leading to the catalytic species shown in Scheme 4. In the benchmark reaction with DEDAM, TONs up to 830 were achieved.[31, 32] All monolith-based catalytic systems summarized here were successfully used as pressure stable catalytic reactors.

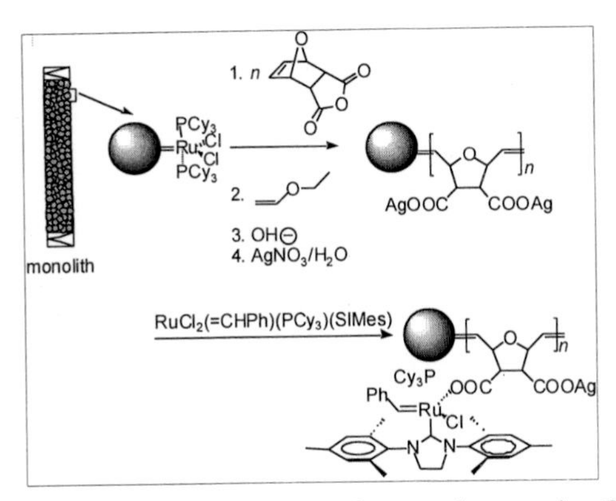

Scheme 4. Immobilization of a second-generation Grubbs-type catalyst via Cl-exchange.

Bleeding was virtually suppressed, leading even in RCM to basically ruthenium-free products with a ruthenium-content far below 0.1 %.

Monolithic Supports for Separation Science

Due to the pure hydrocarbon backbone, monoliths prepared from NBE and DMN-H6 are strongly hydrophobic. Nevertheless, the resulting materials significantly differ from PS-DVB based resins, in that the latter one contains aromatic systems that are capable of forming π-stacks with analytes possessing aromatic groups.

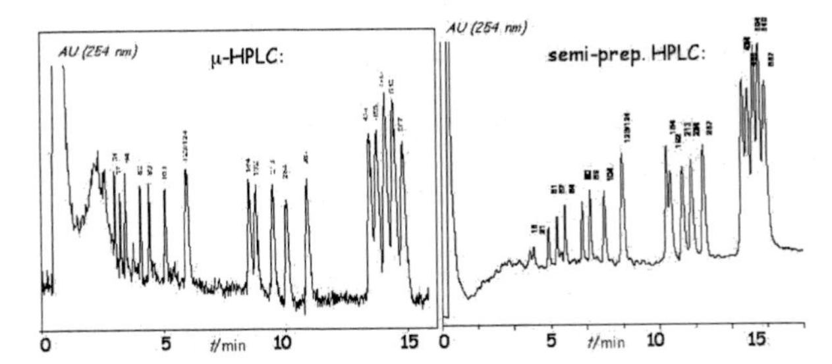

Figure 3. Separation of *ds*-DNA on monolithic supports. A) μ-HPLC, B) semi-preparative HPLC.

The impressive separation capabilities have been demonstrated by the fast separation of biologically relevant compounds such as proteins, double stranded (ds) DNA, oligonucleotides as well as phosphorothioate oligodeoxynucleotides.[3-5, 33] As an example, the separation of 20 base pairs of *ds*-DNA was accomplished on both standard (i. e. 3x100 mm) and microanalytical (200 mm i. d.) monolithic columns (Figure 3).[6, 34]

Summary

Metathesis-based polymerization techniques have certainly found their place in materials science. This has been made possible by adding well-defined and tolerant initiators to the armor of existing polymerization systems. With these initiators, in particular ROMP has had an enormous impact on the development of both surface-modified and polymeric materials. Applications in catalysis and separation science have been added to the more "traditional" ones in optics and electronics. The ongoing developments in organometallic chemistry, polymer chemistry, and in particular in metathesis polymerization will certainly result in the permanent improvement of existing systems and techniques as well as in new applications in many areas of chemistry and materials science.

Acknowledgement

Our work was supported by the *Austrian Science Fund* (START Y-158).

[1] N. B. Afeyan, S. P. Fulton, F. E. Regnier, *J. Chromatogr.* **1991**, *544*, 267.
[2] E. C. Peters, F. Svec, J. M. J. Fréchet, *Adv. Mater.* **1999**, *11*, 1169.
[3] F. Sinner, M. R. Buchmeiser, *Macromolecules* **2000**, *33*, 5777.
[4] F. Sinner, M. R. Buchmeiser, *Angew. Chem.* **2000**, *112*, 1491.
[5] B. Mayr, R. Tessadri, E. Post, M. R. Buchmeiser, *Anal. Chem.* **2001**, *73*, 4071.
[6] S. Lubbad, B. Mayr, C. G. Huber, M. R. Buchmeiser, *J. Chromatogr. A* **2002**, *959*, 121.
[7] S. Lubbad, M. R. Buchmeiser, *Macromol. Rapid Commun.* **2002**, *23*, 617.
[8] M. R. Buchmeiser, *Macromol. Rapid. Commun.* **2001**, *22*, 1081.
[9] M. R. Buchmeiser, *J. Molec. Catal. A: Chemical* **2002**, *190*, 145.
[10] F. Svec, Z. Deyl, Elsevier, Amsterdam, **2002**.
[11] M. Mayr, B. Mayr, M. R. Buchmeiser, *Angew. Chem.* **2001**, *113*, 3957.
[12] M. R. Buchmeiser, F. Sinner, in *European Patent*, Buchmeiser, M., 409 095 (A 960/99, 310599), PCT/EP00/04 768, WO 00/73782 A1, EP 1 190244 B1.
[13] M. R. Buchmeiser, S. Lubbad, M. Mayr, K. Wurst, *Inorg. Chim. Acta* **2003**, *345*, 145.
[14] A. E. Rodrigues, *J. Chromatogr. B* **1997**, *699*, 47.
[15] Y. Xu, A. I. Liapis, *J. Chromatogr. A* **1996**, *724*, 13.
[16] D. Sykora, F. Svec, J. M. J. Fréchet, *J. Chromatogr. A* **1999**, *852*, 297.
[17] C. Viklund, F. Svec, J. M. J. Fréchet, K. Irgum, *Chem. Mater.* **1996**, *8*, 744.
[18] C. Viklund, E. Pontén, B. Glad, K. Irgum, P. Hörstedt, F. Svec, *Chem. Mater.* **1997**, *9*, 463.

[19] Q. C. Wang, F. Svec, J. M. J. Fréchet, *Anal. Chem.* **1993**, *65*, 2243.
[20] M. R. Buchmeiser, in *Monolithic Materials: Preparation, Properties and Applications (J. Chromatogr. Library), Vol. 67* (Eds.: F. Scvec, T. B. Tennikova, Z. Deyl), Elsevier, Amsterdam, **2003**.
[21] I. Halász, K. Martin, *Angew. Chem.* **1978**, *90*, 954.
[22] C. A. Leon y Leon, M. A. Thomas, *GIT Lab. J.* **1997**, *2*, 101.
[23] M. Szwarc, *Makromol. Chem. Rapid Commun.* **1992**, *13*, 141.
[24] K. Matyjaszewski, *Macromolecules* **1993**, *26*, 1787.
[25] S. Lubbad, M. R. Buchmeiser, *Macromol. Rapid Commun.* **2003**, *24*, 580.
[26] M. Scholl, S. Ding, C. W. Lee, R. H. Grubbs, *Org. Lett.* **1999**, *1*, 953.
[27] M. R. Buchmeiser, M. Mayr, B. Mayr, in *Austrian Pat. Appl.*, Buchmeiser, M. R., AT, **2001**.
[28] M. R. Buchmeiser, *Bioorg. Med. Chem. Lett.* **2002**, *12*, 1837.
[29] M. Mayr, B. Mayr, M. R. Buchmeiser, *Designed Monomers and Polymers* **2002**, *5*, 325.
[30] M. R. Buchmeiser, *Chem. Rev.* **2000**, *100*, 1565.
[31] J. O. Krause, S. Lubbad, M. Mayr, O. Nuyken, M. R. Buchmeiser, *Polym. Prepr. (Am. Chem. Soc., Div. Polym. Chem.)* **2003**, *44*, 790.
[32] J. O. Krause, S. Lubbad, O. Nuyken, M. R. Buchmeiser, *Adv. Synth. Catal.* **2003**, *345*, 996.
[33] B. Mayr, M. R. Buchmeiser, *J. Chromatogr. A* **2001**, *907*, 73.
[34] B. Mayr, G. Hölzl, K. Eder, M. R. Buchmeiser, C. G. Huber, *Anal. Chem.* **2002**, *74*, 6080.

Positron Annihilation: A Unique Method for Studying Polymers

Günter Dlubek,[1] Duncan Kilburn,[2] Vladimir Bondarenko,[3] Jürgen Pionteck,[4] Reinhard Krause-Rehberg,[3] M. Ashraf Alam[2]*

[1]ITA Institut für Innovative Technologien, Köthen, Aussenstelle Halle, Wiesenring 4, D-06120 Lieskau (bei Halle/S.), Germany
E-mail: gdlubek@aol.com
[2]H.H.Wills Physics Lab., University of Bristol, Tyndall Avenue, Bristol, BS8 1TL, UK
[3]Martin-Luther-Universität Halle-Wittenberg, FB Physik, D-06099 Halle/S, Germany
[4]Institut für Polymerforschung e.V., Hohe Strasse 6, D-01069 Dresden, Germany

Summary: Positron annihilation is a unique technique for studying the local free volume of polymers. Employing the positron annihilation lifetime spectroscopy (PALS) the size and size distribution of subnanometer size holes which constitute the excess free volume may be studied. In combination with macroscopic volume data the fractional free volume and the number density of holes may be estimated. After presenting the principles of the method, some examples typically for the investigation of the free volume in polymers will be given. Moreover, the study of interdiffusion in demixed polymer blends and further applications are shortly reviewed

Keywords: free volume; glass transition; interdiffusion; microstructure

Introduction

The free volume of polymers is generally defined by $V_f = V - V_{occ}$ where V is the total and V_{occ} the occupied volume. When identifying V_{occ} with the van der Waals volume V_W, $V_{occ} = V_W$, the value of V_f represents the total free volume which is known also in crystals and termed there as interstitial free volume. In amorphous polymers an additional or excess free volume appears due to the (static or dynamic) structural disorder. This excess free volume appears as many small holes and can be calculated assuming $V_{occ} = V_c$, $V_f = V - V_c$, where V_c is the specific volume of the corresponding crystal. Many properties of polymers are related to this type of free (excess or hole) volume, including the diffusion of small molecules through glassy polymers, the sorption of humidity and the dynamics of rubbery polymers. Modern theories of glass transition are more complex than the original free volume concepts but show also the close relation between mobility (dynamic heterogeneity) and free volume (structural heterogeneity).[1] Despite a great deal of interest in the investigation of free volume, there is still only limited information available about its structures: the hole dimensions, number

 DOI: 10.1002/masy.200450602

densities and the size and shape distribution.

A unique tool to probe such holes is the positron annihilation lifetime spectroscopy (PALS).[2,3] In molecular substances a significant fraction of the injected positrons annihilates from the positronium (Ps) bound state. The Ps forms either in the para (p-Ps) or ortho (o-Ps) states with a relative abundance of 1:3. In vacuum, p-Ps has a lifetime of 125 ps and annihilates via 2 γ-photons while o-Ps lives 142 ns and annihilates via 3 γ-photons. In amorphous polymers Ps is trapped by the holes of the excess free volume (Anderson localisation). When within the holes, the o-Ps has a finite probability of annihilating with an electron other than its bound partner (and of opposite spin) during the numerous collisions that it undergoes with the molecules of the surrounding material, a process generally known as the 'pick-off'. The result is a sharply reduced o-Ps lifetime depending on the frequency of collisions. The collision frequency of the Ps with the surrounding molecules will depend on the dimensions of the confining volume. This results in a highly sensitive correspondence of the o-Ps pick-off rate, and therefore the lifetime, to the free volume hole size[2-4]. In this paper we present a short introduction in the positron annihilation method and show typical examples for its applications for studying polymers.

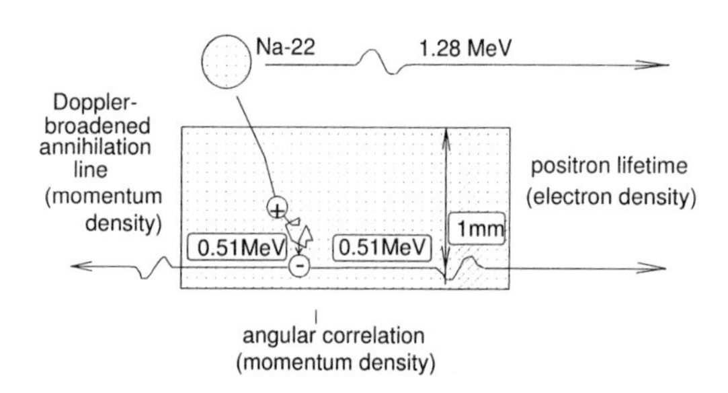

Figure 1. The positron experiments: Positron lifetime, 2γ-angular correlation and γ-spectroscopy.

The Positron Annihilation Techniques

Positrons are supplied by radioactive sources such as ^{22}Na with high kinetic energies (Figure 1).[2,3] If implanted into a solid the fast positrons slow down within a few ps due to ionisation and excitation of molecules. The implantation profile of positrons is an exponentially decreasing function exp($-\alpha x$) where α = 42 cm^{-1} is the positron absorption coefficient (in materials of density 1 g/cm^3) and x is the depth where the positron is stopped. 50, 90 and 99 %

of implanted positrons are stopped at depths of 0.17, 0.55 and 1.1 mm. Thin polymer foils may be stacked to obtain the required thickness of sample. The free (not Ps) positrons annihilate via emission of two (almost co-linear) γ-photons of 0.51 MeV energy. Simultaneously with the emission of the positron the ^{22}Na nucleus emits a 1.3 MeV photon. The time delay between the 1.3 MeV (start γ, positron birth) and the 0. 51 MeV photons (stop γ, positron annihilation), i. e. the lifetime of a positron, can be measured with a positron annihilation lifetime spectrometer (PALS) using two fast scintillation detectors. The momentum distribution of e^+-e^- pairs can be determined either by measuring the Doppler-broadening of the 0.51 MeV annihilation radiation (DBAR) using a Ge detector, or the 2γ angular correlation of annihilation radiation (ACAR, Figure 1).[2-4]

While the lifetime of an individual positron may vary between 0 and ∞, the lifetime spectrum of an ensemble of positrons annihilating from a solitary state is a single exponential $\exp(-t/\tau)$ where τ denotes the characteristic (mean) lifetime of positrons. As shown in Figure 2 for a CR39,[5] typically three lifetime components appear in amorphous polymers:

$$s(t) = \sum(I_i/\tau_i)\exp(-t/\tau_i), \qquad \sum I_i = 1, \qquad (i = 1...3). \tag{1}$$

These lifetimes arise from annihilation of p-Ps (τ_1 = 125 - 200 ps), free positrons (τ_2 = 300 - 400 ps) and o-Ps pick-off (τ_3 = 1 - 5 ns). Only the third component (o-Ps) responds clearly to polymeric material properties: For CR39: τ_3 = 1640 ps (c = 0, T_g = 122 °C) and 2145 ps (content of comonomer c = 75 wt-%, $T_g \approx 0$ °C, Figure 2[5]). Typical specimens in PALS experiments are platelets of 8x8 mm^2 in area and 1.5 mm in thickness. For each experiment, two identical samples are sandwiched around a 5 MBq positron source (^{22}Na), prepared by evaporating carrier-free $^{22}NaCl$ solution on a Kapton foil of 8 μm thickness. One experiment lasts between 1 and 10 hours depending on the total counts accumulated.

Following convolution of $s(t)$ with the experimental resolution function and subtraction of background and source components the spectra can be analysed by a non-linear least squares fit of Equation (1) to the data points N_i. A continuous lifetime distribution may also be analysed using the Laplace-inversion routine CONTIN-PALS2 or the maximum entropy routine MELT. A relatively new analysis routine LT9.0 assumes a log-normal distribution of annihilation rates $\alpha(\lambda)$, $\lambda = 1/\tau$, for some, if not all, annihilation channels and calculates the mass centre and width of this distribution.[3,6]

The size of the free volume holes in which o-Ps annihilates can be estimated using a simple quantum mechanical model which assumes the Ps to be confined in a spherical potential well of radius $r = r_h + \delta r$ and an infinite depth where r_h is the radius of the hole (see ref.[3,4] and references given therein). The Ps has a spatial overlap with molecules within a layer δr of the potential wall.

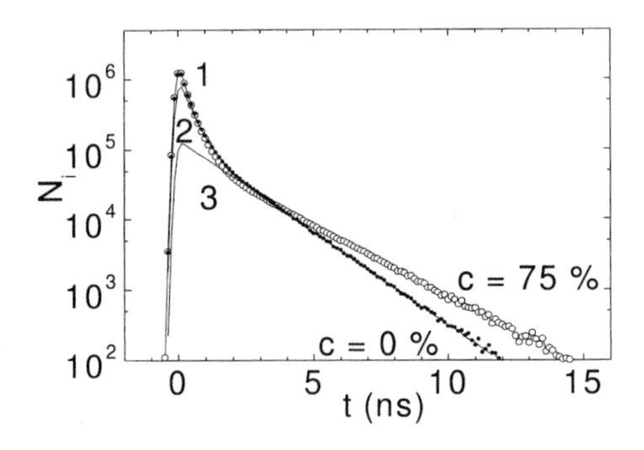

Figure 2. Positron lifetime spectra in CR39-copolymers with 0 and 75 % of comonomer.[6]

This provides a simple but very useful relationship between the o-Ps pick-off lifetime and the hole radius r_h:

$$\tau_{po} = 0.5ns\left[1 - \frac{r_h}{r_h + \delta r} + \frac{1}{2\pi}\sin\left(\frac{2\pi r_h}{r_h + \delta r}\right)\right]^{-1} \tag{2}$$

where 0.5 ns is the inverse of the spin averaged Ps annihilation rate in dense electron systems and δr is empirically derived to be 0.166 nm.[3,4] Similar formulas were derived for cylindrical or cubical holes.[3] Typically o-Ps lifetimes and holes sizes are $\tau_{po} = \tau_3 = 1$ - 5 ns and $r_h = 2$ - 4 Å ($v_h = 4\pi r_h^3/3 = 50$ - 200 Å^3). In polymer crystals Ps is sometimes formed in interstitial free volumes. In this case, the o-Ps pick-off lifetime reflects the packing coefficient C of the crystals ($C \approx 0.70$, $\tau_3 \approx 1$ ns). The Ps yield in crystals is usually lower than in the amorphous phase or can also be completely absent. Based on Equation (2) the mean size of the free-volume holes may be estimated as a function of temperature, pressure, the content of plasticizer, humidity, or the composition of copolymers and blends etc. The lower detection limit of the method is estimated to be $r_h \approx 1.5$ Å ($v_h \approx 20$ Å^3), while in mesoscopic systems τ_3 values of 100 ns corresponding to $r_h \approx 10$ nm have been observed.[3]

Application of PALS as a probe of polymers

Characterisation of the free volume

PALS is able to deliver the mean size and size distribution of free volume holes.[3-5] As an example, we show in Figure 4 the mean hole volume, v_h, calculated from the o-Ps lifetime τ_3, for atactic polystyrene (PS) provided by the BASF AG, Germany (M_n = 175 kg/mol, M_n/M_w = 2.25, T_g = 104 °C). Together with $v_h(\tau_3)$, the boundaries $v_h(\tau_3 - \sigma_3)$ and $v_h(\tau_3 + \sigma_3)$ are shown

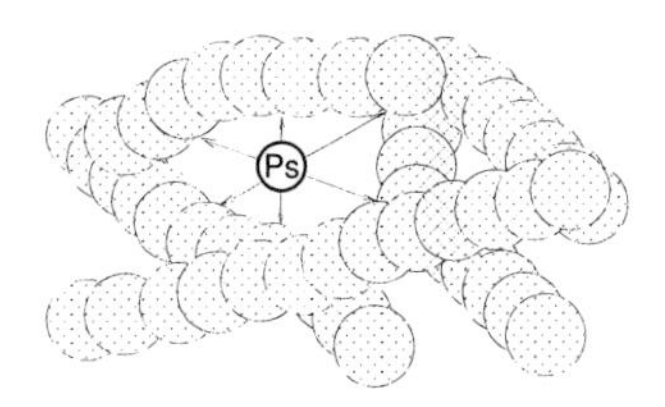

Figure 3. Ps localisation and annihilation in a hole of the (excess) free volume.

where σ_3 is the dispersion (standard deviation) of the lifetime distribution. The mean volume v_h shows a strong increase in its expansivity at T_g. At low temperatures, o-Ps is trapped in local free volumes within the glassy matrix and τ_3, and hence v_h, show the mean size of static holes. The averaging occurs over the hole sizes and shapes. The slight increase of v_h with temperature mirrors the thermal expansion of free volume in the glass due to the anharmonicity of molecular vibrations and local motions in the vicinity of the holes. In the rubbery phase the molecular and segmental motions increase rapidly resulting in a steep rise in the hole size with temperature. Now v_h represents an average value of the local free volumes whose size and shape fluctuate in space and time. The o-Ps lifetime mirrors the mean geometrical hole size as long as the structural relaxation times do not reach the order of magnitude of this lifetime. This is probably the case above a temperature which we have denoted by T_k in Figure 4. The coefficients of thermal expansion of hole volume were estimated to be $\alpha_{hg} = (1.95 \pm 0.2)\times10^{-3}$ K^{-1} ($T < T_g$) and $\alpha_{hr} = (9.5 \pm 0.08) \times10^{-3}$ K^{-1} ($T > T_g$), the hole volume at T_g is $v_{hg} = (121 \pm 2)$ Å^3. The α_{hr} value is larger by a factor of ~15 than of the macroscopic coefficient of thermal expansion, α_r. The number density and volume fraction of holes can not be obtained from PALS alone, however, from a comparison of PALS with the macroscopic volume obtained from pressure-volume-temperature (PVT) experiments, in particular when these are analysed with the Simha-Somcynsky equation of state (S-S eos) (see ref.[7] and references given therein). This theory assumes a lattice with an occupation y of less than 1. The hole fraction, $h = 1 - y = h(T/T^*, V/V^*, P/P^*)$, ($T^*$, V^* and P^* are scaling parameters) is calculated from an equation obtained from the pressure relation $P = -(\partial F/\partial V)_T$, where $F = F(V, T, h)$ is the (Helmholtz) configurational free energy. h can be identified with the fractional (excess) free volume defined by $f = V_f/V \equiv h$ where V and V_f are the specific total and free volumes.

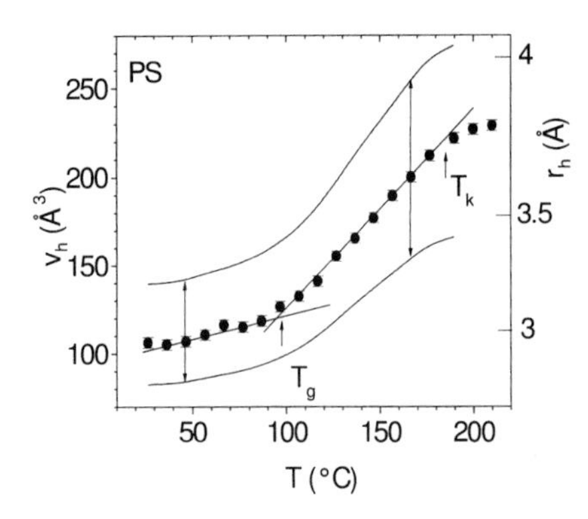

Figure 4. The mean volume $v_h(\tau_3)$ of excess free volume holes and its dispersion ($\pm\sigma$) of PS as a function of temperature T. Above T_k the o-Ps lifetime τ_3 and thus v_h does not anymore represent correctly the true hole volume.

We have determined the specific volume from PVT experiments using a fully automated GNOMIX high-pressure (mercury) dilatometer. Figure 5 shows the specific volume, V, and the specific occupied volume, $V_{occ} = (1 - h)V$, for PS together with the specific free volume, $V_f = hV = V - V_{occ}$ as example. V_{occ} includes the interstitial free volume, V_{if}, $V_{occ} = V_W + V_{fi}$, and has at T_g a value of $V_{occ} = 1.45V_W$ which corresponds well to a typical specific volume for crystalline polymers. The coefficient of thermal expansion of the occupied volume, changes at T_g from $\alpha_{occ,g} = 1.0\times10^{-4}$ K^{-1} ($T < T_g$) to $\alpha_{occ,r} = 0.2\times10^{-4}$ K^{-1} ($T > T_g$). This change seems to be unexpected, but is confirmed by the PALS data (Figure 6, behaviour of V below T_g). As can be observed in Figure 5, the specific volume shows an increase in its coefficient of thermal expansion at T_g from $\alpha_g = 2.25\times10^{-4}$ K^{-1} to $\alpha_r = 6.36\times10^{-4}$ K^{-1}. The corresponding increase for the free volume is from $\alpha_{fg} = 1.8\times10^{-3}$ K^{-1} to $\alpha_{fr} = 8.4\times10^{-3}$ K^{-1}. The fractional free volume at T_g amounts to $f \equiv h = 0.070$.

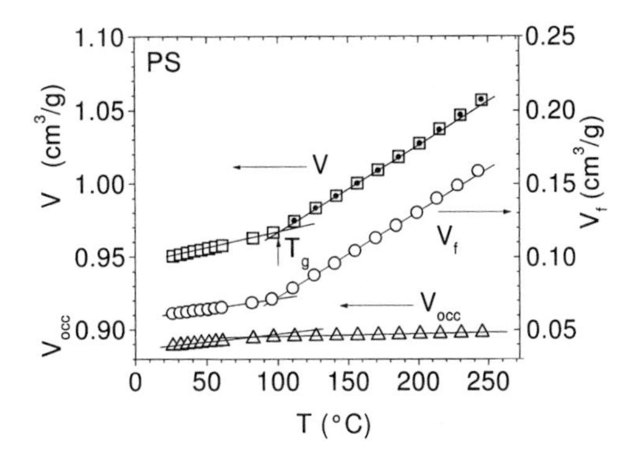

Figure 5. The specific total, V, free V_f, and occupied, V_{occ}, volume of polystyrene (PS) as a function of the temperature T at ambient pressure. Open symbols: experimental data, dots: Simha-Somcynsky equation of state fits in the temperature range $T > T_g$.

An accurate estimate of the mean number of holes per mass unit, N_h', may be obtained from one of the relations[6,8]

$$V_f = N_h' v_h \qquad (3) \qquad\qquad V = V_{occ} + N_h' v_h. \qquad (4)$$

Figure 6 shows the specific total volume, V, and specific free volume, $V_f = hV$, which were plotted vs. the hole volume v_h. One observes that both V and V_f are linear functions of v_h for data points above T_g with slopes of $N_h' = 0.55$ and $0.53(\pm 0.02)\times 10^{21}$ g^{-1}. The values correspond to a hole volume density of $N_h = 0.55$ nm^{-3}. There is almost complete agreement between the N_h' values estimated from both Equations (3) and (4).The constancy of the hole density, N_h', is a surprising and interesting result also confirmed by all other work in the literature[9] related to this question.

The intersection of the V_f vs. v_h lines with the abscissa does not show systematic deviations from zero and varies between -0.005 and 0.008 (±0.005). This shows that the (mean) o-Ps lifetime $\tau_{po} = \tau_3$ mirrors the true mean of the hole size distribution well although eq. (2) is based on a simple quantum mechanical model assuming spherical holes. From comparison of computer simulations of the hole size distribution with PALS results it has been concluded that o-Ps underestimates the hole size since non-spherical holes are detected as too small holes.[5] A possible preference to larger holes may, however, compensate this effect.[3] In the past we have employed PALS for the study of a variety of questions related to the free volume in polymers such as (i) the temperature dependence of the free volume in polyethylene and

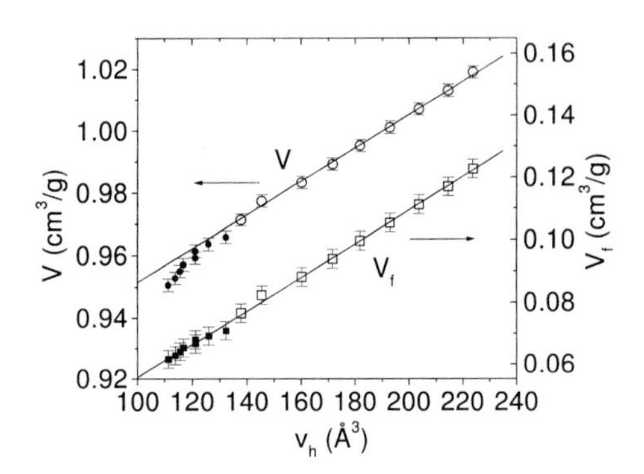

Figure 6. Plots of V and $V_f = hV$, vs. the mean hole volume v_h for PS. The lines are linear fits to the data above T_g and below T_k (open symbols). The filled symbols are data from below T_g.

polytetrafluoroethylene,[10] plasticization and antiplasticization caused by humidity in polyimides[11] and polyamide 6[12], the relation between free volume, T_g, and branching in polyethylene[13] and propylene/α-olefine copolymers,[14] free volume in PA6-ABS blends,[15] ionic (Li^+, ClO_4^-) conduction and free volume in poly(ethylene oxide),[16] the estimation of the fraction of mobile and immobile amorphous phases in semicrystalline ethylene-octene copolymers,[17] and phase transitions in liquid crystals.[18]

Interdiffusion in polymer blends

Recently we have shown an interesting application of PALS which is not directly related to the free volume: the study of the interdiffusion in an originally demixed blend of two miscible polymers.[19,20] The blend was prepared as a particle-matrix system by mixing and pressing powders of the two polymers and annealing at a temperature between the T_g´s of both. DSC experiments were used to characterise the state of the blend. Two systems were investigated: (i) styrene-maleic anhydride-copolymer containing 24 wt.-% maleic anhydride (SMA-24, particle) and poly(methyl methacrylate) (PMMA, matrix),[19] (ii) polyvinylchloride (PVC, particle) and poly (n-butyl methacrylate) (PnBMA, matrix).[20] We presented experimental evidence that the o-Ps intensity I_3 and the average positron lifetime τ_{av} respond on the interdiffusion process. The response is due to the diffusion of inhibitors for Ps formation from one phase into the other. Figure 7 shows that I_3 varies linearly with the weight fraction of the components when the blend is completely demixed. During the annealing diffusion of polymer molecules from the PVC particles into the PnBMA matrix and vice versa occurs which leads to a lowering of I_3. A core-shell model for the description of the PALS response to the local chemical inhomogeneity has been developed which makes use of a calculated

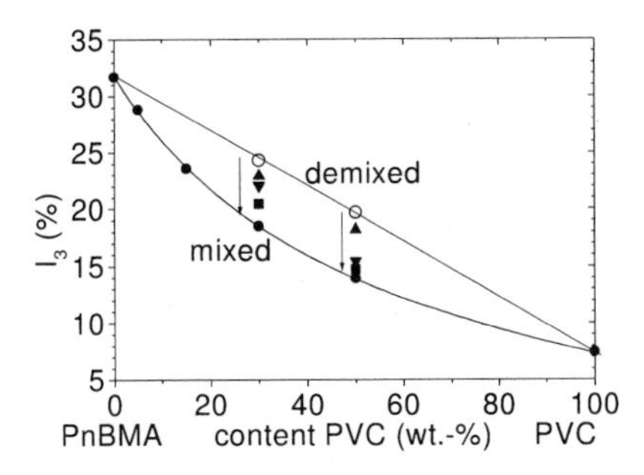

Figure 7. Variation of the o-Ps intensity I_3 as a function of the composition in molecularly mixed (filled symbols) and demixed (empty symbols) blends of PVC (particles) and PnBMA (matrix). The arrows show the change in I_3 due to interdiffusion (T_a = 110 °C, t_a = 3, 10, and 30 min).[20]

concentration-distance profile. The comparison with the experiment allows the estimation of the interface width and of kinetic parameters, such as the exponent k of the time dependence of the interfacial width $\sigma = \text{const.}\ t_{ann}^{\ k}$ and the coefficient of mutual diffusion D.[20] When this model is applied to DSC experiments, kinetic parameters very similar to those derived from PALS are obtained.[21]

Outlook

We have attempted to point out the uniqueness of Positron Annihilation as a technique for polymer studies. The main application of the method lies in the study of the microstructure (hole size distribution) of the free volume. There are several further applications of the method such as the estimation of the chemical surroundings of holes sensed by DBAR[22] or the hole anisotropy in highly crystalline fibres detected by ACAR.[23] A particular interesting field is the study of surface and near-surface properties using a slow, monoenergetic positron beam.[3,24,25] This technique allows a variation of the energy of monoenergetic positrons from a few eV to about 40 KeV and thus the penetration depth of positrons into polymers from a few Å´s up to several µm´s. In past we have used this technique for the study of ion-irradiation effects in polyimides layers.[24] The study of interdiffusion in thin polymer bi-layers is now in progress. Further interesting applications are the study of glass transition at surfaces[25] and degradation of polymer coatings due to weathering.[26]

[1] E.-J. Donth, "*The glass transition: Relaxation dynamics in liquids and disordered materials*", Springer, Berlin 2001.
[2] O. E. Mogensen, "*Positron Annihilation in Chemistry*", Springer-Verlag, Berlin, Heidelberg, New York, 1995.
[3] Y. C. Jean, P. E. Mallon, D. M. Schrader (Eds.*), "Principles and Application of Positron and Positronium Chemistry*", World Scientific, Singapore, 2003.
[4] Y. C. Jean, *Microchem. J.* **1990**, *42*, 72; and in: *"Positron Annihilation"*, Proc. of the 10th Int. Conf., Y. J. He, B.-S. Cao, Y. C. Jean, Eds., *Mat. Sci. Forum* **1995**, *175-178*, 59.
[5] G. Dlubek, J. Stejny, M. A. Alam, *Macromolecules* **1998**, *31*, 4574.
[6] H. Schmitz, F. Müller-Plathe, *J. Chem. Phys.* **2000**, *112*,1040.
[7] L. A. Utracki, R. Simha, *Macromol. Theory Simul.* **2001**, *10*, 17.
[8] G. Dlubek, V. Bondarenko, J. Pionteck, M. Supey, A. Wutzler, T. Krause-Rehberg, *Polymer* **2003**, *44*, 1921.
[9] J. Bohlen, R. Kirchheim, *Macromolecules* **2001**, *34*, 4210.
[10] G. Dlubek, K. Saarinen, H. M. Fretwell, *J. Polymer Sci.Part B: Polym. Phys.* **1998**, *36*, 1513.
[11] G. Dlubek, R. Buchhold, Ch. Hübner, A. Naklada, *Macromolecules* **1999**, *32,* 2348.
[12] G. Dlubek, F. Redman, R. Krause-Rehberg, *J. Appl. Polym. Sci.* **2002**, *84*, 244.
[13] G. Dlubek, J. Stejny, Th. Lüpke, D. Bamford, K. Petters, Ch. Hübner, M. A. Alam, M. J. Hill, *J. Polym. Sci.: Part B: Polym. Phys.* **2002**, *40*, 434.
[14] G. Dlubek, D. Bamford, A. Rodriguez-Gonzalez, S. Bornemann, J. Stejny, B. Schade, M. A. Alam, M. Arnold, *J. Polym. Sci.: Part B: Polym. Phys.* **2002**, *40*, 434.
[15] G. Dlubek, M. A. Alam, M. Stolp, H.-J. Radusch, *J Polymer. Sci.: Part B: Polym. Phys.* **1999**, *37,* 1749.
[16] D. Bamford, A. Reiche, G. Dlubek, F. Alloin, J.-Y. Sanchez, M. A. Alam, *J. Chem. Phys.* **2003,** *118*, 9420.
[17] D. Kilburn, D. Bamford, T. Lüpke, G: Dlubek, T. J. Menke, M. A. Alam, *Polymer* **2002**, *43*, 6973.
[18] G. Dlubek, D. Bamford, I. Wilkinson, K. Borisch, M. A. Alam, C. Tschierske, *Liquid Crystals* **1999**, *26,* 836.
[19] G. Dlubek, C. Taesler, G. Pompe, J. Pionteck, K. Petters, F. Redmann, R. Krause-Rehberg, *J. Appl. Polym. Sci.* **2002**, *84*, 654.
[20] G. Dlubek, J. Pionteck, V. Bondarenko, G. Pompe, Ch. Taesler, K. Petters, R. Krause-Rehberg, *Macromolecules* **2002**, **35**, 6313.
[21] G. Dlubek, G. Pompe, J. Pionteck, A. Janke, C. Kilburn, *Macromol. Chem. Phys.* **2003**, *204*, 1234.
[22] G. Dlubek, H. M. Fretwell, M. A. Alam, *Macromolecules* **2000**, *33*,187.
[23] D. Bamford, M. Jones, J. Latham, R. J Hughes, M. A. Alam, J. Stejny, G. Dlubek, *Macromolecules* **2001**, *34,* 8156.
[24] G. Dlubek, R. Buchhold, Ch. Hübner, A. Nakladal, K. Sahre, *J. Polym. Sci.: Part B: Polym. Phys.* **1999**, *37,* 2539.
[25] G. B. DeMaggion, W. E. Frieze, D. W. Gidley, M. Zhu, H. A. Hristov, *Phys. Rev. Lett.* **1977**, *78*, 1524.
[26] H. Cao, Y. He, R. Zhang, J. P. Yuan, T. C. Sandreczki, Y. C. Jean, B. Nielsen, *J. Polym. Sci.: Part B: Polym. Phys.* **1999**, *37*, 1289.

Sulphonated Poly(ether ether ketone): Synthesis and Characterisation

*V. V. Lakshmi, V. Choudhary, I. K. Varma**
Centre for Polymer Science and Engineering, Indian Institute of Technology Delhi, New Delhi-16, India
E-mail: ikvarma@hotmail.com

Summary: The paper describes the sulphonation of commercially available poly(ether ether ketone) PEEK (GATONE™, Gharda Chemicals Limited, India and VICTREX®, ICI Limited, UK) by using concentrated sulphuric acid. The concentration of GATONE in conc. H_2SO_4 was varied from 4-10 % (w/v) whereas in VICTREX® the concentration was 4 % (w/v). The temperature was varied from 35-55°C and the duration of reaction was 3-7 h. Structural characterisation of sulphonated polymers was done by elemental analysis, FT-IR and ^{1}H- NMR spectroscopy. On the basis of elemental analysis, the extent of sulphonation of GATONE was found to be 57-75%. The extent of sulphonation as determined by ^{1}H-NMR in case of GATONE was in the range of 53-80% and for VICTREX 58-87 %. Thermal analysis, proton conductivity and water uptake of these samples were also studied. Proton conductivity of the films was comparable to the perflourinated polymer (Nafion).

Keywords: fuel cells; membranes; poly(ether ether ketone); proton conductivity; sulphonation

Introduction

Poly (ether ether ketone)s (PEEK) are engineering plastics and find applications in electrical and electronic parts, automotives, aerospace, oil and chemical industries and military equipments.[1] Since PEEK is hydrophobic in nature, it cannot be used for certain applications. Hydrophilicity of these polymers can be increased by sulphonation. The possibility of using these ionic materials as proton exchange membranes (PEM) or direct methanol fuel cell membranes has generated considerable interest in the past. Membranes based on sulphonated PEEK have been reported earlier for fuel cell applications.[2-4]

Jin et al [5] reported that sulphonation takes place only on the phenyl flanked by two ether groups of the PEEK repeat unit. The extent of sulphonation is controlled by the reaction time, acid concentration, and temperature, which can provide a sulphonation range of 30- 100% per repeat unit.

Conc. H_2SO_4

SO_3H

PEEK SPEEK

DOI: 10.1002/masy.200450603

A series of novel sulphonated poly (aryl ketone)s have been developed by Hoechst AG for potential application in PEM fuel cells. The sulphonation of PEEK has been carried out by using 98% sulphuric acid [5-9], mixture of sulphuric acid and methane sulphonic acid[10], sulphur trioxide- triethyl phosphate complex and chlorosulphonic acid.[11] A crosslinked structure was formed by using 100% sulphuric acid due to sulphone formation[12], which is negligible in aqueous sulphuric acid i.e. 97.4% because water decomposes the postulated aryl pyrosulphonate intermediate which is required for sulphone formation. Chlorosulphonated PEEK is also formed when chlorosulphonic acid was used. Sulphonated PEEK (SPEEK) is soluble in dimethyl formamide (DMF), dimethyl acetamide (DMAc), dimethyl sulphoxide (DMSO), pyridine and N- methyl- 2 – pyrrolidine (NMP).[13]

Considerable work has been reported on sulphonation of commercially available PEEK (VICTREX ICI product) but no reports are available on sulphonation of other commercially available poly(ether ether ketone)s such as GATONE of Gharda Chemicals Limited, India. There is a difference in the properties of these polymers (Table 1), which may be attributed to the structural difference. This variation may affect the sulphonation behaviour. Therefore, it was considered of interest to carry out the sulphonation of GATONE and evaluate the properties of sulphonated GATONE and compare with the sulphonated VICTREX.

Experimental

The concentration of GATONE in conc. H_2SO_4 was varied from 4-10 % (w/v) whereas in VICTREX® the concentration was 4 % (w/v). The temperature was varied from 35-55°C and the duration of reaction was 3-7 h. The reaction mixture was stirred using mechanical stirrer & the resulting sulphonated polymers (SPEEK) were precipitated by dropwise addition of solution to 500 ml of ice cooled distilled water. The samples were washed till the excess acid was removed and dried in oven overnight at 70°C. The details of sulphonation along with the letter designations are given in Table 2. The films were obtained by casting a DMF solution of SPEEK (3.3 % (w/v)) on a glass plate.

Structural Characterisation

A Carlo ERBA, EA 1108 elemental analyzer (CHNSO) was used for elemental analysis. FT-IR spectra of polymer samples either in film or in KBr pellets were recorded on NICOLET FT-IR spectrometer. A BRUKER AC300 spectrometer was used to record ^{1}H- NMR using tetramethyl silane as an internal standard and DMSO-d_6 as a solvent.

Table 1. Properties of Polymers (as given in Manufacturers Data Sheet)

S. No	Property	Test method	GATONE™5200P	VICTREX® 450P
1.	Density	ISO1183	1.32 g/cc	
Mechanical properties				
2.	Tensile Strength	ISO527	95 MPa	97 MPa at 23°C 12 MPa at 250°C
3.	Tensile Modulus	ISO527	3800 MPa	
4.	Elongation at Break	ISO527	>50 %	Up to 60 %
5.	Flexural Strength	ISO178	160 MPa	53 MPa at 23°C
6.	Flexural Modulus	ISO178	3800 MPa	4100 MPa
7.	Izod Impact	ISO179	6.0 kJ/m^2 (notched)	-
		ISO180/1A	-	6.4 (notched) and no break (unnotched)
Thermal properties				
8.	HDT at 1.82 MPa	ISO75	150°C	152°C
9.	Glass Transition temperature	DSC	148°C	143°C
10	Melting Point	DSC	338°C	340°C

Thermal behaviour

Thermal characterisation was done using a TA 2100 thermal analyzer having a 951 TG module. Thermogravimetric analysis was carried out in the temperature range of 50°C-850°C in N_2 atmosphere (flow rate = 60cm^3/min) using a heating rate of 20°C/min. A sample weight of 10 ± 2 mg was used.

Water uptake

SPEEK film samples were first dried at 120°C for 20 h. The dried films were weighed and kept immersed in water at 35 ± 2°C. The films surface was dried by a filter paper and the hydrated polymer was weighed after definite time intervals. From the knowledge of difference in weight of dried and wet samples, water uptake was calculated.

Table 2. Reaction Conditions and Sample Designations for the preparation of Sulphonated PEEK Samples in 98% H_2SO_4

S. No	Reaction Time (h)	Reaction temp (°C)	Conc. of PEEK (w/v %)	Sample designation
1	3	35	6.25	G-1
2	3	40	6.25	G-2
3	3	45	6.25	G-3
4	3	45	5.00	G-4
5	3	55	4.00	V-1
6	5	55	4.00	V-2
7	7	55	4.00	V-3

* G_1-G_4 were obtained by using GATONE™ while V_1-V_3 were from VICTREX

Proton conductivity measurements

Proton conductivity of all the samples was measured by impedance spectroscopy using an EG&G PARC potentiostat/galvanostat (model 273) and a Schlumberger 1255HF frequency response analyser unit.

Results and Discussion

VICTREX® used for sulphonation was flaky in nature whereas GATONE was in powder form. The dissolution of GATONE in H_2SO_4 was faster than VICTREX. The sulphonation within the sample (both GATONE and VICTREX) is not homogeneous.

The sulphonated samples obtained after 5h of reaction were either completely soluble in water (when the reaction temperature was 45°C) or were highly swollen (at a reaction temperature of 35°C). Isolation of such samples in dry form was difficult, therefore the results of sulphonated samples obtained after 3h of reaction are given in Table 2. Sulphonated samples were flaky in nature and pinkish in color. The films obtained were transparent with yellowish tinge.

Elemental analysis

The elemental composition of some GATONE samples is given in Table 3. The degree of sulphonation could be calculated from the % sulphur content and was found to be in the range of 57-75%. Higher sulphur content as well as degree of sulphonation was found for samples prepared at 45°C.

Table 3. Results of Elemental Analysis and Degree of Sulphonation (DS %)

S. No.	Sample designation	Carbon (%)	Hydrogen (%)	Sulphur (%)	Empirical formulae	DS (%)
1	G-1	62.96	3.98	4.94	$C_{34}H_{26}O_{11}S_1$	57
2	G-2	63.37	4.44	5.83	$C_{29}H_{24}O_9S_1$	67
3	G-3	62.94	4.62	6.11	$C_{55}H_{48}O_{17}S_2$	70
4	G-4	59.68	4.79	6.54	$C_{49}H_{47}O_{18}S_2$	75

FT-IR

In sulphonated samples, additional absorption bands associated with sulphonic acid groups were observed at 3440, 1252, 1080, 1024 and 709cm^{-1}. The broad band at ~3440 cm^{-1} is assigned to the O-H vibration of ~SO_3H as well as to the absorbed moisture. The other bands are due to sulphur-oxygen vibrations; asymmetric O=S=O stretch (1252 cm^{-1}), symmetric O=S=O (1080 cm^{-1}) stretch, S=O stretch (1024 cm^{-1}), and S-O stretch (709 cm^{-1}).

The aromatic C-C absorption band of PEEK at ~1490 cm^{-1} and 1414 cm^{-1} were split and new bands 1472 and 1402 respectively appeared upon sulphonation. The intensity of the new absorption peak~1472 cm^{-1} increased while that of 1492 cm^{-1} decreased with increase in temperature of the reaction.

Both PEEK and sulphonated PEEK samples had an absorption peak at ~840 cm^{-1} which is characteristic of the out-of-plane bending of two hydrogens of 1,4-disubstituted benzene ring. In addition, the sulphonated samples had a new band at ~867cm^{-1} which is characteristic of the out-of-plane C-H bending of an isolated hydrogen in a 1,2,4-trisubstituted phenyl ring. So, the changes observed in the IR spectra indicate the presence of $-SO_3H$ groups

^{1}H-NMR

The ^{1}H- NMR spectra of SPEEK samples are shown in Fig. 1. In the ^{1}H- NMR spectra of SPEEK samples, the aromatic proton resonance signals were observed in the range of δ = 7-

7.75 ppm. In sulphonated samples, H_B and H_D appear as a doublet at 7.12 ppm & triplet at 7.27ppm respectively. The doublet at δ = 7.87 ppm and triplet at δ = 7.92 ppm correspond to H_A and $H_{A'}$ protons. Presence of sulphonic acid group in the hydroquinone ring of SPEEK leads to a downfield shift of H_E protons compared to H_C or H_D protons. Therefore the observed signal at 7.50 ppm should correspond to H_E protons. The labeling of various magnetically non-equivalent protons in SPEEK samples is shown as,

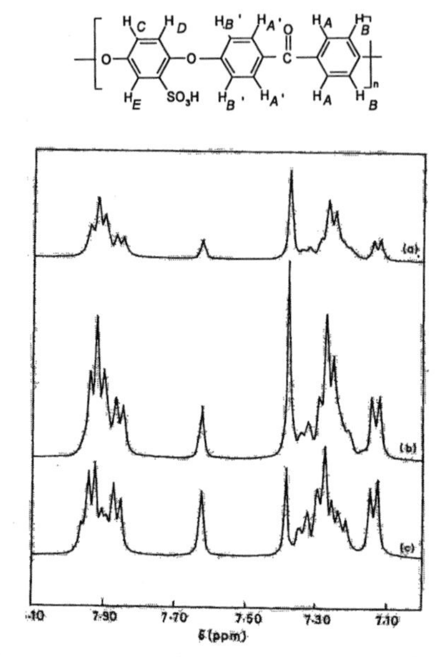

Figure 1. ^{1}H-NMR Spectra of SPEEK Samples (a) G-1 (b) G-3 and (c) G-4

The intensity of H_E signal therefore may be used to estimate for the SO_3H content.[13] The ratio between the peak areas of the signals corresponding to all other aromatic hydrogen atoms may be expressed as,

$$\frac{H}{12-2n} = \frac{AH_E}{\Sigma\, AH_{A,\,A',\,B,\,B',\,C,\,D}}$$

Where 'n' is the number of H_E per repeat unit. An estimate of degree of sulphonation (DS) expressed is = n x 100 (%).

The calculated DS values are given in Table 4. Depending on the temperature and duration of sulphonation reaction, DS values ranging from 53-80 % were obtained in case of GATONE and 58-87 % in case of VICTREX.

Table 4. Degree of Sulphonation as Calculated from ^{1}H NMR

S. No	Sample Designation	DS (%)
1	G-1	53
2	G-2	57
3	G-3	65
4	G-4	80
5	V-1	58
6	V-2	80
7	V-3	87

Thermal behaviour

Thermogravimetric Analysis

The TG traces showed multistep mass loss in sulphonated samples while single step decomposition was observed in PEEK above 400°C (Table 5). Three-step degradation was observed in SPEEK samples i.e. from 50-200°C, 200-450°C and 450-800°C. The mass loss from 50-200°C in SPEEK is due to loss of physically and chemically bound water. The observed mass loss between 200-450°C in SPEEK samples may be attributed to decomposition of sulphonic acid group. Breakdown of the polymer backbone takes place above 450°C.

Water uptake

The presence of water facilitates proton transfer and increases the conductivity of solid-state electrolytes. The enhancement of hydrophilicity by sulphonation of PEEK polymer can be followed by water absorption of SPEEK films as a function of degree of sulphonation (Table 6).

Proton conductivity

The proton conductivity of thin films (60-100μ) was found to be in the range of 1.3×10^{-2} S/cm in VICTREX polymers and 0.04- 0.22×10^{-2} S/cm in GATONE polymers. The conductivity values are comparable with the values reported for Nafion membranes ($\sigma \geq 10^{-2}$ S/cm).[14]

Table 5. Results of TG Analysis of PEEK & SPEEK Samples

S. No	Sample designation	MASS LOSS (%)			
		50-200°C (%)	200°C-450°C (%)	Above 450°C (%)	Char residue (%)
1	GATONE™	-	-	49.8	50.2
2	G-1	7.6	-	39.2	53.2
3	G-2	5.0	14.8	35.4	55.2
4	G-3	10	16.0	30.0	44.0
5	G-4	1.0	18.0	29.6	51.4
6	VICTREX®	----	----	48.9	51
7	V-1	5	14	33	48
12	V-2	6	21		73
14	V-3	6	42		42

Char residue indicate the mass remaining at 800°C.

Table 6. Results of Water Uptake and Proton Conductivity in SPEEK Samples

S. No	Sample designation	Water uptake (%)	Proton conductivity x 10^{-2} (S/cm)
1	G-1	23	0.22
2	G-2	24	0.04
3	G-3	28	0.14
4	G-4	47	0.15
5	V-1	15	-
6	V-2	58	0.71
7	V-3	84	0.35

Conclusions

A series of sulphonated PEEK samples were synthesized by varying the reaction time, temperature and the PEEK concentration in 98% sulphuric acid. The sulphonation was confirmed by FTIR spectra of SPEEK where additional absorption bands associated with sulphonic acid groups were present at 3440, 1252, 1080, 1024, and 709cm^{-1}. Splitting of C-C aromatic bands on sulphonation was also observed. All these structural characterization results are consistent with each other. The extent of sulphonation (measured by elemental analysis and ^{1}H- NMR) increased by increasing temperature of the reaction and 80% sulphonation was attained at 45 and 50°C. An increase in moisture absorption of SPEEK films was observed with an increase in degree of sulphonation. Decomposition of sulphonic acid group was indicated by mass loss in the temperature range of 200°C-450°C in thermogravimetric analysis. In unsulphonated PEEK, no mass loss was observed in this region. The proton conductivities were found to be in the range of 0.34 –2.6 x 10^{-2} S/cm. The major difference between GATONE and VICTREX was in terms of ease of sulphonation, proton conductivity and moisture absorption.

Acknowledgement

Authors thank Gharda Chemicals Limited, India for providing free samples of GATONE, Reliance Industries Limited, India for creating Reliance chair at IIT Delhi (I. K. Varma) and Council of Scientific and Industrial Research (CSIR, Government of India) for providing financial assistance to one of the authors (Muthu Lakshmi. R. T. S), Dr. R. P. Singh of National Chemical Laboratory, Pune for elemental analysis and Dr. Dhathathreyan of SPIC Science Foundation, Chennai for proton conductivity measurements.

1. Martin Hograth and Xavier Glipa, High Temperature Membranes for Solid Polymer Fuel Cells, ETSU/F02/00189/REP; DTI/PUB URN 01/ 893, **2001**.
2. C. Genies, R. Mercier, B. Sillion, R. Petiaud, N. Cornet, G. Gebel, M. Pineri, *Polymer*, **2001**, 42, 5097.
3. P. Genova- Dimitrova, B. Baradie, D. Foscallo, C. Poinsignón, J. Y. Sanchez; *J. Memb. Sci*, **2001**,185, 59.
4. D. J. Jones and J. Roziere; *J. Memb. Sci.*, **2001**, 185, 41.
5. X. Jin, Mathew T. Bishop, Thomas S. Ellis, Frank E. Karash, *Brit. Polym. J.* **1985**, 17, 4.
6. N. Shibuya and Roger S. Porter; *Macromolecules*, **1992**, 25, 6495.
7. N. Shibuya and Roger S. Porter; *Polymer*, **1994**, 35, 15.
8. Muthu Lakshmi R. T. S, I. K. Varma, U. Yugandhar, and T. S. R. Murthy, Proceedings 39th International Symposium on Macromolecules, MACRO 2002, July 7-12, **2002**, Beijing, China.
9. C. Bailly, D. J. Willims, F.E. Karaz, and W. J. Macknight; P*olymer,* **1987**, 28, 1009.
10. T. Okawa and C. S. Marvel; *J. Polym. Sci., Polym. Chem. Edn*, **1985**, 23, 1231.
11. M. Bishop, F. E. Karaz, P. Russo and K. J. Langley; *Macromolecules*, 1985, 18, 86.
12. D. Daoust, J.Devaux, P. Godard A. Jonas and R. Legras; Advanced Thermoplastics and Their Composites. Characterisation and Processing (Ed. H. H. Kausch), Ch.1, Carl Hanser Verlag, Munich, **1992**, 3.
13. S. M. J Zaidi, S. D. Michailenko, G. P. Robertson, M.D. Guiver, and S. Kalaiaguine; *J. Memb. Sci.*, **2000**, 173, 17.
14. G. Alberti, M. Casciola, L. Massinelli. and B. Bauer, Polymeric proton conducting membranes for medium temperature fuel cells (110-160 °C), *J. Memb. Sci.*, **2001**, 185, 73.

Monte Carlo Simulation of Polymeranalogous Reaction in Confined Conditions: Effects of Ordering

Victor V. Yashin,[1] *Yaroslav V. Kudryavtsev,*[2] *Yury A. Kriksin,*[3]
*Arkady D. Litmanovich**[2]

[1] University of Pittsburgh, Department of Chemical Engineering, Pittsburgh, PA 15261, U.S.A.
[2] Topchiev Institute of Petrochemical Synthesis of Russian Academy of Sciences, Leninsky prosp. 29, Moscow B-71, 119991, Russia
E-mail: alit@ips.ac.ru
[3] Institute for Mathematical Modelling of RAS, Miusskaya sq. 4-a, 125047 Moscow, Russia.

Summary: The influence of energetic parameters of the interchain homo- and heterocontacts on a local ordering of Bernoullian copolymers has been studied using Monte Carlo simulations and probabilistic analysis. The results of both methods are in a good agreement. Then simple Monte Carlo procedure was employed to study the ordering in products of a polymeranalogous reaction with accelerating effect of neighboring groups. When the reaction with intra- and interchain acceleration and local ordering proceed simultaneously in confined conditions, the ordering might affect the process so that the formation of certain nano-structures (in particular, not trivial strip-like ones) is possible.

Keywords: Monte Carlo simulation; nanostructures; polymeranalogous reaction; probabilistic analysis; self-assembly

Introduction

Regular di- and multi-block *AB* copolymers easily form ordered structures via microsegregation provided *AB*-contacts are unfavorable.[1] Any disorder in the blocks distribution along the chain hinders an ordering.[2] Just therefore ordering of irregular statistical copolymers sparks the keen interest of theorists. A phase behavior of the copolymers is mainly considered. Recently de Gennes[3] considered the weak segregation of a random *AB* copolymer in a melt using very simple elegant approach. He found that the segregation might proceed at high value of Flory-Huggins parameter $\chi \geq 2$, the result being in accordance with detailed calculations performed earlier[4] using the random phase approximation.[5] Phase behaviour of random copolymers both in melts[6] and in concentrated solutions[7] was studied also by Monte Carlo simulations; a segregation in melts[6] and inhomogeneity of the solutions structure[7] were predicted but again for $\chi \geq 2$.

In the meantime, another aspect is of great interest, namely the local ordering as a component of self-assembly of statistical copolymers leading to a formation of energy-wise advantageous

 DOI: 10.1002/masy.200450604

structure. Let us consider the simplest model system: two stretched *AB* chains of composition p (p is a molar fraction of *A* units). Their interaction energy consists of energies of interchain *AA*-, *BB*- and *AB*-contacts. Potential capability of an ensemble of such chains to ordering is determined by their primary structure, namely by units distribution. Among irregular copolymers, Bernoullian ones are most disordered. So it is especially interestingly to elucidate whether Bernoullian copolymers are capable to a local ordering and, if so, how to estimate such a tendency?

To study the problem, a simple Monte Carlo procedure has been suggested recently.[8]

Monte Carlo procedure and Probabilistic analysis

M Bernoullian chains of the length N were generated and arranged in the NxM rectangle. N and M were varied within 20 – 1000 and 100 – 1000000, respectively. The numbers of interchain *AA*-, *BB*- and *AB*- contacts were calculated. In all cases, an average value of the fraction of the interchain *AB*-contacts, ϕ_{in}, was equal to the pure random value of $2p(1\text{-}p)$. Then the rectangle was closed to a cylinder (with a generatrix M) and each upper ring was rotated over the lower ring to attain a position with minimal fraction of unfavorable *AB*-contacts, ϕ_{min}. The values of ϕ_{min} averaged over the whole ensemble deviate noticeably from the pure random value $\phi_{\text{in}} = 2p(1\text{-}p)$: 0.286 and 0.500, respectively, for $p = 0.50$ and $N = 20$.

For Bernoullian chains, it is possible to elaborate an analytical approach as well.

Let us enumerate units in each chain from 0 to $N - 1$. Now a Bernoullian chain of the length N is represented by a set of independent variables $\{x_k\}$; x_k takes the value A with the probability p and B with the probability $q = 1 - p$. Now join such sequences in an infinite chain, so that $x_{k+N} \equiv x_k$. This is N-periodic Bernoullian *AB* chain $\{x_k\}$. Consider two such jointly independent periodic chains $\{x_k\}$ and $\{y_k\}$. Sliding of the chain $\{x_k\}$ along the chain $\{y_k\}$ with a shift, s, $0 \le s \le N\text{-}1$ is equivalent to the Monte Carlo rotation procedure for two corresponding cycles of the length N.

Let the interaction energy of two units x_k and y_l, $u_{kl} \equiv u(x_k, y_l)$, takes the values ε_{AA}, ε_{AB}, and ε_{BB} with probabilities p^2, $2pq$, and q^2, respectively.

Interaction energy of two N-periodic *AB*-chains $\{x_k\}$ and $\{y_k\}$ and at a shift s (per one contact), u_s, is given by Eq. (1):

$$u_s = \frac{1}{N}\sum_{k=0}^{N-1} u(x_k, y_{k+s}) = \frac{1}{N}\sum_{k=0}^{N-1} u_{k,k+s}, \quad s = 0,1,\ldots,N-1 \tag{1}$$

As $N \to \infty$, the distribution of u_s values tends asymptotically to the normal distribution.[8] Hereafter we will consider variables $\{u_s\}$ as normally distributed.

For any shift s, the mean value of the random variable u_s is given by Eq. (2):

$$\langle u_s \rangle = p^2 \varepsilon_{AA} + 2pq\varepsilon_{AB} + q^2 \varepsilon_{BB} = \varepsilon . \tag{2}$$

The normally distributed vector $(u_0, u_1, ..., u_{N-1})$ is completely characterized by its covariation matrix

$$a_{mn} = \langle (u_m - \langle u_m \rangle)(u_n - \langle u_n \rangle) \rangle = \langle u_m u_n \rangle - \langle u_m \rangle \langle u_n \rangle = \langle u_m u_n \rangle - \varepsilon^2 , \tag{3}$$

which may be directly calculated:

$$a_{mn} = N^{-1} \left((pq(\varepsilon_{AA} - 2\varepsilon_{AB} + \varepsilon_{BB}))^2 \delta_{mn} + 2pq(- p\varepsilon_{AA} + (2p-1)\varepsilon_{AB} + (1-p)\varepsilon_{BB})^2 \right. , \tag{4}$$

where δ_{mn} is the Kronecker symbol.

Let us introduce the variables

$$\chi = \varepsilon_{AB} - (\varepsilon_{AA} + \varepsilon_{BB})/2, \qquad r_w = -p\varepsilon_{AA} + (2p-1)\varepsilon_{AB} + (1-p)\varepsilon_{BB} \tag{5}$$

In this notation, elements of the covariation matrix take the form

$$a_{mn} = N^{-1} (\delta_{mn} (1-a)^2 \chi^2 / 4 + (1-a) r_w^2 / 2) . \tag{6}$$

where $a = (1-2p)^2$.

As follows from Eq. (6), the random variables u_s (s = 0, 1, …, N-1) are independent only at $r_w = 0$. In the general case, we may write

$$u_s = \varepsilon + v_s + w , \tag{7}$$

where ε is the constant introduced above, w and v_s (s = 0, 1, …, N-1) are mutually independent variables that are distributed normally with the zero mean value $\langle v_s \rangle = \langle w \rangle = 0$ and the dispersions

$$\sigma_v^2 = (1-a)^2 \chi^2 /(4N), \qquad \sigma_w^2 = (1-a) r_w^2 /(2N) \tag{8}$$

In order to find the minimum value of the interaction energy

$$u_{\min} = \min_{0 \le n \le N-1} (\varepsilon + v_n + w) = \varepsilon + \min_{0 \le n \le N-1} (v_n + w) \tag{9}$$

let us calculate the probability that $v = \min\limits_{0 \le n \le N-1} (v_n + w)$ is not less than θ, $F(\theta)$:

$$F(\theta) = P(v \ge \theta) = \int_{-\infty}^{+\infty} d\eta [F_v(\theta - \eta)]^N f_w(\eta) \tag{10}$$

where $F_v(\theta - \eta) = \int_{\theta-\eta}^{\infty} dy f_v(y) = \frac{1}{2}\left(1 - erf\left(\frac{\theta - \eta}{|\chi|(1-a)} \sqrt{2N}\right)\right)$ is the probability that v_n is not less than $\theta - \eta$, $f_w(\eta) = \frac{1}{\sqrt{2\pi}\sigma_w} \exp\left(-\frac{\eta^2}{2\sigma_w}\right)$, and $erf(x) \equiv \frac{2}{\sqrt{\pi}} \int_0^x dy \exp(-y^2)$ is the error function.

The mean value of minimum interaction energy of two chains (per one contact) is

$$\langle u_{\min} \rangle = \varepsilon + \langle v \rangle = \varepsilon - \int_{-\infty}^{+\infty} d\theta \cdot \theta \frac{dF(\theta)}{d\theta}. \tag{11}$$

Since $f_w(\eta)$ differs noticeably from zero only in the narrow region of width ~ $N^{-1/2}$ near $\eta = 0$, it is possible to use the approximation[8] $F(\theta) \approx [F_v(\theta)]^N$, $F_v(\theta - \eta) \approx F_v(\theta)$, whence

$$\langle u_{\min} \rangle = \varepsilon + \langle v \rangle = \varepsilon - \int_{-\infty}^{+\infty} d\theta \cdot \theta \frac{d[F_v(\theta)]^N}{d\theta}. \tag{12}$$

For equal energies $\varepsilon_{AA} = \varepsilon_{BB}$, the interaction energy of two chains (per one contact) is related identically to the fraction of *AB*-contacts between those chains, ϕ_{AB}:

$$\phi_{AB} = \frac{u - \varepsilon_{AA}}{\varepsilon_{AB} - \varepsilon_{AA}} \tag{13}$$

Therefore the fraction of *AB*-contacts corresponding to the mean minimum value of the interaction energy for those chains is given by Eq. (14)

$$\langle \phi_{AB\min} \rangle = \frac{\langle u_{\min} \rangle - \varepsilon_{AA}}{\varepsilon_{AB} - \varepsilon_{AA}}. \tag{14}$$

Dependencies of $\langle \phi_{AB\min} \rangle$ on the logarithm of chain length N for unfavorable and favorable *AB*-interaction were calculated using both a set of equations given above and Monte Carlo simulations. It is seen in the Figure 1 that the results of both methods are in a good agreement. Deviations of $\langle \phi_{AB\min} \rangle$ from the pure random value might serve as a measure of the capability of Bernoullian chains to ordering. These deviations decrease when N increases, however they are significant even for long chains: ~25-30 % and ~10 % for N = 100 and 1000, respectively (Figure 1).

Note that estimations of a capability to an ordering given by the rotation procedure are good for a sliding not only of periodic but also for that of true Bernoullian chains.[8] Therefore the simple Monte Carlo procedure seems to be suitable for estimating ordering in copolymers of other classes when an analytical approach is more complicated.

Using this procedure, ordering of *AB* copolymers formed during a polymeranalogous reaction $A \rightarrow B$ with accelerating effect of the neighboring *B* units has been studied. The ratios of the rate constants for central *A* units in triads *AAA*, *AAB* (*BAA*), *BAB* were k_0: k_1: k_2 = 0.02: 0.245: 0.490. An ensemble of *M A*-chains of the length N was generated. Then *A* units transformed into *B* ones with probabilities proportional to corresponding rate constants till the mean value over the whole ensemble for the prescribed the product composition, p, was attained.

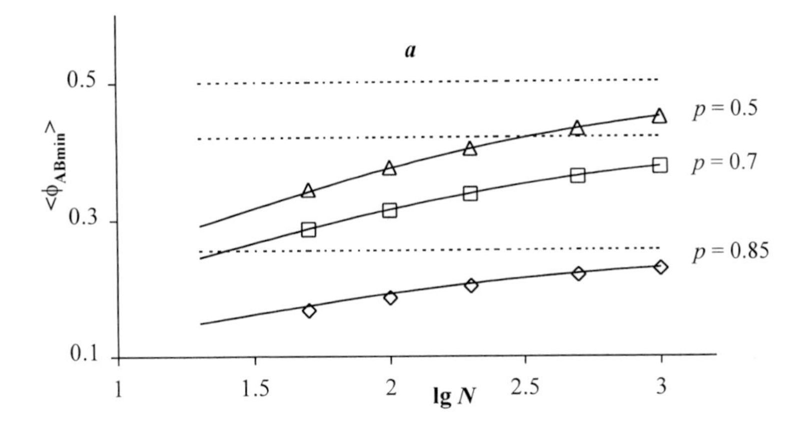

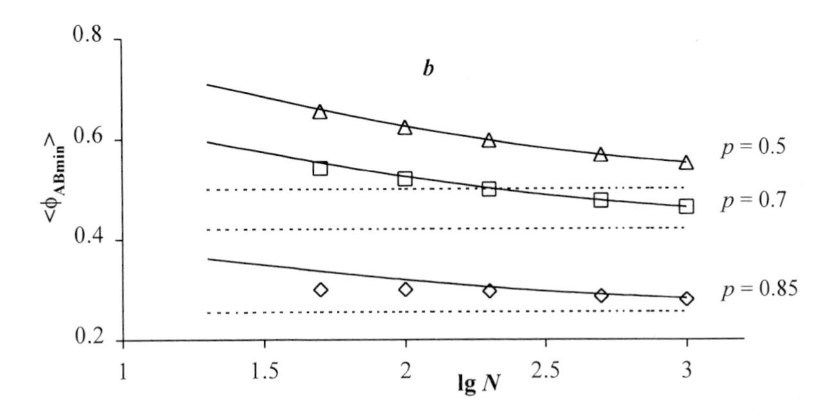

Figure 1. Minimized fraction of interchain *AB*-contacts $\langle \phi_{AB\min} \rangle$ *vs* chain length. (a) unfavorable *AB*-contacts: $\varepsilon_{AB} = -0.1$, $\varepsilon_{AA} = \varepsilon_{BB} = -1$; (b) favorable *AB*-contacts: $\varepsilon_{AB} = -1$, $\varepsilon_{AA} = \varepsilon_{BB} = -0.1$; dashed lines – pure random values for corresponding average composition *p*.

For such copolymers, the deviations $\langle \phi_{AB\min} \rangle$ from ϕ_{in} are also significant. Moreover, the deviations turn out to be greater than the ones for Bernoullian copolymers (see Figure 2). This is due to a tendency to the block distribution typical for the products of reactions with accelerating neighbor effect.[9]

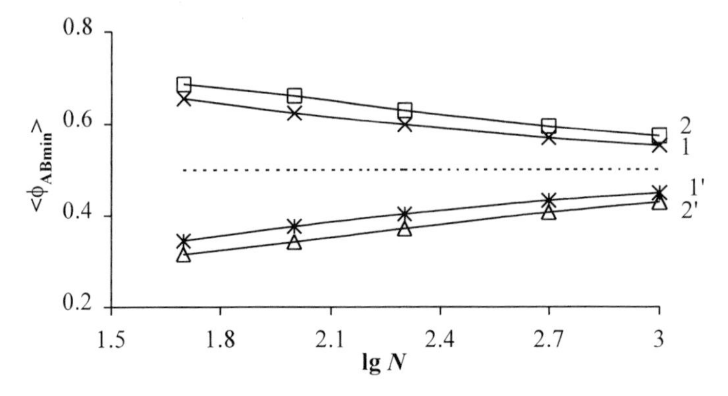

Figure 2. Effect of the chain structure on the ordering: $\langle \phi_{AB\min} \rangle$ *vs* chain length for mean chain composition $p = 0.5$; Bernoullian chains (1,1′) and reaction products (2,2′), rate constant ratios k_0: k_1: k_2 = 0.02: 0.245: 0.490; favorable (1,2) and unfavorable (1′,2′) *AB* contacts; interchain contact energies like in Fig. 1; dashed line – no ordering.

Ordering during Reaction

It is reasonable to assume that a tendency to ordering might affect both reaction kinetics and structure of the product if ordering and a polymeranalogous transformation proceed simultaneously.

Let us consider a two-dimensional system composed of *M* stretched chains *A* of the length *N* where $A \rightarrow B$ reaction with accelerating effect of both inner and external neighboring *B* units takes place. It might be viewed as a very thin layer on a surface or within a narrow slit, so that the chains may slide one along the other while the transversal displacements are hindered. This is an already familiar rectangle, so the effects of ordering were studied using the Monte Carlo procedure – rotation of rings. Let accelerating effects of the internal and external neighbors be equal so that the ratios of the rate constants for *A* units having 0, 1, 2, 3 и 4 *B* neighbors are k_0: k_1: k_2 : k_3: k_4 = 0.020: 0.265: 0.510: 0.755: 1.000. Each Monte Carlo test consists of the following successive operations: a random choice of a unit; testing whether the unit reacts or not (provided the chosen unit is *A*); displacement of the chain containing the unit in accordance with the model studied.

The following reaction models were studied.

Local Ordering - after each Monte Carlo test for $A \rightarrow B$ reaction, the tested chain is shifted to a position with minimal energy of interchain contacts with two neighboring chains. Two cases are considered: favorable and unfavorable AB-contacts, **Local Ordering** $\mathbf{AB_f}$ and **Local Ordering** $\mathbf{AB_{unf}}$, respectively.

For the purpose of comparison, the other models were studied as well.

Immobile chains - the reaction in a system of immobile chains;

Ideal Mixing - random sliding in a system with equal energies of interchain contacts;

No Interchain Effects - the reaction without interchain acceleration and ordering.

The product structure was characterized by composition p – fraction of units A, probability of a boundary between A and B sequencies, R, and a dispersion of compositional distribution, D. The fraction of interchain AB-contacts, ϕ_{ABtransv}, was also calculated.

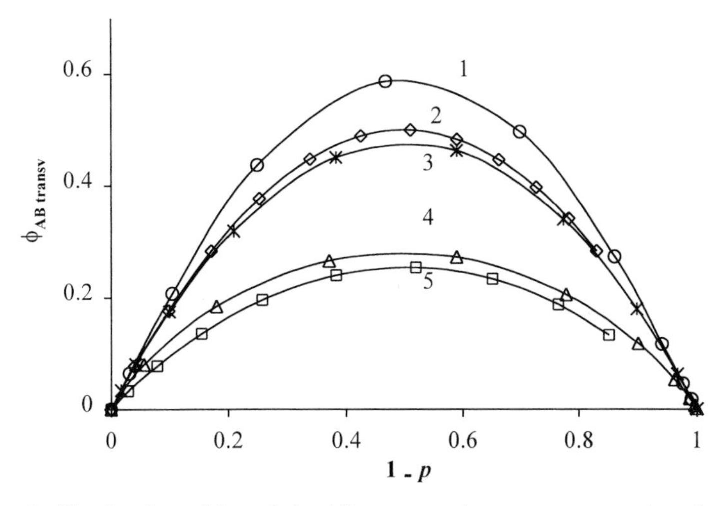

Figure 3. The fraction of interchain AB-contacts, ϕ_{ABtransv}, vs conversion, $1 - p$. Curves correspond to the different reaction models: **Local Ordering AB$_f$** (1), **No Interchain Effects** (2), **Ideal Mixing** (3), **Immobile chains** (4), **Local Ordering AB$_{unf}$** (5).

For the model **No Interchain Effects**, the values of ϕ_{ABtransv} are equal to the pure random values $2p(1\text{-}p)$ (see Figure 3). For **Ideal Mixing**, the values of ϕ_{ABtransv} were found to be close to the latter as well. Ordering significantly affects this quantity. For **Local Ordering AB$_f$**, the value of ϕ_{ABtransv} is considerably greater than $2p(1\text{-}p)$, while for **Local Ordering AB$_{unf}$** it is markedly less than the random values. This kind of data is quite informative. So, ϕ_{ABtransv} for **Local Ordering AB$_{unf}$** and for **Immobile chains** are close to one another. Accordingly, the other characteristics of those systems are quite similar.

During the reaction with an accelerating effect of inner neighbors, relatively long blocks are formed; consequently, the probability of a boundary between A and B sequences, R, is relatively small (Figure 4). Due to interchain acceleration, the rate of generation of new blocks increases and the probability of a boundary increases as well. Naturally, the increase of R is minimal for **Local Ordering AB$_{unf}$** and maximal for **Local Ordering AB$_f$**.

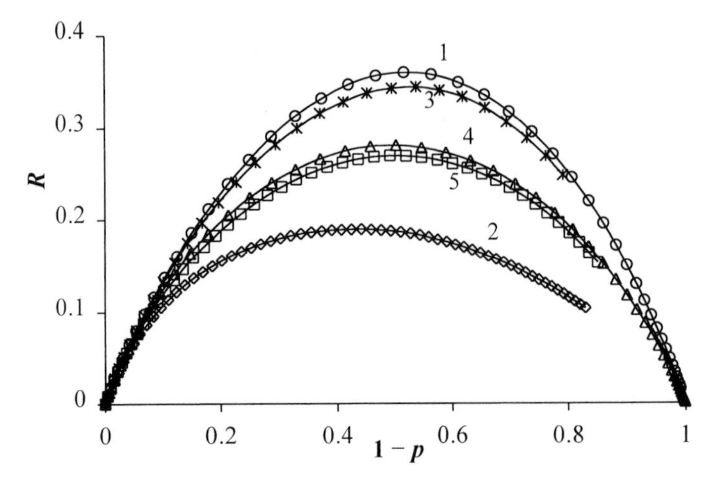

Figure 4. The probability of an AB-boundary, R, vs conversion, $1-p$. Curve numbers are as in Fig. 3.

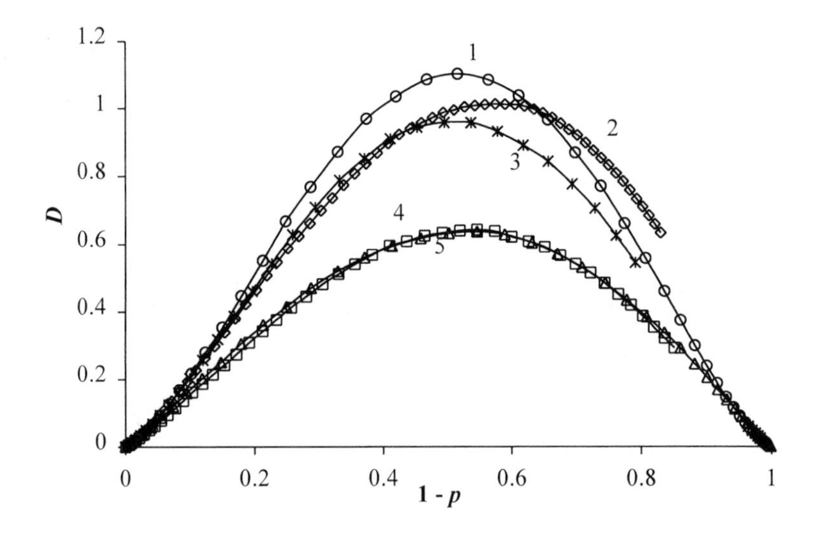

Figure 5. The dispersion of compositional distribution, R, vs conversion, $1-p$. Curve numbers are as in Fig. 3.

For a reaction with an accelerating effect of inner neighbors only, the dispersion of compositional distribution D is significant and its dependence on conversion is asymmetrical (the maximum is shifted to the right).[9] **Ideal mixing** shifts the maximum to the center, the D value decreases slightly at high conversions (Figure 5). The dispersion decreases greatly for

Local Ordering AB_{unf} and **Immobile chains**. In general, the more R the less D. Therefore it is worthy to notice that for **Local Ordering AB_f** the simultaneous maximum values for both R and D are observed in the results of simulations. It indicates a specific effect of ordering on evolution of the reacting system in two dimensions.

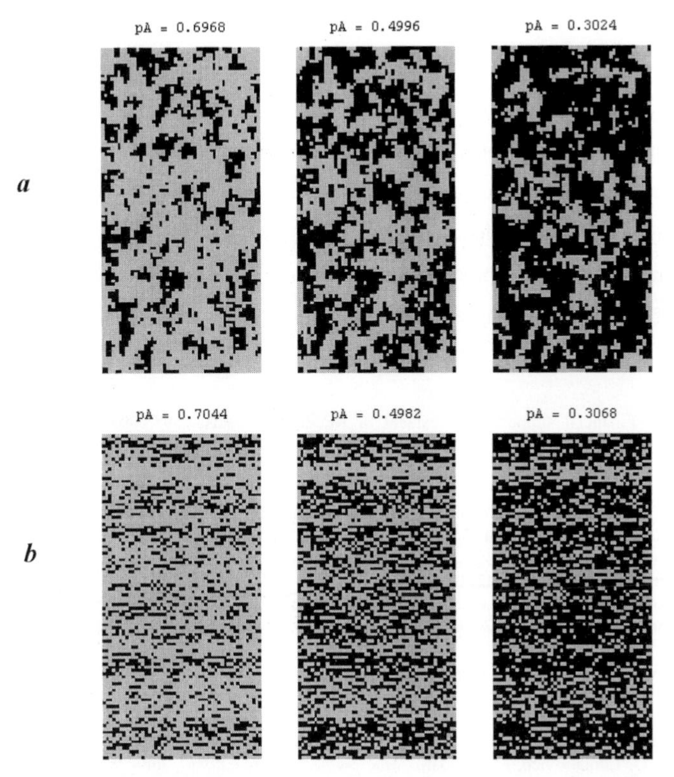

Figure 6. Snap-shots of the system structure for the reaction models **Local Ordering AB_{unf}** (a) and **Local Ordering AB_f** (b). Ensemble consists of M = 150 chains of N = 75 units. Chains are settled along abscissa.

Evolution of the system structure for different models is shown on snap-shots (Figure 6). Chains are settled along abscissa. In case of **Local Ordering AB_{unf}**, the B-blocks, which propagate along a chain, involve the neighboring chains into the reaction; the reaction spreads both in the horizontal and vertical directions, so that black "islands" are formed and grow being distributed relatively uniformly in the white "sea". This is due to the fact that the interchain BB-contacts formed during the reaction are favorable, therefore the chains behave as immobile (for **Immobile chains** similar snapshots "islands in the sea" were obtained). On the contrary, the fraction of favorable AB-contacts increases in the course of ordering in the

Local Ordering AB_f model. As a result, the reaction spreads noticeably along the horizontal axis. This leads to formation of the observed black and white strips, in which the chains significantly differ in the degree of conversion. It corresponds to the greater values of dispersion D.

Thus, local ordering might affect the reaction in a two-dimensional system (under confined conditions) and a formation of certain nano-structures is possible in such systems.

Conclusions

The influence of energetic parameters of the interchain homo- and heterocontacts on a local ordering of both Bernoullian copolymers and products of a polymeranalogous reaction with accelerating neighbor effect proceeding in confined conditions has been studied using Monte Carlo simulations and probabilistic analysis. When the reaction with intra- and interchain acceleration and local ordering proceed simultaneously, the ordering might affect the process so that the formation of certain nano-structures (in particular, not trivial strip-like ones) is possible.

Acknowledgement: Financial support by President of RF grant No НШ-1598.2003.3 for leading scientific schools is greatfully acknowledged.

[1] H. Hasegawa, T. Hashimoto, in "*Comprehensive Polymer Science",* 2nd Suppl. Vol., S. Aggarval, S. Russo, Eds., Pergamon, Oxford 1996, Ch. 14/497 ff.
[2] P.-G. de Gennes, *Faraday Discuss. Chem. Soc.* **1979**, *68*, 96.
[3] P.-G. de Gennes, *Macromol. Symp.* **2003**, *191*, 7.
[4] G.H. Fredrickson, S.T. Milner, L. Leibler, *Macromolecules* **1992**, *25*, 6341
[5] P.-G. de Gennes, "*Scaling concepts in Polymer Phtsics*", 2nd print., Cornell Univ. Press, Ithaca New York 1985.
[6] J. Houdayer, M. Müller, *Europhys. Lett.* **2002**, *58*, 660.
[7] B.W.Swift, M. Olvera de la Cruz, *Europhys. Lett.* **1996**, *35*, 487.
[8] A.D. Litmanovich, Ya.V. Kudryavtsev, Yu.A. Kriksin, O.A. Kononenko, *Macromol. Theory Simul.* **2003**, *12*, 11.
[9] N.A. Platé, A.D. Litmanovich, O.V. Noah, "*Macromolecular Reactions*", John Wiley & Sons, Chichester 1995.

Reactivity of Supramolecular Systems Based on Calix[4]resorcinarene Derivatives and Surfactants in Hydrolysis of Phosphorus Acid Esters

Irina S. Ryzhkina,[*1] Tat'yana N. Pashirova,[1] Wolf D. Habicher,[2] Ludmila A. Kudryavtseva,[1] Alexander I. Konovalov[1]*

[1]A. E. Arbuzov Institute of Organic and Physical Chemistry, Kazan Research Center of the Russian Academy of Sciences, Akad. Arbuzova 8, 420088 Kazan, Russia
E-mail: ryzhkina@iopc.knc.ru
[2]Dresden University of Technology, Institute of Organic Chemistry, Mommsenstr.13, D-01062 Dresden, Germany
E-mail:wolf.habicher@chemie.tu-dresden.de

Summary: The formation of mixed micelles of amphiphilic calix[4]resorcinarenes with aminomethyl (AMC, PAMC), tris(hydroxymethyl)amide (THAC) fragments and the cationic surfactant cetyl trimethylammonium bromide (CTAB) in water and aqueous DMF solutions (10-50 % DMF) leads to the decrease of the critical micelle concentration of the systems and the increase of the size of the mixed micelles in comparison with CTAB micelles. The catalytic activity of the mixed systems in the hydrolysis of phosphorus acid esters is higher than those of CTAB micelle and AMC, PAMC or THAC aggregates.
Keywords: amphiphilic calix[4]resorcinarene; critical micelle concentration; kinetics; micelle formation; mixed micelle; surfactant

Introduction

Micelle formation in mixtures of surfactants is of considerable interest from both fundamental and practical points of view. The behavior of such mixtures, in particular, the catalytic activity is quite different from that of the individual surfactants. Recently the approaches to the syntheses of functionalized amphiphilic compounds are intensivly developed[1,2]. Calix[4]resorcinarenes or resorcarenes (CRA) attracted much attention as cyclooligomes with a stronge inherent amphiphilicity[3,4]. Two feathers are essential for any amphiphilic compounds: the adsorption at the interface and the aggregation in solution. The amphiphilicity of resorcarenes was used for the formation of organized mono- and multi-layers at the air-water interface or on solid surface[3,4]. However, the aggregation of calix[4]resorcinarenes and their derivatives in the solutions and the effect of the nature of surfactants on the aggregate behavior, the properties and the reactivity of calix[4]resorcinarenes has been insufficiently studied until now[5,6].

 DOI: 10.1002/masy.200450605

The aim of this work was to study the micellization in mixed systems based on different calix[4]resorcinarenes (CRA) **1-6** and the cationic surfactant cetyltrimethylammonium bromide (CTAB) in water and aqueous DMF solutions, as well as the study of catalytic activity of mixed micellar systems on the base of these compounds in the hydrolysis of 4-nitrophenyl bis(chloromethyl)phosphinate (NBCP) **7** (Scheme 1). The previously obtained results reveal that aggregates of amphiphilic resorcarenes strongly bind phosphorus acid esters and exhibit high hydrolytic activity at low concentrations of resorcarenes[7,8].

We investigated the aggregation behavior and the reactivity of amphiphilic derivatives of resorcarene with different polar head groups. Two types of resorcarene were included for the investigation. First are resorcarene derivatives with aminomethyl (AMC **1, 2**) or tris(hydroxymethyl)amide (THAC **5, 6**) fragments as hydrophilic groups and long alkyl chain as hydrophobic group. Second are resorcarenes with four strong polar alkylphosphonic acid (PAMC **3,4**) fragments instead of hydrophobic tails of AMC. These are compounds without a clear separation between hydrophilic and hydrophobic parts.

Experimental

AMC, PAMC were synthesized according to literature[9,10]. THAC **5, 6** were obtained from the corresponding derivatives of resorcarene [11] and tris(hydroxymethyl)aminomethan. The structures of compounds **1-6** were proved by means of FTIR, NMR ^{1}H, ^{13}C, ^{31}P spectroscopy and MALDI TOF mass spectroscopy. The aggregation of the compounds was explored by surface tension (σ) and conductivity (χ) measurements. The size of aggregates and mixed micelles was estimated by the NMR FT-PGSE method. The conformation of resorcarenes and the intra- and intermolecular interactions were studied by ROESY NMR. The experiments were carried out in water and water/DMF solutions in presence and in absence of the cationic surfactant CTAB. The kinetics of the reactions was studied by spectrophotometry from the increase in optical density due to the formation of p-nitrophenolate (λ 400 nm) and ^{31}P NMR measurements under pseudo-first order conditions at 30 °C. The quantitative treatment of the experimental kinetic data (the observed rate constant for hydrolysis, k_{obs}) in terms of a pseudo-phase model of micellar catalysis permitted to characterize on the known equation [12] the reactivity of compounds in a micellar microenvironment (the rate constant in micellar phase, k_m), to estimate the binding efficiency of reagents with micelles (the constant of binding of substrate by the aggregates, K_b) and to evaluate the critical micelle concentration (CMC). As AMC are practically water-insoluble, these compounds were studied in aqueous DMF solutions. THAC and PAMC are very readily soluble in aqueous solutions.

Results and discussion

It was found, that AMC **1**, **2** and PAMC **3, 4** are surfactants in a water-DMF and aqueous solutions. The CMC values of PAMC **3** and **4** are practically the same. Table 1 presents the CMC values of studied compounds. The aggregation behavior of THAC **5, 6** surprisingly differs from those of AMC **1, 2** and PAMC **3, 4**. THAC **5** and **6** are not surface active at least at $1 \cdot 10^{-2}$ mol·l^{-1} in water and aqueous media containing 10% of DMF or 30 % of DMSO as far as surface tensions are concerned (Figure 1). Nevertheless an indication of their amphiphilic nature comes from their broadened ^{1}H NMR signals in D_2O.

Indeed combined evidence from NMR FTPGSE method and molecular simulation by HYPERCHEM indicates that THACs form small micelle-like nanoparticles in water. Molecular simulation has shown that the maximum intramolecular distance between protons of THAC **5** molecule in 1,3-alternate conformation equals 28 Å. The study of aggregate formation of THAC **5** by NMR FTPGSE method reveals that the effective hydrodynamic diameter of kinetic particles in water are 66 Å and 77 Å at the concentration of THAC $1 \cdot 10^{-3}$ and $5 \cdot 10^{-3}$ mol·l^{-1}, correspondingly. Taking into account the molecule size of THAC these values of diameters of kinetic particles mean that in warer THAC **5** forms the small aggregates, which are composed of 3-4 molecules of THAC.

Scheme 1. Hydrolysis of 4-nitropheny bis(chloromethyl)phosphinate in the presence of various resorcarenes.

1 - 6 + **7** $\xrightarrow{+H_2O}$ − HO–C$_6$H$_4$–NO$_2$; ^{4}R(^{5}R)P(O)OH + **1 - 6**

AMC -**1**: $R^1 = C_9H_{19}$; $R^2 = H$; $R^3 = CH_2N(CH_3)_2$; **2**: $R^1 = C_{11}H_{23}$; $R^2 = H$; $R^3 = CH_2N(C_2H_5)_2$

PAMC- **3**: $R^1 = CH_2P(O)(OH)(OC_3H_7)$; $R^2 = H$; $R^3 = CH_2N(C_2H_5)_2$; **4**: $R^1 = CH_2P(O)(OH)(OC_4H_9)$; $R^2 = H$; $R^3 = CH_2N(C_2H_5)_2$

THAC- **5**: $R^1 = C_8H_{17}$; $R^2 = CH_2C(O)NHC(CH_2OH)_3$; $R^3 = H$; **6**: $R^1 = C_{11}H_{23}$; $R^2 = CH_2C(O)NHC(CH_2OH)_3$; $R^3 = H$

NBCP- **7**: $R^4 = R^5 = ClCH_2$

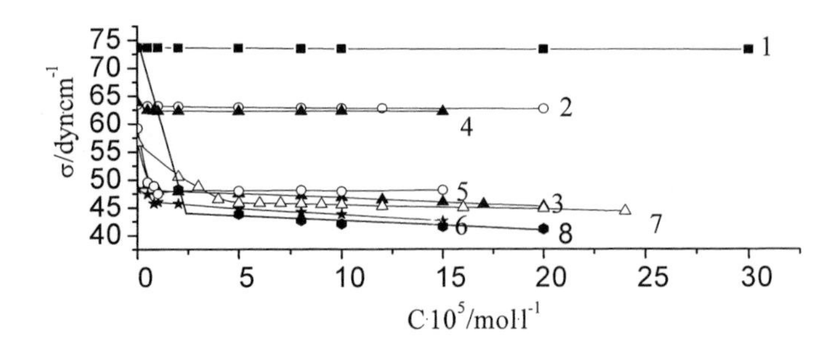

Figure 1. Surface tension of aqueous solutions of CRA as a function of their concentration; reaction conditions: T = 30 °C, medium: 1 = THAC **5, 6** -water, 2 = THAC **5**-10 % DMF, 3 = THAC **5**-30% DMF, 4 = THAC **5**-30% DMSO, 5 = THAC **5** -50 % DMSO, **6** - 30% 1,4-dioxan, 7 = AMC **1**-30% DMF, 8 = PAMC **4**-water.

Recently it was shown[13,14] that analogous resorcarenes having four long alkyl chains (R^1=$C_{11}H_{23}$) and eight saccharide moieties are surface inactive in water. Nevertheless these compounds form unusual stable and essentially irreversible small micelle-like nanoparticals in the solvent. That is, perhaps, aggregates of THAC are irreversible small micelle-like nanoparticals too. THACs nanoparticles shows no surface activity even in concentration $1 \cdot 10^{-2}$ $mol \cdot l^{-1}$ indicating that there is no equilibrium dissociation of an aggregate into monomers in water and in aqueous media containing 10% of DMF or 30 % of DMSO.

However, THAC are the micelle forming surfactants in the solutions with the higher concentration of organic solvent, for example 30% of DMF and 1,4-dioxan or 50 % of DMSO (Figure 1).

Table 1. CMC values ($mol \times l^{-1}$) of CTAB, AMC **1**, PAMC **3, 4** and mixtures of CRA with CTAB in water and in aqueous DMF solutions.

Content of DMF in v. %	CMC/$mol \times l^{-1}$					
	CTAB	AMC **1**	PAMC **3,4**	AMC**1**-CTAB	PAMC **3,4**-CTAB	THAC **5**-CTAB
0	8×10^{-4}	–	2×10^{-5} 1×10^{-4}	–	$^{a}2\times10^{-4}$, $^{b}8\times10^{-4}$	$^{a}2\times10^{-5}$, $^{b}5\times10^{-5}$, $^{c}3\times10^{-4}$, $^{d}1\times10^{-3}$
30	7×10^{-3}	$5\cdot10^{-5}$	2×10^{-5} 1×10^{-4}	$^{a}1\times10^{-4}$ $^{d}3\times10^{-3}$	7×10^{-4}	$^{a}2\times10^{-5}$, $^{c}6\times10^{-4}$, $^{d}1\times10^{-3}$
50	6×10^{-3}	$7\cdot10^{-5}$	6×10^{-4}	3×10^{-4}	8×10^{-4}	-

[a]CMC-1, [b]CMC-2, [c]CMC-3, [d]CMC-4 based on data of figure 2 for PAMC **3** and THAC **5**.

The study of aggregate formation of THAC **5** by NMR FTPGSE method in aqueous DMF (30 vol.% of DMF) allowed us to determine the effective hydrodynamic diameter of kinetic particles in the solutions, wich are 47 Å. Obviously, such organic solvents as DMF or DMSO cause the micelle transition leading to the decrease of micelle size and the appearance of equilibrium dissociation of a aggregate into monomers, that is surface activity of THAC.

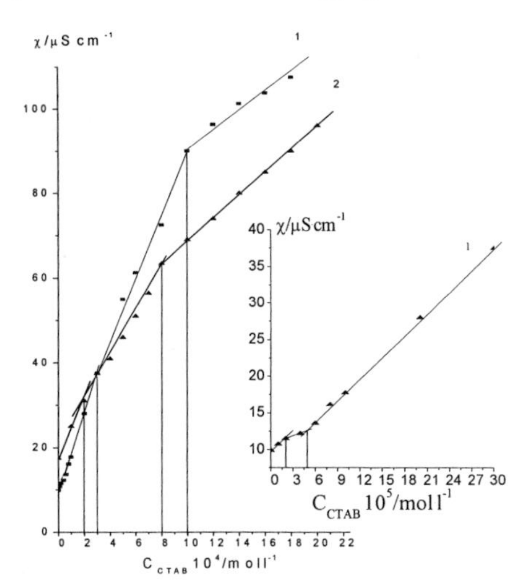

Figure 2. Specific conductance of aqueous CTAB solutions as a function of CTAB concentration in presence of THAC **5** (1) and PAMC **3** (2); reaction conditions: T = 30 °C , C_{THAC} = 2×10^{-5} ; C_{PAMC} = 1×10^{-4} mol×l^{-1}. Insert: Specific conductance of aqueous CTAB solutions as a function of CTAB concentration in presence of THAC **5** (1); reaction conditions: T = 30 °C, C_{THAC} = 2×10^{-5} mol×l^{-1}.

Thus, it was determined that all studied amphiphilic resocarenes are micelle forming amphiphiles and surfactants in the water-DMF solutions. PAMC are the micelle forming amphiphiles and the surfactants in water and aqueous DMF solutions. Micellization of macrocyclic resocarene surfactants is characterized by CMC values, which are less than conventional surfactant such as CTAB (Table 1).

The study of self-organization of resorcarenes **1-6** in the presence of CTAB confirms the formation of mixed micelles based on these compounds. The mixed aggregate formation is accompanied by the change of CMC values of the individual systems (Table 1). Another confirmation of the formation of mixed aggregates is the peculiarity of the aggregation behavior of mixtures observed at different ratios of components.

Figure 2 shows the availability of several transition points corresponding to the CMC values summarized in Table 1. The interaction of resocarenes with CTAB yields mixed aggregates of variable composition depending on the molar ratio of components. The first CMC points (CMC-1) appear at the ratio resorcarene: CTAB in region from 1:1 to 1:5 depending on the nature of resorcarene, in particular, 1:1 for THAC and 1:5 for AMC. The increase of concentration of CTAB leads to the formation of larger aggregates (CMC-2).

Thus, resorcarene forms aggregates with a monomeric form or cations of CTAB (CMC-1, CMC-2), if the surfactant is in deficiency. The increase of concentration of CTAB leads to a penetration of resorcarenes into the CTAB micelles at the ratios of resorcarene: CTAB approximately from 1:10 to 1:50 (CMC-3, CMC-4). Such aggregation behavior of mixed systems indicates a gradual formation of CRA-CTAB mixed aggregates.

The kinetic study of the reaction of 4-nitrophenyl bis(chloromethyl)phosphinate (NBCP) **7** with CRA was carried out in the absence and in the presence of the surfactant CTAB (Table 2).

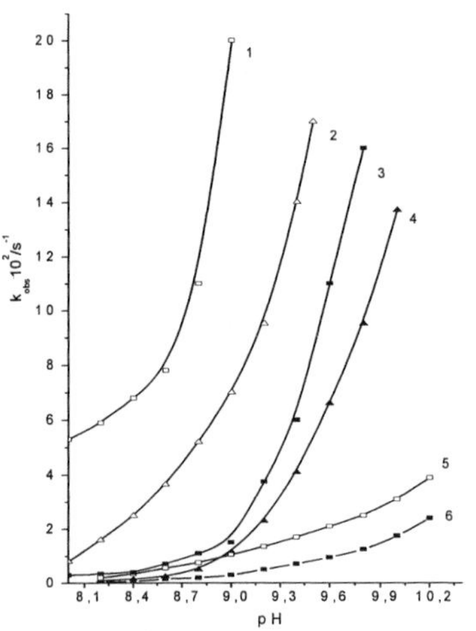

Figure 3. Observed rate constants of the reaction (k_{obs}) of 7 with THAC **5** and PAMC **4** as a function of pH in water and in aqueous DMF (30% vol. of DMF) solution in absence and in presence CTAB; reaction conditions: T = 30 °C, C_{PAMC}= 6×10^{-4} mol×l^{-1}, C_{THAC}= 4×10^{-4} mol×l^{-1}, C_{CTAB}= 5×10^{-3}, medium: 1 = PAMC-CTAB-water, 2 = THAC-CTAB-water, 3 = PAMC-water, 4 = THAC-water, 5 = THAC-CTAB-DMF-water, 6 = THAC-DMF-water.

The kinetic study in depending of pH has shown that the reactivity of water-soluble PAMC and THAC is slightly different in neutral or weak basic aqueous solutions (Figure 3). In theory the reactivity of PAMC have to be higher then THAC due to presence of strong nucleophilic phenoxyl and amino groups, but in reality is lower. Furthermore, mixed system PAMC **4**/CTAB/DMF/water differs from those based on AMC **2** by less efficiency of substrate binding and the less catalytic activity (Table 2). Apparently, the some functional groups of PAMC take part on the self-organization or a mixed aggregate formation. This leads to decreases the reactivity of PAMC aggregates and their mixed micelles. The presence of DMF decreases the reactivity of both water-soluble resorcarenes, but aggregates of THAC are more sensitive to addition of DMF (Figure 4 and Figure 5).

The unusual stable or irreversible assemblies of THAC have a higher substrate binding efficiency compared with both AMC and PAMC. The study of intermolecular interaction between THAC and p-nitrophenol by ROESY NMR revealed the very strong interaction between aromatic protons of p-nitrophenol and methylene groups of alkyl chains of THAC.

Table 2. Parameters of the reaction of substrate **7** with AMC **2**, THAC **5**, PAMC **4**, and CTAB in water and aqueous DMF solution (30% vol. of DMF).

It clearly confirms the formation of a stable substrate-micelle complex. The formation of mixed aggregates is accompanied by a sharp increase in the catalytic activity of the systems, a decrease in their critical micelle concentration in comparison with the CTAB micelles and a shift of reactivity from weak basic to neutral pH of the reaction mixture (Figure 3-5, Table 2). Mixed micelles of THAC are characterized by the low CMC value, a high substrate binding efficiency and a high reactivity in hydrolysis of PAE in weak basic media (Table 2). By analogy with resorcinarene aggregates the reactivity of mixed micelles decreases in presence of DMF. This process occurs very sharply in a case of mixed micelles of THAC (Figure 5).

Table 2.

Systems	CMC/mol×l^{-1}	K_b/l×mol^{-1}	k_m/s^{-1}	pH
AMC **2**-CTAB-DMF-water $C_{CTAB}=5\times10^{-3}$ mol×l^{-1}	2×10^{-4}	1700	6.7×10^{-2}	8
AMC **2**-CTAB-DMF-water $C_{CTAB}=1\times10^{-2}$ mol×l^{-1}	2×10^{-4}	1745	1.5×10^{-1}	8
THAC **5**-DMF-water	3.6×10^{-4}	4600	1.2×10^{-2}	9,2
THAC **5**-CTAB-DMF-water, $C_{CTAB}=5\times10^{-3}$ mol×l^{-1}	2.7×10^{-4}	4000	2.4×10^{-2}	9,2
THAC **5**-water	1×10^{-4}	4500	3.4×10^{-2}	9,2
THAC **5**-CTAB-water, $C_{CTAB}=5\times10^{-3}$ mol×l^{-1}	1×10^{-5}	6000	2.1×10^{-1}	9,2
PAMC 4-water	3×10^{-4}	2500	3.2×10^{-2}	9.0
PAMC **4**-DMF-water	6×10^{-4}	1300	2.2×10^{-2}	9.0
PAMC **4**-DMF(50%)-water	5×10^{-4}	2400	4.2×10^{-3}	9.0
PAMC **4**-CTAB-water $C_{PAMC}=4\times\cdot10^{-4}$ mol×l^{-1}	8×10^{-4}	420	3.6×10^{-2}	8.0
PAMC **4**-CTAB-DMF-water $C_{PAMC}=4\times10^{-4}$ mol×·l^{-1}	1.5×10^{-3}	90	1.3×10^{-2}	8.0
CTAB-DMF-water	6×10^{-3}	100	1×10^{-2}	8.0

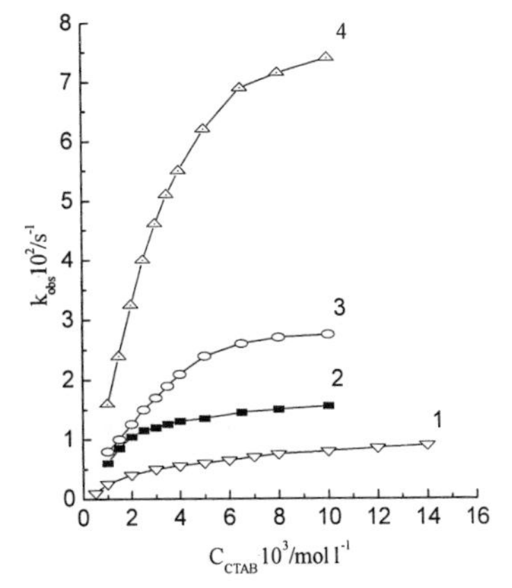

Figure 4. Observed rate constants of the reaction (k_{obs}) of 7 with PAMC **4** in water (2-4) and in aqueous DMF solution (1) (30% vol. of DMF) as a function of CTAB concentration; reaction conditions: T = 30 °C, pH = 8, 1 = C_{PAMC}= 4×10^{-4} mol×l^{-1}, 2 = C_{PAMC}= 2×10^{-4} mol×l^{-1}, 3 = C_{PAMC}= 4×10^{-4} mol×l^{-1}, 4 = C_{PAMC}= 6×10^{-4} mol×l^{-1}.

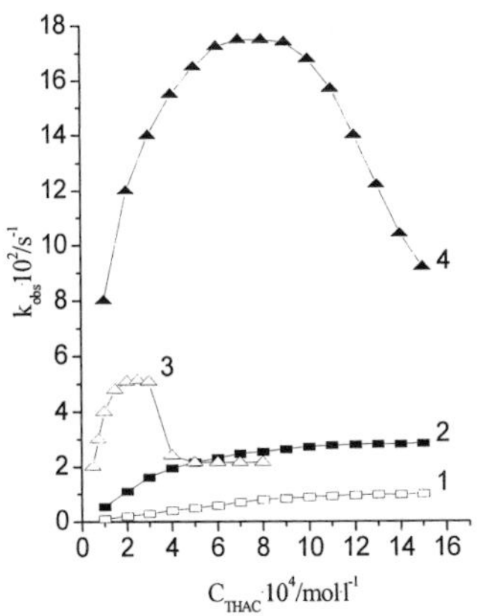

Figure 5. Observed rate constants of the reaction (k_{obs}) of 7 with THAC **5** as a function of THAC concentration in water (2-4) and aqueous DMF solution (1) (30% vol. of DMF) in absence (1, 2) and in presence of CTAB; reaction conditions: T = 30 °C, pH= 9.2, 3 = C_{CTAB}= 1×10^{-3} mol×l^{-1} , 4 = C_{CTAB}= 5×10^{-3} mol×l^{-1}.

Conclusion

1. Self-organization of amphiphilic derivatives of resorcarenes leads to the formation of surface inactive in water irreversible micelle-like nanoparticals, to surface-active aggregates in water and water-organic media and to the formation of mixed aggregates in water-surfactant and in water-organic-surfactant systems. Micellization of the macrocyclic calixarene surfactants is characterized by CMC values, which are less than those of amphiphilic phenols and conventional surfactants such as CTAB. The mixed aggregate formation is accompanied by a decrease of CMC values of CTAB and an increase of the micelles size.

2. The resorcarene aggregates and mixed micelles based on these compounds and CTAB are the catalytic systems, which provide a specific microenvironment for substrate binding and subsequent hydrolysis of phosphorus esters in water and water-DMF media. The formation of mixed aggregates is accompanied by a sharp increase in the catalytic activity of the systems, a decrease in their critical micelle concentration compared with the CTAB micelles.

The work was financially supported by the Russian Foundation for Basic Research (Project no. 03-03-32953) and the DFG within the "Sonderforschungsbereich 387".

[1] F. M. Fendler, V.A. Migulin, *J. Org. Chem.* **1999**, 64, 8916.
[2]Handbook of surfactants and related products/ Ed. M. Yu. Pletnev, Clavel Pablishing, 2002, 768.
[3] *Calixarenes 2001.* Eds. Z. Asfari, V. Bohmer, J. Harrowfield, J. Vicens. Kluwer Academic Pablishers. Dordrecht-Boston-London, 2001, 683.
[4] *Calixarenes in Action.* Eds L. Mandolini, R Ungaro. Imperial College Press, London, 2000, 271.
[5] I. S. Ryzhkina, L. A. Kudryavtseva, A. R. Burilov, E. Kh. Kazakova, A.I.Konovalov, *Izv. Akad. Nauk, Ser. Khim.*, **1998**, 275 [*Russ. Chem. Bull.* , Int. Ed., **1998**, 47, 269].
[6] I. S. Ryzhkina, Ya. A. Babkina, S.S. Lukashenko, K.M. Enikeev, L.A.Kudryavtseva, A.I. Konovalov *Izv. Akad. Nauk, Ser. Khim.*, **2002**, 2026[*Russ. Chem. Bull.* , Int. Ed., **2002,** 51, 2183].
[7] I.S. Ryzhkina, V. V. Yanilkin, V. I .Morozov, L. A. Kudryavtseva, A.I.Konovalov, *Zh.. Fizich. Khim.,* **2003**, 77, 491[*Russ. J. Phys. Chem.,* **2003,** 77, 426 (Engl. Transl.)].
[8] I.S. Ryzhkina, L.A.Kudryavtseva , Ya. A. Babkina, K.M. Enikeev, M.A. Pudovik, A.I. Konovalov, *Izv. Akad. Nauk, Ser. Khim.*, **2000**, 1361[*Russ. Chem. Bull.* , Int. Ed., **2000**, 49, 1355].
[9] Y. Matsuskita, T. Matsui, *Tetrahedron Lett.*, **1993**, 34,7433.
[10] E.V. Popova, A.R. Burilov, M.A. Pudovik, W.D. Habicher, A.I. Konovalov, *Zh..Obshch. Khim*, **2002**, 72, 1049 (*Russ. J. Gen. Chem.*, **2002**, 72, (Engl. Transl.)].
[11] E. U. Toden van Velzen, J.F.J. Engbersen, D.N. Reinhoudt, *Synthesis,* **1995**, 989.
[12] E. Fendler, J. Fendler, in *Advances in Physical Organic Chemistry*, Ed. V. Gold, Academic Press, London, 1970.
[13] O.Hayashida, K.Mizuki, K. Akagi, A. Matsuo, T. Kanamori, T. Nakai, S. Sando, Y. Aoyama, *J.Am.Chem.Soc.***2003,** 125, 594
[14] Y. Aoyama, T. Kanamori, T. Nakai, T. Sasaki, S. Horiuchi, S. Sando, T. Niidome *J.Am.Chem.Soc.***2003,** 125, 3455.

Studies on the Copolymerisation of N-Isopropyl-Acrylamide with Glycidyl Methacrylate

Veena Choudhary, Sonal Rajya, Dipti Singh*

Centre for Polymer Science and Engineering, Indian Institute of Technology, Delhi, Hauz Khas, New Delhi 110016, India
E-mail: veenach@hotmail.com

Summary: The paper describes the homopolymerisation and copolymerisation of N-isopropyl acrylamide (NIPAM) with glycidyl methacrylate (GMA) in solution at 60°C using azobisisobutyronitrile (AIBN) as an initiator and dioxan as solvent. Copolymers were synthesized by varying the mol fraction of GMA in the initial feed from 0.025-0.125. All the polymerization reactions were terminated at low % conversion (10-15%) and the copolymer composition was determined by measuring the epoxy content. Percent epoxy content was determined by titration method using pyridine- HCl mixture. The reactivity ratios determined using Fineman-Ross method were found to be 0.94±0.05 (r_1, NIPAM) and 1.05±0.08 (r_2, GMA). All the polymers have high molecular weights with wide molecular weight distribution as determined by gel permeation chromatography (GPC) i.e. M_n in the range of 3.7×10^4 - 7.8×10^4 and M_w in the range of 1.2×10^5 - 4.1×10^5 with a polydispersity index in the range of 2.3-5.3. Lower critical solution temperature (LCST) of NIPAM homopolymer and copolymers was determined by recording DSC scans of polymers in aqueous solution. Incorporation of GMA in the poly(NIPAM) backbone resulted in a decrease in the LCST.

Keywords: copolymerization; glycidyl methacrylate; LCST; N-isopropyl acrylamide

Introduction

Copolymerisation is the most useful method for tailor making a polymer product with specifically desired properties. Copolymerisation modifies the symmetry of the polymer chain and modulates both intermolecular and intramolecular forces, so properties such as lower critical solution temperature (LCST), solubility etc of the smart polymer may be varied within wide limits. Polymers having transition temperature ranging from 3-150°C by using N-substituted acrylamides or methacrylamide as a monomer in the feed has been reported in the literature.[1-4] The cloud point of such polymers was strongly dependent on the polymer structure as well as on

DOI: 10.1002/masy.200450606

the copolymer composition. The greatest number of publication has been devoted to the polymers based on N-isopropylacrylamide, NIPAM, as both homo and copolymers of NIPAM can be easily obtained by free radical polymerization in aqueous and organic solvents. Although copolymerisaton of NIPAM with a variety of vinyl monomers is well reported in the literature however, no systematic study on the copolymerisation of NIPAM with functional monomer is reported.

In recent years, copolymers based on glycidyl methacrylate (GMA) have received increasing attention. The interest in these copolymers is largely due to the ability of pendant epoxide groups to enter into a large number of chemical reactions[5-7], thus offering the opportunity for chemical modification of the parent copolymer for various applications. For example, the copolymers based on glycidyl methacrylate has been used for binding enzymes and other biologically active species[8-9] and in electronic applications as negative electron beam resist. The high reactivity of the epoxide group is primarily due to the considerable strain in the three membered ring. The properties of homopolymer of GMA are largely affected by the polymerization method as well as by the microstructure in the copolymers. The objective of the present work was to combine the useful properties of NIPAM and functional monomer such as glycidyl methacrylate.

The present work was undertaken with an aim to investigate systematically the copolymerisation of NIPAM with varying mole fractions of GMA. Several copolymer samples were prepared by taking 0.025, 0.05, 0.075, 0.1 and 0.125 mole fractions of GMA in the feed. The rate of homopolymerisation of NIPAM /or GMA and copolymerisation of NIPAM: GMA (mole ratio 0.9:0.1) was investigated as a function of time, monomer concentration and initiator concentration. Copolymer composition was determined by measuring the epoxy group content in copolymers. LCST of these copolymers as determined by recording DSC scans in the presence of water decreased with increasing mole fraction of GMA in the copolymers. Thermal stability of the copolymers was determined by recording thermogravimetric traces in nitrogen atmosphere.

Experimental

Materials: N-isopropylacrylamide (NIPAM)(Aldrich), glycidyl methacrylate (GMA)(Fluka), petroleum spirit (Qualigens), methanol (Qualigens), N, N'- methylene bisacrylamide

(BAM)(Aldrich), potassium hydroxide (Qualigens) and sodium metal (S.D Fine chemicals) were used as such.

Azobisisobutyronitrile (AIBN) (High Polymer labs) and benzoyl peroxide (CDH) were recrystallised before use.

Homopolymerisation and Copolymerisation of NIPAM with GMA

Homopolymerisation and copolymerisation in nitrogen atmosphere of NIPAM and GMA at 60 ± 1°C was carried out in solution using AIBN as an initiator and dioxan as solvent.

A 20% w/v solution of NIPAM / GMA/ or required amounts of NIPAM : GMA mixture in dioxan was placed in a flask equipped with a magnetic stirrer, nitrogen inlet tube and reflux condenser. The temperature of the reaction mixture was raised to 60°C while passing nitrogen through it and then 0.5% (w/w) AIBN was added and the reaction was carried out for a desired interval of time, so as to keep the percent conversion below 15. The polymer was precipitated by slowly pouring the solution in to a beaker containing petroleum spirit. The polymer was isolated by filtration, purified by reprecipitating using dioxan /petroleum spirit. The polymers were then dried in vacuum oven at 60°C.

Five copolymer samples were prepared by varying the mole fraction of glycidyl methacrylate (GMA) from 0.025-0.125 in the initial feed. The samples have been designated as PNG followed by numerical suffix indicating the mole fraction of GMA multiplied by 100. For example, copolymers obtained by taking 0.025, 0.05, 0.075 and 0.125 moles of GMA have been designated as PNG–2.5, PNG-5, PNG-7.5 and PNG-12.5 respectively. Homopolymer of GMA and NIPAM have been designated as PG-100 and PN-100 respectively.

Rate of polymerisation

In order to determine the rate of hompolymerisaton and copolymerisation, solution polymerization reactions were carried out in nitrogen atmosphere at 60 ± 1°C for different intervals of time using 0.5% (w/w) AIBN as an initiator using dioxan as a solvent.

For copolymerisations, molar ratio of NIPAM : GMA (0.9:0.1 i.e sample PNG-10) were taken in the initial feed in three necked flask equipped with nitrogen inlet tube and water condenser. Dioxan was used as a solvent. After passing nitrogen for ~ 10 min and heating to 60°C, 0.5 % (w/w) AIBN was added to the continuously stirred reaction mixture. After desired interval of time, the contents of the flask were poured with continuous stirring into excess of petroleum spirit. The precipitated polymers were collected by filtaration followed by drying in vaccum oven at 50-60°C till constant weight. The weight of the polymer was used to calculate the % conversion.

The homopolymerisation and copolymerisation reactions were also carried out at 60 ± 0.1°C in N_2 atmosphere using different initiator concentrations i.e. 0.25, 0.5, 0.75 and 1.0 % (w/w) for studying the effect of initiator concentration on the rate of polymerization.

The effect of monomer concentration on the rate of polymerization was also investigated by keeping initiator concentration constant (0.5 % w/w) and taking 20 % (w/v) and 40 % (w/v) monomer concentration in the initial feed.

Characterization

Structural characterisation of the copolymers was done by recording FT-IR spectra in thin films using Perkin-Elmer 580B infra-red spectrophotometer.

The copolymer compositions were determined by estimating the epoxy group using pyridine-HCl method.

Intrinsic viscosity of the homopolymer and copolymers was determined in dioxan at 30 ± 0.1°C using Ubbelohde suspension level viscometer. Waters gel permeation chromatograph was used to determine molecular weight and molecular weight distribution.

Thermal characterization

DSC traces of homopolymers and copolymers in presence of water were recorded using TA 2100 thermal analyzer having 910 DSC module. A sample weight of 4±1 mg and 2 mg of water in DSC crucible (which were sealed to prevent the evaporation of water) were used for recording DSC traces at a heating rate of 5°C /min. The lower critical solution temperature (LCST) was recorded as the endothermic shift in the base line.

Thermal stability of the homopolymers and copolymers was evaluated by recording TG/DTG traces in nitrogen atmosphere in the temperature range of 50-800°C. A Dupont 2100 thermal analyzer having 951 TG module and a heating rate of 20°C /min was used for recording TG/DTG traces in nitrogen atmosphere. A sample weight of 10 ± 2 mg was used in each experiment.

Results and Discussion

Rate of polymerisation

For kinetic studies, homopolymerisation and copolymerisation of NIPAM: GMA (0.9:0.1,sample PNG-10) was carried out in nitrogen atmosphere for different intervals of time or at varying concentrations of initiator. The rate of polymerization may be influenced by (a) structure of monomers/comonomers (b) monomer concentration (c) initiator concentration and (d) temperature.

Effect of time and monomer concentration

The rate of homopolymerisation of NIPAM /GMA and copolymerisation of NIPAM:GMA (molar ratio 0.9:0.1 sample PNG-10) was followed gravimetrically. A linear increase in % conversion was observed in the homopolymerisation of NIPAM (sample PN-100), GMA (sample PG-100) and PNG-10 monomer concentrations of 20%(w/v) and 40%(w/v). As expected the rate of polymerization was found to be higher at 40%(w/v) monomer concentration in all the systems (Table 1).

Effect of initiator concentration

The effect of initiator concentration on the rate of polymerization was investigated by taking 0.25, 0.50, 0.75 and 1.0 % w/w of AIBN in the initial feed. The monomer concentration (20%) (w/v) and the reaction time (30 min) was kept constant to investigate the effect of initiator concentration. A plot of R_p vs $In^{1/2}$ for homopolymerisation of NIPAM or GMA and their copolymerisation (sample PNG-10) was found to be linear indicating thereby that the rate of polymerization was dependent on the half power of the initiator concentration. This implies that the termination was either by combination or disproportionate. No departure from polymerization kinetics explains the absence of side reactions especially primary radical termination.

Characterization of copolymers

In the FT IR spectra of copolymers a characteristic absorption band due to amide group of NIPAM at 1652 ± 1cm^{-1} and 1521 ± 1cm^{-1} and due to carbonyl and epoxy group of glycidyl at 1732 ±1cm^{-1} and 909 ± 1 cm^{-1} respectively was observed. The epoxy content in the copolymers

Table 1. Effect of monomer concentration on the rate of polymerization

Sample Designation	Weight of monomers in feed (g)		Dioxan (ml)	Rp % conversion/min
	NIPAM	GMA		
PN-100	4	-	10	0.96
PN-100*	4	-	20	0.61
PG-100	-	4.0	10	0.33
PG-100*	-	4.0	20	0.281
PNG-10	4.068	0.568	11.6	0.72
PNG-10*	4.068	0.568	23.2	0.45

** represents 20% (w/v) monomer concentration*

determined quantitatively by titration method was used to calculate the copolymer composition (Table 2). From the knowledge of copolymer composition, reactivity ratios were calculated using Fineman Ross and Kelen Tudos method and it was found to be 0.94 ± 0.05 (r_1) and 1.05 ± 0.08(r_2).

Molecular characterization

Intrinsic viscosity increased with increasing GMA content in the copolymers (Table 3). [η] is a measure of hydrodynamic volume and depends on the molecular weight as well as on the size of the polymer coil in solution. Increase in [η] with inceasing GMA content could be due to the better solvating power of dioxan. All the homopolymers and copolymers have high molecular weight i.e. M_n in the range of 3.7 x 10^4 – 7.8 x 10^4 and M_w in the range of 1.2 x 10^5 – 4.1 x 10^5 with polydispersity index in the range of 2.3-5.3.

Table 2. Details of feed composition and copolymer composition

Sample designation	Mole fraction of monomers in Feed		Epoxy content %	Mole fraction of GMA (m_2)
	NIPAM (M_1)	GMA (M_2)		
PNG 2.5	0.975	0.025	0.98	0.026
PNG 5.0	0.950	0.050	2.32	0.062
PNG 7.5	0.925	0.075	3.27	0.088
PNG 10	0.900	0.100	4.50	0.122
PNG 12.5	0.875	0.125	5.24	0.143

Thermal characterization

Fig 1 and 2 show the DSC scans of PN 100* of varying molecular weight and copolymers of varying composition respectively in the presence of water. An endothermic shift in base line due to phase transition, known as lower critical solution temperature (LCST) was observed in all the DSC scans. The endothermic transition was characterized by noting T_i (onset temperature), T_{mid} (LCST) and T_f (final temperature) and the results are summarized in Table 3. LCST of PN-100 was found to be in the range of 30.4-31.5 °C and was marginally affected by the molecular weight. LCST decreased with increasing amounts of GMA in the copolymers. This can be attributed to the hydrophobic nature of GMA which increases the hydrophobicity of the copolymers and lowers the phase transition temperature An increase /decrease in LCST upon copolymerisation of NIPAM with hydrophilic or hydrophobic monomers has already been reported in the literature. [4,5]

Table 3. Results of intrinsic viscosity and LCST (DSC traces recorded at a rate of 5°C/min)

Sample designation	[η] mL/g	T_i (°C)	T_{mid} (°C)	T_f (°C)
PN-100	12	29.9	30.9	31.5
PN-100	27	29.3	31.5	32.2
PN-100	29	29.1	31.1	32.0
PNG-2.5	14	24.4	28.63	30.20
PNG-5	21	24.0	27.27	30.00
PNG-7.5	48	23.10	27.00	29.90
PNG-10	62	14.60	20.00	23.00
PNG-12.5	133	-	-	-

Thermogravimetric analysis

The relative thermal stability of samples was compared by comparing the initial decomposition temperature (IDT), temperature of maximum rate of weight loss (T_{max}) and the final decomposition temperature (FDT) and the results are summarized in Table 4. A typical TG/DTG trace PNG-10 is shown in Fig 3. A weight loss of 5-13% was observed in the temperature range of 50-300°C in all the samples which may be due to the presence of absorbed and bound water. Major weight loss was observed in the temperature range of 300 to 800°C. The effect of molecular weight on the thermal stability of NIPAM homopolymer and NIPAM-GMA

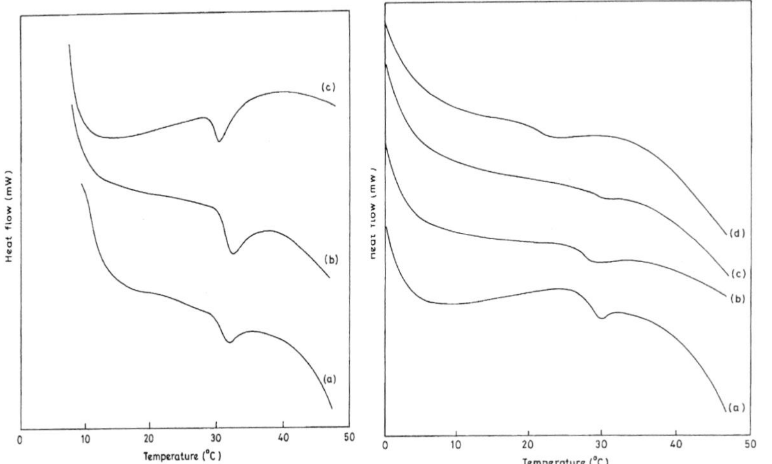

Figure1. DSC traces of PN*- 100 at varying % conversion (a)3.4 (b) 29.4 (c) 38.1

Figure 2. DSC traces of copolymers (a) PNG-2.5 (b) PNG-5 (c) PNG-7.5 (d) PNG-10

copolymer i.e. PNG-10 was also investigated. As expected thermal stability of homopolymer was independent of molecular weight, whereas in copolymer (PNG-10), an increase in IDT, T_{max} and FDT was observed. An increase in these temperatures with % conversion in copolymer could be due to the change in copolymer composition or due to the microstructural changes. The varying mole fraction of GMA showed a similar thermal behaviour with slight variation of IDT, T_{max} and FDT values.

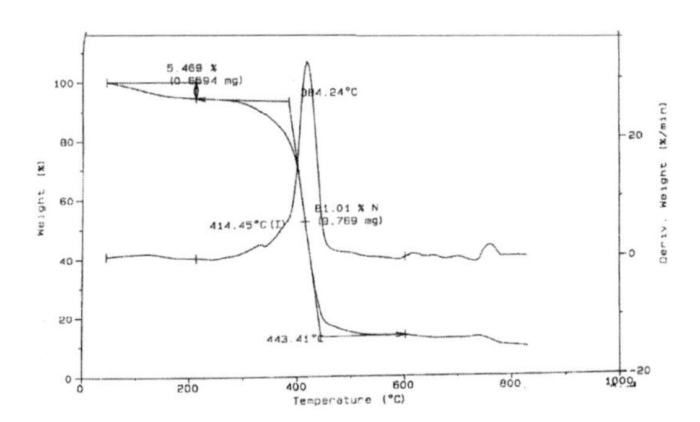

Figure3. TG/DTG trace of PNG-10

Table 4. Results of thermogravimetric analysis of homopolymers and copolymers (Heating rate 20°C/min, N_2 atmosphere)

Sample designation	% Conversion	[η]	IDT (°C)	T_{max} (°C)	FDT (°C)
PN-100	27.23	22.9	415.4	437.5	450.3
PN-100	65.69	50.0	417.1	440.6	449.9
PG-100	10.42	39.0	406.5	425.0	441.9
PG-100	29.82	91.5	411.6	428.1	446.6
PNG-2.5	8.76	14.0	395.2	418.8	433.4
PNG-5.0	10.92	21.0	396.1	428.1	449.9
PNG-7.5	11.56	48.0	389.1	415.6	445.1
PNG-10	12.12	54.0	384.2	413.3	443.4
PNG-10	22.16	62.0	408.9	425.0	438.6
PNG-10	32.15	70.4	414.3	435.5	445.0
PNG-12.5	14.85	133.2	411.5	431.3	445.2

Conclusion

Incorporation of small amounts of GMA can be used to introduce functional group in the polymer backbone without much affecting the LCST.

[1] X. Kono, K. Kawakami, K. Morimoto, T. Takagishi, *J. Appl. Polym. Sci.*, **1999**, *72*,1763.
[2] K. S. Oh, J. S. Oh, H. S. Choi, Y. C. Bae, *Macromolecules*, **1998**, *31*, 7328.
[3] H. Y. Liu, X. X. Zhu, *Polymer*, **1999**, *40*, 6985.
[4] X. Y. Shi, J. B. Li, C. M. Sun, S. K. Wu, *J. Appl. Polym. Sci.*, **2000**, *75*, 247.
[5] J. Kalal, F. Svec, V. Marousek, *J. Polym. Sci. Symp.*, **1974**, *47*, 155.
[6] H. Lee, K. Newille, *Handbook of Epoxy Resins*, McGrawHill, New York 1967.
[7] D. A. Tomalia, *"Functional Monomers"*, R. H. Yocum, E. B. Nuquist, Eds., Dekker, New York 1974, Vol. 2/Ch. 1.
[8] J. Kalal, *J. Polym. Sci., Polym. Symp.*, **1978**, *62*, 251.
[9] J. Korav, M. Navratilova, L. Shursky, J. Drobnic, F. Svec, *Biotech. Bioeng.*, **1982**, *24(4)*, 837.

Heterotelechelic Poly(oxazolidine-acetals) and Their Functionalization

Jolanta Maślińska-Solich, Edyta Gibas*

Department of Physical Chemistry and Technology of Polymers, Silesian University of Technology, ul. M. Strzody 9, 44-100 Gliwice, Poland
E-mail: jms1acet@zeus.polsl.gliwice.pl

Summary: Telechelic poly(1,3-oxazolidine-acetal)s with $-CH_2OH$ and –CHO groups were synthesized by polycondensation of the 2-amino-2-hydroxy-1,3-propanediol (**1**) (TRIS) with terephthaldehyde (**2**). The degree of polymerization (DP) was controlled by the ratio of **1** to **2** at the given reaction time. Characterization was achieved by ^1H and ^{13}C NMR and IR spectroscopy. The distribution of oxazolidine-acetal units in the polymer chain has been performed using ESI-MS. The activities of telechelic poly(oxazolidine-acetal) were determined in reaction oxidation (4-chloroperbenzoic acid), reduction (CH_3MgCl) and nucleophilic substitution (acylation, alkylation).

Keywords: nucleophilic substitution; oxazolidine; oxidation; polycondensation; reduction

Introduction

The introduction of a particular functional group, reagent or catalyst to a polymer supported by chemical modification of the latter is by far the most commonly used method. This approach has the advantage of not requiring the synthesis of an elaborate comonomer.

In this context, some attempts have been made to use functionalized macromolecules with N,O- and O,O-acetal moieties for facile preparation of grafting and branched structures.

Experimental

Measurements

The NMR spectra were recorded using a UNITY INOVA 300 MHz (Varian Associates Inc.) multinuclear spectrometer. The ^1H and ^{13}C NMR spectra were run in DMSO-D_6 and in deuterated chloroform using TMS as an internal standard.

ESI-MS experiments were performed using a Finnigan MAT TSQ 700 triple stage quadrupole mass spectrometer equipped with an electrospray ionization (ESI) source (Finnigan, San Jose, CA, US).

 DOI: 10.1002/masy.200450607

EPR spectrum was recorded at room temperature on a RADIOPAN type SE/X – 2543 spectrometer with compatible computer acted as the control center of this system, operating at 9.3 GHz microwave frequency. The modulation amplitude was 0.1 mT, the attenuation 10 dB, the time constant was 0.3 s, the receiver gain $1*10^4$ (50% diode current) and the scan range was 20 mT.

Polycondensation

Poly(oxazolidine-acetals) **3** were obtained by polycondensation of **1** with **2** (at molar ratio 1:1) at 80 – 110 °C in solution (benzene, toluene, benzene DMSO, benzene DMF) at 80-110 °C in the presence of an acidic catalyst (p-toluenesulfonic acid (TsOH) or its complex with poly(4-vinylpyridine-divinylbenzene) PVP TsOH).

Chemical modification

*N-Alkylation of polycondensation products (**4**)*

Polymer samples **3** (0.5-1.0g) were alkylated with an excess of alkyl iodide (methyl, ethyl, butyl and decyl) at 30 to 40 °C in a sealed tube (5-14 hr) under ultrasonication. The semi-solid product was washed with water, acidified with a diluted solution of acetic acid and extracted with chloroform. The organic layer was first washed with a 10 % solution of NaOH, then with water, finally it was dried over anhydrous $MgSO_4$. Chloroform was evaporated and the residue was dried under reduced pressure. The N-alkylated polymer **4** was purified by precipitation from tetrahydrofurane (THF) solution with methanol. The degree of N-alkylation (60-89 %) was determined from ^1H NMR spectra.

*N-Acetyl derivative of poly(oxazolidine-acetal) (**5**)*

The polymer sample **3**, 1.5g in 10 ml of pyridine and 2.5 ml of acetic anhydride was stirred at room temperature for 14 day. The reaction mixture was then suspended in water ice. The resulting solid polymer was neutralized with sodium hydrocarbonate and then washed with a sufficient amount of water. Finally the polymer was dissolved in THF and after precipitating using ethanol, it was dried under vacuum at 80 °C for several hours. The amount of N-acyloxazolidine units calculated from the ^1H NMR spectrum of **5** corresponds closely to the almost of oxazolidine units in polymer **3** (calc. CH_3CO 20.0%; found 23.5%).

*Reduction with Grignagd reagent (**6**)*

A sample of 1.16g (5.3 mmol) of polymer **3** dissolved in 10 ml of THF was added dropwise to a stirred 2M solution of CH_3MgCl (53 mmol) in THF at 0 °C under argon atmosphere. The reaction mixture was left to warm to ambient temperature during 8 hours and than quenched with saturated aqueous ammonium chloride solution (10 ml). The aqueous phase was separated and extracted with chloroform (4x10 ml). The combined organic extracts were dried ($MgSO_4$) and evaporated. The degree of reduction (57 %) was determined from ^{1}H NMR spectrum.

*Oxidation (**7**)*

The polymer sample **3**, 1g and 2g of 4-chloroperbenzoic acid in 20 ml of dry chloroform were stirred at room temperature for 12 hrs. The chloroform layer was washed with cold 5 % sodium bicarbonate solution and dried over anhydrous $MgSO_4$. The solvent was removed under vacuum and the product **7** was analyzed by EPR (Figure 4).

Results and discussion

Synthesis and characterization

The polycondensation of **1** with **2** was performed in solution (benzene, toluene, benzene DMSO, benzene DMF) at 80-110 °C in the presence of an acidic catalyst.[1] The examination of condensation products allows to follow the influence of several parameters on the formation of macromolecules. It was found that the stepwise reaction between the functional groups of reactants **1** and **2** with the molar ratio 1:2 proceeds almost immediately after the start of reaction to form the bis-oxazolidine compound. The O,N acetal units exist predominately in solution as oxazolidine-imine tautomers according the appropriate resonances in their ^{1}H and ^{13}C NMR spectra, as demonstrated in Table 1.

Table 1. The dynamic equilibrium of oxazolidine-imine system in DMSO-D_6.

time	% imine	% oxazolidine
0	-	100
3 h	6.7	93.3
4 days	11.86	88.13
5 days	12.56	87.44

The condensation of **1** with **2** at a molar ratio 1:1 conducted in the presence of an acidic catalyst (TsOH or PVP TsOH) leads to the formation of products with M_w~9600 g · mol^{-1}. An excess of one of the reagents (TRIS) lead to a mixture of oligomers or a crosslinked structure.[1]

The IR and NMR spectra indicate the presence of O-C-O and O-C-N units and also aldehyde groups in the polymers. The IR spectrum of the polymer shows typical absorption peaks at 1080 - 1180 cm^{-1} assigned to -O-C-O and O-C-N groups as well as distinct absorption due to -OH groups at 3340-3450 cm^{-1}. The absorption band at 1645 cm^{-1}(with weak intensity) may be assigned to the CH=N group in the polymer. The ^{1}H NMR spectrum of the polymer showed broad peaks centered at 3.3, 3.65, 3.88. 3.81 ($C\underline{H}_2O$), 5.02, 5.54 (N-C$\underline{H}$-O and O-C$\underline{H}$-O), 6.8-7.9 (aromatic protons) and a weak peak at 10 ppm. By monitoring the intensities of the peaks due to Ar-$\underline{H}$, OC$\underline{H}$O, OC$\underline{H}$N, and $C\underline{H}_2O$ protons, it was possible to show that the polymers exist in solution as oxazolidine-acetal macromolecules. The determination of differences in the chemical structure of the end groups was possible only by ESI-MS investigation.[2]

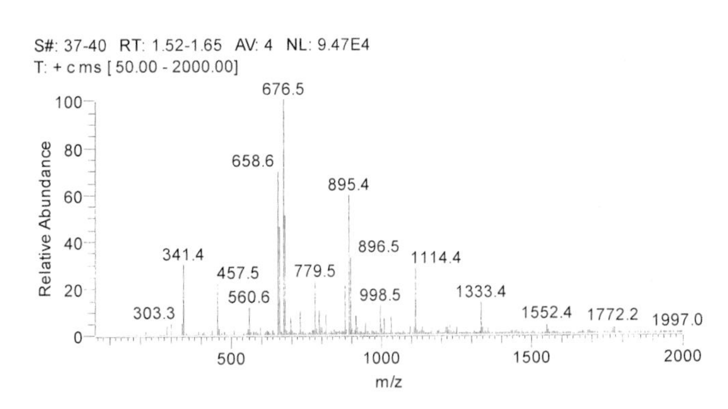

Figure 1. The ESI-MS spectrum of polymer **3**.

The ESI mass spectrum (in positive-ion mode) of the sample **3** (Mn = 2300 g · mol^{-1}) is presented in Figure 1. Analysis of this ESI mass spectrum revealed the presence of protonated molecules with m/z values corresponding to three kinds of macromolecular chains containing different end groups, i.e., CHO and OH groups (A), only OH groups (B), and oligomers with no end-groups, i.e., macrocyclic compounds (C). The number of repeating units in the MH^+ ions varied from A_n = 3-8, B_n = 2-3, C_n = 3-5. The signals in the spectrum show a peak-to-peak mass increment of 219 Da, which is equal to the molecular weight of mono-oxazolidine as the repeating unit in polymer **3**.

Chemical modifications of polymer **3**

The telechelic poly(1,3-oxazolidine-acetal)s were tested in oxidation, reduction and nucleophilic substitution (alkylation, acylation) reactions, as depicted on Scheme 1.

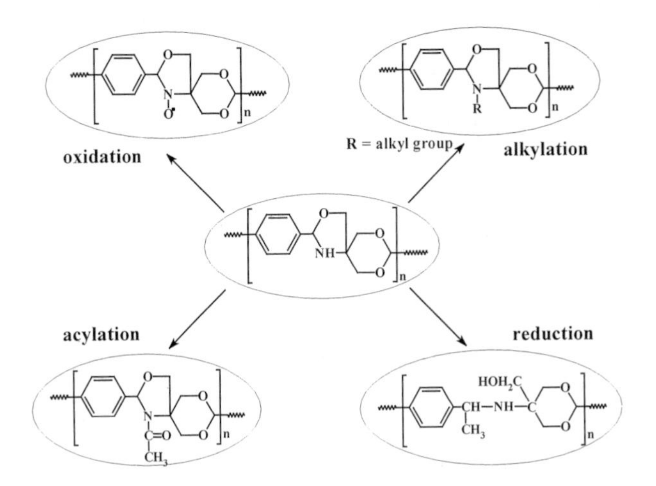

Scheme 1. Chemical modification of new heterotelechelic poly(oxazolidine-acetal).

The presence of CHO groups in the polymer **4** after alkylation was observed in IR and NMR spectra. The signal at 191.2 ppm increased in their intensity when compared to the polymer before alkylation (Figure 2). During alkylation the oxazolidine end groups of macromolecules are isomerized to the imine form, which then is partially hydrolyzed into secondary amino-triol and macromolecules with aldehyde groups at the chain end.[1, 3]

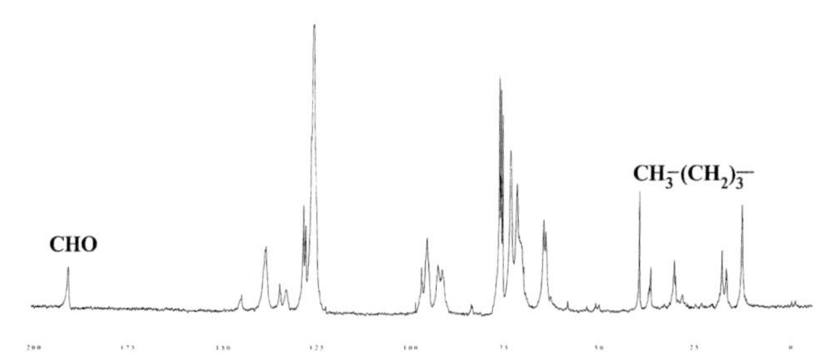

Figure 2. ^{13}C NMR (75 MHz) ($CDCl_3$) spectrum of N-alkylated polymer **4** with buthyl iodide.

Acylation of the polymer **3** with acetic anhydride occurs with an effective N-substitution of O,N-acetal units. Figure 3b presents the spectrum of N-acetylated polymer **5**. The observed difference in the chemical shifts of CH_3CO- protons (1.89 and 2.06 ppm) may be accounted for the existence of both, the envelope and twist conformations of oxazolidine rings in the polymer chain.

The Grignard reagents (RMgX) are known to react with 1,3-oxazolidines.[4] The addition of methylmagnesium chloride (MeMgCl) was carried out in THF at room temperature. Reduction of O,N-acetal units of polymer **3** occurs with forming poly(amine-acetal). The ^{1}H NMR spectrum of polymer **6** shows the appearance of a new band centered at 1.25 ppm (-NH-), 1.8 ppm (CH_3-CH) and 4.18-4.23 ppm (-CH-N). Based on the ^{1}H NMR spectrum it was found that 57 % oxazolidine units were converted into the amine groups. This reaction is presented in Scheme 1 and Figure 3c.

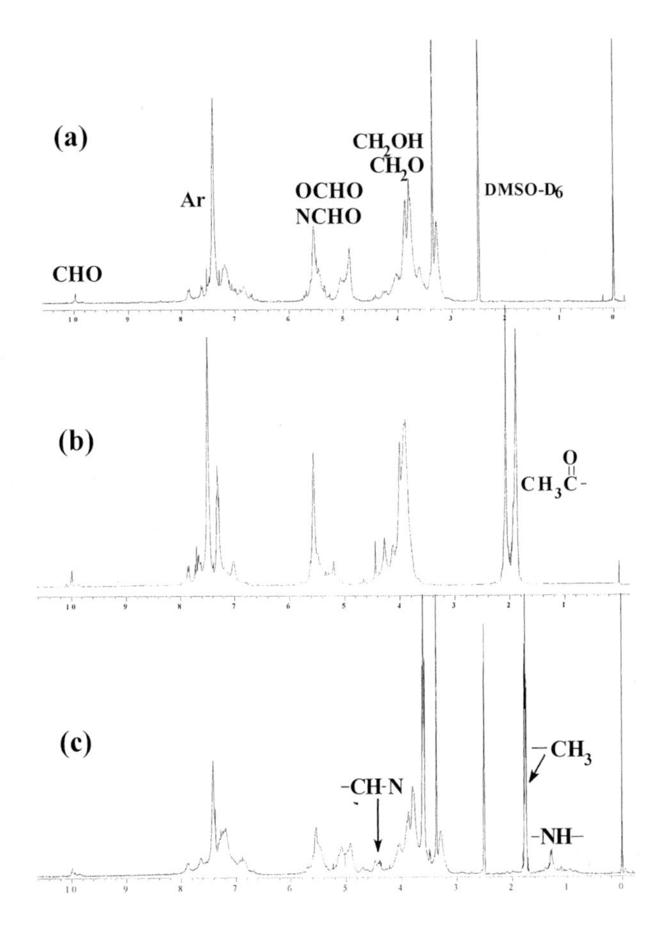

Figure 3. ^{1}H NMR (300 MHz) (dmso-d_6) spectra of: a) polymer **3**; b) ($CDCl_3$) N-acylated polymer **5**; c) and after reduction with CH_3MgCl **6**.

The spectroscopic data of all new compounds are in accordance with their structural assignments.

New stable nitroxide free radicals were generated in the oxidation reaction of oxazolidine units of polymer **3** with m-chloroperbenzoic acid. The radicals gave EPR spectrum, similar to TEMPO, with a triplet hyperfine splitting due to the nitrogen nucleus (see Figure 4). Studies on polymerization of model graft polymers with the nitroxide free radicals are in progress.

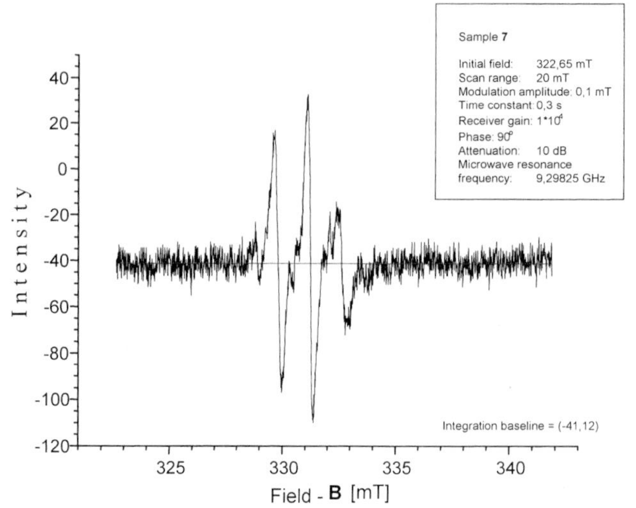

Figure 4. The EPR spectrum of the nitroxide free radicals in polymer chain.

Conclusion

The polycondensation of terephthalaldehyde with 2-amino-2-hydroxymethyl-1,3-propanediol under acidic catalysis leads to the formation of linear polymers containing different end groups, i.e., CHO and OH.

The activities of poly(oxazolidine-acetal)s were investigated in oxidation, reduction and nucleophilic substitution (acylation, alkylation) reaction. It was demonstrated that new nitroxide radicals were easily prepared in the reaction polymer **3** with 4-chloroperbenzoic acid.

[1] J. Maślińska-Solich, E. Kudrej-Gibas, *Design.Monom.&Polym.* **2003**, 3, 31.
[2] J. Maślińska-Solich, E. Kudrej-Gibas, *Rapid Commun. Mass Spectrom.* **2003**, 17, 1769.
[3] J. Maślińska-Solich, E. Kudrej-Gibas, to be published.
[4] K. Higashiyama, H. Inoue, H. Takahashi, *Tetrahedron Lett.* **1992**, 33, 235.

Synthesis of Functional Polymers with Acetal-Monosaccharide Moieties

Jolanta Maślińska-Solich, Sylwia Kukowka*

Department of Physical Chemistry and Technology of Polymers, Silesian University of Technology, ul. M. Strzody 9, 44-100 Gliwice, Poland
E-mail: jms1acet@zeus.polsl.gliwice.pl

Summary: The polycondensation of methyl α-D-mannopyranoside (**1**) with 1,n-bis(formylphenoxy)alkanes (**2**), using acidic catalysts, leads to the formation of linear polymers and macrocyclic compounds. The structure of the polymer and macrocycles was determined by ^{1}H, ^{13}C NMR spectroscopy and ESI-MS analysis.

Keywords: acetal; dialdehydes; macrocycle; methyl α-D-hexopyranoside; polycondensation

Introduction

The monosaccharides are probably the most densely functionalized naturally occurring organic molecules and are undoubtedly among the most important molecules in the biochemistry of all living things. Synthetic polymers combined with carbohydrates are of great interest in the applications as polymeric protecting groups, polymer support in the asymmetric synthesis or polymer-supported phase transfer catalysts.[1]

Recently, our investigations have been focused on the polyacetals derived from methyl D-hexopyranosides with terephthaldehyde[2] or 1,4-bis(2-formylphenoxy)butane.[3] It was found that in the polymer syntheses the equilibrium of macrocycles and linear macromolecules are occurred. Polymer of **1** with terephthaldehyde was tested as a polymeric phase transfer catalyst in the polyetherification reactions.[4]

In this paper, we report the results of studies on the equilibration reactions in the polycondensation of methyl α-D-mannopyranoside (**1**) with 1,n-bis(formylphenoxy)alkanes (**2**).

 DOI: 10.1002/masy.200450608

Experimental

Measurements

^{1}H and ^{13}C NMR spectra were recorded using a Varian Unity/Inova spectrometer (300 MHz and 75 MHz, respectively) in deuterated chloroform with tetramethylsilane as an internal standard.
Gel permeation chromatography (GPC) analyses were performed in tetrahydrofuran using Waters System equipped with a refractive index detector. Two (300 x 7.5) PL-gel mixed E (3 μm) columns were used and maintained at 40 ^{0}C. Polystyrene standards (Polymer Laboratories) were used to calibrate the system. ESI MS experiments were carried out using a Finnigan MAT TSQ 700 triple stage quadrupole mass spectrometer equipped with an electrospray ionization (ESI) source (Finnigan, San Jose, CA, US).
Optical rotation measurements were made in chloroform solution using a Carl-Zeiss Jena Polamat A spectropolarimeter with a sensitivity ±0.01°.

Polycondensation of 1 with 2

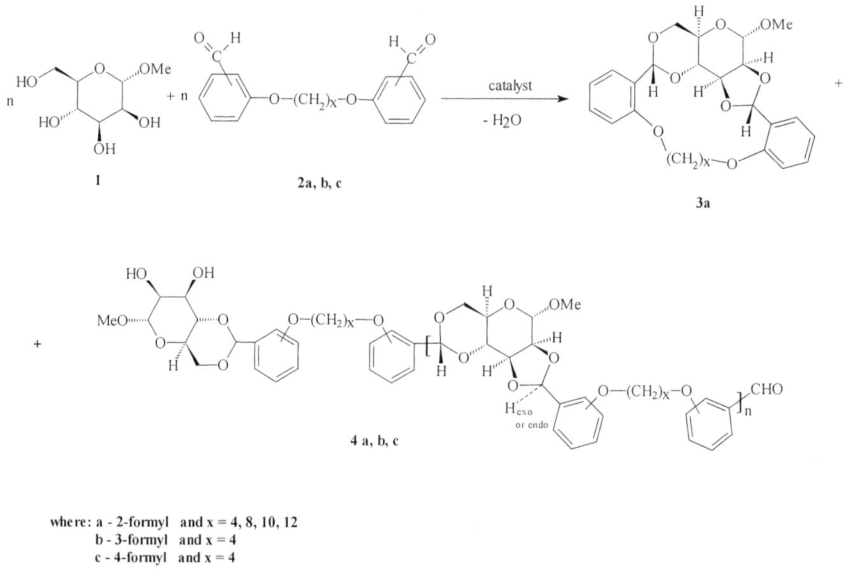

where: a - 2-formyl and x = 4, 8, 10, 12
b - 3-formyl and x = 4
c - 4-formyl and x = 4

Scheme 1. Polyacetalising of methyl α-D-mannopyranoside (**1**) with 1,n-bis(formylphenoxy)alkanes (**2**)

General procedure:

A mixture of **1** (0.05 mol) and **2(a,b,c)** (0.05 mol) in solution (benzene/dioxane, benzene/DMF, benzene/DMSO) containing TsOH (0.2-0.05 g), or PVP-TsOH (0.6-1.5 g) was subjected to azeotropic distillation (Dean-Stark). After 8-27 h the catalyst was filtered off (or deactivated with $CaCO_3$) and the solvent was removed under reduced pressure. The residue was extracted with chloroform, the organic layer was washed with a solution of $NaHSO_3$, water (3-4 times), and dried over anh. $MgSO_4$ and filtered off. The chloroform was removed under reduced pressure and the residue was analyzed by TLC, NMR. The reaction mixture of compounds was fractionated on silicagel (flash chromatography). The polymeric materials were purified by reprecipitation in THF/ ethanol.

For **4a** (where x = 4), $[\alpha]^{20}_{546}$ (c. 1, $CHCl_3$)= +16.5 deg · dm^{-1} · g^{-1} · cm^3.

^{1}H NMR ($CDCl_3$): δ [ppm] = 1.70–2.20 (m, -C**H₂**-); 3.30–3.50 (m, -OC**H₃**); 3.60–4.71 (m, -OC**H**-, -OC**H₂**-); 4.76–5.20 (m, -OC**H**O $_{anom}$); 5.80–6.04 (m, -OC**H**O-, H-2 dioxan-2-yl); 6.12–6.36 (m, -OC**H**O-, H-2 dioxolan-2-yl, *exo*), 6.54–6.68 (m, -OC**H**O-, H-2 dioxolan-2-yl, *endo*), 6.70–7.90 (m, $\mathbf{H_{Ar}}$, Ar-O**H**), 10.40–10.60 (m, -C**H**O)

^{13}C NMR ($CDCl_3$): δ[ppm] = 24.0–27.6 (-**C**H_2-); 55.1–82.0 (-O**C**H_3, -**C**H_2O-, -O**C**H-); 96.0–102.0 (-O**C**HO-acetal); 110.0–136.0 ($\mathbf{C}_{Ar}$); 155.0–160.0 ($\mathbf{C}_{Ar}$-O-); 190.0 (-**C**HO).

For **4b** (where x = 4), $[\alpha]^{20}_{546}$ (c. 2.5, $CHCl_3$) = -2.5 deg · dm^{-1} · g^{-1} · cm^3.

^{1}H NMR ($CDCl_3$): δ [ppm] = 1.80–2.10 (m, -C**H₂**-); 3.30–3.50 (m, -OC**H₃**); 3.63–4.66 (m, -OC**H**-, -OC**H₂**-); 4.70–5.08 (m, -OC**H**O $_{anom}$); 5.44–5.63 (m, -OC**H**O-, H-2 dioxan-2-yl); 5.85–5.94 (m, -OC**H**O-, H-2 dioxolan-2-yl, *exo*), 6.09–6.25 (m, -OC**H**O-, H-2 dioxolan-2-yl, *endo*), 6.84–7.48 (m, $\mathbf{H_{Ar}}$, Ar-O**H**), 9.92–10.00 (m, -C**H**O).

^{13}C NMR ($CDCl_3$): δ[ppm] = 26.0–26.4 (-**C**H_2-); 55.1–80.6 (-O**C**H_3, -**C**H_2O-, -O**C**H-); 99.9–104.1 (-O**C**HO-acetal); 112.0–140.4 ($\mathbf{C}_{Ar}$); 159.0–159.8 ($\mathbf{C}_{Ar}$-O-); 192.4 (-**C**HO).

For **4c** (where x = 4), $[\alpha]^{20}_{546}$ (c. 2.5, $CHCl_3$) = -17.6 deg · dm^{-1} · g^{-1} · cm^3.

^{1}H NMR ($CDCl_3$): δ [ppm] = 1.84–2.04 (m, -C**H₂**-); 3.31–3.42 (m, -OC**H₃**); 3.64–4.64 (m, -OC**H**-, -OC**H₂**-); 4.76–5.06 (m, -OC**H**O $_{anom}$); 5.45–5.60 (m, -OC**H**O-, H-2 dioxan-2-yl); 5.86–5.90 (m, -OC**H**O-, H-2 dioxolan-2-yl, *exo*), 6.11–6.24 (m, -OC**H**O-, H-2 dioxolan-2-yl, *endo*), 6.80–7.84 (m, $\mathbf{H_{Ar}}$, Ar-O**H**), 9.93–9.92 (m, -C**H**O)

^{13}C NMR ($CDCl_3$): δ[ppm] = 25.8–26.0 (-**C**H_2-); 55.0–80.6 (-O**C**H_3, -**C**H_2O-, -O**C**H-); 98.7–

104.1 (-OCHO-acetal); 114.1–132.0 (**C**$_{Ar}$); 164.0 (**C**$_{Ar}$-O-); 190.8 (-**C**HO).

Results and discussion

The polycondensation of **1** with dialdehyde **2** (**a,b,c**) was performed in solution (benzene/dioxane, benzene/DMF, benzene/DMSO) in the presence of a catalytic amount of p-toluenesulfonic acid (TsOH) or its complex with poly(4-vinylpyridine) PVP-TsOH under azeotropic removal of water.

The formation of macrocycles and polymers was followed by ^{1}H NMR spectroscopy.

In case of a 1:1 molar ratio of comonomers **1** and **2a** (**x** = **4, 8, 10, 12**), in benzene/DMSO as solvent, the major products were cyclic diacetals **3a (**where **x** = **4, 8, 10, 12)** and linear polymers **4a** (see Table 1).

Table 1. Formation of macrocycles as a function of dialdehyde **2** structure.

-(CH$_2$)$_x$-	**Macrocycles 3a [%]**	**Polymer 4a [%]**
4	69.8 [1+1] [2+2]	30.2
8	58.9 [1+1]	41.1
10	50.1[1+1]	48.1
12	29.7 [1+1]	70.3

[M$_1$]= [M$_2$]=0.1 mole/dm^3; benzene-DMSO (4:1),TsOH as catalyst; 12h

The evidence of the cyclic diacetal **3a** is demonstrated in the NMR spectrum (Figure 1, for **3a** where x = 8).

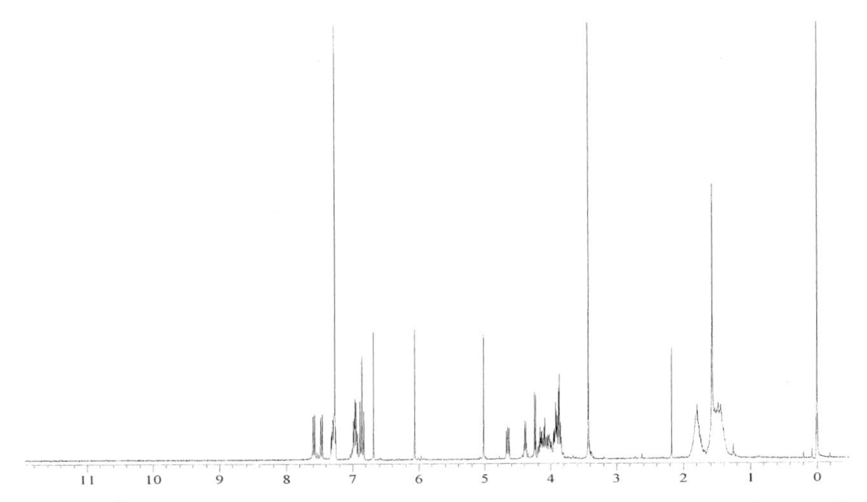

Figure 1. ^{1}H NMR (300 MHz, $CDCl_3$) spectrum of the macrocyclic compound **3a** (x = 8).

The characteristic signals due to methoxy protons at 3.41 ppm, acetal protons in the region of 5.89–6.06 and 6.54–6.68 ppm and aromatic protons at 6.55-7.52 ppm, confirmed that the condensation product **3a** consists of two cyclic acetal rings bridged by a di-O-[2,2'(1,n-alkoxy)]phenylidene unit (Table 2).

Table 2. Chemical shifts of acetals and C^1 (anomeric) protons of selected macrocyclic compounds **3a**.

	Chemical shifts [ppm]		
	2,3-acetal	4,6-acetal	OCHO anom.
Methyl 2,3:4,6-di-O-[2,2(1,4-butoxy)]phenylidene-α-D-mannopyranoside	6.54 H *endo*	5.89	5.08
Methyl 2,3:4,6-di-O-[2,2(1,8-octoxy)]phenylidene-α-D-mannopyranoside	6.68 H *endo*	6.06	5.01
Methyl 2,3:4,6-di-O-[2,2(1,10-decoxy)]phenylidene-α-D-mannopyranoside	6.65 H *endo*	6.01	5.02
Methyl 2,3:4,6-di-O-[2,2(1,12-dodecoxy)]phenylidene-α-D-mannopyranoside	6.67 H *endo*	6.04	5.01

A detailed ^{1}H NMR and X-ray analysis of **3a** (x = 4) confirmed that the conformation of five-membered 1,3-dioxolane rings is different in two symmetry-independent molecules: in one of the molecules it is close to an envelope, while the other one has a distorted half chair. Both symmetry-independent molecules are H-*endo* isomers. Whereas the substituent at C-2 (in 1,3-dioxane) is in equatorial orientation with respect to the chair-shaped dioxane ring fused to a tetrahydropyran ring.[3]

Moreover the presence of macrocycles were detected by ESI mass spectrometry (Figure 2). Analyses of ESI MS **3a** (where x = 4) revealed the presence of potassium adduct ions with m/z values 495.5 and 950.9 that correspond to the individual macrocycles [1+1] and [2+2].

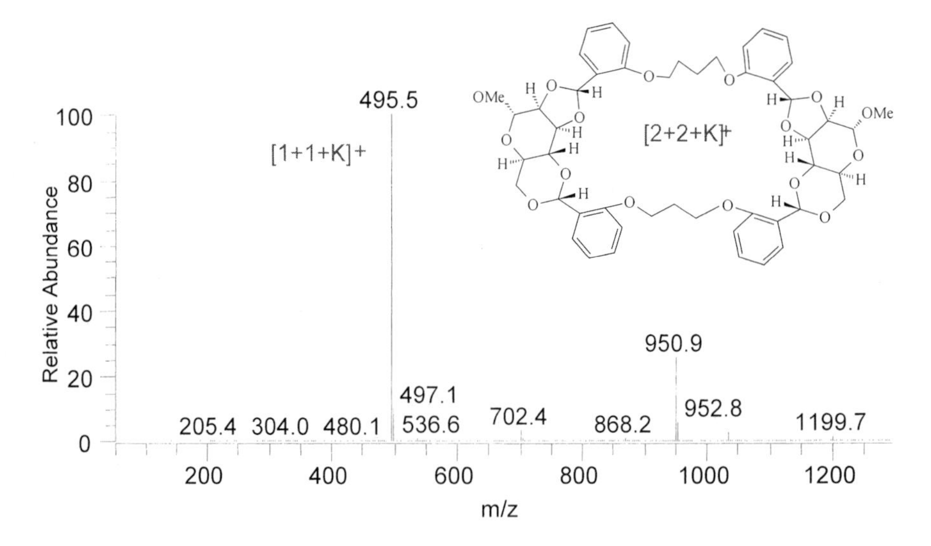

Figure 2. ESI MS macrocycles [1+1] and [2+2] of **3a** (x = 4).

It has shown the ability of co-monomers to form also the linear macromolecules that co-exit with macrocyclic structure. The polymers were soluble in THF, $CHCl_3$, DMSO but insoluble in alcohols, aliphatic and aromatic hydrocarbons. The polymers were purified by repeated precipitation with ethanol from solution in THF. It was noted that the specific rotation, the molecular weight and molecular weight distribution depends on the structure of 1,n-bis(formylphenoxy)alkanes (**2**).

The polycondensation of **1** with 1,4-bis(3-formylphenoxy)butane **2b** or 1,4-bis(4-formylphenoxy)butane **2c** leads to the formation only of linear polymers with molecular range from 2500 to 4500 g · mol^{-1} (M_w, GPC).

The NMR data of polymer have been used to establish the configuration of a five-membered acetal ring at C (2) and C (3) of **1** in the polymer chain of **4**.

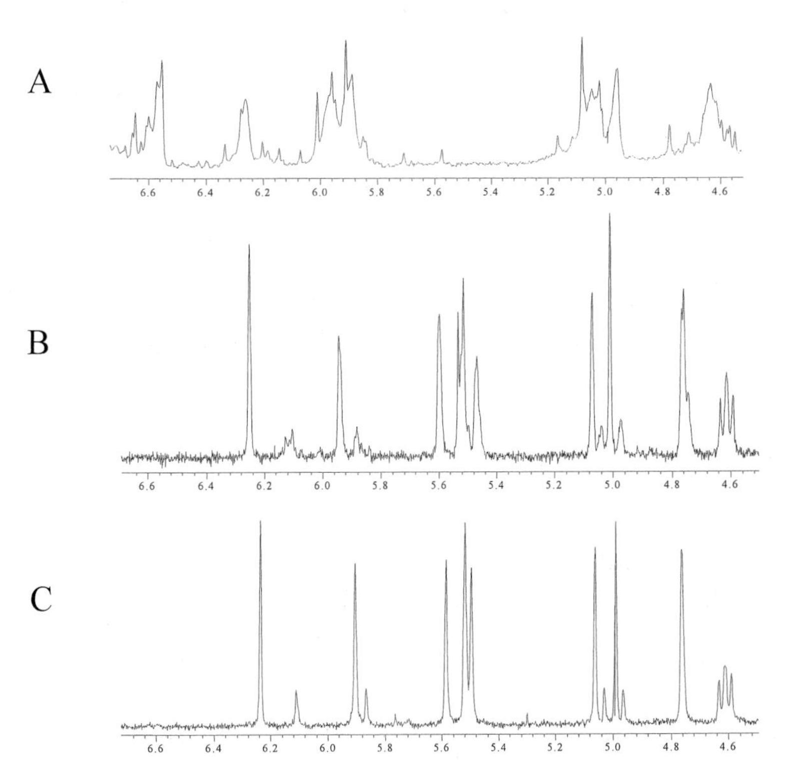

Figure 3. ^{1}H NMR (300 MHz, $CDCl_3$) spectra (expansion of the region 4.6 to 6.6 ppm) of polymers A - **4a,** B - **4b,**C - **4c**.

^{1}H NMR spectra (expansion region of OC<u>H</u>O protons) show characteristic signals of anomeric C-1 protons at 4.7-5.2 ppm, the acetal protons at 5.8-6.0 ppm (1,3-dioxane), 6.1-6.4 ppm (H-*exo* 1,3-dioxolane) and 6.5-6.7 ppm (H-*endo* 1,3-dioxolane) for polymer **4a** (x = 4).

The influence of steric factors caused by five-membered acetal rings (which adopt a suitable half-chair or twist conformation) with six-membered acetal rings in a sugar unit has been considered as an interpretation of multi-set signals of acetal protons (**4a**, Fig. 3 A). However, slight sterically effect in polymers **4b** and **4c** is shielded the acetal protons as can be seen in Figure 3(B and C). The characteristic signals due to the OCHO protons appear at 5.4-5.6 ppm (1,3-dioxane), 5.8-5.9 ppm (H-2 *exo* 1,3-dioxolane) and 6.1-6.3 ppm (H-*endo* 1,3-dioxolane).

It is known that methyl 2,3:4,6-di-O-benzylidene-α-D-mannopyranoside exists in C1 conformation where the 1,3-dioxolane (at C-2 and C-3) system shows a strong preference for the *endo*-H (*exo*-phenyl) isomer.[5]. The *exo*-H diastereomer was only formed by benzylidenation of **1** with benzylidene bromide under basic-catalyzed (potassium tert-butoxide) conditions.[6] Clode[7] postulated that the acid-catalyzed acetal formation might be divided into a kinetic and a thermodynamic phase. Then, *endo*-phenyl diastereomer of methyl 2,3:4,6-di-O-benzylidene-α-D-mannopyranoside was formed acid catalyzed under kinetically controlled conditions.

The formation of **3a** (x = 4, 8, 10, 12) may need thermodynamically rather than kinetically controlled conditions for the given reaction. Thus, although there are theoretically four possible intramolecular condensations of **1** with **2a**, only one compound **3a** (highly favored H-2 *endo*-5-membered ring and H-2 *axial*-6-membered ring) is produced.

The presence of *endo*-H and *exo*-H isomers of five-membered rings in the polymer chains **4** can be explained by less sterically interactions between methyl α-mannopyranosyl and arylidene groups in the kinetically controlled polycondensation.

All these results confirm that the polymer **4** is constructed of the cyclic acetal rings (five and six-membered) of dialdehyde **2** and methyl α-D-mannopyranoside. The chemical analysis has also established the presence of macromolecular chains containing formyl and hydroxyl end groups.

From the experimental data one can conclude that macrocycle formation was efficient under thermodynamic conditions (*endo*-H isomer), whereas the polymer was formed under kinetically controlled conditions which led to a mixture of *exo*-H and *endo*-H (in 1,3-dioxolane) units.

The relative quantities of *endo*-H and *exo*-H (in 1,3-dioxolan-2yl ring) units in the linear macromolecules are equal 1:1. This relationship is valid for the polymers with molecular weights $M_w \sim 2000\ g \cdot mol^{-1}$.

In spite of the moderate stability of these new compounds, their transformation into functional polymers has been achieved in thermal or chemical conditions. Further investigations are in progress in order to explore the synthetic potential of this new class of sugar polymers as chiral catalysts.

Conclusion

The results of this work indicate that the polycondensation of methyl α-D-mannopyranoside (**1**) with 1,n-bis(formylphenoxy)alkanes (**2**) in the presence of acidic catalyst yields polyacetals. It was demonstrated that the intramolecular cyclization reactions [1+1] of **1** with 1,n-bis(2-formylphenoxy)alkanes decrease with the elongation of an alkane spacer between phenoxyaldehyde units. The formation of linear macromolecules with *endo*-H and *exo*-H isomers of five-membered rings in the polymer chain supports the regioselective character of this reaction.

[1] P.Hodge, D.C.Sherrington (eds.) "*Polymer-supported Reactions in Organic Synthesis*", Wiley, New York, 1980, p.188, 319; 303.
[2] J. Maślińska-Solich, *Macromol. Biosci.* **2001**, *1*, 312.
[3] J. Maślińska-Solich, N.Kuźnik, M.Kubicki, S. Kukowka, *Chem. Commun.* **2002**, 984.
[4] J. Maślińska-Solich, S. Kukowka, M. Demarczyk, "*Polymers for the 21st Century*", *Proc. 5th Inter. Polym. Sem.* Gliwice **2003**, 111.
[5] J. Gelas, *Adv. Carbohydr. Chem. Biochem.* **1981**, *39*, 71.
[6] N. Baggett, J. M. Duxbury, A. B. Foster, J. M. Webber, *Carbohydr. Res.* **1965**, *67*, 22.
[7] D. M. Clode, *Chem. Rev.* **1979**, *79*, 491.

Macromol. Symp. **2004**, *210*, 77-84

ω-Substituted Long Chain Alkylphosphonic Acids - Their Synthesis and Deposition on Metal Oxides and Subsequent Functional Group Conversion of the Deposited Compounds

*Michael Eschner,**[1] *Ralf Frenzel,*[1] *Frank Simon,*[1] *Dieter Pleul,*[1] *Petra Uhlmann,*[1] *Hans-Jürgen Adler*[2]

[1]Institut für Polymerforschung Dresden e. V, Hohe Straße 6, 01069 Dresden, Germany
E-mail: eschner@ipfdd.de
[2]Technische Universität Dresden, Institut für Makromolekulare Chemie und Textilchemie, Mommsenstr. 13, D-01062 Dresden, Germany

Summary: The crystal structure and the surface of alumina layers on silicon wafers were investigated by means of GIWAXS, WAXS, and AFM. Self-assembled monolayers of ω-substituted long-chain alkylphosphonic acids were deposited on alumina and the layer thickness and homogeneity were determined by ellipsometry revealing a dependency of thickness and homogeneity on the nature of the substituent. During the adsorption process surface etching by the phosphonic acid was observed causing an increase in surface roughness.
Furthermore, ex post functional group conversions were carried out yielding surface bound azides and rhodanides.
Keywords: alkylphosphonic acid; alumina; atomic force microscopy; self-assembled monolayer; surfaces

Introduction

This article is part of our research on scope and limitation of our concept of ex post chemical SAM (self-assembly monolayer) modification. First, a homogeneous SAM is generated on a surface using readily available anchor groups, which bear reactive functional groups. With this highly functionalized SAM modified surface on hand further chemical modifications can be carried out. Generally reactions with high conversion and high selectivity are applied. But, even if conversion is low, e. g. as in case of introduction of bulky substituents, this concept is advantageous over SAM formation of highly substituted anchor groups because bulky substituents always cause great distances between neighbored molecules due to spatial reasons. Thus, there will be empty space between the anchor moieties, which might be filled up by any

 DOI: 10.1002/masy.200450609

sort of compounds, so to speak there will be plenty of space for permanent surface contamination. Co-adsorption of two different anchor molecules may lead to phase separation instead of formation of a homogeneous mixed layer. In contrast, our concept of ex post SAM modification forms a homogeneous SAM first which will be modified afterwards yielding a dense mixed layer.

The work described here aims at the development of surface modifiers for metal oxides. Such compounds will render the chemical properties of oxydic surfaces of metallic materials. These compounds can be used for corrosion inhibitors [1], anchor groups for dyes, polymers, bio-polymers, e. g. antibodies, enzymes, poly-sugars, and the like, peptide mimetics, additional applications are electronic devices and molecular recognition etc., simply by adopting the functional head-groups to the users needs. Phosphonic acids are versatile compounds for these purposes. They adsorb well on metal oxides like titanium oxide [2], iron oxide, and aluminum oxide [1] forming self-assembly layers on the surfaces of metal oxides.
Furthermore, alkylphosphonic acids as well as ω-substituted alkylphosphonic acids are readily accessible via Michaelis-Arbusov reaction.

Surface Characterization [3]

As substrates for this research silicon wafers coated with aluminum were used. The aluminum layer had been deposited by PVD (physical vapor deposition). Afterwards, the aluminum surface was oxidized in an oxygen plasma.
To gain a deeper insight into the structure of the metal oxide layers used in this investigation the surface of the applied alumina was characterized in detail with AFM, WAXS (Wide Angle X-Ray Scattering), and GIWAXS (Grazing Incidence Wide Angle X-Ray Scattering). The latter methods revealed dense layers of aluminum oxide (5.05 nm) and metallic Al (45.6 nm) on top of natural silicon dioxide (2 nm) followed by bulk silicon. The aluminum oxide is contaminated with some silicon dioxide. No crystalline alumina was found (Fig. 1), hence, the obtained aluminum oxide layer is amorphous.

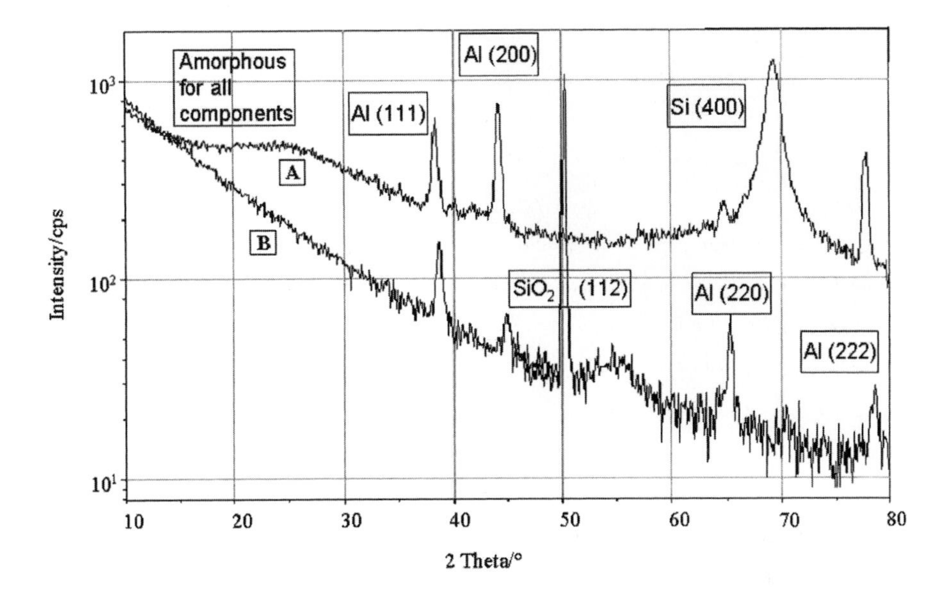

Figure 1: Comparison of WAXS (**A**) and GIWAXS ($\omega = 0.2°$, **B**) of an aluminum oxide bearing aluminum coated silicon wafer

Aluminum, which had been deposited on a silicon wafer by PVD, was oxidized in an oxygen plasma. Afterwards, the obtained aluminum oxide layer was investigated by various methods. AFM (atomic force microscopy) imaging revealed a surface covered with numerous little but singular peaks, which are depicted in white. Treatment of the Si-wafer with an aqueous ethanolic solution of phosphonic acids caused not only formation of a self-assembly layer, but also an increase in surface roughness and a pronounced increase in the number of peaks due to surface etching during the adsorption process (Fig.2).

Surface Modification [3]

Various ω-substituted dodecylphosphonic acids had been dissolved in ethanol/water 1:1 at concentrations of 1 mmol/l and were adsorbed on alumina.

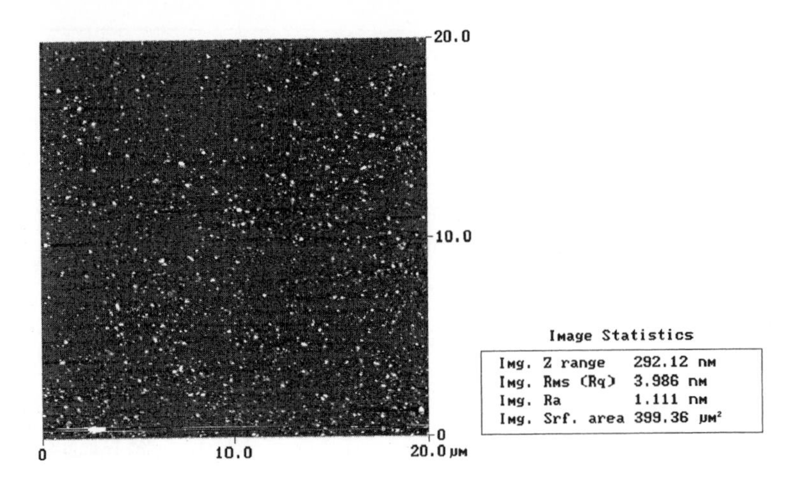

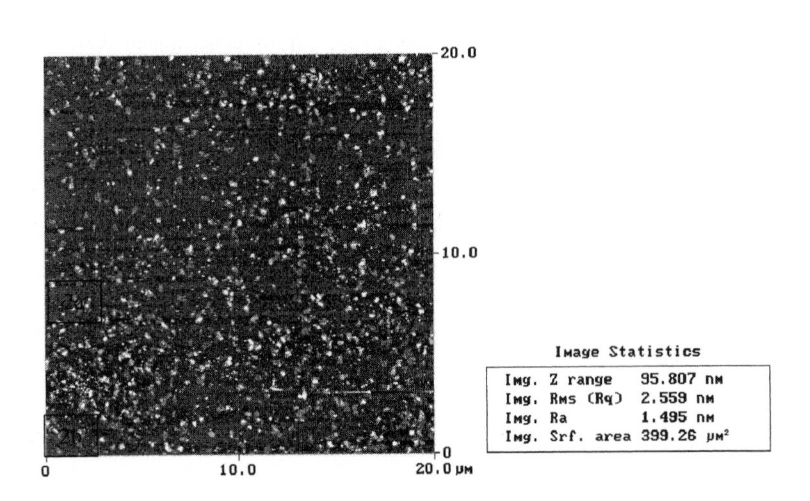

Figure 2: Time dependence of etching of the aluminum oxide layer on an aluminum coated silicon wafer. 2a) after 1 h, 2b) after 20 h

AFM investigation of 12-bromo and 12-mercapto dedecylphosphonic acid showed intense surface etching. The surface roughness increased from 1.1 nm for the untreated material to 3.1 nm for the coated wafer in the case of 12-bromo dodecylphosphonic acid and to 1.5 nm for the 12-mercapto dodecylphosphonic acid. Both, 12-amino and 12-ethylamino dodecylphosphonic acid showed much less surface etching and a surface roughness in the same order of magnitude as the untreated material. Because of the low solubility of the latter compounds due to internal salt formation the concentration in solution is to low to cause intense etching. Hence, only small amounts of these phosphonic acids were deposited on the alumina surface as confirmed by ellipsometric measurements. The determined layer thicknesses and the calculated lengths of the molecules are summarized in table 1; the error of the ellipsometric data given in table 1 is 15 %. It should be kept in mind that ellipsometry is an analytical tool with only low lateral resolution. Thus, it integrates the layer thickness over a distinct surface area. Any inhomogeneities in surface coating result in an increase or decrease in layer thickness. Therefore, layer thicknesses below the molecular length mean an imperfect coarse layer or a layer with low coverage due to angle formation between the alkyl chain and surface normal. Any tilt angle will cause a layer thickness less than the molecular length, too. Furthermore, the space filling ethyl amino group prevents formation of a dense layer, hence, the SAM of 12-amino dodecylhosphonic acid shows an even lower layer thickness than its unsubstituted counterpart.

In contrast to the amino functionalized phosphonic acids the 12-mercapto dodecylphosphonic acid forms layers more than twice as thick than the calculated molecule length. This finding can be interpreted in terms of disulfane formation during the adsorption process due to oxidation of the mercapto group or in terms of oxidative disulfane formation in the stock solution prior to adsorption.

The unexpected high layer thickness of the 12-bromo dodecylphosphonic acid can be explained by multilayer formation [4]. May be the tremendous increase in overall surface roughness influenced the measurements, too.

Table 1. Ellipsometric determined layer thicknesses of various ω-substituted dodecylphosphonic acids adsorbed on aluminum oxide.

ω-substitued dodecyl-phosphonic acid R:	Measured layer thickness/nm	Calculated molecule length/nm [7]
-H 1 h adsorption time	1.4	1.65
-H 20 h adsorption time	2.4	1.65
-Br	4.7	1.72
$-NH_2$	1.3	1.76
-NHEt	0.7	2.01
-SH	4.8	1.80

Despite the little amount of adsorbed 12-amino and 12-ethylamino dodecylphosphonic acid the obtained contact angles of the SA-layers (Table 2) show a tremendous increase in surface hydrophilicity for both compounds in comparison to dodecylphosphonic acid. The advancing contact angle decreases from 111.2° for dodecylphosphonic acid to 71.3° for ω-amino dodecyl-phosphonic acid.

With increasing hydrophilicity of the substituent the advancing contact angle decreases. The little contact angle hysteresis, which was found for dodecyl and 12-bromine dodecylphosphonic acid, suggests formation of a dense SA layer on the aluminum oxide, otherwise a greater difference between advancing and receding contact angle would have been observed.

Adsorption of alkylphosphonic acids performs via salt formation due to an acid-base reaction between the phosphonic acid and surface bound aluminum hydroxide moieties. This reaction leads to insoluble aluminum salts of phosphonic acids [5].

Table 2. Contact angles of various aluminum oxide covered Si-wafers, which are coated with ω-substituted dodecylphosphonic acids.

Substituted dodecyl phosphonic acid R:	θ_a/°	σ	θ_r/°	σ
-H 1 h adsorption time	111.2	0.3	86.4	3.8
-H 20 h adsorption time	112.5	0.6	84.1	4.9
-Br	90.4	1.4	61.9	0.8
$-NH_2$	71.3	4.2	----	----
-NHEt	74.1	1.8	----	----
-SH	74.2	0.9	----	----

θ_a: advancing contact angle, θ_r: receding contact angle, σ: standard deviation
---- Values could not be determined

SAM Modifications by Interfacial Reactions

The synthesis of functionalized alkylphosphonic acids started with the Michalis-Arbuzov reaction of 1,12-dibromododecane followed by further derivatization [6]. Details on the preparation of these and other derivatives will be described in a forthcoming paper. After deposition on aluminum oxide further modification had been carried out at the interface between solvent and solid. Here we present the nucleophilic substitution of terminal bromine by azide and rhodanide ions [7]. Successful substrate conversion was proved with diffuse reflection IR-spectroscopy (Fig. 3). After substitution of bromine by KSCN an new absorption band arose at 2155 cm^{-1}, which was attributed to the rhodano moiety [8]; after replacement of Br with NaN_3 a new absorption band appeared at 2099 cm^{-1}, which was assigned to the azide group [8].

XPS (X-ray photoelectron spectroscopy) gave further information on the reaction. In neither of the XPS spectra of both supported compounds, azide and rhodanide, residual bromine could be detected; hence, complete conversion had occurred.

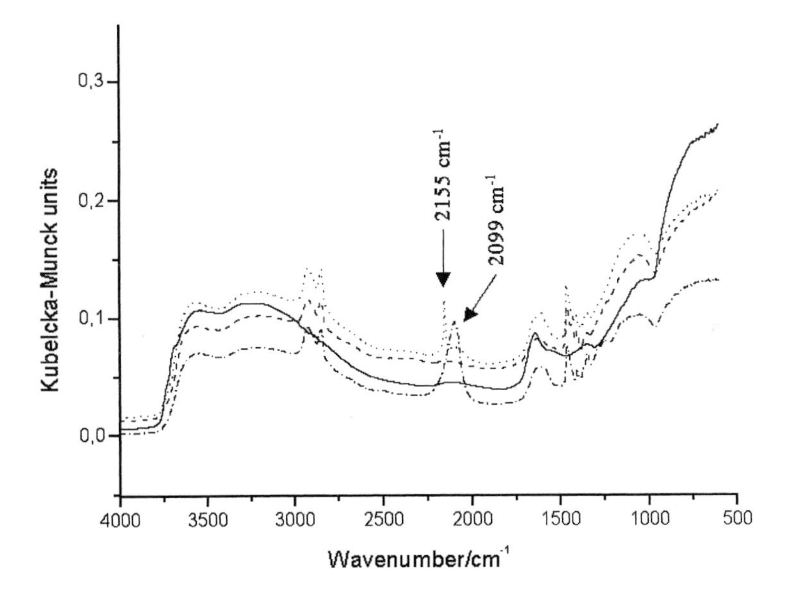

Figure 3: Comparison of the IR-spectra of supported ω-substituted dodecylphosphonic acids (Al_2O_3: ——— , $Br(CH_2)_{12}P(O)(OH)_2$: ------ , $SCN(CH_2)_{12}P(O)(OH)_2$: ········ , $N_3(CH_2)_{12}$-$P(O)(OH)_2$: -·-·-·-)

Conclusion

We investigated aluminum oxide surfaces on aluminum deposited on silicon wafers. WAXS investigations confirmed the inhomogeneity of the metal oxide layer. AFM revealed a smooth surface with scattered peaks. Treatment with phosphonic acids caused etching of the aluminum oxide surface which lead to an additional formation of peaks, as was confirmed by AFM. Due to the low resolution of AFM it was impossible to detect the organic matter.

Indirectly, contact angle measurements confirmed the presence of ω-substituted phosphonic acids. With increasing hydrophilicity of the terminal functional group the contact angle dropped from 112° for dodecylphosphonic acid to 71° for 12-amino dodecylphosphonic acid.

Ellipsometric measurements, too, confirmed the existence of a thin organic layer on the aluminum oxide surface.

We successfully synthesized ω-substituted phosphonic acids with long alkyl chains. Furthermore, we could exchange terminal bromine of the adsorbed by ω-substituted phosphonic azide and rhodanide groups. Both functional groups allow further conversions and introduction of other functional moieties at the interface.

Proving functional group conversion in SA-layers requires concerted application of sophisticated analytical techniques like XPS, AFM, and diffuse reflection IR-spectroscopy.

Acknowledgements

We thank the DFG for funding this research in the framework of Sonderforschungsbereich 287.

We thank to Dr. Klaus Eichhorn and Roland Schulze for performance and intepretation of the ellipsometric measurements, Dr. Karina Grundke and Kathrin Pöschel for contact angle measurements and discussions, and Dr. Dieter Jehnichen for performance and interpretation of WAXS measurements

[1] I. Maege, E. Jaehne, A. Henke, H.-J. Adler, C. Bram, C. Jung, M. Stratmann, Self-assembling adhesion promoters for corrosion resistant metal polymer interfaces, *Progress in Organic Coatings* 1997, *34* (1-4), 1-12

[2] S. Pawsey, K. Yach, L. Reven, *Langmuir* 2002, *18*, 5205-5212

[3] P. Uhlmann, unpublished results

[4] D. L. Allara, R. Nuzzo, *Langmuir* 1985, *1*, 45-52

[5] I. Mäge, Dissertation, Dresden 1998

[6] R. Frenzel, unpublished results

[7] M. Eschner, unpublished results

[8] H. Günzler, H. M. Heise, *IR-Spektroskopie Eine Einführung*, 3. Neubearbeitete Auflage, Weinheim 1996, VCH, ISBN 3-527-28759-0, S. 223

Reactive Polymers on the Basis of Functional Peroxides

R. I. Fleychuk, A. M .Kohut, O. I. Hevus, S. Voronov*

Lviv Polytechnic National University, S.Bandera Str. 12, 79013 Lviv, Ukraine
E-mail: fleychuk@polynet.lviv.ua

Summary: The copolymerization constants of new monomer containing ditertiary peroxide groups with styrene have been determined. The peroxide monomers were prepared by the acylation of 3-(tert-butylperoxy)-3-methyl-1-butanol with methacryloyl chloride or maleic anhydride in the presence of tertiary amines. Peroxide containing copolymers were obtained by copolymerization of peroxyalkyl methacrylate and peroxyalkyl maleate with styrene.

Keywords: copolymerization; macroinitiators; monomers; peroxides; reactive polymers

Introduction

Peroxide copolymers are of the most efficient classes of the reactive modifiers for macro- and microsurfaces. Depending on peroxide group placement, these copolymers could be classified as those containing peroxide group: in the terminus of macrochain (telechelic); along the skeleton; and in the side substituents of polymer macrochain. These copolymers are obtained by the copolymerization of peroxide monomers (PM) with other ones (styrene, α-methylstyrene, vinyl acetate etc.),[1-4] by the reaction of dihydroperoxides with bifunctional compounds (chloroanhydrides of dicarboxylic acids, ditertiary diols, diisocyanates etc.)[5-7] or by the interaction of maleic anhydride copolymers with hydroxy peroxides.[8] Polymers having peroxide groups in side substituent of macrochain are of especial interest since they are able to initiate free radical processes, e.g. polymerization, grafting, structurizing etc. while retaining macrochain due to the homolysis of O-O bond under the influence of transition metals or heating. The peroxide derivatives of 2-methyl-5-hexen-3-yne-2-ol are the most studied monomers of this kind. Polyreactive plastics, rubbers, latexes, and peroxide surfactants were obtained on the basis of these monomers. However, the combination of the chemical-physical properties and the reactivity (solubility, activity in copolymerization reactions, temperature range of free radicals generation, and the activity of the radicals formed) limits partially the application of known PM sometimes.

The purpose of this work is a synthesis of reactive polymers on the basis of the new kind of peroxide monomers incorporating double carbon-carbon bond and peroxide group separated by alkyl spacer.

 DOI: 10.1002/masy.200450610

Experimental

Materials

3-tert-Butylperoxy-3-methyl-1-butanol was synthesized according to the technique reported before.[9] Maleic anhydride from Aldrich was purified by vacuum distillation and kept under argon. Methacryloyl chloride from Fluka was used as supplied. Styrene from Aldrich was distilled under reduced pressure just before use. The initiator, 2,2'-azobis(isobutyronitrile), AIBN, from Merck was purified by recrystalization from anhydrous methanol for three times.

Analysis Methods

Individuality of the synthesized compounds was verified by thin layer chromatography with plates of Silica gel 60 F_{254} from Merck. The plates were developed in iodine vapour as well as by reflux with solutions of reagents forming coloured derivatives: N,N-dimethyl-1,4-phenylenediamine – with peroxide group, cerium ammonium nitrate – with hydroxyl one. IR spectra were recorded in thin film (for methacrylate **2**) or in tetrachloromethane solution (for maleate **3**) with a RS 1000 FT-IR spectrometer (UNICAM Analytische System GmbH). ^{1}H-NMR spectra were recorded using a Brucker 150 spectrometer at working frequency of 300 MHz, substance concentrations of 5...10%, internal standart – hexamethyldisiloxane. The average molecular weights of obtained copolymers were determined with a gel-permeation chromatograph "Waters" (200). The active oxygen content in the synthesized peroxide compounds and prepared polymers was determined by means of iodometry according to the known technique.[10] Functional groups in the synthesized substances were determined using the following techniques: carboxylic – potentiometric titration;[11] carbon-carbon double bonds – addition of $Hg(ClO_4)_2$ with following back titration by 0.01 N solution of thioglycolic acid.[12]

Synthesis of peroxide monomers

3-tert-Butylperoxy-3-methyl-1-butyl methacrylate **2** was obtained in petroleum-ether solution by the reaction of 4.4 g (0.025 mol) of hydroxy peroxide (**1**) with 2.6 g (0.025 mol) of methacryloyl chloride in the presence of 2.6 g (0.025 mol) of triethylamine at 0…5°C according to the technique described elsewhere.[9] 5.4 g amount of target methacrylate was obtained (yield 89%). IR spectrum, cm^{-1}: 830 (O-O), 955 (COOC), 1016…1110 and 1220 (C-O), 1210 (C-O), 1680 (C=C), 1722 (C=O). ^{1}H-NMR (CDCl$_3$), δ, ppm: 1.15 s (9H, t-Bu), 1.25

s (6H, $2CH_3$), 1.93 t (2H, CH_2), 1.94 s (3H, CH_3), 4.2 t (2H, OCH_2), 5.56 s and 6.07 s (2H, CH_2).

3-tert-Butylperoxy-3-methyl-1-butyl maleate **3** was obtained by the reaction of 17.62 g (0.10 mol) of 3-tert-butylperoxy-3-methyl-1-butanol and 9.8 g (0.10 mol) of maleic anhydride according to the technique reported before.[9] 23.8 g (yield 87%) amount of peroxyalkyl maleate was obtained. IR spectrum, cm^{-1}: 830 (O-O), 960 (COOC), 1210 (C-O), 1168 and 1280 (C-O), 1658 and 1640 (C=C), 1720 (C=O), 2500…3200 (HO). ^{1}H-NMR ($CDCl_3$), δ, ppm: 1.15 s (9H, t-Bu), 1.25 s (6H, $2CH_3$), 1.93 t (2H, CH_2), 4.21 t (2H, OCH_2), 6.42 and 6.47 (1H, CH), 6.36 d and 6.41 d (1H, CH), 10.34 s (1H, OH).

Copolymerisation procedure

Copolymerization of peroxyalkyl methacrylate ***2*** *with styrene in solution.* The copolymerization was performed in ampoules under argon in which benzene, 3-tert-butylperoxy-3-methyl-1-butyl methacrylate, styrene, and 2,2'-azobis(isobutyronitrile) (2 wt % with respect to monomer mixture) were charged. The volume ratio of the solvent-monomer mixture was of 5 : 1, respectively. The ampoule contents were freezed, vacuumized and charged with argon for several times in turn. Ampoules were sealed after that and heated to 60°C for 2 h. The prepared polymers were isolated and purified by precipitation from benzene to methanol. Copolymer composition was determined from elemental analysis data and active oxygen content after drying in vacuum at room temperature up to constant weight.

Copolymerization of peroxyalkyl maleate ***3*** *with styrene in solution.* The copolymerization was performed in ampoules under argon, the volume ratio of the solvent-monomer mixture was of 5 : 1, respectively (solvent – benzene). 2,2'-Azobis(isobutyronitrile) was utilized as initiator (2 wt % with respect to monomer mixture). The ampoule contents were freezed, vacuumized and charged with argon for several times in turn. Ampoules were sealed after that and heated to 60°C for 2 h. The polymerization proceeded till monomer conversion of 13…18%. The prepared polymers were isolated and purified by precipitation from benzene to methanol. The polymers were dried in vacuum at room temperature up to constant weight and copolymer yields were calculated. Copolymer composition was determined from elemental analysis data, active oxygen content and acid number value.

Results and discussion

Obtaining of peroxide monomers

Peroxide monomer – methacrylic acid derivative – was synthesized by acylation of 3-tert-butylperoxy-3-methyl-1-butanol **1** with equimolar amount of methacryloyl chloride (Scheme 1, reaction 1). The reaction was carried out in organic solvent at 0…5°C in the presence of triethylamine as hydrogen chloride acceptor. The yield of 3-tert-butylperoxy-3-methyl-1-butyl methacrylate **2** was of 89%.

$HO{-}CH_2{-}CH_2{-}C(CH_3)_2{-}OO{-}C(CH_3)_3$ **1**

$H_2C{=}C(CH_3){-}C(O)Cl$ — reaction 1 → $H_2C{=}C(CH_3){-}C(O){-}O{-}(CH_2)_2{-}C(CH_3)_2{-}OO{-}C(CH_3)_3$ **2**

maleic anhydride — reaction 2 → $HOOC{-}CH{=}CH{-}C(O){-}O{-}(CH_2)_2{-}C(CH_3)_2{-}OO{-}C(CH_3)_3$ **3**

Scheme 1

It was failed to obtain peroxyalkyl methacrylate **2** by the esterification of 3-tert-butylperoxy-3-methyl-1-butanol with methacrylic acid in the presence of sulphuric acid as a catalyst or by interesterification of methyl methacrylate with hydroxy peroxide **1** at satisfactory yields due to the heterolysis of peroxide group catalyzed by mineral acids.

Hydroxy peroxide **1** and maleic anhydride were utilized as initial substances to prepare peroxide monomers – maleic acid derivatives. The interaction of equimolar amounts of 3-tert-butylperoxy-3-methyl-1-butanol and maleic anhydride was performed for 18 h at 30…40°C in the presence of catalytic amount of triethylamine (Scheme 1, reaction 2). The yield of 3-tert-butylperoxy-3-methyl-1-butyl maleate **3** was of 87%.

The structure of the synthesized substances has been confirmed by data of elemental and functional analyses, IR and NMR H^1 spectra. The characteristics of the obtained products are listed in Table 1.

Table 1. Characteristics of peroxide monomers – derivatives of methacrylic and maleic acids.

Product Number	Yield, %	d_4^{20}	n_D^{20}	MR_D Obtained/ Calculated	Obtained / Calculated, % C	H	Formula	M Obtained/ Calculated
2	89	0.9388	1.4320	67.504/67.728	63.78/63.91	9.86/9.90	$C_{13}H_{24}O_4$	248/244.33
3[a)]	87	--	--	--	56.78/56.92	8.21/8.08	$C_{13}H_{22}O_6$	281/274.31

a) melting point 60.0...61.5°C.

The peroxide monomer **2** is a colourless oily liquid, and the monomer **3** is a crystalline substance. The peroxide monomers synthesized are safe for use and capable to be kept without the appreciable loss of active oxygen content for a long time. The monomer **3** has an essential advantage over the other kinds of known PM: it does not undergo self-initiated homopolymerization.

Obtaining of reactive copolymers

Copolymerisation of peroxyalkyl methacrylate with styrene

The copolymerization of synthesized peroxyalkyl methacrylate **2** with styrene in solution has been investigated in order to study its polymerizability (Scheme 2). The copolymerization was carried out for five monomer ratios (Table 2).

Unsaturated peroxide loss for side reactions is known to do not exceed 2...3% during copolymerization at 60-70°C.[13] It permits the utilization of known techniques for determination of copolymerization constants under these conditions.

The polymerization constants were calculated according to the Mayo-Lewis method using the composition equation in the integrated form:[14] $r_1 = 0.54$, $r_2 = 0.43$. Figure 2 represents the composition curve of the copolymer. One can see from Figure 2 that both copolymerization constants are less then 1 during copolymerization of peroxyalkyl methacrylate with styrene, i.e. every comonomer is added more readily to radical of another than to its own one. It is probably explained by the difference in polarities of double bonds of peroxyalkyl methacrylate and styrene. The product of r_1 and r_2 is of low value, that tesifies to the tendency of this system to form alternate copolymer.

$$n\, CH_2{=}C(CH_3){-}C({=}O){-}O{-}(CH_2)_2{-}C(CH_3)_2{-}OOBu\text{-}t + m\, C_6H_5CH{=}CH_2 \longrightarrow {-}(CH_2{-}CH(C_6H_5))_m{-}(CH_2{-}C(CH_3)(C({=}O){-}O{-}(CH_2)_2{-}C(CH_3)_2{-}OOBu\text{-}t))_n{-}$$

2   **4**

Scheme 2

The polymers of linear structure are formed as a result of the performed copolymerization. The low value of average molecular weight of the prepared copolymers (about 12,500) testifies to this fact. Furthermore, 3-tert-butylperoxy-3-methylbutyl methacrylate undergoes mainly chain propagation reaction without peroxide group decomposition under copolymerization proceeding conditions as it is indicated by active oxygen content of the final copolymers (see Table 2). The mentioned results coincide with data on copolymerization of other kind of ditertiary peroxide monomers reported in literature before.[15]

Table 2. Copolymerization of peroxyalkyl methacrylate (M_2) with styrene (M_1).

Monomer mixture composition, $[M_1] : [M_2]$	Elemental composition of copolymers, % w/w			Copolymer composition, $[m_1] : [m_2]$	Active oxygen content in copolymer, %	Average molecular weight of copolymer
	C	H	O			
0.7955 : 0.0483	79.23	8.72	11.97	0.7358 : 0.2642	2.99	11,200
0.6000 : 0.4000	74.53	9.09	16.37	0.5846 : 0.4154	4.09	12,100
0.3978 : 0.6022	71.25	9.33	19.39	0.4512 : 0.5488	4.84	12,600
0.1976 : 0.8024	68.13	9.57	22.27	0.2913 : 0.7087	5.56	13,100
0.0988 : 0.9012	66.27	9.72	24.00	0.1761 : 0.8239	5.99	13,200

Copolymerisation of peroxyalkyl maleate with styrene

The copolymerization of peroxyalkyl maleate **3** with styrene (Scheme 3) in solution was carried out for three monomer ratios (Table 3). Figure 3 represents the composition diagram of the copolymer.

n HO–C(=O)–CH=HC–C(=O)–O–$(CH_2)_2$–C(CH_3)(CH_3)–OOBu-t + m $CH=CH_2$(C₆H₅) ⟶ –(CH(C_6H_5)–CH_2)$_m$–(CH(C(=O)OH)–CH(C(=O)–O–$(CH_2)_2$–C(CH_3)(CH_3)–OO–Bu-t))$_n$–

3 **5**

Scheme 3

The copolymerization constants of peroxyalkyl maleate **3** with styrene calculated according to the method[14] were: $r_1 = 0.15$ (styrene), $r_2 = 0.04$ (peroxyalkyl maleate). One can see from the obtained copolymerization constants that the addition of the peroxide monomer **3** molecule to its own radical is hindered during the copolymerization with styrene because of steric factors. That is why, styrene mainly is added to the maleate radical as it is indicated by low copolymerization constant value ($r_1<1$). The average molecular weight of obtained copolymer **5** amounts to about 15,700.

Table 3. Copolymerization of peroxyalkyl maleate (M_2) with styrene (M_1).

Monomer mixture composition, $[M_1] : [M_2]$	Elemental composition of copolymers, % w/w			Copolymer composition, $[m_1] : [m_2]$	Active oxygen content in copolymer %	Acid number, mg KOH/g	Average molecular weight of copolymer
	C	H	O				
0.9046 : 0.0954	73.46	15.41	18.59	0.6988 : 0.3012	3.09	108.71	17,000
0.7562 : 0.2438	67.39	7.97	24.60	0.5260 : 0.4740	4.10	143.87	16,500
0.4829 : 0.5171	67.29	7.97	24.69	0.5227 : 0.4773	4.11	144.44	13,700

Peroxyalkyl maleate link content in the copolymers calculated from acid number value coincides with copolymer composition determined from active oxygen content. It points out the fact that copolymerization of peroxyalkyl maleate (**3**) with styrene undergoes without the appreciable decomposition of peroxide groups under process proceeding conditions. The composition of the synthesized copolymers calculated from functional analyses data corresponds to the composition determined from the results of elemental analyses. The formed copolymers are linear. The low value of average molecular weight and retaining peroxide groups under copolymerization proceeding conditions testify to such structure of the copolymers.

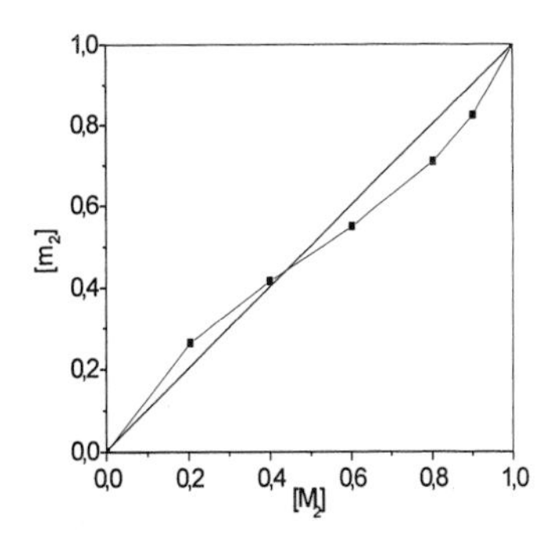

Fig. 1. The composition curve of the copolymer **4** of peroxyalkyl methacrylate (M_2) with styrene (M_1).

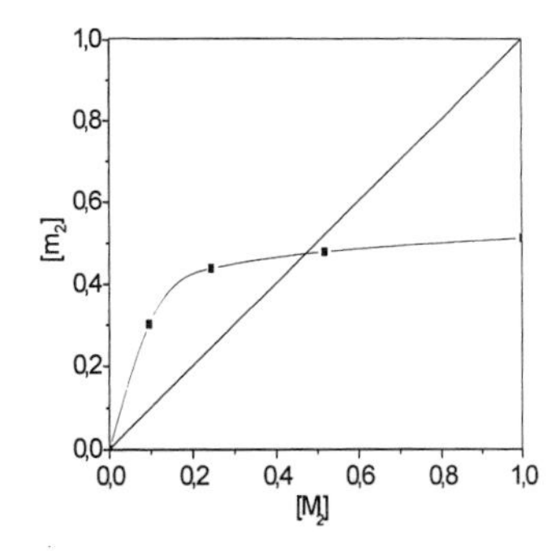

Fig. 2. The composition curve of the copolymer **5** of peroxyalkyl maleate (M_2) with styrene (M_1).

Conclusions

New peroxide monomers on the basis of unsaturated carboxylic acids have been synthesized by the interaction of hydroxy peroxide with derivatives of methacrylic and maleic acid. New reactive peroxide polymers with peroxide groups in side substituents of macrochain have been prepared by the copolymerization of 3-tert-butylperoxy-3-methylbutyl methacrylate with styrene. The features of the radical copolymerization of peroxyalkyl maleate with styrene were studied and copolymerization constants were determined.

[1] Yurzhenko T.I., Puchin V.A., Voronov S.A. *Dop. Acad. Nauk SSSR.* **1965**. Vol. 164. No. 6. P. 1335.
[2] Voronov S., Tokarev V., Bednarska O., Pich A. *Bereiner Polymerentage.* **1997**. 9. 11.X. P. 62.
[3] S. Voronov, V. Tokarev, G. Petrovska, *"Heterofunctional Polyperoxides. Theoretical Basis of Their Synthesis and Application in Compounds"*. State University Lvivska Polytechnica, 1994.
[4] Voronov S., Minko S., Puchin V., Dykyj M. *Izv. Vuzov. Chim. i chim. Technol.* **1980**. Vol. 26. No. 4. P. 492.
[5] Ivanchev S.S., Budtov V.P., Romantsova O.N. e.a. *Vysokomolek. Soyed.* Ser. A. **1976.** Vol. 18. No. 5. P. 1005.
[6] Ivanchev S.S., Romantsova O.N., Romanova O.S. *Vysokomolek. Soyed.* Ser. A. **1975**. T.17. No 11. P. 2401.
[7] Pat. 4494 (Japan). The method of obtaining graft-copolymers. Itinoe Seintiro, Opozuki Mitio. Publ. 1963.[8] Fedorova V.O., Horin Ya.I., Chuyko L.S. *Voprosy Khimii i Khim. Tekhnologii*, **1988**, P. 108.
[8] Fedorova V.O., Selivanova I.M. *Dopovidi·NAN Ukrayiny.* **1997**. No. 5. P. 143.
[9] R.I.Fleychuk, O.I.Hevus, S.A.Voronov, *Russ. J. Org. Chem.* **2003**, *39*, accepted.
[10] V.L.Antonovskiy, M.M.Buzlanova, *"Analytical Chemistry of Organic Peroxide Compounds"*, Khimiya, Moscow, 1978.
[11] Toroptseva A.M., Belogorodskaya A.M., Bondarenko V.M. *"Laboratory training in chemistry and technology of high molecular weight compounds"*, Khimiya, Moscow, 1972.
[12] N.D.Cheronis, T.S.Ma, *"Organic Functional Group Ananlysis by Micro and Semimicro Methods"*, John Wiley & Sons, New York, 1964.
[13] Tokarev V.S. Ph.D. Thesis. Lviv, 1980. P. 189.
[14] T.Alfrey, J.J.Bohrer, H.Mark. *„Copolymerization"*, John Wiley & Sons, New York, 1952.
[15] Voronov S.A. Dr.Sc. Thesis. Lviv, 1981. P. 351.

Ring-Expansion Polycondensation of 2-Stanna-1,3-dioxepane (or 1,3-dioxepene) with Dicarboxylic Acid Chlorides

*Hans R. Kricheldorf,**[1] *Saber Chatti,*[2] *Gert Schwarz*[1]

[1] Institut für Technische und Makromolekulare Chemie, Bundesstr. 45A, 20146 Hamburg, Germany
E-mail: kricheld@chemie.uni-hamburg.de
[2] Institut National de Recherche et d'Analyse Physico-chimique, Pôle Technologique-Sidi Thabet-2020 Ariana, Tunisia
E-mail : sabchatti@yahoo.com

Summary: 2,2-Dibutyl-2-stanna-1,3-dioxepane 1 (or 1,3-dioxepene 2) were prepared from 1,4-butane (or 1,4-butene) diol and dibutyltin dimethoxide. They were polycondensed at 80°C in *n*-heptane with adipoyl-, suberoyl, sabacoyl chloride and with decane-1,10-dicarbonyl chloride. In the case of suberoyl chloride and 2,2-dibutyl-2-stanna-1,3-dioxepane reaction time, temperature and stoichiometry were varied to optimize both the molecular weight and the fraction of cyclic polyesters. With a slight excess of the dicarboxylic acid chlorides, only macrocyclic polyesters were obtained in all cases. The resulting cyclic polyesters were characterized by viscosity measurements, by ^{1}H and ^{13}C NMR and by MALDI-TOF mass spectrometry.

Keywords: cyclization; MALDI-TOF; polyesters; ring-expansion polycondensation

Introduction

In a previous publication [1] the ring-opening polycondensation in bulk of 2,2-dibutyl-2-stanna-1,3-dioxepane (**1**, DSDOP) with various aliphatic dicarboxylic acid chlorides was reported. The polycondensations of **1** were particularly remarkable, because it was demonstrated for the first time that the products of such a ring-opening polycondensation were mainly cyclic.[2] These results raised the question if donor-acceptor interactions between Bu_2SnO- and -COCl chain ends is responsible for the cyclization reactions or if the formation of cycles is the normal consequence of a kinetically controlled polycondensation proceeding in a homogeneous phase and in the absence of side reactions.[3-4] In almost all textbooks the theory of step-growth polymerisation is based on the work of Carothers and Flory and describes the formation of linear polymers from linear monomers via linear oligomers. However, the older "Ruggli-Ziegler Dilution Method" designed for synthesis of macrocycles assumes a permanent competition between chain growth and

 DOI: 10.1002/masy.200450611

cyclization. Increasing concentration favors, of course, higher molecular weights, but the tendency of cyclization does not automatically vanish. A quantitative description of this aspect was published almost three decades ago [3-4] by two groups of british authors but it was not substantiated by experiments and it was never discussed in textbooks. Recently, we have demonstrated by numerous experiments that, in fact, cyclic oligo- and polyesters are main reaction products of any clean polycondensation proceeding in a homogeneous phase.[5]

In this connection the present work had two aims. Firstly, the reproducibility of our first study [1] concerning the polycondensation of DSDOP with various dicarboxylic acid chlorides in *n*-heptane at 80°C to optimise the formation of cyclic oligo- and polyesters (Scheme3 1). Secondly, the optimal conditions obtained with DSDOP will be used for the polycondensation of 2,2-dibutyl-2-stanna-1,3-dioxepene (DSDEN) and various dicarboxylic acid chlorides. The ring formation should provide an easy access to unsaturated cyclic oligo- and polyesters based on 1,4-butene diol (Scheme 3). These unsaturated macrocycles may be of interest for a variety of postreactions (even when containing small amounts of linear species), for instance, ring-opening polymerizations such as the metathesis copolymerization with cyclic olefins.

Results and discussion

A) Polycondensation of 2-Stanna-1,3-dioxepane (DSDOP) with Dicarboxylic Acid Chlorides

The first cyclic monomer used in this work (DSDOP, **1**) was easily prepared by condensation of dry 1,4-butane diol with dibutyltin dimethoxide. A complete conversion which can be monitored by ^{1}H-NMR spectroscopy of the reaction mixture is essential for the success of this synthesis because unreacted starting materials are difficult to separate from the product by distillation over a short-path apparatus. The DSDOP was isolated as a colorless liquid in yields of 85-90%. The polycondensations were performed in a simple way just by dropwise addition of aliphatic dicarboxylic acid chlorides (ADADs) in *n*-heptane to a solution of DSDOP in *n*-heptane with rapid mechanical stirring (Figure 1).

$[CH_3(CH_2)_3]_2Sn(O\text{-}CH_2\text{-}CH_2)_2$ (cyclic) **1** + $Cl\text{-}CO\text{-}(CH_2)_m\text{-}CO\text{-}Cl$, m = 4, 6, 8, 10 $\xrightarrow[\text{temperature}]{n\text{-heptane}}$ cyclo-$[O\text{-}(CH_2)_4\text{-}O\text{-}CO\text{-}(CH_2)_m\text{-}CO]_n$, m = 4, 6, 8, 10 + $[CH_3(CH_2)_3]_2SnCl_2$

Figure 1. Polycondensation of DSDOP with various dicarboxylic acid chlorides

The first reactions were performed with equimolecular quantities of DSDOP and suberoyl chloride at 40°C. After 24 hours of reaction time, the MALDI mass spectrometry analysis of the resulting polyesters indicated the presence of three structures, one cyclic polyester (**C** = 55%) and two linear species corresponding to the linear structures **Lb** = 21%, having the 1,4-butane diol as chain end and **Lc** = 24% having the suberoyl chloride as chain end. This result was confirmed by ^{1}H-NMR where we clearly evidenced the presence of CH_2-OH and CH_2-COOH groups as chain ends (Figure 2).

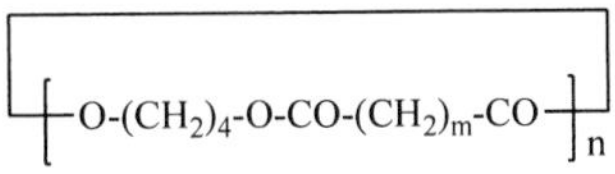

$$\left[\text{O-(CH}_2\text{)}_4\text{-O-CO-(CH}_2\text{)}_m\text{-CO} \right]_n \text{ (cyclic)}$$

C

$$\text{H}\left[\text{O-(CH}_2\text{)}_4\text{-O-CO-(CH}_2\text{)}_m\text{-CO} \right]_n \text{O-(CH}_2\text{)}_4\text{-OH}$$

Lb

$$\text{HO-CO-(CH}_2\text{)}_m\text{-CO}\left[\text{O-(CH}_2\text{)}_4\text{-O-CO-(CH}_2\text{)}_m\text{-CO} \right]_n \text{OH}$$

Lc

Figure 2. Structures of different polyesters obtained as a result of the polycondensation reaction between DSDOP and suberoyl chloride at 40°C during 24 hours

When +1% excess of acid dichloride was used, a significant improvement of the MALDI spectrum was noticed, and in this case only two structures were obtained, one of the cyclic polyester **C** and the other one of the linear polymer **Lb**. This mixture of cyclic and linear products has an inherent viscosity of 0.48 dL\g. When the reaction temperature was raised to 80°C with +1% excess of acid dichloride and a reaction time of 24 hours, the MALDI mass spectrum shows the exclusive presence of the cyclic polyester **C,** with a higher inherent viscosity of 0.76 dL\g. Considering these results, we noticed that when we worked above the melting point of the polyester based on DSDOP and suberoyl chloride (Tm = 58°C)[1], we have a better and higher reactivity of the active centers and, as a consequence, less secondary reactions. Under these conditions a quasi-total conversion and 100% cyclic polyester were obtained (Figure 3A and 3B) in agreement with our new theory of step growth-polymerization.

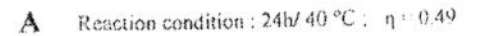

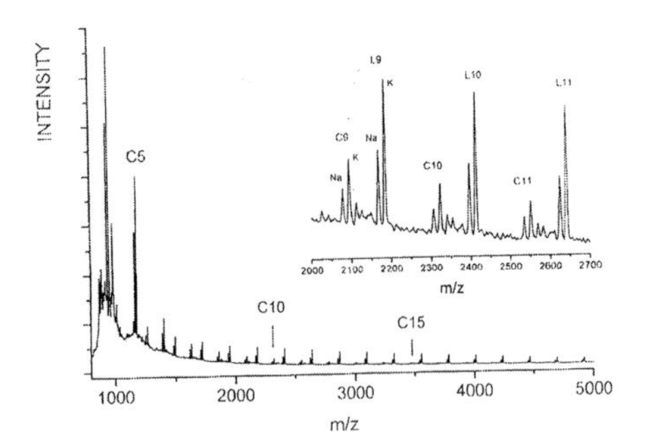

B Reaction condition : 24h/ 80 °C ; η = 0.76

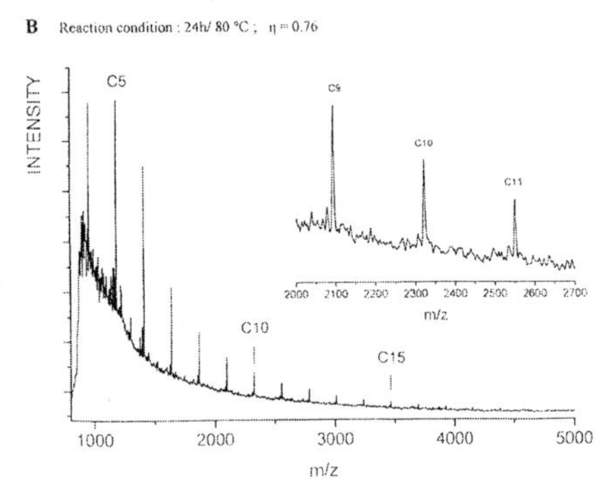

Figure 3. MALDI-TOF spectra showing the influence of reaction conditions on formation of cyclic polyesters derived of DSDOP and suberoyl chloride

The optimum conditions found with suberoyl chloride were subsequently used for the polycondensation of DSDOP with other acid dichlorides (Table 1).

Table 1. Reaction conditions and results of the polycondensations conducted with DSDOP and various dicarboxylic acid chlorides

exp. no.	acid dichloride (1% excess)	time hours	temperature °C	yields %	η_{inh} [a)] dL/g	results of MALDI-TOF intensi % C	La
1	adipic [b)]	24	80	87	0.64	100.0	0.0
2	suberic	24	40	91	0.48	15.0	85.0
3	suberic	24	80	96	0.76	100.0	0.0
4	sebacic	24	80	81	0.18	77.0	23.0
5	dodecane-dioic	24	80	91	1.22	100.0	0.0

[a)] Measured at 20°C with c = 2 g/L in CH_2Cl_2
[b)] 3% excess

For the poycondensation of DSDOP with adipoyl chloride with +1% excess of the acic dichloride, the MALDI mass spectrum indicated the presence of a mixture of cyclic (**C**) and linear (**Lb** and **Lc**) polyesters in equal proportions. When +3% excess of adipoyl chloride was used, the MALDI spectrum indicated the presence of the cyclic polyester **C** with an inherent viscosity of 0.64 dL\g.

The same optimum conditions were used in the case of sabacoyl chloride, but a mixture of cyclic polyester (**C** = 77 %) and linear polyester with hydroxyl chain ends (**Lb** = 23%) were obtained. This mixture of cyclic and linear polyesters has a low inherent viscosity of only 0.18 dL/g. This result is a bit surprising, because viscosity should be higher than those obtained with suberoyl chloride, where we have an inherent viscosity of 0.76 dL/g. We noticed that neither a longer reaction time nor a higher excess of acid dichloride did not allow a significant improvement in percentage of cyclic polyesters or molecular weights.

Finally the polycondensation of DSDOP with decane-1,10-dicarbonyl chloride gave only cyclic polyesters (**C** = 100%) with a higher inherent viscosity of 1.22 dL/g.

B) Polycondensation of 2-Stanna-1,3-dioxepene (DSDEN) with Dicarboxylic Acid Chlorides

The second cyclic monomer used in this work (DSDEN, **2**) was easily prepared by condensation of dry 1,4-butene diol with dibutyltin dimethoxide under the same procedure used with DSDOP. The DSDEN was isolated as a yellow liquid in yields of 80-85%.

The polycondensations were performed again by dropwise addition of dicarboxylic acid dichlorides (ADADs) in *n*-heptane to a solution of DSDOP and *n*-heptane with rapid mechanical stirring (Figure 4, Table 2).

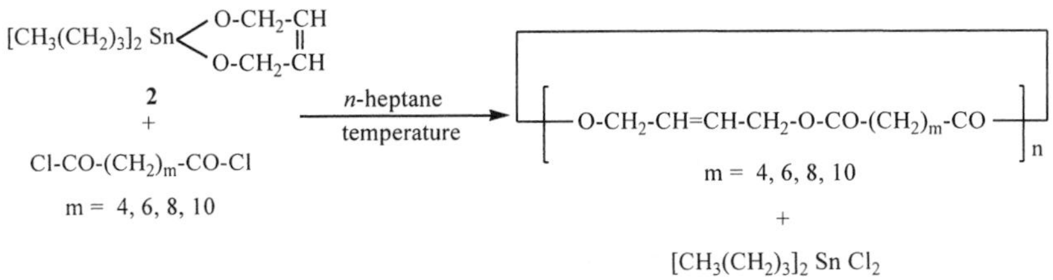

$[CH_3(CH_2)_3]_2\,Sn(O\text{-}CH_2\text{-}CH{=}CH\text{-}CH_2\text{-}O)$ (**2**) + $Cl\text{-}CO\text{-}(CH_2)_m\text{-}CO\text{-}Cl$, m = 4, 6, 8, 10 $\xrightarrow[\text{temperature}]{n\text{-heptane}}$ cyclic $[\text{-O-}CH_2\text{-}CH{=}CH\text{-}CH_2\text{-O-CO-}(CH_2)_m\text{-CO-}]_n$, m = 4, 6, 8, 10 + $[CH_3(CH_2)_3]_2\,Sn\,Cl_2$

Figure 4. Polycondensation of DSDEN with various dicarboxylic acid chlorides

Table 2. Reaction conditions and results of the polycondensations conducted with DSDEN and various dicarboxylic acid chlorides

exp. no.	acid chloride (1% excess)	time hours	temperature °C	yields %	η_{inh} [a)] dL/g	results of MALDI-TOF intensity %		
						C	Lb	Lc
1	adipic	24	80	88	0.57	89.0	0.0	11.0
2	suberic	24	80	89	0.49	60.0	40.0	0.0
3	sebacic	24	80	93	0.58	25.0	76.0	0.0
4	dodecane-dioic	24	80	95	0.65	93.0	0.0	07.0

[a)] Measured at 20°C with c = 2 g/L in CH_2Cl_2

We proceeded in the same manner as in the case of DSDOP, that is using +1% excess of acid dichloride at 80°C during 24 hours. The best results were obtained in the case of adipoyl chloride and decane-1,10-dicarbonyl chloride. Practically only unsaturated cyclic polyesters were obtained (**C** = 93-89%) with inherent viscosities between 0.57 and 0.65 dL/g. In Figure 5 is presented the MALDI mass spectrometry spectrum obtained for the polycondensation of DSDEN with adipoyl chloride. We can notice the majority presence of cyclic polyesters (**C** = 89%) and a minority presence of the linear polyester (**Lc** = 11%).

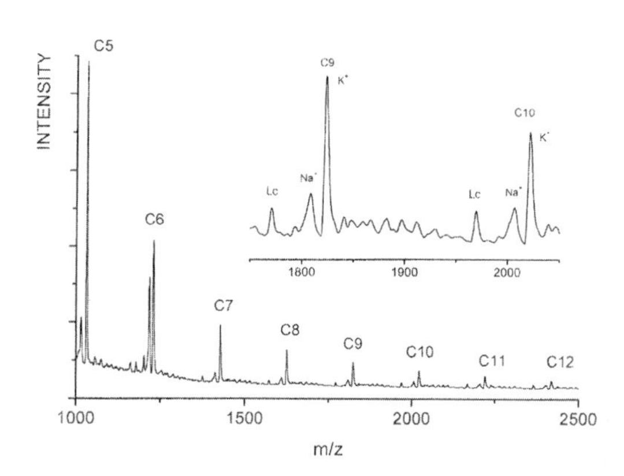

Figure 5. MALDI-TOF spectrum obtained after polycondensation of DSDEN with adipoyl chloride a 80°C during 24 hours

Due to the fact that we have an acidic chain end, we performed the same reaction with an equimolecular mixture of DSDEN and adipoyl chloride at 80°C during 24 hours. Unfortunately, the analysis of the reaction product by MALDI mass spectrometry showed a mixture of cyclic (**C** = 70%) and linear (**Lb** + **Lc** = 30%) polyesters. An increase of the reaction time did not lead in any improvement concerning the percentage of cyclic polyesters. For the polycondensation of DSDEN with suberoyl and sabacoyl chloride we did not succeed in obtaining a high percentage of unsaturated cyclic polyesters. The best results were obtained in the case of suberoyl chloride, 60% of cyclic polyesters **C** and 40% of linear polyesters **Lb**. The ^{1}H-NMR spectrum of this compound showed the presence of CH_2-OH group as chain end and confirmed the result of the MALDI mass spectroscopy. In order to improve these results we varied the excess of acid chloride between +2% and +5%, but we noticed the constant presence of a non-negligible quantity of linear polyesters (20-30%) with acidic chain ends **Lc**. In the same time, higher reaction times (7 days) did not determined a significant increase of the percentage of cyclic polyesters or of the molecular weight of these unsaturated polyesters derived from DSDEN.

Conclusion

The results of this work represent a satisfactory reproducibility of the previous study dealing with polycondensations of DSDOP (**1**). Firstly, only cyclic oligo- and polyesters (**C = 100%**) were formed in all polycondensations of DSDOP (**1**) with aliphatic dicarboxylic acid chlorides when we are working in *n*-heptane at 80°C. Particularly important is the observation that any optimisation of the reaction conditions favouring higher molecular weights enhances the fraction of cyclic products at the expense of the linear species. Therefore, the cyclizations become themselves a decisive limiting factor of the chain growth. This finding is in perfect agreement with our interpretation of kinetically controlled polycondensation according to the "Ruggli-Ziegler Dilution Method" which is based on the assumption of a permanent competition between ring-closure and propagation. Secondly, the optimal conditions obtained with DSDOP were used for the polycondensation of 2,2-dibutyl-2-stanna-1,3-dioxepene (DSDEN) and various dicarboxylic acid chlorides. Large fractions of cyclic oligo- and polyesters (approximately **C = 90%**) were formed in all polycondensations of DSDEN (**2**) with aliphatic dicarboxylic acid chlorides. These unsaturated macrocycles may be of interest for a variety of postreactions (even when containing small amounts of linear species), for instance, ring-opening polymerizations such as the metathesis copolymerization with cyclic olefins.

Acknowledgement

The authors greatfully acknowledge the financial support of the Alexander von Humboldt Foundation.

[1] H.R. Kricheldorf, D. Langanke, *Macromolecules* **1999**, *32*, 3559
[2] H.R. Kricheldorf, *Macromol.Rapid Commun.* **2000**, *21*, 528
[3] M. Gordon, W.B. Temple, *Macromol.Chem.* **1972**, *160*, 262 and 270
[4] J.L. Stanford, R.F.T. Stepto, D.R. Waywell; *J.Chem.Soc. Faraday Trans.* **1975**, *71*, 1308
[5] H.R. Kricheldorf, G. Schwarz, *Macromol. Rapid. Commun.* **2003**, *26*, 359.

Macromol. Symp. **2004**, *210,* 101-110

Hyperbranched Thermolabile Polycarbonates Derived from a A_2+B_3 Monomer System

Arnulf Scheel, Hartmut Komber, Brigitte Voit*

Institute of Polymer Research Dresden e.V., Hohe Strasse 6, 01069 Dresden, Germany

Summary: Hyperbranched polycarbonates were synthesized successfully via the A_2 + B_3 route by the reaction of a bis(carbonylimidazolide) with triethanolamine. These polymers containing the carbonate group as thermolabile moiety are decomposing into volatile products at around 200°C. The polymers were characterized with $^{1}H/^{13}C$ NMR spectroscopy, SEC, DSC and TGA techniques.

Keywords: degradation; hyperbranched; NMR; polycarbonate; thermal properties

Introduction

Developing thermolabile polymers is of interest for several applications like imaging[1] or for use as porogen in polymer matrices.[2] Nanoporous polymers have a high potential e.g. as ultra low dielectric constant material for use in microelectronic devices.[3,4] In this application the thermolabile polymer used as porogen has to fulfill several conditions. To achieve voids in nanometer scale the porogen or template has to be miscible with the polymeric matrix material. Furthermore, its decomposition temperature has to be below the glass transition temperature (T_g) and above the crosslinking temperature of the matrix materials. Hyperbranched polymers are of interest as pore template due to their globular structure and molecular dimensions and the high density of functional groups in the structure. They are usually much easier synthesized than dendrimers but exhibit a not perfect, irregular branched structure and a broad molar mass distribution. Due to the possibility of incorporating many thermally labile parts in the hyperbranched structure the polymer should decompose into small volatile products which can diffuse out of the matrix leaving the desired voids. In previous studies, photo- and thermally labile triazene units could be incorporated into hyperbranched polyesters[5] and the material was

 DOI: 10.1002/masy.200450612

successfully incorporated and decomposed in a tetramethyldivinylsiloxane-bisbenzocyclobutane (DVS-BCB) matrix leading to reduced dielectric constants.[6]

Polycarbonates are known to have sharp decomposition temperatures depending on the structural environment of the carbonate group.[7, 8, 9] Polycarbonates derived from tertiary alcohols have lower decomposition temperatures as those derived from primary alcohols and the decomposition temperature can be further reduced even below 100°C in the presence of acids.[1] Aromatic polycarbonates are thermally quite stable. Until now only hyperbranched polycarbonates derived from phenolic structures have been synthesized[10, 11] which seem not to be suitable for use as porogen.

Experimental

Materials: Dry toluene (Fluka), potassium hydroxide (Fluka, 86%), 2,5-dimethyl-2,5-hexanediol (Fluka, 97%) and carbonyldiimidazole (Merck, 98%) were used as received. Dichloromethane (Acros) and triethanolamine (Fluka, 99%) were dried before use over molecular sieve (4 Å).

Measurements: The NMR measurements were performed with a Bruker DRX 500 NMR spectrometer at 500.13 MHz for ^{1}H NMR spectra and at 125.75 MHz for ^{13}C NMR spectra. DMSO-d_6 was used as solvent for all NMR experiments. For internal calibration the solvent peaks of DMSO were used: δ (^{13}C) = 39.60 ppm; δ (^{1}H) = 2.50 ppm. Signal assignment was done by ^{1}H-^{1}H-COSY, ^{1}H-^{13}C-HMQC and ^{1}H-^{13}C-HMBC 2D NMR experiments using standard pulse sequences provided by Bruker. Thermal analysis were carried out with a Perkin Elmer TGA 7 instrument at a heating rate of 10 K/min for the dynamic measurement and a DSC 7 at a scan rate of 20 K/min. SEC was performed with MERCK LiChrogel PS40-column with a flow rate of 1.0 ml/min with linear polystyrene as standard and a Knauer RI-detector. Melting points were determined with Mettler Toledo FP62.

Synthesis of 2,5-dimethyl-2,5-hexanedicarbonylimidazole (3)

A solution of 8.53 g (58 mmol) 2,5-dimethyl-2,5-hexanediol and 20.11 g (120 mmol) carbonyldiimidazole (CDI) were dissolved in 150 ml dry toluene under argon atmosphere. To the solution 40 mg (1 mmol) KOH were added. The mixture was stirred for 5 hours at 60°C. After cooling to room temperature the reaction mixture was poured onto ice. After separation of the organic layer the aqueous layer was extracted twice with 20 ml toluene. The organic layers were unified and washed until neutralization. The organic layer was dried over anhydrous sodium sulfate. After evaporation of the solvent the crude product was recrystallized from a hexane/ethylacetate mixture to give colorless crystals (yield: 94%). mp: 93.8°C

^{1}H NMR (ppm, DMSO-d_6): δ = 8.17 (s, 2H, H_1); 7.51 (s, 2H, H_3); 7.06 (s, 2H, H_2); 2.00 (s, 4H, H_7); 1.58 (s, 12H, H_8).

^{13}C NMR (ppm, DMSO-d_6): δ = 146.74 (C_4); 137.13 (C_1); 130.11 (C_3); 117.44 (C_2); 87.07(C_5); 33.78(C_7); 25.41(C_6).

Synthesis of the hyperbranched polycarbonate 5 (selected example)

4 g (12 mmol) of **1** and 1.16 g (8 mmol) triethanolamine **(4)** were dissolved in 20 ml dichloromethane under argon atmosphere. 14 mg (0.2 mmol) KOH were added and the mixture was refluxed for 12 h. After cooling to room temperature the mixture was poured onto ice. The organic layer was separated and the aqueous was extracted two times with 20 ml dichloromethane. The organic layers were unified and washed with water till neutralization. The organic layer was dried with sodium sulfate and evaporation of the solvent gave 2 g crude colorless oily product. For purification the product was dissolved in dichloromethane and precipitated in cold ether to give 0.98 g of polymer. (M_w: 10400 g/mol; M_n: 3300 g/mol)

^{1}H NMR (ppm, DMSO-d_6): δ = 8.16 (H_{14}); 7.51 (H_{16}); 7.03 (H_{15}); 4.25 (H_1); 4.07 and 3.99 (H_5, also in cyclic structures); 3.41 (H_2); 2.75 and 2.71 (H_4, also in cyclic structures); 2.58 (H_3); 1.88 (H_9 and H_{10}); 1.85 (H_{17} in $[a_2b_2]_c$); 1.74 (H_{17}); 1.55 (H_{12}); 1.39 – 1.34 (H_8 and H_{18}, also in cyclic structures).

^{13}C NMR (ppm, DMSO-d_6): δ = 152.85, 152.78, 152.71 (all C_6); 146.6 (C_{13}); 137.1 (C_{14}); 130.0 (C_{16}); 117.4 (C_{15}); 86.9 (C_{11}); 82.8 (C_7); 64.8 – 64.3 (several signals, C_5, also in cyclic structures); 59.40 (b_2**B**), 59.34 (b$\mathbf{B}_2$), 58.70 ($[a_2b_2]_c$**B**) (all C_2); 56.96 (b$\mathbf{B}_2$), 56.71 (b_2**B**), 58.70 ($[a_2b_2]_c$**B**) (all C_3); 53.45 ($[a_2\mathbf{b}_2]_c$B), 53.30 ($[a_2\mathbf{b}_2]_c$b), 53.06 (**b**B_2), 52.90 ($\mathbf{b}_2$B), 52.8 – 52.6 ($\mathbf{b}_3$), 51.52 $[a_2b_2]_c\mathbf{b}$)

(all C_3); 34.20 (C_{10}); 33.7 (C_{17}); 33.2 (C_9); 31.70 and 31.68 (C_{17} in $[a_2b_2]_c$); 26.59 (C_{18} in $[a_2b_2]_c$); 25.55 and 25.50 (C_8); 25.4 (C_{18}); 25.28 (C_{12}).

Results and Discussion

There are two ways to build up hyperbranched structures: first the classical AB_n (with $n \geq 2$) route which can not lead to crosslinked products according to Flory [12] and second the $A_n + B_m$ (with $n = 2$ and $m \geq 3$) approach. The second approach offers an easier accessibility of monomers but during the polymerization crosslinking might take place and the resulting structures are far more complex. The $A_2 + B_3$ approach was followed for the synthesis of the hyperbranched polycarbonates. The A_2 monomer **(3)** was obtained analogue to Rannard et al.[13] from the reaction of 2,5-dimethyl-2,5-hexandiol **(1)** with CDI **(2)** in toluene with a catalytic amount of KOH (Scheme 1).

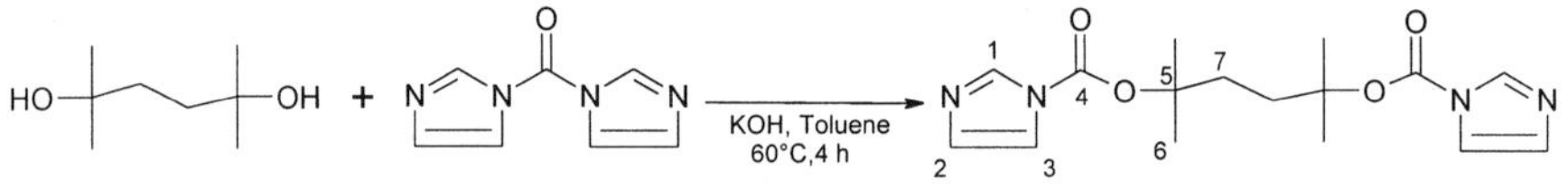

Scheme 1: Synthesis of $\mathbf{A_2}$

The polymerization reaction was carried out using different ratios n : m (with n = 3 to 1 and m = 1 to 2) of A_2 and B_3 and different concentrations (0.01 wt% to 15 wt%) of **3** in dichloromethane under reflux (Scheme 2). Using higher concentrations of **3** leads to insoluble products. Not converted B_3 monomer was removed during work up by washing with water and remaining A_2 monomer was removed by precipitating the crude polymer from dichloromethane into cold diethylether. With the precipitation also a fractionation of the polymer took place resulting in a high molecular weight fraction which was not soluble in diethylether and a low molecular weight fraction which could be recovered from the diethylether (Fig. 5). The resulting polymers are viscous colorless substances.

In the following the results obtained for a hyperbranched polycarbonate prepared in a 15 wt%

solution of **3** in dichloromethane with a ratio $A_2 : B_3$ of 3:2 are discussed.

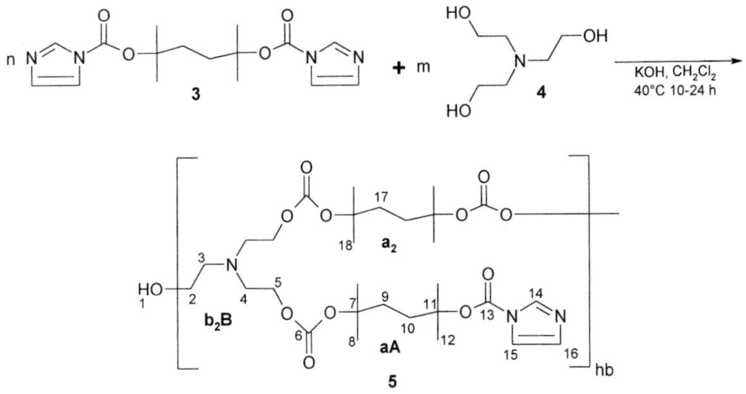

Scheme 2: Polymerization reaction leading to polycarbonate **5**

NMR Analysis

NMR analysis proved to be highly effective for the structural characterization of hyperbranched structures.[5,14] The ^{1}H NMR spectrum of polymer **5** (Fig. 1) proves the presence of all the expected structural units and allows the quantification both of unreacted A and B units. More structural information can be obtained by ^{13}C NMR spectroscopy looking at specific reaction mixtures. Depending on the different degree of conversion of B units it can be distinguished between the B_3 (monomer), bB_2 and b_2B structures from the ^{13}C signals of the B unit (Fig. 2a). In contrast, the signals of the b units are less sensitive on different substitution patterns. Furthermore, it can be differentiated between unreacted A_2 monomer, half reacted aA where one carbonylimidazole group has reacted with one hydroxy group and a_2 where both carbonylimidazole groups have reacted with hydroxy groups as shown in the spectrum in Figure 2b, obtained from a reaction mixture with a high excess of A_2. In conclusion, the ^{13}C NMR spectra prove the existence of branched structures.

Additional signals were observed besides the ones of the regular structure shown in Scheme 2. These are caused by further structural units within the branched macromolecule. From the observed chemical shift effects, signal intensity ratios and from 2D NMR it can be concluded that these signals are caused by the cyclic structures $[a_2b_2]_cB$ or $[a_2b_2]_cb$ (Scheme 3), which are formed when both A functionalities of an A_2 monomer react with the same B_3 monomer. As expected the content of this structure increases with decreasing concentration of monomers the feed. Whereas this smallest cyclic system can still be identified by NMR spectroscopy, larger ring systems result in the same chemical shifts as non-cyclic structures.

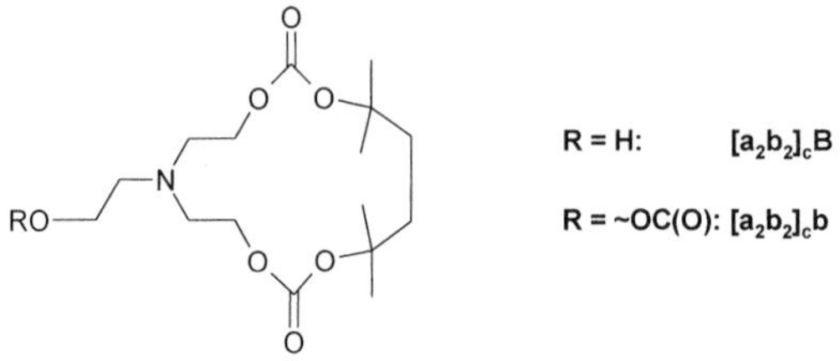

R = H: $[a_2b_2]_cB$

R = ~OC(O): $[a_2b_2]_cb$

Scheme 3: Ring formation

Ring formation has consequences for the determination of the number of terminal groups and the calculation of the degree of branching. The structure $[a_2b_2]_cb$ (Scheme 3) appears with respect to the B_3 monomer as dendritic unit but is a terminal unit and terminates the polymer growth. Similar considerations are possible for larger cyclic structures. So the calculation of a meaningful degree of branching is not possible for this polymer system.

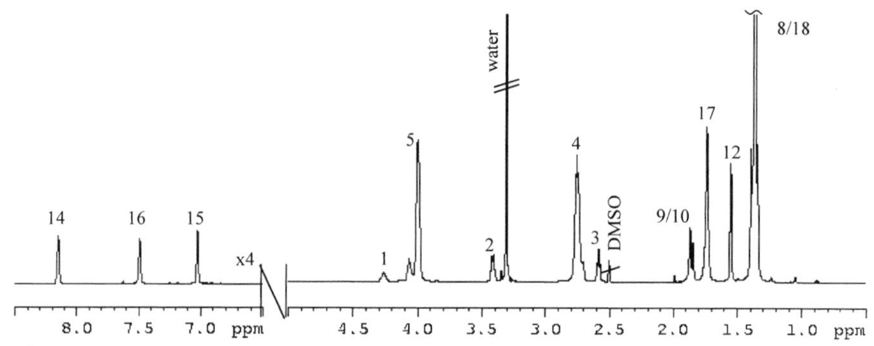

Fig. 1: ^{1}H NMR spectra of hb polycarbonate **5** (A_2:B_3 = 3:2, 15 wt% in CH_2Cl_2) in DMSO-d_6 (for peak assignment see Scheme 2)

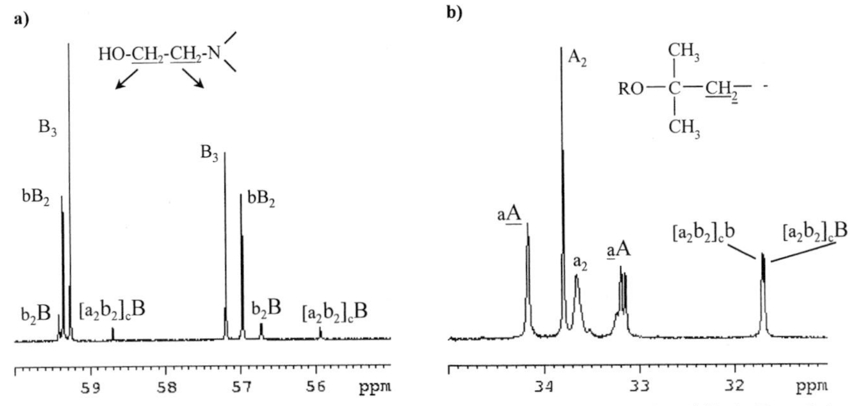

Fig. 2: ^{13}C NMR spectra regions of different reaction mixtures: a) A_2:B_3 = 1:2 and b) A_2:B_3 = 2:1 showing characteristic signals for differently substituted B_3 units a) and A_2 units b) in DMSO-d_6

Thermal behavior

The thermal behavior was studied by differential scanning calorimetry (DSC) and thermogravimetric analysis (TGA). The DSC of the polymer **5** shows a T_g of -14°C. The dynamic TGA measurement indicates that the polymer decomposes in three steps. The main process was found at 221°C and the decomposition is completed at 250°C (Fig. 3). With isothermic TGA measurements it is possible to determine the decomposition velocity at a certain temperature. The isothermal measurement at the beginning of the main decomposition step at 200°C shows that the decomposition goes near to completion in about 15 min (Fig. 4). Therefore the polycarbonate should be suitable to function as porogen in matrices.

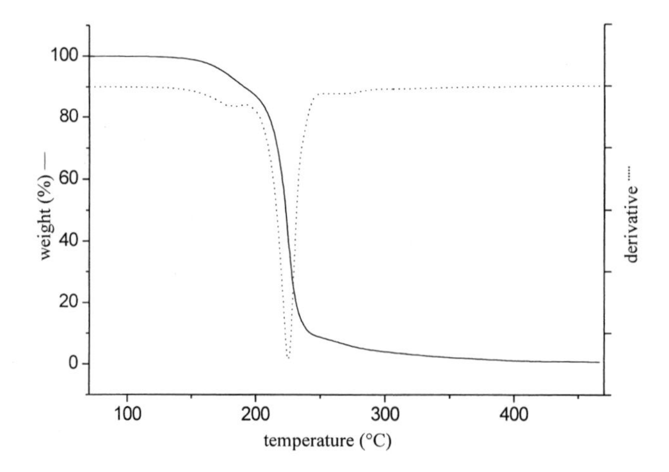

Fig. 3: Dynamic TGA-measurement of A_2+B_3 polycarbonate **5** at a heating rate of 10 K/min

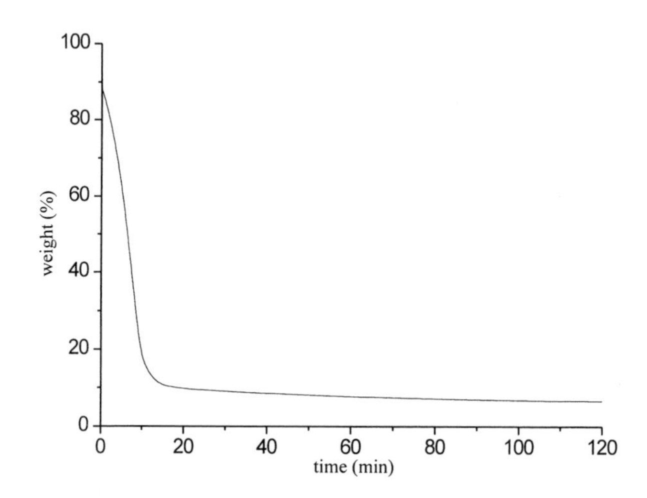

Fig. 4: Isothermic TGA-measurement of A_2+B_3 polycarbonate **5** at 200°C

Size Exclusion Chromatography - Analysis

In general size exclusion chromatography (SEC) is not suitable for molar mass determination of hyperbranched polymers. The SEC is calibrated with linear polystyrene as standard and, due to the more compact structure of hyperbranched polymers in comparison to linear polymers, the molar mass should be higher than the results indicated by SEC. Therefore the measured molar masses of M_n = 3300 g/mol and M_w = 10400 g/mol of the precipitated fraction of hb polycarbonate **5** should be considered as probably too low compared to the real molar mass. The broad polydispersity of 3.2 is usual for hyperbranched polymers obtained by A_2 + B_3 polymerization.

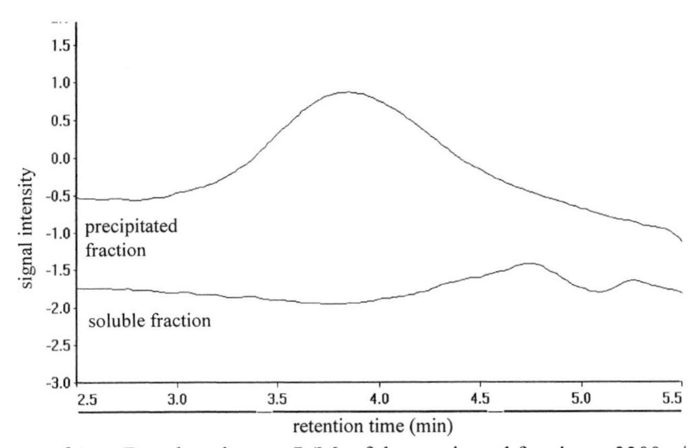

Fig.5: SEC curves of A_2 + B_3 polycarbonate **5** (M_n of the precipated fraction = 3300 g/mol)

Conclusions

Fully soluble hyperbranched polycarbonates were synthesized successfully via the A_2 + B_3 route by the reaction of a bis(carbonylimidazolide) with triethanolamine. SEC molar masses (M_w) of up to 10000 g/mol were achieved. NMR analysis proved the existence of the branched structure but also the presence of internal cyclics. These low T_g polymers (T_g = -14 °C) containing the carbonate group as thermolabile moiety are decomposing completely into volatile products at about 200°C. Therefore the use of these thermolabile highly branched structures as globular polymeric template for the preparation of nanoporous polymeric materials seems possible.

Acknowledgement

We would like to thank L. Häußler for the TG and DSC measurements and P. Treppe for the SEC characterization. We also would like to thank the DFG (SFB 287) and the DAAD for the financial support.

[1] J. M. J. Frechet; F. Bouchard; E. Eichler; F. M. Houlihan; T. Iizawa; B. Kryczka; C. G. Willson, *Polymer Journal* **1987**, *19* (1), 31-49
[2] Q. R. Huang; W. Volksen; E. Huang; M. Toney; C. W. Frank; R. D. Miller, *Chemistry of Materials* **2002**, *14*, 3674-3685
[3] C. V. Nguyen; K. R. Carter; C. J. Hawker; J. L. Hedrick; R. L. Jaffe; R. D. Miller; J. F Remenar; H.-W. Rhee; P. M. Rice, M. F. Toney; M. Trollsas; D. Y. Yoon, *Chemistry of Materials* **1999**, *11*, 3080-3085
[4] D. Mecerreyes; E. Huang; T. Magbitang; W. Volksen; C. J. Hawker; V. Y. Lee; R. D. Miller; J. L. Hedrick, *High Performance Polymers* **2001**, *2*, 11-19
[5] M. Eigner; H. Komber; B. Voit, *Macromolecular Chemistry and Physics* **2001**, *202* (2), 245-256
[6] M. Eigner; B. Voit; K. Estel; J. Bartha, *e- Polymers* **2002**, *28*, 1-18
[7] V. van Speybroeck; Y. Martele; M. Waroquier; E. Schacht, *Journal of the American Chemical Society* **2001**, *123* (43), 10650-10657
[8] Y. Martele; V. van Speybroeck; M. Waroquier; E. Schacht, *e-Polymers* **2002**, *049*, 1-14
[9] V. van Speybroeck; Y. Martele; E. Schacht; M. Waroquier, *The Journal of Physical Chemistry* **2002**, *106*(51), 12370-12375
[10] D. H. Bolton; K. L. Wooley, *Macromolecules* **1997**, *30*, 1890-1896
[11] D. H. Bolton; K. L. Wooley, *Journal of Polymers Science: Part A: Polymer Chemistry* **2002**, *40*, 823-835
[12] P. J. Flory, *Journal of the American Chemical Society* **1952**, *74*, 2718-2723
[13] S. P. Rannard; N. J. Davis, *Organic Letters* **1999**, *1* (6), 933-936
[14] H. Komber; B. Voit; O. Monticelli; S. Russo, *Macromolecules* **2001**, *34* (16), 5487-5493

Synthesis of Boc Protected Block Copolymers Based on para-Hydroxystyrene via NMRP

Martin Messerschmidt,[1] *Liane Häußler,*[1] *Brigitte Voit,**[1] *Tilo Krause,*[2] *Wolf-Dieter Habicher*[2]

[1] Institute of Polymer Research Dresden e. V., Hohe Strasse 6, 01069 Dresden, Germany
[2] Institute of Organic Chemistry, Technical University of Dresden, Mommsenstrasse 13, 01062 Dresden, Germany

Summary: Nitroxide mediated free radical polymerization (NMRP) was used for the preparation of orthogonally protected block copolymers based on para-hydroxystyrene. The polymers have a low polydispersity and an active chain end. By a series of polymer analogous reactions, a partly deprotected block copolymer was synthesized consisting of a block with unprotected phenolic OH groups and a further block which is protected by the thermolabile Boc group.

Keywords: diblock copolymers; p-hydroxystyrene; macroinitiators; NMRP; protecting groups

Introduction

A characteristic feature of block copolymer systems is the repulsion between unlike blocks causing microphase separation at mesoscopic length scales.[1] The size and type of ordering can be controlled varying the molecular weight, chemical structure, molecular architecture and composition of the block copolymers. Understanding and controlling the morphology of a block copolymer is essential for any application.[2]

Here we employ block copolymers based on p-hydroxystyrene. They consist of a block with protected phenolic OH groups and a further block with unprotected phenolic OH groups. The big difference in the polarity of the two blocks causes a strong incompatibility which result in a phase separation even for blocks of low molecular weight. The gradual removal of a protecting group leads to a change of the composition of the block copolymer which should also affect the morphology. This coherence is of great interest especially for phase investigations in thin films (thickness <100 nm). The formation of a defined nanostructure requires usually block copolymers

 DOI: 10.1002/masy.200450613

with a very low polydispersity.[2,3] Therefore the synthesis of the block copolymer was carried out applying nitroxide mediated free radical polymerization (NMRP).[4] As the phenolic OH group act as a transfer reagent the direct polymerization of unprotected p-hydroxystyrene is not practicable.[5] To overcome this problem orthogonally protected p-hydroxystrene derivatives were employed. Selective removal of only one protecting group leads to the desired partly deprotected, phase-separated block copolymer. Of special interest was the preparation of block copolymers with one Boc protected block since those protecting groups can be removed by simple thermal treatment.

Experimental

Characterization:

^{1}H NMR measurements were performed in 5 mm diameter tubes with a Bruker DRX 500 NMR spectrometer at 500.13 MHz. Acetone-d_6 has been used as solvent for the polymers as well as for internal calibration of the spectra.

GPC measurements were carried out on a Knaur modular equipment with two different columns and solvents. The homopolymers and the protected block copolymers were measured with a MERCK LiChrogel PS 40 column and with chloroform as solvent. Polystyrene standards were applied for calibration. The partly deprotected block copolymers were measured with two Zorbax Trimodal-S columns in dimethyl acetamide which contained also 2 vol% water and 3.0 g/L LiCl. Here PVP standards were used for calibration.

TGA measurements were performed on a Perkin Elmer TGA 7 with a heating rate of 10 K/min.

Materials

When not specified all employed chemicals were purchased form Fluka, Aldrich and Merck and were also of analytical grade.

Synthesis of monomer and polymers

The synthesis of the monomer TBDMS-OSt and the procedure of the polymerization was performed similar as already described in literature.[6]

Synthesis of the initiators and polymerization

Initiator 1 and initiator 2 have been synthesized similar to literature.[7-10] The homopolymers as well as the block copolymer **A** were synthesized via NMRP similarly as described previously.[6]

Polymer analogous reactions

Synthesis of block copolymer **B**

In a mixture of 60 mL dioxane and 12 mL methanol 2 g of block copolymer **A** was dissolved. After addition of 4 mL of hydrazine monohydrate the reaction was stirred for 48 h. The polymer was precipitated in 400 mL water, recovered by filtration, dissolved in a mixture of ethyl acetate and methanol (1:3) and precipitated in water again. After filtration, the polymer was dried at 70 °C in a vacuum oven and amounted to 1.4 g of block copolymer **B.**[12]

Synthesis of block copolymer **C**:

1 g of block copolymer **B** was dissolved in 10 mL ethyl acetate followed by the addition of 1.8 g potassium carbonate and 2 g of di-tert-butyl dicarbonate. After stirring for 24 h the polymer was precipitated in 400 mL methanol and recovered by filtration. The obtained precipitate was purified by reprecipitation in methanol and dried in vacuum at 40 °C to give 1.2 g of block copolymer **C.**[13,14]

Synthesis of block copolymer **D**:

0.8 g of block copolymer **C** and 4 mL of a 1 M solution of tetra(n-butyl) ammonium fluoride in THF were mixed in 6 mL of dry THF at room temperature. After stirring for 4 h, the polymer was precipitated in 400 mL of a 1:1 mixture of water and methanol. Precipitation was repeated several times to remove impurities, yielding 0.5 g of block copolymer **D.**[13]

Polymerization of protected p-hydroxystyrene derivatives via NMRP

For this study, the following orthogonally protected p-hydroxystyrene derivatives were used as monomers for NMRP: p-acetoxystyrene (Ac-OSt), p-tert-butyldimethylsilyloxystyrene (TBDMS-OSt) and p-tert-butoxycarbonyloxystyrene (Boc-OSt).

Figure 1 shows the applied monomers and also the appropriate reagent for the selective removal of the protecting group.

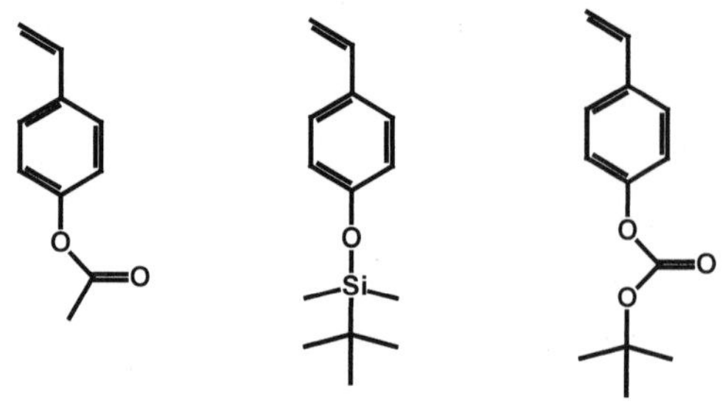

Abbreviation:	Ac-OSt	TBDMS-OSt	Boc-OSt
Removal by:	hydrazine monohydrate	fluoride	heat or H^+

Figure 1: Protected p-hydroxystyrene derivatives

Figure 2 show the two different initiators which were applied for NMRP. They were synthesized very similar as already described in literature[7-10] and are known as "Hawker-adducts". In contrast to initiator 1, initiator 2 contains three hydroxyl groups which are able to stabilize the free nitroxide radical. Hence it follows that the polymerization can be conducted at much lower temperatures. This can be very important for polymerizations of monomers which contain a thermolabile group like the Boc group.

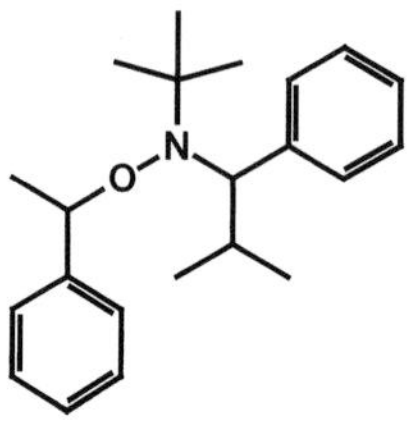

initiator 1

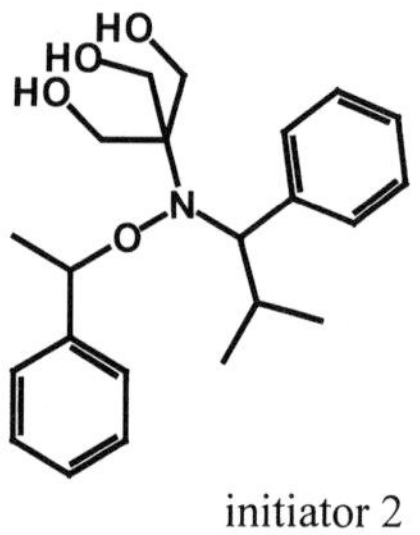

initiator 2

Figure 2: Applied initiators for NMRP

Table 1 summarizes some results of the polymerizations carried out with initiator 1 and initiator 2. All polymerizations with initiator 1 were run in bulk at 120 °C for approximately 16 hours. In respect of the amount of the initiator two equivalents of acetic anhydride were additionally added as it works as a rate accelerating additive.[11] In contrast to Ac-OSt and TBDMS-OSt, which resulted in well controlled homopolymers with narrow molar mass distribution, Boc-OSt could not be polymerized under these conditions.

Polymerizations with initiator 2 were run without addition of acetic anhydride as it most likely reacts with the OH groups of the initiator which results in a high polydispersity (PD). Firstly we polymerized Ac-OSt also at 120 °C but only over a period of 195 minutes. Despite the short reaction time we achieved already 56 % conversion and the molecular weight of the obtained poly(Ac-OSt) is 9900 g/mol. This is in good agreement with the theoretical value ($M_{n,cal}$ = 9100 g/mol, conversion implemented). Moreover the PD of poly(Ac-OSt) is only 1.16 which is comparable with those obtained for the polymerizations with initiator 1. These results underline the high efficiency of this initiator for NMRP. Due to the thermal lability of the Boc group, the polymerization of Boc-OSt was conducted at only 85 °C but over a period of 9 days. Under these conditions it was possible to polymerize Boc-OSt in a controlled fashion but about 7 % of the Boc groups were degraded. These results told us that the preparation of Boc group containing block copolymers cannot be accomplished by the direct polymerization of Boc-OSt. In order to achieve our goal anyhow we decided to apply a series of polymer analogous reactions.

Table 1: Results of the polymerization of protected p-hydroxystyrene derivatives

Initiator	**Homopolymer**	**M_n [g/ mol]**	**$M_{n,cal}$ [g/ mol]**	**PD**	**Yield [%]**
1	poly(Ac-OSt)	11150	11600	1.18	85
1	poly(TBDMS-OSt) macroinitiator (**MI**)	9800	11900	1.16	82
2	poly(Ac-OSt)	9900	9100	1.16	56
2	poly(Boc-OSt)	16500	6600	1.38	26

Polymer analogous reactions on block copolymers based on p-hydroxystyrene

The starting point was the preparation of the block copolymer poly(TBDMS-OSt)-b-poly(Ac-OSt) **A**. In order to achieve this, the homopolymer poly(TBDMS-OSt) was taken as macroinitiator **MI** and the second block was formed by sequential monomer addition of Ac-OSt. The block copolymer formation was also conducted at 120 °C and also with the addition of two equivalents of acetic anhydride in respect to the amount of macroinitiator. Diglyme was additionally added to dissolve all poly(TBDMS-OSt). After polymerization for approximately 24 hours we obtained block copolymer **A** in 81 % yield and with a molar mass of 42700 g/mol and a polydispersity of 1.19. Scheme 1 gives an overview over the subsequent single reaction steps while the corresponding ^{1}H NMR spectra are presented in Figure 3. After formation of block copolymer **A** all acetyl groups were removed using hydrazine monohydrate giving block copolymer **B.**[12] Block copolymer **C** was obtained by the reaction of block copolymer **B** with di-tert-butyl dicarbonate.[13, 14] The following treatment of **C** with tetrabutylammonium fluoride (TBAF) resulted in the orthogonal removal of the TBDMS groups leading to the desired block copolymer **D.**[13]

Table 2: Results of the preparation of the orthogonally protected block copolymers

block copolymers	macroinitiator MI		block copolymers		
	Mn [g/mol]	PD	M_n [g/mol]	$M_{n,cal}$ [g/mol]	PD
Poly(TBDMS-OSt)-b-poly(Ac-OSt) **A**	9800	1.16	42700	40000	1.19
Poly(TBDMS-OSt)-b-poly(H-OSt) **B**	—	—	46900	40000	1.63
Poly(TBDMS-OSt)-b-poly(Boc-OSt) **C**	—	—	48700	40000	1.20
Poly(H-OSt)-b-poly(Boc-OSt) **D**	—	—	51500	40000	1.52

All ^{1}H NMR (Figure 3) spectra were recorded in acetone-d_6 as solvent and show clearly that all polymer analogous reactions are orthogonal and highly effective.

In Table 2 the molecular weight (M_n) and the polydispersity (PD) of the macroinitiator and the derived block copolymers **A**, **B**, **C** and **D** are summarized. The molecular weight and the polydispersity was determined by GPC. The GPC measurements of the protected block copolymers **A** and **C** and the macroinitiator **MI** were carried out in chloroform while for the partly deprotected block copolymers **B** and **D** dimethyl acetamide was used. However in the GPC curve of the partly deprotected block copolymer **B** and also in the curve of block copolymer **D** a shoulder in the range of increasing molecular weight is clearly discernible. This means that under the conditions of the GPC measurement a fraction of a higher molecular weight is present. Since this higher molecular fraction is not observable in the GPC curve of block copolymer **C** we assumed that this fraction is caused by a simple physical association of at least two single molecules. This assumption is also corroborated by the fact that hydrogen bonding caused by phenolic OH groups are significantly stronger than those of normal aliphatic alcohols. Therefore the preparation of a molecular dispersed solution of a partly deprotected and thus, amphiphilic block copolymer like **B** and **D** is presumably much more difficult than for protected polymers like **A** and **C**. This would also explain the higher polydispersity of the block copolymers **B** and **D** compared with those of **A** and **C**.

In Figure 4 the GPC eluation peaks of the macroinitiator poly(TBDMS-OSt) **MI** and the derived block copolymers **A** and **C** are presented. All GPC curves are monomodal and have no shoulders. This implies that the initiation by the macroinitiator is highly efficient and the formation of well controlled and pure block copolymers is verified.

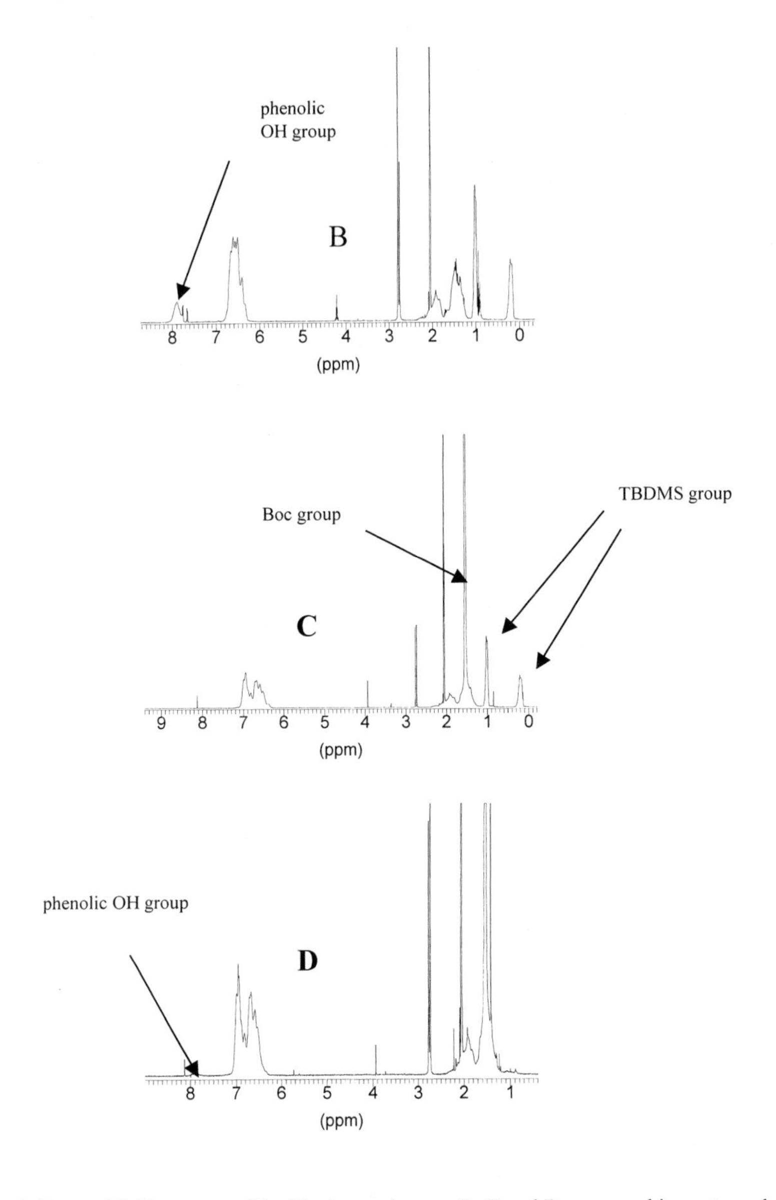

Figure 3: Proton NMR spectra of the block copolymers **B**, **C** and **D** measured in acetone-d_6

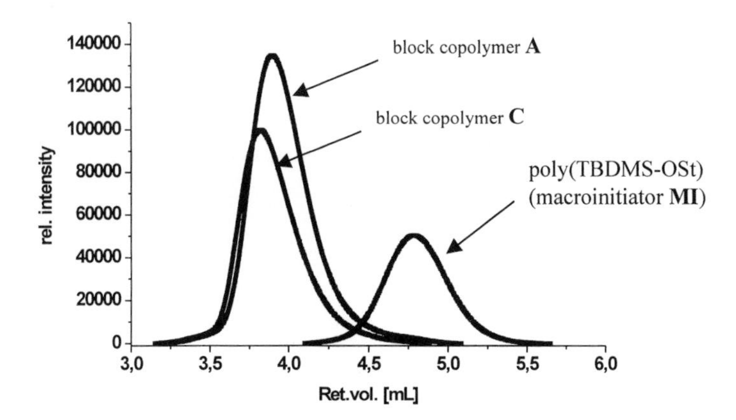

Figure 4: GPC elution curves of the macroinitiator poly(TBDMS-OSt) **MI** and the derived block copolymers **A** and **C** measured in chloroform

TGA investigations of the block copolymers C and D

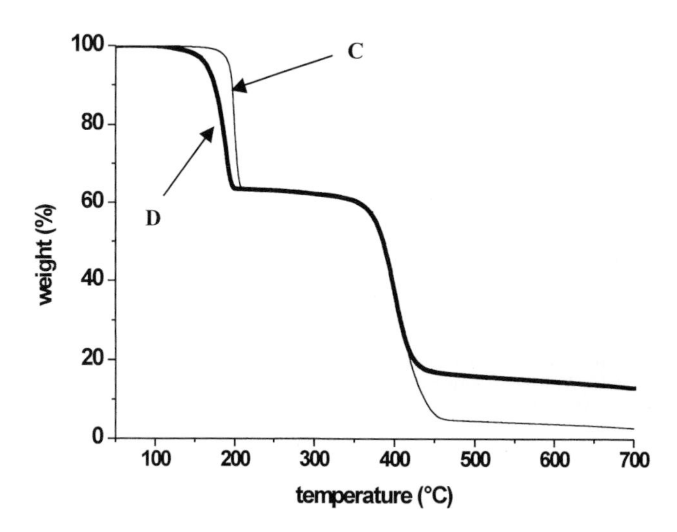

Figure 5: TGA curves of the polymer **C** and **D**

Figure 5 shows the TGA curves of the block copolymers **C** and **D**. While heating up with 10 °C per minute all Boc groups were removed near to 200 °C which is in good agreement with results already presented in literature.[15] But it is worthwhile to mention that the degradation of the Boc groups in the partly deprotected block copolymer **D** (thick line) started at least 10 K earlier than in block copolymer **C** (thin line).

Conclusion

Orthogonally protected block copolymers based on p-hydroxystyrene were successfully prepared with high control over molar mass and narrow molar mass distribution using NMRP. A quantitative and orthogonal removal of the acetyl groups, the TBDMS groups and the Boc groups was accomplished by employing hydrazine monohydrate, TBAF and heat, respectively. Using series of polymer analogous reactions a partly deprotected block copolymer consisting of a block with unprotected phenolic OH groups and a second block protected with Boc groups was synthesized.

Acknowledgement

The authors thank Ms. Treppe and Mr. Voigt for GPC measurements and Dr. Komber for NMR interpretation. Financial support of the DFG in the frame of the SFB 287 is gratefully acknowledged.

[1] F. S. Bates, G. H. Fredrickson, *Annu. Rev. Phys. Chem.* **1990**, *41*, 525.
[2] E. Huang, L. Rockford, T. P. Russell, C. J. Hawker, *Nature* **1998**, *395*, 757.
[3] G. Kim, M. Libera, *Macromolecules* **1998**, *31*, 2569-2577.
[4] C. J. Hawker, A. W. Bosman, E. Harth, *Chem. Rev.* **2001**, *101*, 3661.
[5] S. Nakahama, A. Hirao, *Prog. Polym. Sci.* **1990**, *15*, 299.
[6] A. Leuteritz, M. Messerschmidt, B. Voit, M. Yin, T. Krause, W.-D. Habicher, *Polymer Preprints* **2002**, *43*, 283.
[7] E. Harth, B. Van Horn, C. Hawker, *Chem. Comm.* **2001**, 823.
[8] S. Marque, H. Fischer, E. Baier, A. Studer, *J. Org. Chem.* **2001**, *66*, 1146.
[9] D. Benoit, V. Chaplinski, C. J. Hawker, R. Braslau, *J. Am. Chem. Soc.* **1999**, *121*, 3904.
[10] A. Studer, *Angew. Chem.* **2000**, *112*, 1157.
[11] E. Malmström, R. D. Miller, C. J. Hawker, *Tetrahedron* **1997**, *53*, 15225-15236.
[12] X. Chem, K. Jankova, J. Kops, W. Batsberg, *J. Polym. Sci. Part A* **1999**,*37*, 627.
[13] H. Ito, A. Knebelkamp, S. B. Lundmark, C. V. Nguyen, W. D. Hinsberg, *J. Polym. Sci. Part A: Polym. Chem.* **2000**, *38*, 2415-2427.
[14] F. Houlihan, F. Bouchard, J. M. J. Fréchet, and C. G. Willson, *Can. J. Chem.* **1985**, *63*, 153.
[15] J. M. J. Fréchet, E. Eichler, H. Ito and C. G. Willson, *Polymer* **1983**, *24*, 995.

High Energy Binders: Glycidyl Azide and Allyl Azide Polymer

I.K. Varma

Reliance Emeritus Professor, Centre for Polymer Science and Engineering, Indian Institute of Technology Delhi, Hauz Khas, New Delhi-110016, India
E-mail: ikvarma@hotmail.com

Summary: Hydroxy-terminated azido polymers such as poly(glycidyl azide), poly bis(azidomethyl oxetane) and poly(azidomethyl methyloxetane) have been investigated in the past in propellent formulations and as fuels in rocket technology. The high energy released upon the decomposition of the azido group is responsible for their specialized application as high-energy binders. The present paper describes the synthesis and characterization of new low molecular mass azido polymer i.e. poly(allyl azide). The curing reaction was carried out by using 1,3-cyclic dipolar addition reaction. The dipolarophiles, such as dimethylene glycol dimethacrylate (EGDMA) and addition polyimides (bismaleimides, bisnadimides and bisitaconimides) were used for curing of azido polymers. The curing reaction was monitored by FT-IR and differential scanning calorimetry. Curing was carried out at 40°C for 16 h (EGDMA) or 2 days (bismaleimide) and then at 60°C by using different phr of dipolarophiles. The heat of exothermic transition, due to decomposition of azide groups and thermal polymerization of addition polyimides, was very high and an improvement in thermal stability of cured resins was observed.

Keywords: bismaleimides; cycloaddition; ethylene glycol dimethacrylate; poly (allyl azide); poly (glycidyl azide)

Introduction

High-energy solid rocket propellants are composite materials having a binder [hydroxy terminated polybutadiene (HTPB)], and high-energy additives [e.g. ammonium perchlorate (AP)] and pyrolants (metallic powder). HTPB is an inert binder, which has been used in cast-cure propellant systems. Substitution of HTPB by more energetic binders may lead to an enhancement in performance of such propellants. Azido polymers have attracted researchers attention for the past two decades and compatible propellants with acceptable physical properties could be formulated from azide-containing ingredients.[1-4] The polymeric azides include poly (glycidyl azide) (GAP) and its copolymers, poly (*bis*-azidomethyl oxetane) [poly (BAMO)] and its copolymers, and poly (azidomethyl methyl oxetane)[poly (AMMO)]. [3] Although extensive studies have been reported on glycidyl azide polymers, no studies on poly (allyl azide) (PAA)

 DOI: 10.1002/masy.200450614

have been carried out. In this paper we report the synthesis, characterization, curing behaviour and thermal stability of poly (allyl azide).

The hydroxyl-terminated azido prepolymers are cured by reacting with isocyanates. There are certain problems associated while using isocyanates as curing agents for hydroxy- terminated binders. An alternative approach is the use of cycloaddition reaction with appropriate dipolarophiles. Curing of glycidyl azide polymers with multifunctional acrylic or acetylenic esters has been reported in the literature. [5,6] In this paper we present the results of our studies on curing of poly(allyl azide) using ethylene glycol dimethacrylate and addition polyimides such as bismaleimides, bisitaconimides or bis (*endo*-5-norbornene-2,3-dicaroximide) (bisnadimide) end-capped resins.

Experimental

Titanium tetrachloride (anhydrous) (Spectrochem. Pvt. Ltd); aluminum chloride (anhydrous) (Spectrochem. Pvt. Ltd); ferric chloride (anhydrous) (S.d.fine chem. Ltd); aluminium powder (Loba Chemie Co.); sodium azide (CDH); dimethyl sulphoxide (Merck); chloroform (Qualigens) and ethylene glycol dimethacrylate (EGDMA) (Merck) were used as such. Allyl chloride, b.p. 44-48°C (Lancaster) was distilled before use.

Poly (allyl azide)[7] and bismaleimides[8-10] i.e. 4,4'- bis(maleimidodiphenyl) ether (BE), 4,4'-(bismaleimidodiphenyl) sulphone (BS), and 1,6- bismaleimido hexane (BH) were synthesized according to the method reported earlier. The m.pt of the bismaleimides- BE, BS and BH were 174-175°C, 252°C and 138°C respectively. Synthesis of PAA was carried out by first preparing the poly(allyl chloride) followed by azidation.

The curing behaviour of poly (allyl azide) with varying amounts (10-30 phr) of bismaleimides or ethylene glycol dimethacrylate (5-45 phr) was evaluated by using TA 2100 thermal analyzer having 910 DSC module. A heating rate of 5-10°C/min in static air atmosphere and a sample mass of 4±1 mg were used. The exothermic transition observed in the temperature range of 150-280°C was characterised by determining the temperatures of (a) onset of exotherm (Ti), (b) extrapolated onset (To), (c) end of exotherm (Te) and exothermic peak temperature (Tp). Heat of the reaction (ΔH) was determined from the area under the exothermic transition.

Studies on isothermal curing of PAA with bismaleimides was also carried out in an air oven at 40°C (2 days) or 16 h with EGDMA and 60°C for several days. DSC was used for the evaluation of changes in the exothermic transition in these samples after heating at 40°C for two days and then after keeping at 60°C for 0-6 days.

A TA 2100 thermal analyser having 951 TG Module was used for thermal characterisation of isothermally cured poly (allyl azide) in nitrogen atmosphere (flow rate 60 mL/min). A sample mass of 5-6 mg was used and the rate of heating was 5-10 °C/min. The relative thermal stability of these samples was estimated by comparing initial decomposition temperature (IDT), final decomposition temperature (T_f), temperature of maximum rate of mass loss (T_{max}) and char yield at 600°C.

Characterisation Techniques

FT-IR spectrometer (Bio-Rad Digilab FTS-40/ Nicolet 5PC) was used for recording the IR spectra of PAC and PAA. Molecular mass of the polymers was determined using Knauer Vapor Pressure Osmometer K-7000 and benzil as calibration standard. A 2 % polymer solution in chloroform was used for molecular mass determination.

Structural changes taking place in the polymer during isothermal curing were studied by FT-IR. For this purpose, poly (allyl azide) was mixed with 20 phr of EGDMA and a thin film was coated on KBr disc, which was then heated in an air oven at 60°C for several days. FT-IR spectra were recorded at regular intervals of time for 5 days.

Results and Discussion

Polymerisation of allyl chloride

Poly(allyl chloride) (PAC) with different molecular masses could be obtained by changing catalyst and monomer concentration. A PAC sample having molecular mass of 1872 was used for carrying out further studies. Percentage of chlorine in PAC samples was found to be in the range of 35-40 and was much lower than the theoretical value (46.05 %). On the other hand, carbon content was higher than the expected value.

Thermal behaviour of poly (allyl azide)

The DSC scan of poly (allyl azide) showed an exothermic transition in the temperature range of 155-274°C, with exothermic peak temperature at 231°C. The energy liberated was 1099 J/g (Fig. 1a). The TG trace of poly (allyl azide) showed an initial mass loss of 7 %, which may be attributed to the moisture and low molecular mass oligomers. The main decomposition proceeded in two stages. The first step of decomposition occurred in the temperature range of 160-322°C, which was in good agreement with the temperature range of exothermic transition in the DSC scan. A mass loss of 28.3 % was observed (Fig. 1b). The exothermic transition observed in DSC accompanied by a mass loss (28.3 %) has been attributed to breakdown of azido group, by a mechanism reported earlier for glycidyl azide polymer. [11-12]

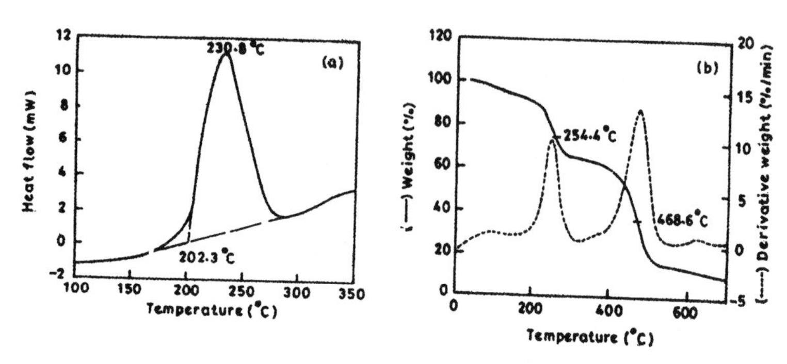

Figure 1. DSC and TGA Traces of poly(allyl azide)

Theoretically the decomposition of the azido group in a linear PAA should have resulted in a mass loss of 33.7 %. The poly (allyl azide) thus had lower azido content. This may be due to the branched structure of poly (allyl chloride), which resulted in lower chlorine content of the polymer. Since poly (allyl azide) was prepared by azidation of PAC, therefore the azido content will be decreased.

Major mass loss was observed above 400°C (~59 %) and is due to breakdown of the polymer backbone leading to the formation of hydrogen, carbon monoxide, carbon dioxide, methane, ammonia, hydrogen cyanide gas and other higher hydrocarbons. A char residue of 12.5 % at 600°C and 7 % at 800°C was obtained and, may be due to the formation of cross-linked structure. [10]

Curing studies

Poly (allyl azide) and EGDMA

Samples of PAA containing upto 25 phr of EGDMA were flexible. A further increase in EGDMA content resulted in a brittle product. In the DSC scans of PAA cured in presence of 25/

45/ 75 phr of EGDMA, an exothermic transition was obtained above 70°C and may be due to cycloaddition reaction of EGDMA with PAA and thermal polymerisation of EGDMA. The Ti values were 70.4, 93.4 and 100°C and the peak exothermic temperatures (Tp) were 148.5, 149.1 and 151.4°C for samples containing 25, 45 and 75 phr of EGDMA respectively. The heat of curing also increased from 136.5 to 212.4 J/g with the increase in the EGDMA content.

An exotherm was observed after the curing exotherm and this may be due to decomposition of residual azido groups in the samples. The ΔH for the decomposition exotherm of neat PAA was 1099 J/g, while for PAA containing 45 phr of EGDMA, it was 181.1 J/g.

Monitoring of residual cure in isothermally cured samples

The DSC scans of EGDMA (20 phr) and PAA were recorded after isothermal curing. A significant decrease in heat of curing was observed on increasing the duration of isothermal curing. The heat of curing at 40°C changed from 128.6 to 47.2 J/g in 12 days whereas at 60°C after 7 days (ΔH = 13.4 J/g) extent of cure was much higher thereby resulting in a significant decrease in the residual cure. The exothermic peak temperature also decreased marginally. Gradual curing of poly (allyl azide) can therefore be carried out with EGDMA at 40°C by keeping the samples for several days whereas one can accelerate the curing process by carrying out the reactions at 60°C.

Structural changes during curing

A shift in the absorption peak due to ester carbonyl of EGDMA from 1721 to 1740 cm^{-1} was observed after 24 h of heating at 60°C. This shift indicates the removal of the double bond, (which was conjugated with the carbonyl group) by the formation of triazoline ring. A new absorption peak was observed at 1135 cm^{-1} and can be attributed to the ring breathing vibrations of 1,2,3-triazolines. [13] These studies thus confirm the cyclic dipolar addition reaction of PAA with EGDMA.

No appreciable decrease in the intensity of azido group at 2099 cm^{-1} was observed. This absorption peak shifted from 2099 to 2123 cm^{-1}. Increasing the phr of EGDMA did not affect the intensity of azido absorption band. This indicates a low conversion of azido group to triazoline ring even at higher ratio of EGDMA at 60°C.

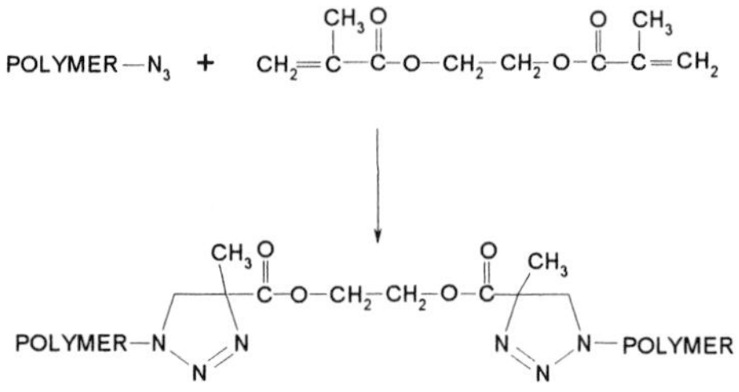

$POLYMER{-}N_3 + CH_2{=}C(CH_3){-}C(O){-}O{-}CH_2{-}CH_2{-}O{-}C(O){-}C(CH_3){=}CH_2$

Scheme 1. Curing of PAA using EGDMA

Poly (allyl azide) and bismaleimides

The curing of PAA was done with three different bismaleimides containing (a) flexible aliphatic units (hexamethylene) (BH), (b) aromatic rings having flexible ether (BE) linkage or (c) a strong electron withdrawing group sulphone (BS) as the bridging units between maleimide end caps. The bismaleimide content was varied to evaluate the effect of concentration on curing (10-30 phr). This corresponds at 10 phr to an azide:maleimide ratio of 1:0.06, 1:0.47 and 1:0.02 for BH, BE and BS respectively. Thus, azide groups were much higher in concentration than maleimide under investigation. This will control the cross-link density of the cured product.

The blends of PAA and bismaleimide were initially sticky mass irrespective of the composition of blends. Isothermal heating at 40°C for 2 days resulted in a marginal change in physical characteristics. When isothermal heating was done at 60°C for several days, the sticky mass converted to a non-sticky elastic product.

The PAA blends with bismaleimides BH, BE or BS have been designated as PH, PE and PS respectively. The phr of the bismaleimides is indicated by writing A (10 phr), B (15 phr) and C (30 phr) after the letter designation of concerned blend. The duration of isothermal curing i.e. 0, 1, 5 days etc are indicated by writing the number after letter designation of the formulation. Thus, a blend of PAA and BE (10, 15, 30 phr) and cured for 5 days is designated as PEA-5, PEB-5, and PEC-5, respectively. The 1,3- cyclic dipolar addition with bismaleimides can be depicted according to the scheme 2.

The exothermic peak position of PAA was 231°C but addition of structurally different bismaleimides reduced it by 20-30°C. The ΔH was 1099 J/g for PAA but in most blend samples ΔH value was either equal to or higher than this value. Slightly lower values were observed only in PEC-6, PEB-0, PHC-1 and PHC-5 samples. The observed exotherm in PAA bismaleimides samples arises due to two reactions- (a) loss of N_2 from PAA to yield nitrenes and (b) curing of bismaleimides. The ΔH values of bismaleimides (150-200 J/g) are lower than the heat evolved during decomposition of azide groups.

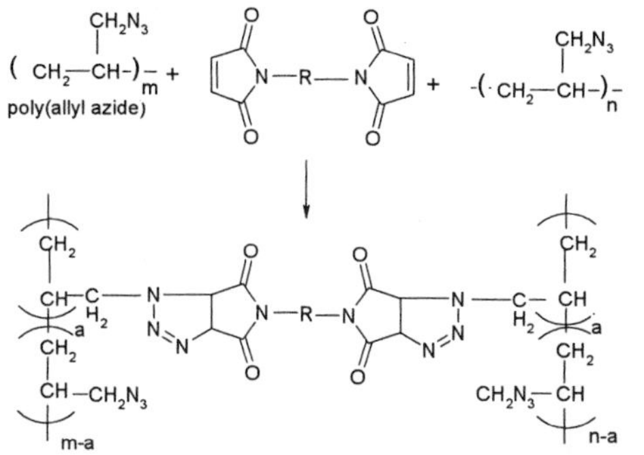

Scheme 2. Curing of Poly (allyl azide) with Bismaleimides

These results thus indicate that only few azide groups had reacted with bismaleimides at 60°C and a large proportion was available for decomposition of these energetic binders. Bismaleimide did not act as an inert material (as in case of isocyanates) but contributed towards the heat of curing in the same temperature range in which azide group undergoes decomposition. Thus the over all heat evolved in the temperature range of 150-277°C was either equal to neat PAA or greater than that.

The results of TG studies are summarized in Table 1. A two-step decomposition was observed. A mass loss of 18 – 23% was observed in temperature range 150- 280°C. In comparison to pure PAA, a reduction in mass loss of about 5-10 % is observed. Curing of bismaleimides is an addition reaction without evolution of any volatiles whereas PAA eliminates N_2 in this temperature range. Increase in phr of bismaleimides reduced the mass loss in this temperature region. The second step decomposition was observed in the temperature range of 290-310°C with a mass loss of 57-64 %. Compared to PAA, the char residue at 600°C was higher in these resins. This is due to the network formation of cured bismaleimides, which favour condensation and char formation at high temperatures.

Table 1. Thermal behaviour of isothermally cured PAA with BH resin (heating rate 10°C/min)

Sample designation	IDT (°C)	T_{max} (°C)	T_f (°C)	Mass loss (%)	Y_c (%, 600°C)
PHA-1	155.4	209.4	280.0	22.0	23.4
	297.7	482.1		63.2	
PHA-5	153.5	201.5	271.4	27.6	22.5
	296.4	459.8	610.7	57.5	
PHB-1	124.4	213.8	291.1	22.0	14.8
	300.0	477.0	600.0	63.2	
PHB-5	144.4	210.3	291.1	21.9	20.0
	304.4	468.3	555.6	58.5	
PHC-1	142.8	221.5	275.0	18.2	21.6
	289.3	466.5	628.6	58.4	
PHC-5	132.1	194.0	264.3	18.1	20.7
	282.1	461.8	600.0	60.3	
PEA-1	135.6	211.8	286.7	24.2	13.5
	295.6	472.4	580.0	62.5	
PEA-6	125.0	210.0	282.1	22.7	19.4
	292.8	459.3	610.7	59.5	
PEB-1	128.9	220.8	297.8	24.5	15.2
	297.8	467.6	573.3	60.3	
PEB-5	135.7	207.7	282.1	20.4	22.0
	285.7	463.6	607.1	60.0	
PEB-1	111.2	203.4	302.1	18.7	25.8
	311.0	464.2	635.5	59.2	
PEB-6	142.8	202.7	282.1	18.7	21.6
	292.8	460.7	592.9	62.6	
PSB-1	124.4	213.8	291.1	22.0	14.8
	300.0	477.0	600.0	63.2	
PSB-5	146.5	212.9	275.5	22.0	19.4
	297.7	460.6	532.2	62.3	

Conclusion

Curing of poly(allyl azide) can be carried out under mild conditions using bismaleimides (10-30 phr) as the dipolarophiles. The cured resins were flexible and retained their elastic behaviour for several weeks. The ΔH value for the exothermic decomposition of azido groups was significantly high. This may be due to exothermic curing of bismaleimides and the decomposition of azido groups in almost same temperature range. In contrast to isocyanates, which are generally used for curing of hydroxyl-terminated azido polymers, bismaleimides do not act as inert curing agents. An increase in char residue of cured resin at 600°C was also observed.

Acknowledgement

The Armament Research Board, Government of India, Ministry of Defence is gratefully acknowledged for providing financial assistance for carrying out this work. The work could not have been completed with out the contributions made by my colleague Prof. Veena Choudhary and students Dr. Bharti Gaur, Ms. Bimlesh Lochab, S. Oberoi and Mr. R. R. Golla.

1. N. Kubota and T. Kuwahara, *Propellants, Explosives, Pyrotechnics*, **1991**,16, 51.
2. Y. J. Lee, C-J. Tang, G. Kudva and T. A. Litzinger, *J. Propulsion and Power*, **1998**,14, 57.
3. Bharti Gaur, Bimlesh Lochab, V. Choudhary and I. K. Varma, *J. Macromol. Sci. Revs.*, **2003**, C-43, 503.
4. C. N. Divekar, S. N. Asthana and H. Singh, *J. Propulsion and Power*, **2001**,17, 58.
5. A. P. Manzara, U. S. Pat. 5, 681, 904, **1997**.
6. M. B. Frankel, E. R. Wilson, D. O. Woolery, U. S. Pat. 4,96,2213, **1990**.
7. Bharti Gaur, Bimlesh Lochab, V. Choudhary and I. K. Varma, *J. Therm. Anal. Cal*, **2003**, 71, 467.
8. I. K. Varma and V. B. Gupta "Thermosetting Reins-Properties" in 2nd volume of Comprehensive Composite Materials, R. Talreza and J-A. E. Manson, Ed. A. Kelly and C. Zweben, Eds-in-chief, Elsevier, **2000**, 1.
9. I. K. Varma, G. M. Fohlen and J. A. Parker, *J. Polym. Sci: Polym. Chem.Ed.*, **1982**, 20, 283.
10. I. K. Varma, Sangita and D. S. Varma, *J. Polym. Sci: Polym. Chem.Ed.*, **1984**, 22, 1419.
11. S. K. Sahu, S. P. Panda, D. S. Sadafule, C. G. Kumbhagar, S. G. Kulkarni, J. V. Thakur, *Polym. Degrad. Stabil.*, **1998,** 62, 495.
12. M. S. Eroglu, O. Guven, J. Appl. Polym. Sci, **1996**, 61, 201.
13. A. T. Balaban, S. Badilsen, I. I. Badilsen in "Physical Methods in Heterocyclic Chemistry", R. R. Gupta, Ed, Wiley Interscience, **1984**, 46.

Pyrrole-Substituted Alkyl Silanes as New Adhesion Promoters on Oxidic Substrates

Xuediao Cai, Evelin Jaehne, Hans-Juergen P. Adler*

Institute for Macromolecular Chemistry and Textile Chemistry, Dresden University of Technology, 01069 Dresden, Germany

Summary: ω-(Pyrrol-1-yl alkyl) dimethylchlorosilanes with different chain length were synthesized and characterized with IR, NMR and elemental analysis. The growth of self-assembled monolayers of ω-(pyrrol-1-yl alkyl) dimethylchlorosilanes on oxide surfaces has been investigated. The order gradually improved with adsorption time and highly ordered SAMs were obtained for nearly 2 days adsorption time. Characterization of the films has been performed with contact angle measurements, ellipsometry, SPR and grazing incident FTIR. The chemical deposition of polypyrrole on the modified surface was investigated. The thickness of polypyrrole layer was influenced by the concentration of monomer, the deposition time, and the ratio of monomer to oxidant.

Keywords: monolayer; polypyrrole; ω-(pyrrol-1-yl alkyl) dimethylchlorosilanes

Introduction

Many varieties of conducting polymers have been described in the last decades [1]. Especially, the electrochemically synthesized conducting polymer films have been extensively investigated because of their wide range of useful applications such as microelectronic devices [2], chemical sensors, biosensors [3, 4], and corrosion protection [5]. Polypyrrole films are one of the important conducting polymers and their films are prepared by electrochemical polymerisation using highly conducting electrodes [6]. However, in most cases, adhesion between the polymer and the conducting substrate (electrode) is poor because of weak physical interactions between them. Generally, polypyrrole can be peeled from the substrate to give freestanding films [7, 8]. Researchers have tried to use adhesion promoter molecules to modify the electrode to increase the adhesion between the substrates and the polymer films [9]. Wrighton and co-workers used N-(3-trimethoxysilyl) pyrrole as a surface modification agent to improve adhesion of polypyrrole films and n-type Si [10]. F. Faveralle et. al reported the chemical polymerisation of polypyrrole

 DOI: 10.1002/masy.200450615

on glass fibre modified with pyrrole-substituted organotrialkoxysilane coupling agents as adhesion promoter between polypyrrole and glass fibre [11]. Recently, the synthesis and application of several substituted pyrroles as adhesion promoters were described, e.g., pyrroles with thiol group on Au [12-17]. Nevertheless, none of these researchers systematically investigated the effect of alkyl chain length on the properties of the monolayer of pyrrolyl-alkyl dimethylchlorosilanes and the chemical deposition of conducting polypyrrole on surface modified inorganic substrates.

In this paper, the properties of the monolayer were studied with various techniques including contact angle measurements, ellipsometry, surface plasmon resonance spectroscopy (SPR), grazing incident FTIR and X-ray photoelectron spectroscopy (XPS). The chemical deposition of polypyrrole on the modified substrates was described and the layer thickness could be adjusted.

EXPERIMENTAL

Substances

Several ω-(pyrrol-1-yl alkyl) dimethylchlorosilanes with different chain length were used as adhesion promoters for grafting polypyrrole layers onto silicon. The synthesis details will be described elsewhere.

6-(pyrrol-1-yl hexyl)dimethylchlorosilane	PMCS-6
8-(pyrrol-1-yl octyl)dimethylchlorosilane	PMCS-8
11-(pyrrol-1-yl undecyl)dimethylchlorosilane	PMCS-11
16-(pyrrol-1-yl hexadecyl)dimethylchlorosilane	PMCS-16

Surface modification of substrates with the silane compounds

Silicon substrates were first treated with 3:1(v/v) of concentrated H_2SO_4 and 30% H_2O_2 at 90° for 30min to clean the surface and produce terminal hydroxyl groups [18]. The acid treated substrates were then carefully washed with water and blown dry with argon. The clean substrate was dipped in a solution of silane compound in bicyclohexyl (10mM) for 48h, washed with chloroform, and thoroughly ultrosonicated for 20min and dried with argon.

Physicochemical studies

KRÜSS DSA10 goniometer (drop shape analysis) was used for contact angle measurements. Ellipsometric studies were preformed with a computer interfaced DRE GmbH ELX-02C ellipsometer over the wavelength 623nm and an incident angle of 70°. XPS studies were carried out using a Physical Electronics PHI 5700 ESCA system. UV-VIS spectra were obtained on a Perkin-Elmer Lambda 35 UV/VIS spectrometer. Nanoscope Dimension 3100 from Veeco/USA was used for AFM measurements, which was run in tapping mode. The thickness of polypyrrole films were measured with Dektak surface profile measuring system.

Chemical deposition of polypyrrole films on silane-modified substrates

In a typical reaction, the modified substrate was dipped in 1.7M oxidant /H_2O solution, and a two times volume of 0.1M pyrrole-MeOH solution was added. The mixture was stirred at room temperature for 1h and then the substrate was removed from solution. It was washed with solvent to remove the loose polymer attached on the film surface, and blown dry with argon.

RESULTS AND DISCUSSION

The strategy for deposition of a conducting polypyrrole film on an insulating substrate is shown in Scheme 1.

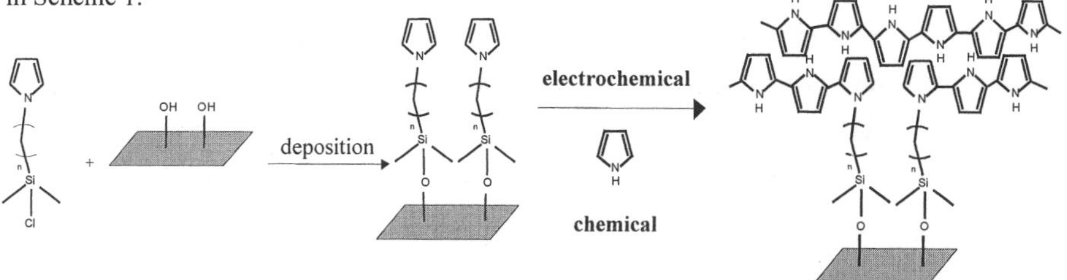

Scheme 1. Polymerisation of pyrrole via template modified surface

The deposition of silane compounds on the oxide substrate was achieved by dipping the clean substrate in 10mM ω-(pyrrole-1-yl alkyl) silanes/bicyclohexyl solution. The static contact angle showed that the longer alkyl chain of ω-(pyrrole-1-yl alkyl) silanes lead to high hydrophobicity of the surface. Optical ellipsometry was used to measure the thickness of the silane compounds on the oxide surface. It was found that the thickness of ω-(pyrrole-1-yl alkyl) silanes on silicon dioxide surface is not proportional to the chain length of the alkyl group and that they are smaller than the size of ω-(pyrrole-1-yl alkyl) silanes calculated from molecular models (Tab.1).

Table 1. Contact angles and thickness of monolayer.

Compounds	PMCS-6	PMCS-8	PMCS-11	PMCS-16
Contact angle [°]	86.7	98.1	101.7	103.4
Layer thickness [nm]	0.95	1.22	1.47	2.59
Calculated length [nm]	1.42	1.53	2.05	2.69
Tilt angle [°]	48.2	37.9	44.2	15.8

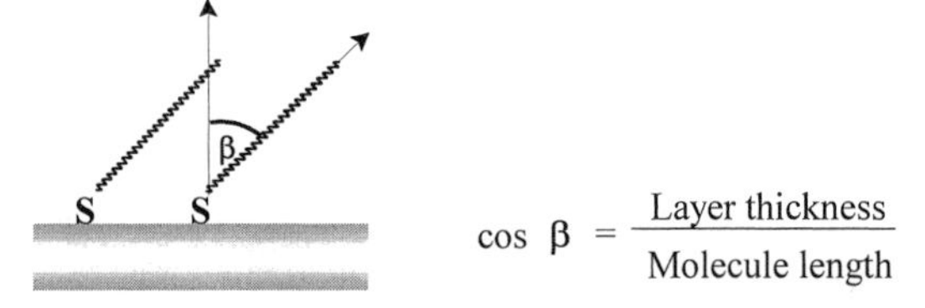

$$\cos\beta = \frac{\text{Layer thickness}}{\text{Molecule length}}$$

Fig. 1: Definition of tilt angle of adsorbed monolayer

These results indicated that ω-(pyrrole-1-yl alkyl) dimethylchlorosilane molecules are not closely packed on the oxide surface, due to two methyl groups at the silicon atom except for the derivative with 16 carbons spacer length. Grazing incident FTIR was measured to study the order of the monolayer. The IR spectra are shown in Figure 1. The order in the PMCS-8 layers can be assessed from the position of the CH_2-stretching vibration. As shown in previous studies [19-22], the frequency of the CH_2-stretching vibration is characteristic for long-chain SAM's indicating close packed structures. For completely disordered structures, the frequency of the CH_2-stretching is close to that of a liquid alkane ($\nu_{as}\sim$ 2924 cm^{-1}). For well-ordered SAMs, the frequency is shifted to lower wavenumbers, and it is close to that of a crystalline alkane ($\nu_{as}\sim$ 2915-2918 cm^{-1}) [23]. The data of Fig. 2 show that CH_2-stretching of 8-(pyrrol-1-yl octyl) dimethylchlorosilane is 2918 cm^{-1}. This indicates SAMs with a high degree of order.

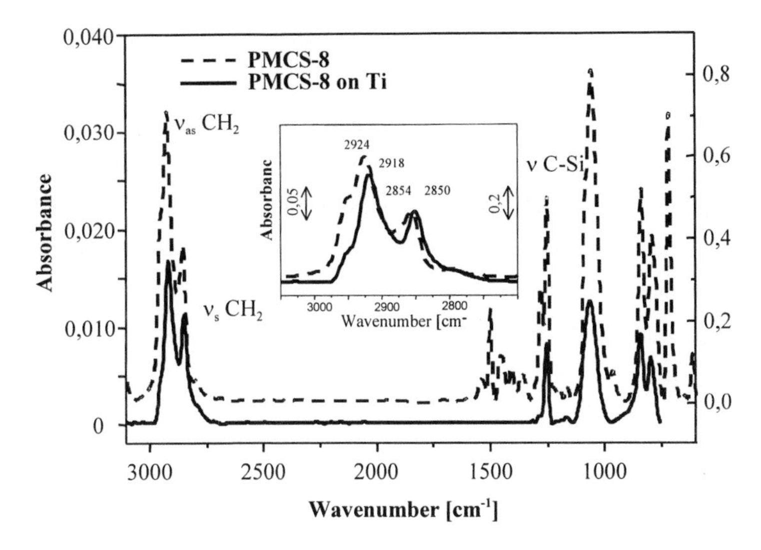

Fig.2: Grazing incident FTIR spectra of PMCS-8

We found that $\nu_{as}CH_2$ is decreasing with the increasing of concentration of solution or adsorption time (Fig. 3). During the adsorption, the early stages of the reactions can be pictured as isolated grafted molecules randomly distributed on the substrate. As the surface coverage increases, the order in the monolayer gradually increase ($\nu_{as}(CH_2)$ decrease), approaching the final highly ordered or the closely packed state.

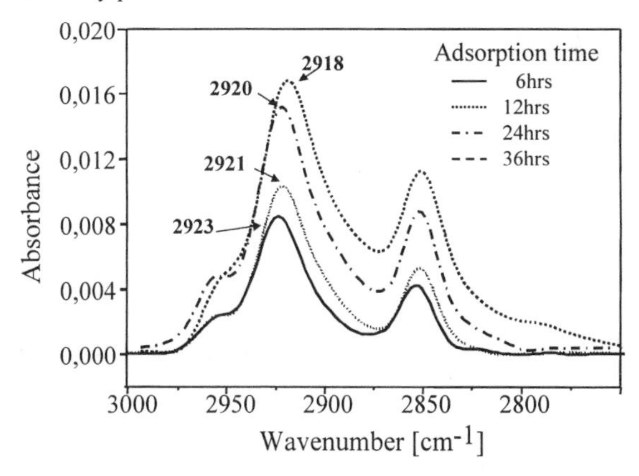

Fig. 3: Grazing incident IR spectra of PMCS-11 on Ti at different adsorption times

Angle-dependent XPS measurements were done to study the orientation of the adsorbed ω-(pyrrole-1-yl alkyl) silane on Ti oxide substrate. The measurements were done at 5°, 20°, 45°, and 85° angles. By varying the detector angle we got information from different depths of the adsorbed layer. Therefore, special marker atoms were used: N from the pyrrole group and Si from the silane group. We calculated I_{Si}/I_N to get an indication how the molecule is oriented on the surface relatively to the detection angle. At a low detector angle more N atoms were observed and at high detector angles more Si atoms. Though, an increase in the Si/N ratio with increasing detector angle indicated that the molecules are oriented with the pyrrole group on top and the silane group attached to the surface (see Fig. 4).

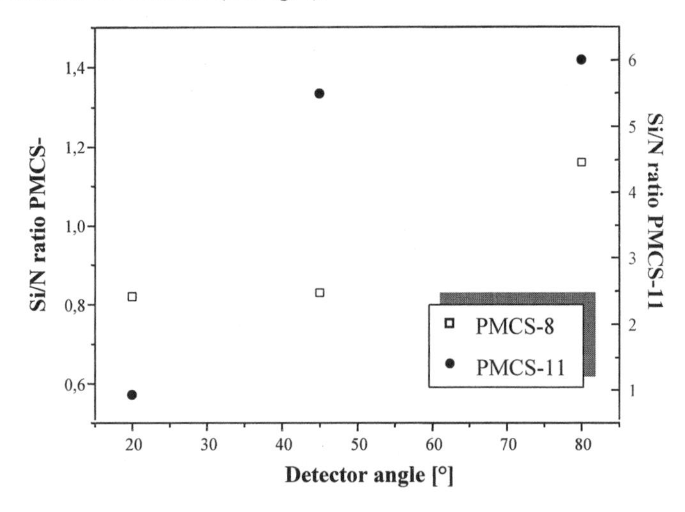

Fig. 4: AR-XPS measurements of adsorbed ω-(pyrrole-1-yl alkyl) silanes: Si/N ratios

There are two methods for deposition of polypyrrole films on the modified substrates, chemical deposition and electrochemical deposition. In this paper, only the results of chemical deposition are given.

Polypyrrole films chemical deposited on a silane-modified substrates show much better adhesion compared to those on unmodified substrates. Thin films up to 400nm in thickness adhered tightly to the silane-modified substrates. On the other hand, when unmodified substrates were used, only loosely attached particles were found on the surface, which could be washed off easily with the solvent. AFM measurements revealed that polypyrrole films deposited over ω-(yrrol-1-yl alkyl) silane with long alkyl chain showed a smoother morphology than that of films deposited over ω-(pyrrole-1-yl alkyl) silane with shorter alkyl chains (Fig.5). Contact angle and ellipsometry data also showed that ω-(pyrrole-1-yl alkyl) silane with long alkyl chains formed more compact and ordered SAMs on the oxide surfaces. The formation of pyrrole molecules on the inorganic surface increased the interaction between the organic polymer and the substrate, and therefore, increased the adhesion between them.

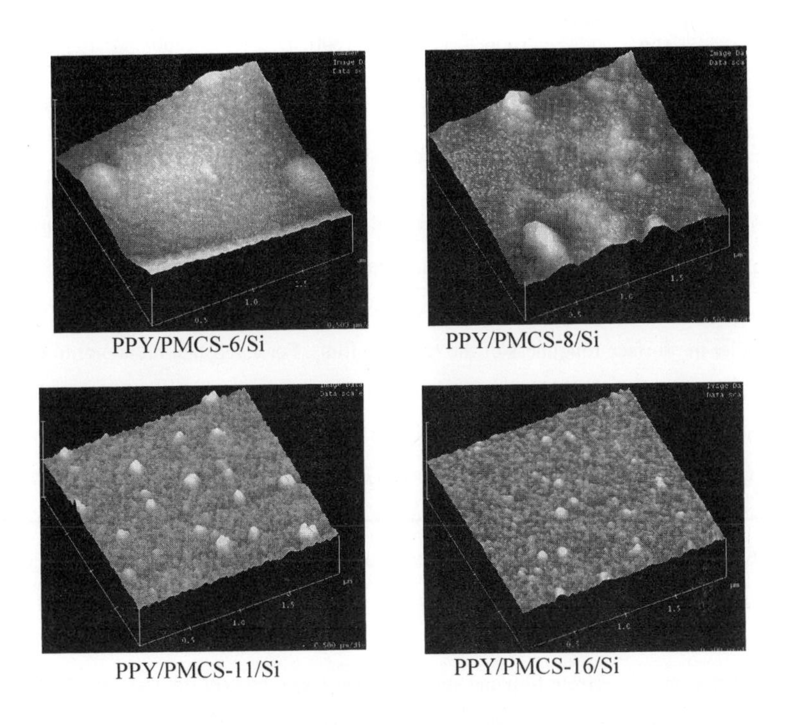

Fig. 5: AFM of polypyrrole deposited on different chain length modified substrate

The formation of polypyrrole was also identified by Raman Spectroscopy (Fig. 6) and UV-VIS data (not shown here), both spectra showed the characteristic absorption patterns of polypyrrole.

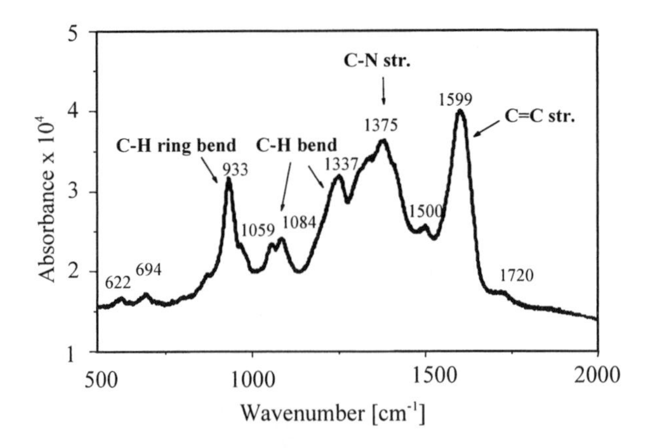

Fig. 6: Raman spectrum of polypyrrole deposited on a silane modified substrate

The thickness of the film increased as the deposition time increased during the first 4 hrs and then reached a plateau, due to complicated equilibrium processes between polypyrrole chains deposited on and removed from the substrate. It was found that the lower the monomer to oxidant ratio, the smaller the surface roughness of the resulting films. For a constant monomer to oxidant ratio, it was found that the polymer films prepared at a concentration of 0.1M have the highest thickness and smallest surface roughness (Fig. 7). At lower monomer concentration, the polypyrrole chains have less chance for deposition on the surface while at high concentration, the reaction rate is too fast to allow the growth of a dense film on the substrate. However, due to the limited solubility of $Na_2S_2O_8$ in water, the concentration of $Na_2S_2O_8$ can not be higher than 0.2M. The reaction atmosphere and temperature did not have any obvious effect on the properties of the resulting films.

The adhesion between polypyrrole film and substrates was tested by peel-out test and ultrasonical test. In every case the film had a good adhesion on the substrate surface.

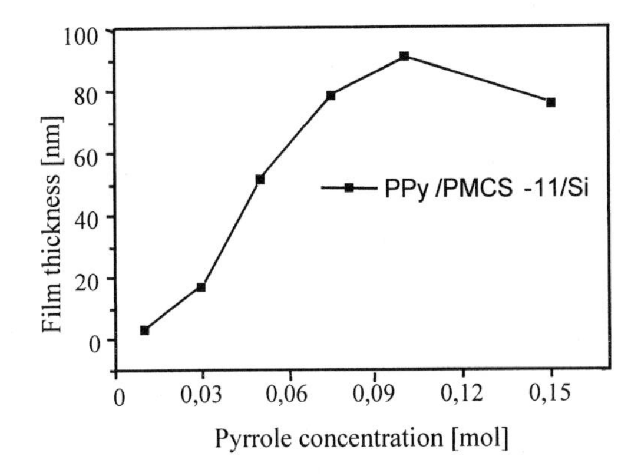

Fig. 7: Effects of polymerisation conditions on the thickness of polypyrrole films (polymerisation time 1h; oxidant $Na_2S_2O_8$; solvent $MeOH/H_2O$ 2:1 v/v; reaction atmosphere air; pyrrole to oxidant ratio 0.17)

CONCLUSION

In this paper, the characterization of ω-(pyrrol-1-yl alkyl) silanes with various alkyl chain lengths, as well as the adsorbed monolayers on oxide substrates and the chemical deposition of polypyrrole over the modified substrates were described. Contact angle measurements showed that hydrophobicity increased with increasing chain length of the adsorbed molecules. Ellipsometry data confirmed that ω-(pyrrol-1-yl alkyl) silane layers formed monolayers on the substrates. The calculation of the tilt angles of the adsorbed molecules revealed that the long chain derivatives formed more compact and well-ordered SAMs on Si substrate. The thickness of the surface polymerised polypyrrole films was influenced by deposition time, monomer to oxidant ratio and monomer concentration. The thickness of polypyrrole films was increased with increasing deposition time during the first 4 hours. At low monomer to oxidant ratios, a small surface roughness and thin polypyrrole films were obtained. ω-(pyrrol-1-yl alkyl) silanes improve the adhesion between polypyrrole layer and substrate. Furthermore, polypyrrole layers deposited over ω-(pyrrol-1-yl alkyl) silanes with long alkyl chains have a smoother morphology than that of layers deposited over ω-(pyrrol-1-yl) alkyl silanes with shorter alkyl chains.

Acknowledgement

The authors are thankful to Dr. Shanglin Gao (AFM measurements) and Dr. Mingtai Wang (ellipsometry measurements) from Institute for Polymer Research Dresden (IPF). We also thank Mr. Gernot Busch (SPR, XPS measurements) and MS. Martina Dziewiencki (grazing incident IR) from Institute for Macromulecular Chemistry and Textile Chemistry at TU-Dresden. We are grateful to Europe Graduate School 720 " Advanced Polymeric Materials" and the German Research Foundation (DFG), the SFB 287 "Reactive Polymers" for financial support.

1. *Handbook of Conducting Polymers*, T. A. Skotkeim, Ed.; Dekker, New York, 1986.
2. S.F. Bond, A. Howie, R. H. Friend, *Surf. Sci.* 1995, 333, 196.
3. J. W. Gardner, P. N. Bartlett, *Sens. Actuators. B.* 1993, 8, 211.
4. B. P. J. De Lacy Costello, P. Evans, N. Guemion, N. M. Ratcliffe, P. S. Sivanand, G. C. Teare, *Syn. Met.* 2000, 114, 181.
5. M. Duprat, A. Shiri, Y. Derbali, N. Pebere, *Electrochemical Methods in Corosion Research, Materials Science Forum;* M. Duprat Ed.; Trans. Tech. Aedermannsdorf 1986, Vol. 8, p 207.
6. B. Sun, D. P. Schweinsberg, *Synth. Met.*, 1994, 68, 49
7. K. K. Kanazawa, A. F. Diaz, R. H. Geiss, W. D. Gill, J. F. Kwak, J. A. Logan, J. Rabolt, G. B. Street, *J. Chem. Soc., Chem. Commun.* 1979, 854.
8. A. F. Diaz, J. M. Vasquez Vallejo, A. Martinez *Duran IBM J. Res. Dev.* 1981, 25, 42.
9. Z. Mekhalif, P. lang, F. Garnier, J. Electroanal. Chem. 1995, 399, 61.
10. R. A. Simon, A. J. Ricco, and M. S. Wrighton, *J. Am. Chem. Soc.* 1982, 104, 2031.
11. F. Faveneralle, A. J. Attias, B. Bloch, P. Audebert, and C. P. Andrieux, *Chem. Mater.* 1998, 10, 740.
12. R. J. Willcut, R.L. McCarley, *J. Am. Chem. Soc.* 1994, 116, 10823.
13. D. B. Wurm, S. T. Brittain, Y.-T. Kim, *Langmiur,* 1996, 12, 3756.
14. C. N. Sayre, David M. Collard, *Langmiur* 1995, 11, 302.
15. E. Smela, G. Zuccarello, H. Kariis, B. Liedberg, *Langmiur,* 1998, 14, 2970.
16. D. B. Wurum, Y.T. Kim, *Langmiur,* 2000, 16, 4533.
17. C.-G. Wu, S.-C. Chiang and C.-H. Wu, *Langmiur*, 2002, 18, 7473.
18. S. R. Wasserman, Y.-T. Tao, G. M. Whitesides, *Langmiur*, 1989, 5, 1074.
19. M. D. Porter, T. B. Bright, D. L. Allara, C. E. D. Chidsey, *J. Am. Chem. Soc.* 1987, 109, 3559.
20. A. N. Parikh, B. Leidberg, S. V. Atre, M. Ho, D. L. Allara, *J. Phys. Chem.* 1995, 99, 9996.
21. K. Kojio, S. Ge, A. Takahara, T. Kajiyama, *Langmiur*, 1998, 14, 971.
22. D. W. Britt, V. Hlady, Langmiur, 1999, 15, 1770.
23. R. G. Snyder, H. L. Straus, C. A. Elliger, *J. Phys. Chem.* 1982, 86, 5145.

Selective Resins, Synthesis and Sorption for Precious Metals

Dorota Jermakowicz-Bartkowiak, Bożena N. Kolarz*

Institute of Organic and Polymer Technology, Wrocław University of Technology,
50-370 Wrocław, Wybrzeże Wyspiańskiego 27, Poland
E-mail: d.bartkowiak@ch.pwr.wroc.pl

Summary: Resins based on vinylbenzyl chloride (VBC) and divinylbenzene (DVB) copolymer were synthesised and used for preconcentration and separation of Au, Pt and Pd from hydrochloric acid solutions. Resulted resins show functionality concentration up to 5,8 mmol/g. The acidity and interference of other ions on the resins sorption were discussed. The sorption capacities of gold, platinum and palladium from hydrochloric solutions reaches to 85, 100 and 60 mg/g and distribution coefficients achieve 50 000 value. Recovery of noble metals revealed average of 60-98 % from mulitcomponent solutions.

Keywords: gold; palladium sorption; platinum; selective resins

Introduction

The increase of the price for precious metals over the years and their expanding use in areas such as manufacturing of automobile catalysts, organic catalysts, jewellery, microelectronics and cancer therapy along with lack of availability of these metals has led to development of mining from low grade ores or from secondary raw materials. The recovery of precious metals: platinum, palladium and gold from different souurces require their separation from other metals. A very promising approach to overcome this problem is to use ion exchange and chelating resins for selective recovery of these metals from multicomponent solutions [1-4]. A series of new polymers bearing amino and guanidine functional group were synthesised, characterised and their abilities for noble metals recovery and sorption were determined. Currently, a new class of materials: chelating and ion exchange resins have been developed. They contain aminoguanidine and guanidine functionality with increased basicity (compared with conventional weak base type resin) and therefore capable for gold complexation from alkali (as dicyanoaurate anion) and acidic (as tetrachloroaurate anion) solutions [5-6].
The aim of our work is to present the abilities of resin to recover gold, platinum and palladium from acidic solution.

 DOI: 10.1002/masy.200450616

Experimental

An expaned gel copolymer, of vinylbenzyl chloride/divinylbenzene (2%) (VBC/DVB), A, was prepared in the presence of toluene. Polymer was obtained in the suspension polymerisation [5]. Resins were obtained by exchange of chlorine atom by diamines: 1,2-diaminoethane and 1,6-diaminohexane, and subsequent reaction with cyanamide. More details concerning synthesis are present in papers [5-7]. Water regain is measured by centrifugation technique. Nitrogen content is determined by Kiejdahl's method. Anion exchange capacity is determined according to Hecker's method and used to calculate ligand concentration.

Batch method was applied to sorption procedure i.e.: resin in swollen form was contacted for 48 hours with hydrogen tetrachloroaurate (III), hydrogen hexachloroplatinate (IV) and sodium tetrachloropalladate (II) in hydrochloric acid solutions. One-component solution contains 0.24 mM of selected metals i.e. (50 mg Au/L, 50 mg Pt/L and 26 mg Pd/L, while multicomponent solutions contain 0,24 mM of noble metals and 2,4 mM or 20 mM of coexisting metals. Metal sorption was determined using atomic absorption spectrophotometry, type Aanalyst 100.

Results

The synthetic route to A (amino functional resin) and AG (guanidino functional resin) can be described schematically in Figure 1. The first step of the reaction is the exchange of chlorine with diamine resulting in A resin. The next step is modification using cyanamide to AG resin. Characteristic of investigated polymers is shown in Table 1. Obtained resins show higher water content than starting VBC/DVB polymer. The higher ligand content presents AG2 resin up to 2.9 mmol/g.

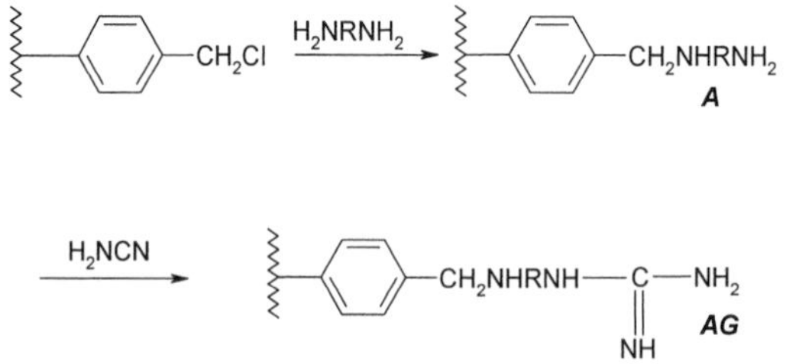

CH_2Cl $\xrightarrow{H_2NRNH_2}$ CH_2NHRNH_2 **A**

$\xrightarrow{H_2NCN}$ $CH_2NHRNH-C(=NH)-NH_2$ **AG**

Where R is $(CH_2)_n$ and n=2 and 6.

Figure 1. Scheme of A and AG resin preparation.

Table 1. Characteristics of investigated polymers

Sample	Water regain g/g	Nitrogen content mmol/g	Ligands conc. mmol/g
Copolymer	0.24	-	-
A2	6.00	6.0	2.8
AG2	3.08	11.6	2.9
A6	2.56	5.2	2.6
AG6	1.90	8.3	2.1

Sorption studies under static conditions were performed with mixture of precious metals and base metals to study the selectivity of resins and thus to judge their application in hydrometallurgical or separation procedures. Multicomponent solutions containing Au(III), Pt(IV), Pd(II) (0,24 mM of each, *S1*) and Fe(III), Cu(II), Ni(II) (2,4 or 20 mM of each, *S2, S3*) were used in sorption experiment. Table 2 and 3 show the influence of each of foreign ions on the noble metals sorption and recovery. It is clear that a large amount of Cu, Fe and Ni (80-fold excess in solution *S3*), caused little interference. The total sorption of Au, Pt and Pd from multicomponent solution *S1, S2* and *S3* are present in Figure 2. The best sorbatibility of investigated noble metals shows AG2 resin, 290 mg/g from *S1* solution and 250 mg/g from *S3* solution with strong fold excess of foreign metals.

Table 2. Precious metals sorption from multicomponent solution *S1*, *S2* and *S3**, (mg/g)

Resin	Foreign ion**	mM	Solution	Au	Pt	Pd	Resin	Au	Pt	Pd
A2	Fe+ Cu+ Ni	0	*S1*	103	93	60	AG2	119	107	66
	Fe+ Cu+ Ni	2.4	*S2*	103	93	60		117	102	67
	Fe+ Cu+ Ni	20	*S3*	90	74	60		100	85	67
A6	Fe+ Cu+ Ni	0	*S1*	82	93	46	AG6	76	77	43
	Fe+ Cu+ Ni	2.4	*S2*	78	78	44		78	70	45
	Fe+ Cu+ Ni	20	*S3*	69	64	34		70	61	41

* Concentration of Au, Pt, Pd is 0.24 mM
**Concentration of each foreign metal

Table 3. Influence of foreign ions on recovery of Au, Pt and Pd*(%)

Resin	Foreign ion** mM		Solution	Au	Pt	Pd	Resin	Au	Pt	Pd
A2	Fe+ Cu+ Ni	0	S1	92	82	98	AG2	95	89	98
	Fe+ Cu+ Ni	2.4	S2	93	83	99		95	82	99
	Fe+ Cu+ Ni	20	S3	81	66	98		80	67	97
A6	Fe+ Cu+ Ni	0	S1	81	91	84	AG6	93	93	96
	Fe+ Cu+ Ni	2.4	S2	77	76	80		90	80	95
	Fe+ Cu+ Ni	20	S3	69	63	61		81	70	87

* Concentration of Au, Pt, Pd is 0.24 mM

**Concentration of each foreign metal

Figure 3 shows the sorption isotherms for Au(III), Pt(IV) and Pd(II) at 0,1 M HCl obtained from multicomponent solutions (Au+Pt+Pd) in the range of 26-1000 mg/L Au, Pt and Pd concentration. The best sorption reveals AG2 resin. As the Au, Pt and Pd concentration increase maximum sorption of noble metals approach constant values of 300 mg/g; 450 mg/g and 200 mg/g, respectively.

The effect of the HCl concentration on sorption of Au(III) was investigated from gold solution in the range of 50-1000 mg/L on AG2 resin. As presents in Figure 4 resin AG2 shows particularly strong preferences for Au(III) reaches 300 mg Au/g. The sorption from 2-3 M HCl solution is lower and achieves level of 250 mg Au/g.

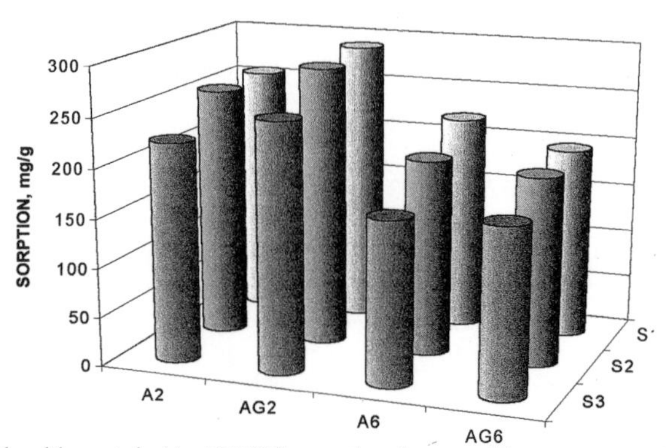

Figure 2. Total noble metals (Au+Pt+Pd) sorption from multicomponent solutions *S1, S2* and *S3*.

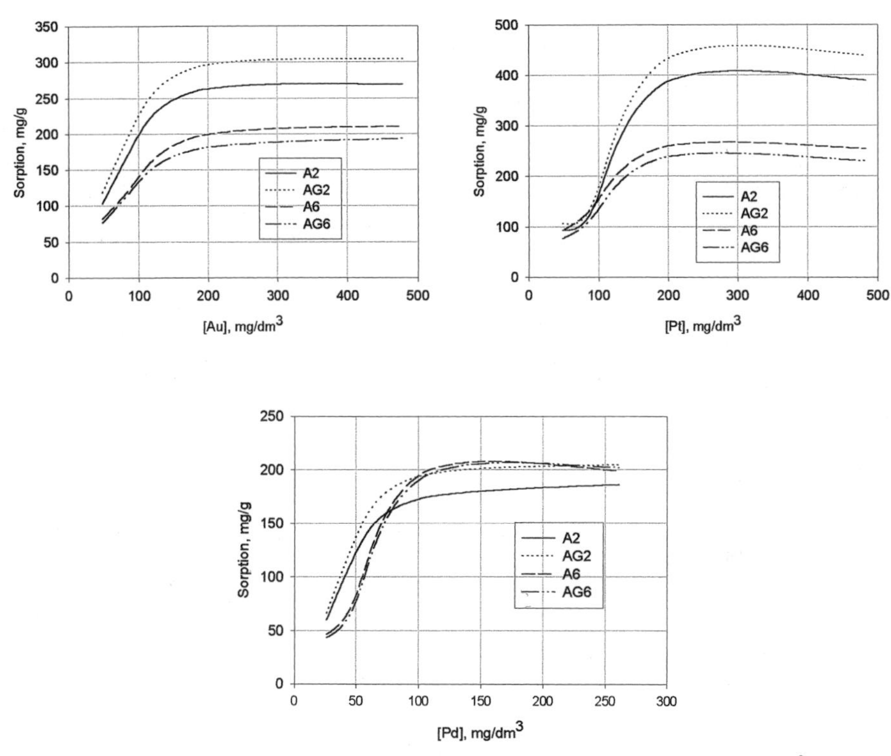

Figure 3. Isotherms of Au(III), Pd(II) and Pt(IV) sorption on resins (0,1 M HCl, 20°C).

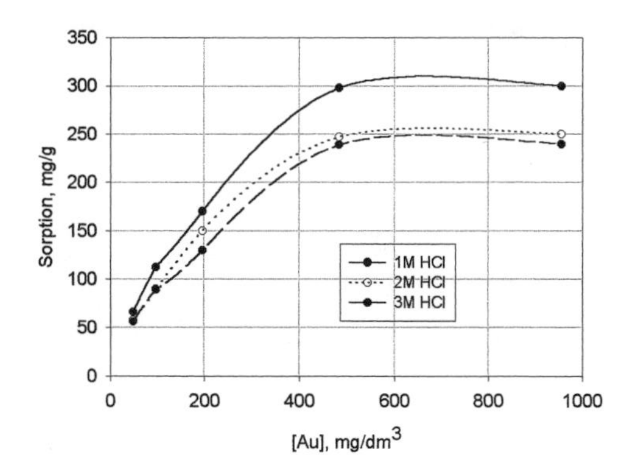

Figure 4. Effect of HCl concentration on the sorption of Au(III) on AG2 resin.

From the results described in this paper, it can be concluded that investigated resins are suitable for the preconcentration and separation of Au, Pt and Pd. The most important characteristics of investigated resins are their high selectivity for these three elements.

Conclusion

1. Incorporation of alkyl diamine ligands to copolymer matrix leads to series of RESINS - A type. The modification of A type RESINS with cyanamide results in the series of resins with guanidyl end groups (AG type).
2. Resins are useful for sorption of tetrachloroaurate, tertachloropalladate and hexachloroplatinate anions from acidic solution. Sorption from single solution of noble metals reaches up to 100 mg/g Au(III) and Pt(IV); sorption of Pd(II) is near 50 mg/g.
3. The best sorption of Au(III) and Pt(IV) from multicomponent solution S3 of (Au+Pt+Pd+Cu+Ni+Fe) reveals RESINS AG2, (100 mg of Au, 85 mg of Pt, for 1 gram of dry resin, respectively.

Acknowledgements

The Polish Committee of Scientific of Research supported this work under grant
7 T09B 119 21.

[1] J.M. Sanchez, M. Hidalgo, M. Valiente, V. Salvado, *J. Polym. Sci, Part A*, **1999,** *38* 269
[2] A. Warshawsky, N. Kahana, V. Kampel, I. Rogachev, E. Meinhardt, R. Kautzmann, J.L. Cortina, C. Sampaio, *Macromol. Mater. Eng.* **2000,** *283* , 103
[3] C.F. Vernon, P.D. Fawell, C. Klauber, *React. Polym.* **1992,** *18*, 35
[4] Chen Yi Yong, Yuan Xing Zhong, *React. Polym.* **1994,** *23* ,165
[5] Jermakowicz-Bartkowiak D., Kolarz B.N. *Eur. Polym .J.* **2002,** *38(11)*, 2239
[6] Kolarz B.N., Trochimczuk A.W., Jermakowicz-Bartkowiak D., Jezierska J., *Polymer*, **2002,** *43,*1061
[7] Jermakowicz-Bartkowiak D., Kolarz B.N, submitted

Macromol. Symp. **2004**, *210*, 147-155

Synthesis of Halogen-Free Amino-Functionalized Polymethyl Methacrylate by Atom Transfer Radical Polymerization (ATRP)

Veera Bhadraiah Sadhu, Jürgen Pionteck, Dieter Voigt, Hartmut Komber, Brigitte Voit*

Institut für Polymerforschung Dresden e.V., Hohe Strasse 6, 01069 Dresden, Germany
E-mail: pionteck@ipfdd.de

Summary: In order to obtain amino-terminated polymethyl methacrylate ($PMMA-NH_2$) free of halogen we used the atom transfer radical polymerization (ATRP) to polymerize methyl methacrylate (MMA) in presence of an initiator containing an alkyl bromide unit and a protected amine functional group. The use of CuBr / N,N,N',N",N"-pentamethyldiethylenetriamine (PMDETA) as co-catalyst system results in a polymer free of halogen due to hydrogen transfer from PMDETA to the growing polymer chain. However, side reactions occur affecting the typically "living" character of the ATRP. The measured molecular weights are consistently higher than the theoretical ones and the molecular weight distributions are relatively broad.

Keywords: amino-containing initiator; atom transfer radical polymerization (ATRP); functionalization of polymers; polymethyl methacrylate (PMMA)

Introduction

Atom transfer radical polymerization (ATRP) offers the possibility to obtain polymers with well-defined functionality, controlled molecular weight, and low polydispersity.[1,2] Typically, alkyl halides, which may carry a functional group, are used as initiators leading to polymers with an (functionalized) alkyl group at the one (starting) end and a "living" halide functionality at the other end. The halogen end groups may be used for further functionalization, but this is often connected with undesired side reaction.[3] Halogen caused side reactions (elimination, radical processes) may also occur under processing conditions and therefore it is advantageous to remove the halogen before further processing. For this, a wide range of methods are available. E.g. it was possible to convert all terminal halogens into hydrogen end groups by treating with trialkyltin hydride.[4] Recently, the synthesis of halogen-free acrylate macromonomers by ATRP in a one-pot reaction using a catalytic system containing CuBr and an excess of N,N,N',N",N"-pentamethyldiethylenetriamine (PMDETA) has been reported.[5] Due to the replacement of the halogen by hydrogen atoms via hydrogen transfer from PMDETA at the end of the polymerization halogen free products have been

 DOI: 10.1002/masy.200450617

obtained. However, in ref. [6] it was shown that the halogen-hydrogen exchange starts already in early stages of the polymerization, even under mild conditions, but to a degree which does not hinder the complete conversion of the studied acrylic monomers and retaining the controlled character of the ATRP.

We studied the applicability of this method for the bulk synthesis of amino-terminated PMMA, which is an interesting material for reactive processing, e.g. in blends. To hinder amino-caused side reactions and interfering of the amino-group with the catalyst system, the amino functionality of the initiator had to be protected before use. In order to obtain low molecular weight polymers, which can be analysed with sufficient accuracy, a relatively high initiator/monomer ratio has been used. The transformation of the terminal bromide into hydrogen and side reactions have been analysed by MALDI TOF MS and ^{1}H NMR spectroscopy.

Experimental Section

Materials. Methyl methacrylate (MMA, 99 %, Aldrich) was vacuum distilled and stored over molecular sieve under argon. Copper (I) bromide (99.999 %, Aldrich) was purified as follows: CuBr was stirred with glacial acetic acid for 24 h and washed consecutively with acetic acid, ethanol and diethyl ether for several times and dried under vacuum at 40°C for 3 days and stored under argon atmosphere. N,N,N',N",N"-pentamethyldiethylenetriamine (PMDETA) (99 %, Aldrich), anisole (99.7 % anhydrous, Aldrich), bromoisobutyryl bromide (Merck), diaminoethane (Merck), dichloromethane (Acros) dried over molecular sieve, tetrahydrofuran (THF) (Acros), triethylamine (Fluka), alumina (Merck), trifluoroacetic acid (99 %, Aldrich), and di-tert-butyl dicarbonate (Fluka) were used as received.

tert-Butyl-N-(2-amino-ethyl) carbonate. A method similar to that reported in ref. [7] was used. A solution of 17.5 g (0.08 mol) di-tert-butyl dicarbonate in 180 ml of 1,4-dioxane was slowly added into a stirred solution of 36 g (0.6 mol) diaminoethane in 180 ml of 1,4-dioxane over a period of 3 h at room temperature. After two days, the precipitate formed was filtered off and the 1,4-dioxane and excess diaminoethane were removed in vacuum from the filtrate. 300 ml water were added to the residue and the water-insoluble bis(N,N'-tert-butyloxycarbonyl)-1,2-diaminoethane was removed by filtration. From the aqueous solution saturated with sodium chloride the product was extracted with dichloromethane. The collected

organic phase was dried over sodium sulfate and eventually evaporated under reduced pressure to give 90-95 % of tert-butyl-N-(2-amino-ethyl) carbonate as colourless oil.
^{1}H NMR (DMSO-d_6): δ 1.37 ((C**H**$_3$)$_3$C-), 2.52 (-N**H**$_2$), 2.73 (-C**H**$_2$NH$_2$), 2.9 (-C**H**$_2$NH-), 5.74 ppm (-N**H**-).

Synthesis of the Initiator (I). tert-Butyl-N-(2-amino-ethyl) carbonate (17.7 g), triethylamine (11.33 g), and dry dichloromethane (180 ml) were placed in a three-neck round-bottomed flask. 2-Bromoisobutyryl bromide (25.4 g) was added slowly with stirring at 0°C. The reaction was left for 48 h with continuous stirring at room temperature. Triethylammonium bromide was formed as white precipitate and filtered off. After removal the solvent in vacuum from the filtrate a yellow solid is left. The product was dissolved in methanol and precipitated into water saturated with Na_2CO_3. Yield: 90-95 %. mp 94-96°C.
^{1}H NMR (DMSO-d_6): δ 1.38 ((C**H**$_3$)$_3$C-), 1.86 ((C**H**$_3$)$_2$C-), 3.02 (-C**H**$_2$-NH-COO-), 3.12 ((-C**H**$_2$-NH-CO-), 6.8 (-CH$_2$-N**H**-COO-), 8.0 ppm (-CH$_2$-N**H**-CO-).

Polymerization. Most samples were prepared by bulk polymerization. CuBr was placed in a dried Schlenk tube. The tube was evacuated and flushed with argon three times. Then the degassed methyl methacrylate, PMDETA, and the initiator were added to the tube and stirred until the system became homogeneously green. The resulting solution was freed from oxygen by performing three freeze-pump-thaw cycles and kept under argon.. The tube was tempered in a bath at the desired temperature and the ATRP was started by adding the initiator. After the polymerization was completed the polymerization mixture was dissolved in THF. The catalyst was removed by passing the solutions through a column of alumina/silica. The polymer was precipitated in n-hexane and dried at 50 °C in vacuum. The BOC protecting group was removed by treating with trifluoroacetic acid (25 % by volume) in dichloromethane for 2 days at room temperature. The solution was neutralized with triethylamine, subsequently precipitated into n-hexane, and vacuum dried.
The polymerization in solution was done with a solvent/MMA ratio of 1/1 by weight.

Analysis. The ^{1}H NMR spectra were recorded on a Bruker DRX 500 spectrometer operating at 500.13 MHz. Gel permeation chromatography (GPC) measurements were performed with a modular chromatographic equipment (KNAUER) containing a refractive index detector at ambient temperature. A single PL Mixed-B column Hibar PS 40 (Merck) was used. The sample concentration was c = 2 g/l and the injection volume was 20 µl. Chloroform was used as eluent at room temperature with a flow rate of 1 ml/min. Linear PMMA standards (Polymer Laboratories) were used for calibration. MALDI TOF mass spectra were acquired on a HP G2030A MALDI TOF MS system (Hewlet Packard). The desorption/ionization was induced by a pulsed N_2 laser. Dihydroxy benzoic acid and sodium trifluoroacetate / KCl were used as matrix and cationizing agent, respectively. All spectra were obtained at an accelerating potential of 24 kV in linear mode and positive polarity using a TLF-unit.

Results and Discussion

The CuBr / PMDETA catalyst system was selected for the atom transfer radical polymerization of MMA with a two-fold excess of ligand relative to the initiator in order to obtain halogen free polymers in a one-pot reaction.[5] The initiator containing a protected amine group was synthesized according to Scheme 1 that also describes the expected polymerization mechanism.

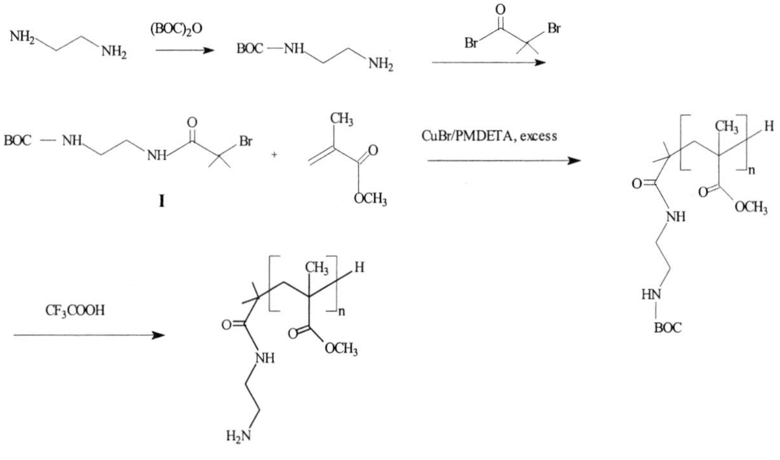

Scheme 1. Synthesis of amino-terminated PMMA by ATRP

The desired reaction product is a polymer with the protected amine group at the one end and devoid of terminal bromine at the other end. The used concentration of the initiator was relatively high in order to obtain low molecular weight polymers. This allows to analyse the structure of the terminal sites with sufficient accuracy. Representative compositions, polymerization conditions and characteristics of the resulting products are summarized in Table 1.

Table 1. Conditions and analytical data for the ATRP of MMA in bulk and solution with I (Scheme 1) as initiator and CuBr/PMDETA as catalyst ($[CuBr]_0/[PMDETA]_0 = 1/2$).

No.	$[CuBr]_0/[I]_0$	solvent	T	$t_{pol.}$	$[MMA]_0/[I]_0$	convn.	$M_{n,calc}$ a)	$M_{n,GPC}$	f =	M_w/M_n
	mol/mol		°C	min	mol/mol	%	g/mol	g/mol	$M_{n,cal}/M_{n,GPC}$	
1	1		60	10	20/1	82	1870	5200	0.36	2.86
2	1		25	30	100/1	53	5540	22400	0.24	2.18
3	0.5		25	60	100/1	80	8240	7750	1.06	1.79
4	0.5		0	240	100/1	n.d.	n.d.	24950	n.d.	2.55
5	0.3		25	90	100/1	43	4530	25700	0.18	2.45
6	0.2		25	180	100/1	22	2430	6050	0.40	2.06
7	0.1		60	180	100/1	42	4430	26050	0.17	1.53
8	0.1		25	360	100/1	n.d.	n.d.	23350	n.d.	1.62
9	1	toluene	25	300	100/1	42	4430	8150	0.54	2.20
10	0.5	toluene	60	180	20/1	98	2190	5350	0.40	2.40
11	1	anisole	90	360	200/1	58	11800	66000	0.18	2.61
12	1	anisole	60	360	200/1	45	9240	49150	0.18	2.23
13	1	anisole	25	1440	200/1	75	15200	28400	0.53	2.21
14	1	anisole	25	1440	100/1	92	9440	18600	0.50	2.50
15	1	anisole	25	30	20/1	66	1550	2200	0.70	2.16.

a) $M_{n,cal}$ (g/mol) = $[M]_0/[I]_0$ x conversion (%/100) x 100.12 + molecular weight of the initiator;
n.d.: not determined

The data in Table 1 indicate that the polymerization of MMA in bulk as well as in solution has an uncontrolled character resulting in rather broad molecular weight distribution and higher experimental M_n-values than the calculated ones. The initiator efficiency f (in ideal case: f = 1) is low (Note: The conversion was calculated from the yield, which was determined gravimetrically, not considering possible fractionation).

The polymerization is highly exothermic. To improve the control over the polymerization conditions the catalyst/initiator molar ratio was decreased up to 0.1, the polymerization temperature reduced to 0°C, and the polymerization mixture diluted. However, even under mild conditions no controlled character of the ATRP was observed. This is likely caused by the fact that the coordination complexes between copper and aliphatic amines have a lower redox potential than the typically used copper/bipyridine complexes[8] resulting in a higher concentration of active radicals which favours undesired side reactions. In ref. [8] also deviations from the ideal living character of the ATRP of MMA have been observed. Additionally, in our system interactions of the initiator with the catalyst system can not be excluded, even if the amino-containing initiator is used in its protected form. Furthermore, the propagation rate constants of MMA are very high.[9] When using an initiator with a slow initiation rate, especially in presence of the highly active catalyst system CuBr/PMDETA for the propagation reaction, a broad molecular weight distribution can be expected. With forming the first active initiator radicals the fast propagation will consume a large quantity of the monomer molecules, while active initiator species formed at later times can not participate in the polymerization process. This explains also the low initiator efficiency.

To study the reasons for the uncontrolled character of the ATRP we analysed the structure of the terminal PMMA groups. Under the polymerization conditions used the expected termination reaction is the bromine exchange by hydrogen from the PMDETA forming polymers devoid of halogen, proven for different acrylates for similar polymerization conditions.[5,6] Figure 1 clearly shows the dominance of the halogen-free amino-terminated polymethyl methacrylate (NH_2-PMMA-H) in the MALDI TOF mass spectra. The main peaks in the NH_2-PMMA-H spectrum correspond to NH_2-PMMA-H (m = 129 + n (100.12) + 1 + M^+), whereas the peaks of NH_2-PMMA-Br should appear at m = 129 + n (100.12) + 79/81 + M^+.

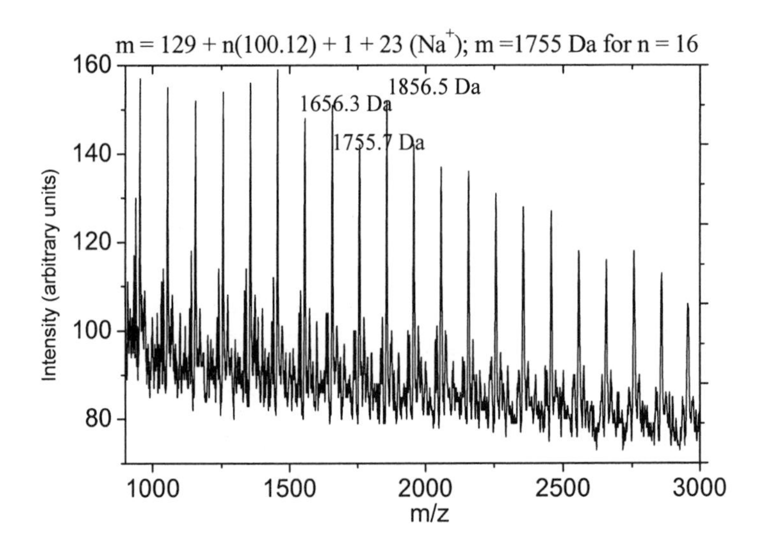

Fig. 1. MALDI TOF MS of PMMA-NH_2 (sample 1, Table 1). The dominant series of peaks correspond to the amino-functionalized hydrogen-terminated polymer chains.

Beside the main product, peaks corresponding to Br-terminated polymers and a third series of minor peaks properly caused by combination of two radicals and corresponding to m = 129 +

n (100.12) + 129 + M^+ could be observed, especially at short polymerization times. Figure. 2 give a detailed peak assignment to the different reaction products.

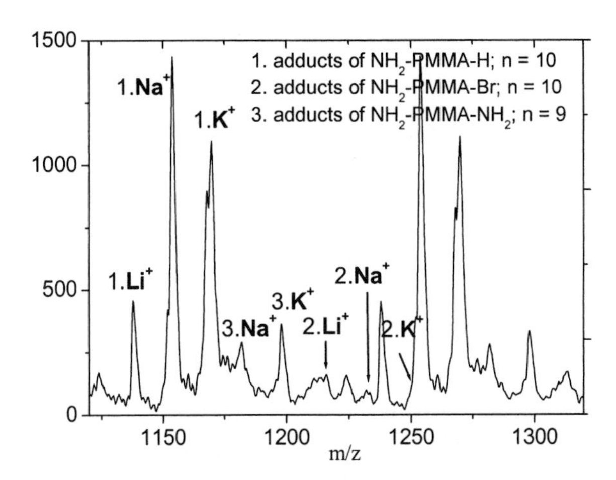

Fig. 2. Detailed peak assignment of main and side products in the sample 15 (Table 1), taken after 30 min. (MMA conversion 66 %).

Another side reaction is detected by ^{1}H NMR spectroscopy. Figure 3 shows the NMR spectra of protected and deprotected amino end group functionalized PMMA devoid of Br (sample 1 in Table 1). The peaks at 5.45 and 6.21 ppm could be assigned to the olefin double bonds formed due to dehydrobromination. According to the signal intensities about 30 % of the polymer chains underwent this elimination reaction. The product is not easy to prove by MALDI TOF MS since the molar mass difference between hydrogen terminated polymers and dehydrobrominated ones is 2 Da only, and appears as small left-hand shoulder of the main product peaks.

To evaluate the stability of the amino-terminated PMMA we analyzed the structure before and after processing in a DACA Micro Compounder (DACA Instruments). The shear rate is comparable to conventional extrusion processes. After processing at 180°C no changes in the NH_2-PMMA-H structure could be observed. At higher temperatures side reactions occur forming a main product of unknown structure (with 50 Da molar mass difference compared to NH_2-PMMA-H, Fig. 4) and increased molecular weight.

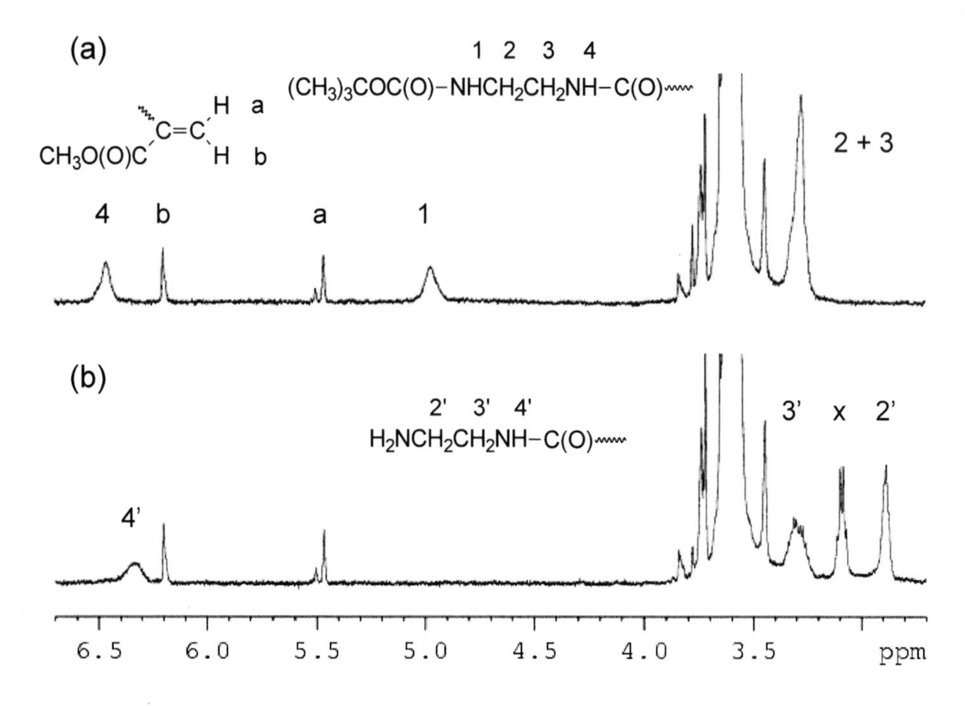

Fig. 3. ¹H NMR spectra of the protected (a) and deprotected (b) amino end-group functionalized PMMA; Region showing the signals of the R-NH-CH_2CH_2-NH-R' moiety and of the olefinic group formed by HBr elimination (x: CH_2 signal of Et_3N traces; truncated signal at ~ 3.6 ppm: OCH_3 group of PMMA).

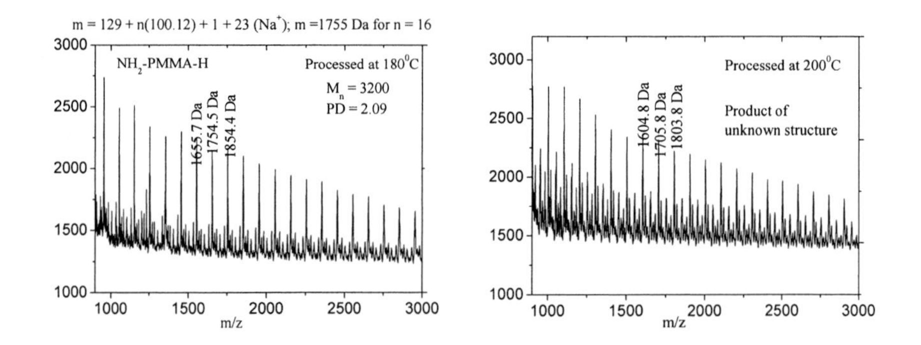

Fig. 4. MALDI TOF mass spectra of PMMA-NH_2 processed in a DACA Micro Compounder for 10 min. with 100 rpm at 180 °C (left) or 200 °C (right)

Conclusions

Atom transfer radical polymerization using the CuBr/PMDETA catalyst system and an initiator containing a protected amino functionality is a suitable technique to synthesize amino-functionalized PMMA devoid of halogen. However, the ATRP of MMA under these conditions proceeds in uncontrolled fashion. The initiator efficiency is smaller than 1, the number average molecular weight of the product higher than the theoretical one, and an untypical high polydispersity is observed, even under mild polymerization conditions. The reasons for this are different terminating reactions. While the bromine exchange with hydrogen results in the desired halogen free product, further side reactions occur as dehybromination and bimolecular radical termination.

However, the main product contains one amino end group and is free of halogen. During melt processing of the product at 180°C no structural changes could be observed, so that it can be used for reactive melt processing , e.g. in blends.

Acknowledgements

The authors are grateful to the Deutsche Forschungsgemeinschaft (Collaborative Research Centre 287) for the financial support, Mrs. P. Treppe for GPC measurements, and Mr. H. Kunath for technical assistance.

[1] K. Matyjaszewski, J.S. Wang, *Macromolecules* **1995**, 28, 7901.
[2] M. Kato, M. Kamigaito, M. Sawamato, T. Higashimura, *Macromolecules* **1995**, 28, 1721.
[3] H. Malz, H. Komber, D. Voigt, I. Hopfe, J. Pionteck, , *Macromol. Chem. Phys.* **1999**, 200, 642.
[4] V. Coessens, K. Matyjaszewski, *Macromol. Rapid Commun.* **1999**, 20, 66.
[5] F. Schön, M. Hartenstein, A.H.E. Müller, *Macromolecules* **2001**, 34, 5934.
[6] M. Bednarek, T. Biedron, P. Kubisa, Macromol. *Chem. Phys.* **2000,** 201, 58.
[7] C. Duan Vo, D. Kuckling, H.-J. P. Adler, M. Schönhoff, *Colloid Polym. Sci.* **2002**, 280, 400.
[8] J. Xia and K. Matyjaszewski, *Macromolecules* **1997**, 30, 7697.
[9] K. Matyjaszewski, J. L. Wang, T. Grimaud, D. A. Shipp, *Macromolecules* **1998**, 31, 1527.

Reversible Switching of Protein Uptake and Release at Polyelectrolyte Multilayers Detected by ATR-FTIR Spectroscopy

M. Müller,[*1] B. Kessler,[1] H.-J. Adler,[2] K. Lunkwitz*[1]

[1] Institute of Polymer Research Dresden e.V. (IPF)
Hohe Str. 6, D-01069 Dresden, Germany
[2] Institute of Macromolecular and Textile Chemistry, Technical University Dresden (TUD), Mommsenstr. 13, D-01062 Dresden, Germany
E-mail: mamuller@ipfdd.de

Summary: The reversible switching of uptake and release of the proteins lysozyme (LYZ, IEP = 11.1) and human serum albumin (HSA, IEP = 4.8) at the surface attached polyelectrolyte multilayer (PEM) consisting of poly(ethylene-imine) (PEI) and poly(acrylic acid) (PAC) is shown. Protein adsorption could be switched by pH setting due to electrostatic interaction. Adsorption of positively charged LYZ at PEM-6 took place at pH = 7.3, where the outermost PAC layer was negatively charged. Complete desorption was obtained at pH = 4, where the outermost PAC layer was neutral. Additionally the charge state of the last adsorbed PAC layer in dependence of the pH of the medium could be determined in the ATR-FTIR difference spectra by the ν(COO^-) and ν(C=O) band due to carboxylate and carboxylic acid groups. Adsorption of negatively charged HSA at PEM-7 was achieved at pH = 7.3, where the outermost PEI layer was positively charged. Part desorption was obtained at pH = 10, where the outermost PEI layer was neutral. PEM of PEI/PAC may be used for the development of bioactive and bionert materials and protein sensors.

Keywords: ATR-FTIR spectroscopy; human serum albumin; isoelectric point; lysozyme; multilayer; polyelectrolytes; protein adsorption; surface modification

Introduction

Protein adsorption at polymer surfaces is an important phenomenon in biomedical, food and pharmaceutical research. On the one hand it is desired for bioactive applications (e.g. uptake of collagenized implants) and on the other it should be prevented for bioinert purposes (e.g clotting on medical devices, membrane fouling) blocking or delaying further bioadhesion cascades. Up to now the correlation between polymer (surface) structure and protein adsorption is not fully solved. In that framework Norde has raised important rules concerning soft and hard properties of the protein and dispersed substrates like polymer latices or inorganic particles [1]. In further fundamental research on this topic often thin films, which

 DOI: 10.1002/masy.200450618

are accessible by a variety of in-situ analytical techniques, are used to mimick the properties of the surface of the bulk polymer [2, 3]. However, the correlation between the properties of the thin film or other models and the real surface of commercial bulk polymer materials is a difficult task. This has been reflected e.g. by industrial work of Pudleiner and cooperators, aiming at the correlation between thrombogenicity and polymer structure, especially shown for extruded copolymers of polysulfone and polyurethanes with PEO soft segments [4].

Moreover, the thin film itself can be also used in practical applications to modify bulk polymer materials with convenient e.g. mechanical, processing or aging properties. Thereby the given mechanical or other properties of commercial polymers can be preserved and new surface properties can be achieved by simply coating the material or device with a thin often nanoscopic functional layer. Consecutively adsorbed polyelectrolyte multilayers (PEM) introduced by Iler [5] and Decher [6] may be used as a surface modification concept in that respect, since they can be anchored on a variety of materials forming versatile platforms for the exposure of defined structural elements on the modified surface via the outermost polyelectrolyte layer. Such elements could be charged groups, hydrophobic or hydrophilic moieties of blockcopolyelectrolytes or even different conformations of charged polypeptides [7]. Here we report on fundamental studies on reversible pH mediated electrostatic interactions between multilayers consisting of poly(ethyleneimine) (PEI) and poly(acrylic acid) (PAC) and the model proteins lysozyme (LYZ) and human serum albumin (HSA). ATR-FIR spectroscopy is used to monitor the reversible uptake/release of proteins upon modulating the surface charge properties of PEMs. These studies shall demonstrate that PEMs are potentially useful as biomaterial coatings or protein sensor devices.

Experimental

Polyelectrolytes, proteins

All commercial polyelectrolyte (PEL) samples were used without further purification. Poly(ethyleneimine) (PEI, pH = 9) and poly(acrylic acid) (PAC, pH = 4) were dissolved in Millipore water at PEL concentrations c_{PEL} = 0.01 M. The ′as is′ pH values are given in brackets.

Lysozyme (LYZ) and human serum albumin (HSA) were dissolved in phosphate buffer saline (PBS, 1 mg/ml) yielding pH = 7.3. As buffered rinsing solutions PBS (1 mg/ml, pH = 7.3), citrate (CIT, 2 mg/ml, pH = 4) and lysine (LYS, 1 mg/ml, titrated to pH = 10) were applied.

Surfaces

Immediately before the usage the silicon internal reflection elements (IRE) were cleaned in piranha solution (30% H_2O_2, 70% H_2SO_4) under ultrasonification and rinsed in Millipore water followed by a plasma treatment (PCD 100, Harrick, New York) under reduced pressure.

Polyelectrolyte multilayers

PEM-X were prepared in the in-situ-ATR-sorption cell (described below) by consecutively exposing the Si support to solutions of cationic PEI (0.01 M) and to anionic PAC (0.01 M) solutions in defined numbers of cycles X = 6 and 7.

ATR-FTIR spectroscopy

The characterization of the deposited PEMs was performed by in-situ-ATR-FTIR spectroscopy, whose principle is given in Fig. 1a. Additionally, the SBSR concept [8], shown in Fig. 1b, was used to obtain well compensated ATR-FTIR spectra. Thereby, the IR beam is guided alternately through the IRE below the sample and the reference compartment, respectively. Typically, the sample compartment is filled with the solution of the adsorbing or binding substance and the reference compartment with the respective solvent. The ATR-FTIR attachment was operated on the IFS 55 Equinox spectrometer (BRUKER-Saxonia, Leipzig) equipped with globar source and MCT detector. Typically, polycation, rinsing, polyanion and protein solutions were injected and cycled through the *in-situ* ATR cell (IPF Dresden) by a peristaltic pump in combination with an addressable valve system (IPF Dresden) operated under computer control as shown in Fig. 1a.

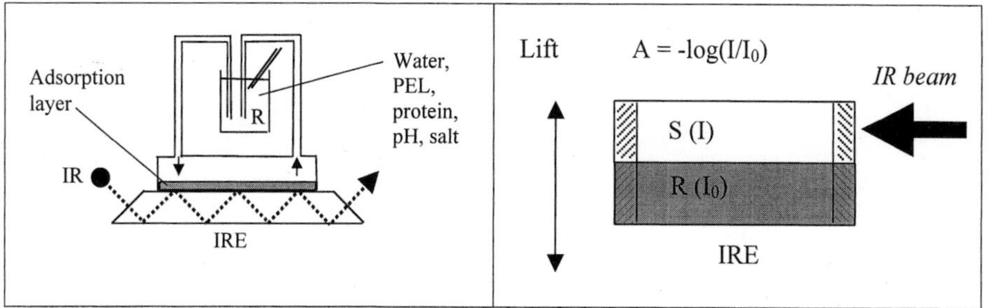

Fig. 1a. Cycling of PEL or protein solutions from a reservoir (R) through the in-situ-ATR-FTIR cell above the Si-IRE.

Fig. 1b. SBSR principle according to [8] applied for reproducible spectral background compensation in the in-situ-ATR-FTIR spectra.

Results

The polyelectrolyte multilayer (PEM) system composed of PEI and PAC has been developped to become a standard system for reproducible deposition of a defined adsorbed amount and the formation of a laterally homogeneous surface coverage. Both PELs are commercially available. PEM of PEI/PAC have been especially used, to study the influence of electrostatic interactions on protein adsorption [9, 10, 11]. Based on those results here we report on the reversible switching of the charge state of the outermost layer as well as on the related adsorption/desorption cycles at the PEM-6 and the PEM-7 of PEI/PAC.

1. Modulation of the charge state of PEM-6

The dissiociation degree and thus the charge density of weak polyacids like PAC is dependent on the pH of the aqueous medium. Therefore the electrostatic interaction between proteins and a polyacid layer can be tuned by the pH setting. In our case the polyacid layer was the terminating PAC layer of a PEM-6 composed totally of three PEI and three PAC layers. In the Fig. 2a in-situ-ATR-FTIR spectra of a PEM-6 of PEI/PAC in contact to citrate buffer (CIT, lower curve) and to phosphate buffer (PBS, upper curve) are shown.

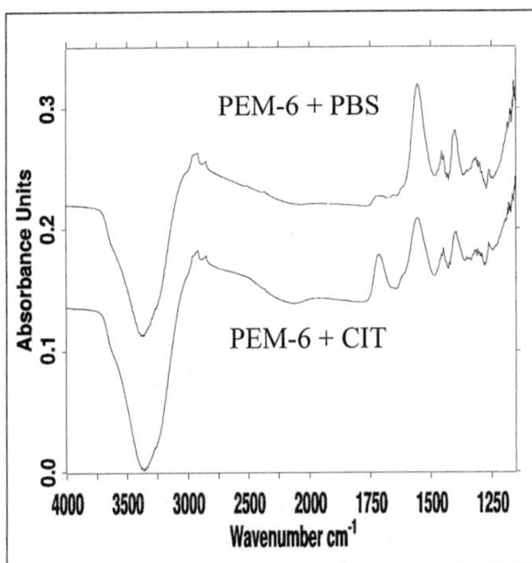

Fig. 2a. ATR-FTIR spectra monitoring the dissociation state of the outermost PAC layer of PEM-6 of PEI/PAC. Upper curve: PEM-6 in the presence of PBS buffer (PBS, pH = 7.3). Lower curve: PEM-6 in the presence of citrate buffer (CIT, pH = 4.0).

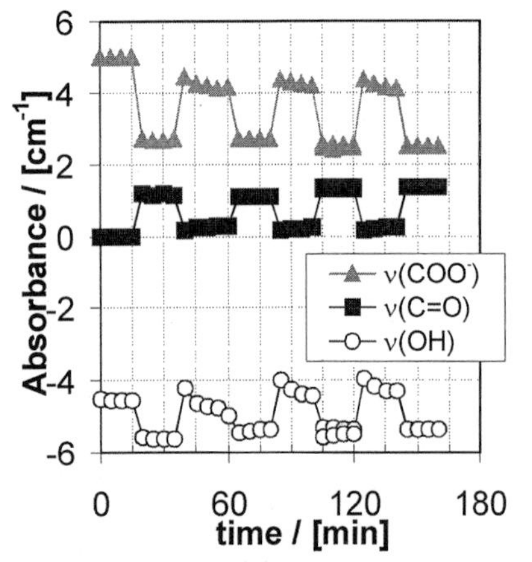

Fig. 2b. Switchable courses of the band integrals of the ν(C=O), ν(COO^-) and the ν(OH) band (factor: 0.1) of PAC within the PEM-PEI/PAC modulated by the pH of the aqueous medium. The pH was switched between 7.3 (PBS) and 4.0 (CIT) starting with pH = 7.3 (first 15 min).

Significantly, the pH elevation from pH = 4 to pH = 7.3 caused changes of the ν(C=O) and ν(COO^-) intensities due to the concentrations of COOH and COO^--groups, respectively. In Fig. 2b the courses of the ν(OH), ν(C=O) and ν(COO^-) band integrals are plotted in dependence of the time, during which the pH value of the aqueous medium was switched subsequently between pH = 7.3 (pH at the beginning) and pH = 4. In detail the ν(C=O) band is lowered and the ν(COO^-) band increased at the pH = 7.3 proving elevation of the number of COO^- groups and thus of charge carriers. The opposite is the case at pH = 4: the ν(C=O) is increased and the ν(COO^-) is lowered proving elevation of the number of COOH groups resulting in diminuation of charge carriers. So with that it can be clearly shown, that the negative surface charge potential can be easily switched by pH setting. As it was shown recently one can also calculate the dissociation degree based on the integated areas A of the ν(COO^-) and ν(C=O) according to $\alpha_{IR} = A_{\nu(COO-)} / [A_{\nu(COO^-)} + F * A_{\nu(C=O)}]$ knowing the ratio $F = \varepsilon_{\nu(COO^-)}/\varepsilon_{\nu(C=O)}$ of their absorption coefficients ε [12]. Furthermore the modulated course of the ν(OH) band intensity reflects a certain pH dependent swelling/shrinking behavior of the PEM-6, which was discussed for comparable systems therein [12].

2. Reversible electrostatic binding of lysozyme at PEM-6

This effect of switching the charge state of the outermost PAC layer between ´neutral´ and ´negative´ can be used for fundamental studies or applications on the electrostatic binding and release of proteins at PEM composed of PEI/PAC. We have chosen lysozyme (LYZ) as a model protein, since it exhibits an isoelectric point of IEP = 11.1 and has therefore at both applied values of pH = 7.3 and pH = 4 a positive net charge. Hence we expected that changing the pH should have the major influence on the outermost PAC layer as shown in the previous section an a minor one on the net charge state of the protein. ATR-FTIR spectroscopy was used to monitor protein binding and release, since the amide I band is a highly sensitive and diagnostic sensor for proteins. If the concentration in the solution is sufficiently small ($c \leq 1$ mg/ml), the detected amide I signal is only due to the formation of a protein layer on the PEM and can therefore be used for monitoring protein sorption. The ATR-FTIR spectra recorded during exposure of LYZ solution (15 min) to a PEM-6 of PEI/PAC are shown in Fig. 3a, whereby at pH = 7.3 (upper curve) a strong increase and at pH = 4 a strong decrease of the amide I signal can be obtained. For quantification the courses of the amide I band integrals are shown in Fig. 3b, which have been obtained by the follwing two steps:

(1) LYZ adsorption at pH = 7.3 (PBS buffer)

(2) LYZ release at pH = 4 (CIT buffer) followed by rinsing with PBS.

The modulated uptake of LYZ can be clearly seen, which correlates well with the switching of the charge state of the last adsorbed PAC layer between ´negative´ and ´neutral´ as it is shown in Fig. 2b. The continuous drop of the adsorbed amount is due to the repeated exposure of the PAC terminated PEM to the same LYZ solution, which was used in the step before and successively depleted by increasing adsorption steps.

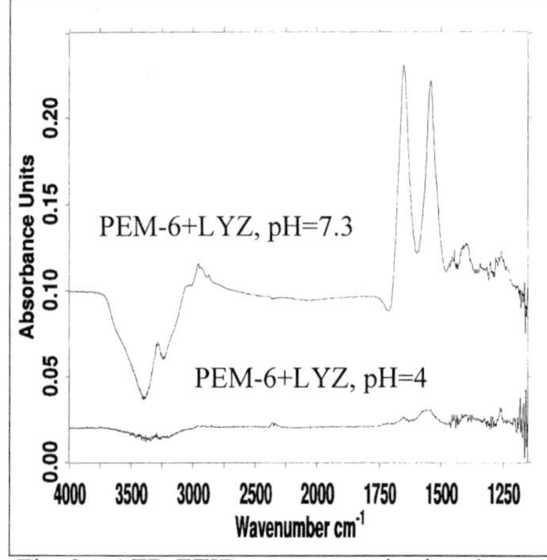

Fig. 3a. ATR-FTIR spectra monitoring the adsorption of LYZ at pH = 7.3 (PBS) (upper curve) and the release at pH = 4 (CIT) (lower curve) at the PEM-6 of PEI/PAC. (Data of the first cycle were taken.)

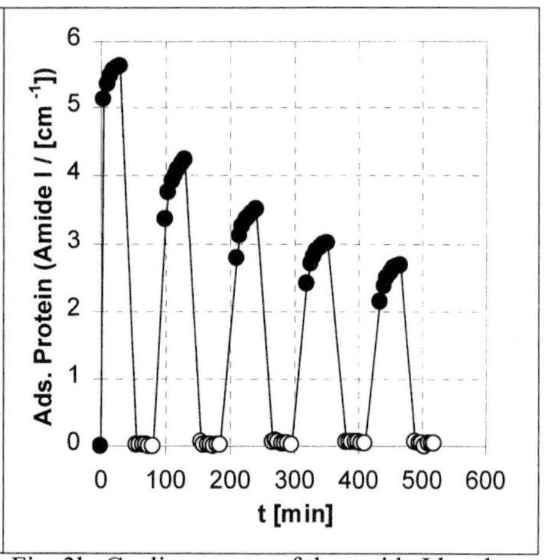

Fig. 3b. Cyclic courses of the amide I band integrals due to the adsorbed LYZ amount at PEM-6 of PEI/PAC modulated by the pH settings: pH = 7.3 (full circles), pH = 4.0 (open circles).

3. Reversible electrostatic binding of serum albumin at PEM-7

As another electrostatically interacting system the PEM-7 was generated simply by adsorbing one more PEI layer on top of the PEM-6. In that case the interaction of the acidic protein human serum albumin (HSA) was studied at two pH values, where HSA (IEP = 4.8) has a negative net charge. The ATR-FTIR spectra recorded during exposure (20 min) of HSA solution to a PEM-7 of PEI/PAC are shown in Fig. 4a, whereby at pH = 7.3 (upper curve) a large and at pH = 10 (lower curve) a smaller amide I signal can be obtained. Analogously to Fig. 3b the courses of the amide I band integrals are shown in Fig. 4b, which have been obtained by the follwing two steps:

(1) HSA adsorption at pH = 7.3 (PBS buffer)

(2) HSA release at pH = 10 (LYS buffer) followed by rinsing with PBS.

Again the modulated uptake of HSA can be clearly seen, which correlates well with the switching of the charge state of the last adsorbed PEI layer between ′negative′ and ′neutral′. Unlike the PEM-6 interaction with LYZ in the case of PEM-7 the HSA could not be fully desorbed by pH = 10. Presumably, at pH = 10 the ammonium groups of PEI were not completely deprotonated to neutral primary or secondary amino groups. Hence positively charged groups remained which are able to bind HSA to a certain extent. A further pH increase was not considered, since solutions of pH > 10 could cause damage of the PEM of PEI/PAC as well as of the silicon support. Further studies will be performed in that direction.

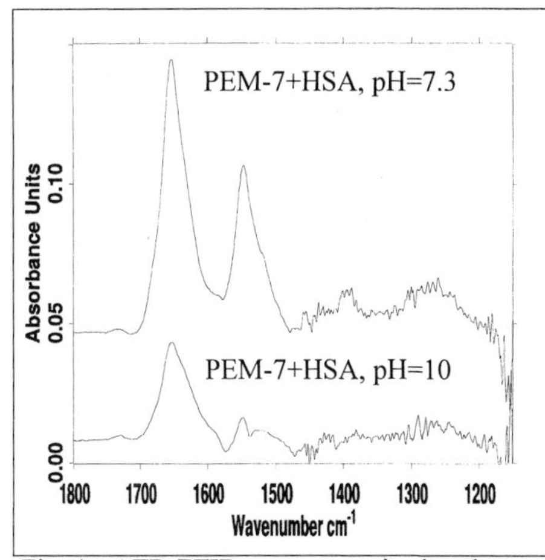

Fig. 4a. ATR-FTIR spectra monitoring the adsorption of HSA at pH = 7.3 (PBS) (upper curve) and the part release at pH = 10 (LYS) (lower curve) at the PEM-7 of PEI/PAC. (Data of the first cycle were taken.)

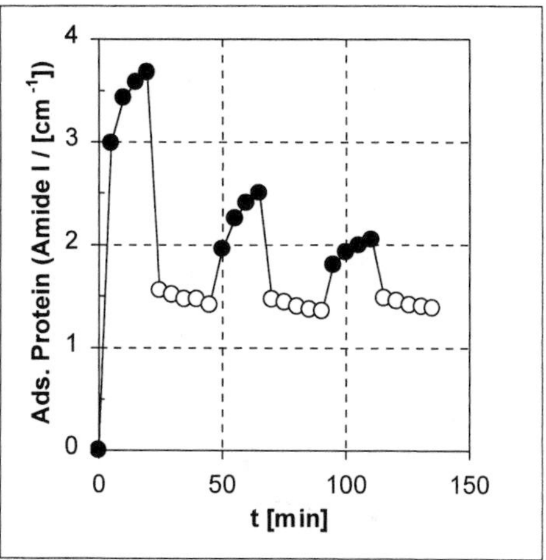

Fig. 4b. Cyclic courses of the amide I band integrals due to the adsorbed HSA amount at PEM-7 of PEI/PAC modulated by the pH settings: pH = 7.3 (full circles), pH = 10 (open circles).

Conclusion

- PEMs composed of PEI/PAC are switchable platforms for the reversible uptake and release of proteins by electrostatic interaction. They could be switched by the pH of the surrounding aqueous medium. The outermost PAC layer of the PEM-6 is fully dissociated at the pH = 7.3 and approximately neutral at pH = 4, whereas the outermost PEI layer is fully protonated at pH = 7.3 and approximately neutral at pH = 10.

- This was used to modulate the binding/release of lysozyme (LYZ), which is a basic protein (IEP = 11.1), by electrostatic interaction: At pH = 7 LYZ was adsorbed at a high amount and at pH = 4 nearly all LYZ could be removed from the PEM-6, due to the decreased negative charge of the outermost PAC layer, as it is shown in Fig. 5a.
- Analogously, at the PEM-7 acidic human serum albumin (HSA, IEP = 4.8) could be adsorbed at pH = 7.3 and partly desorbed at pH = 10 due to the charged and the neutral state of the outermost PEI layer, respectively, as iti shown in Fig. 5b.
- The uptake and release of these proteins could be switched in infinite cycle numbers, which makes PEM of PEI/PAC potentially interesting as protein sensors or also as coatings to prevent biosensor membrane fouling [13].

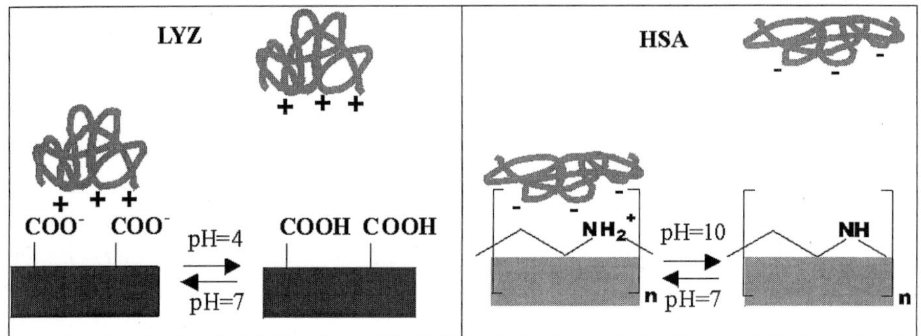

Fig. 5a. Scheme of reversible LYZ interaction to PEM-6 of PEI/PAC.

Fig. 5b. Scheme of reversible HSA interaction to PEM-7 of PEI/PAC.

Acknowledgements

We thank the Deutsche Forschungsgemeinschaft (DFG) for financial support (SFB 287, B5).

1. C.A. Haynes, W. Norde, *Coll. Surf. B*, 2, 517 (1994)
2. J.D. Andrade and V. Hlady, *Adv. Pol. Sci.*, 79, 1 (1986)
3. C.-G. Gölander and E. Kiss, *J. Coll. Interf. Sci.*, 121 (1) 240 (1988)
4. R. Dujardin, H. Pudleiner, J. Goossens and U.T. Seyfert, in *'Werkstoffe für die Medizintechnik'*, J. Breme (ed.), pp. 185, DGM-Informationsgesellschaft Verlag, Frankfurt (1997)
5. R.K. Iler, *J. Coll. Interf. Sci.*, 21, 569 (1966)
6. G. Decher, *Science*, 277, 1232 (1997)
7. M. Müller (manuscript in preparation)
8. U.P. Fringeli, in *'Encyclopedia of Spectroscopy and Spectrometry'*, J.C. Lindon, G.E. Tranter, J.L. Holmes (eds.), Academic Press (2000)
9. M. Müller, T. Rieser, K. Lunkwitz, S. Berwald, J. Meier-Haack, D. Jehnichen, *Macromol. Rapid Commun.*, 19, 333 (1998)
10. M. Müller, M. Briššová, T. Rieser, A.C. Powers, K. Lunkwitz, *Mat. Sci. Eng. C,* 8-9, 167 (1999)
11. M. Müller, T. Rieser, P. Dubin, K. Lunkwitz, *Macromol. Rapid Commun.*, 22, 390 (2001)
12. M. Müller, *in:* "*Handbook of Polyelectrolytes and Their Applications*", S.K. Tripathy, J. Kumar, H.S. Nalwa (eds.), Vol. 1, pp. 293, Stevenson Ranch: American Scientific Publishers (2002)
13. N. Wisniewski, M. Reichert, *Coll. & Surf. B*, 18, 197 (2000)

Creation of Crosslinkable Interphases in Polymer Blends

Veera Bhadraiah Sadhu, Jürgen Pionteck, Petra Pötschke, Lothar Jakisch, Andreas Janke*

Institut für Polymerforschung Dresden e.V., Hohe Strasse 6, 01069 Dresden, Germany
E-mail: pionteck@ipfdd.de

Veera Bhadraiah Sadhu is dedicating this work to his teacher Prof. A. Varada Rajulu, Sri Krishnadevaraya Univ. Anantapur, on the occasion of achieving 'the best teacher' award in 2003.

Summary: A new type of bi- and trifunctional coupling agents containing 2-oxazoline and/or 2-oxazinone as well as hydrosilane moieties has been prepared by hydrosilylation of the corresponding allyl ether containing precursors with poly(methylhydro)siloxanes. In heterogeneous model blends based on mono-carboxylic acid terminated polystyrene (PS) and mono-amine terminated polyamide 12 (PA), the oxazoline and oxazinone units can selectively react with the carboxylic groups or amino groups, respectively. Under this mixing conditions the hydrosilane partially crosslinks.
The morphology development of the three-component blends under melt mixing conditions is a rather complex process. We have shown that the coupling agents are immiscible with the polymers and form their own phase. Under proper processing conditions they locate at least partially at the PS/PA interface and can be used for further modification of the blend interphase, e.g. for crosslinking by hydrolysis. This crosslinking can be accelerated by the addition of a Pt-catalyst during the melt mixing.

Keywords: blends; crosslinking; interface; morphology; polysiloxanes

Introduction

Most polymer blends are thermodynamically immiscible and produce multiphase morphological structures.[1] To improve the compatibility between the blend components and therefore the blend properties, reactive compatibilization has become significant commercial interest.[2] The formation of block or graft copolymers by coupling reaction between functional polymers is the key point for the compatibilization of different immiscible polymers and also for the improvement of the adhesion between polymer-polymer interfaces.[3] These in situ formed copolymers are required to locate at the interface between two phases to act as efficient compatibilizer by lowering the interfacial tension and stabilizing the morphology against

DOI: 10.1002/masy.200450619

coalescence. Though, the in-situ formed copolymers may be easily pulled out of the interfacial region,[4-6] depending on the kinetics of the interfacial reaction.[7,8] This is due to the thermodynamic and hydrodynamic instability caused by the amount of accumulated block or graft copolymers at the interface.[7]

In this paper we focus on the creation of crosslinkable blend interphases by the use of novel types of poly(methylhydro)siloxane-containing coupling agents. Carboxylic acid terminated polystyrene (PS) and amine terminated polyamide 12 (PA) were chosen as model blend system. The new coupling agents contain oxazoline and/or oxazinone sites, which can react with the acid and amine functionalities, respectively, of the blend components. The preparation was done by hydrosilylation of the corresponding allyloxy-modified coupling agents[9,10] with poly (methylhydro)siloxanes. If the coupling agents react with both polymers of the blend, the remaining SiH bonds should locate at the interface and may be used for further modification reactions, e.g. for the creation of crosslinked and stable interphases. We studied the crosslinkability of the coupling agents, their reactivity to the blend components, their distribution in the blends, and their effect on the resulting blend morphology under different processing conditions.

Experimental

Materials

Carboxylic acid terminated polystyrene (PS, Mn = 28000 g/mol, PD = 1.68, functionality = 0.7) was prepared by the TEMPO-mediated free radical polymerisation using 4,4'-azo-bis-(4-cyanopentanoic acid) as an initiator.[11] To stabilize the product during processing the TEMPO living end group was removed by oxidation with m-chloroperbenzoic acid.[12] Mono-amine functionalised polyamide 12 (PA, M_n = 5000 g/mol, functionality of 250 mmol/Kg = 1) was received from Degussa AG. Poly(methylhydro)siloxane-co-dimethyl siloxane (1900-2000 g/mol, 25-30 mole% of MeHSiO), poly(methylhydro)siloxane (1500-1900 g/mol), and platinum-divinyltetramethyldisiloxane complex in xylene (2.1-2.4 % platinum concentration) were received from Gelest ABCR.

The allyloxy-containing coupling agents (Scheme 1) used in hydrosilylation reaction have been synthesized according to [9,10].

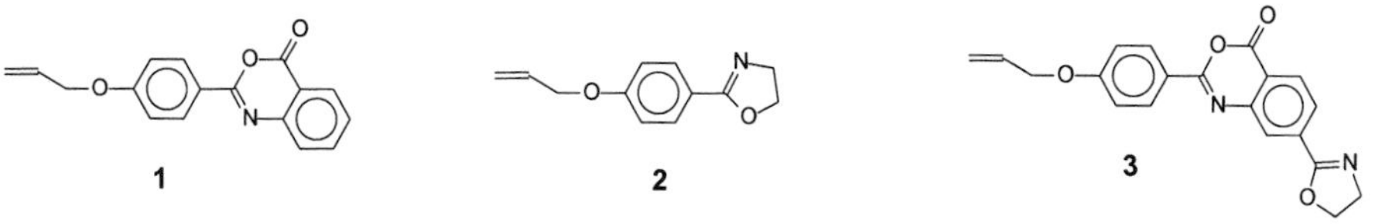

Scheme 1. Structure of the allyloxy-modified coupling agents used for hydrosilylation

Preparation of Polysiloxane-Containing Coupling Agents (SCA)

The hydrosilylated coupling agents are named SCA1 to SCA4. They were prepared according to Scheme 2.

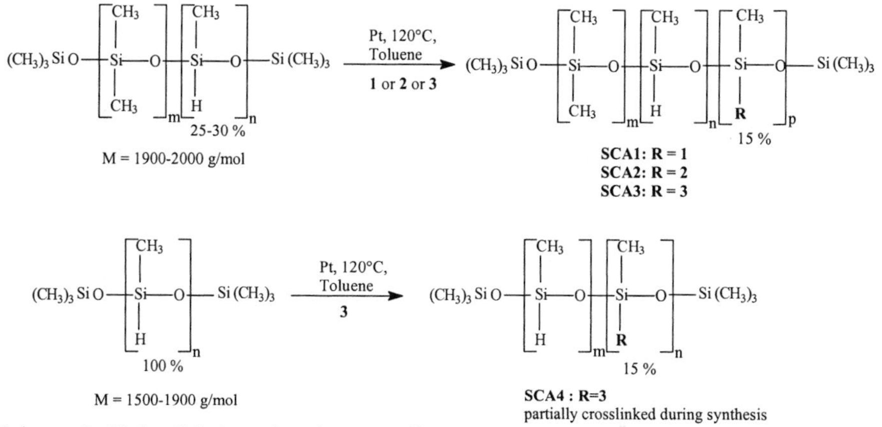

Scheme 2. Hydrosilylation of various coupling agents (Scheme1: 1, 2 and 3) with different polysiloxanes

The synthesis of SCA3 proceeded as follows. To a dry three-neck round bottom flask fitted with a reflux condenser 9.6 g poly(methylhydro)siloxane-co-dimethyl siloxane and 3.2 g (15 mol % to Si-H) of the coupling agent 3 (Scheme 1) were added. The mixture was dissolved in 800 mL toluene and a catalytic amount of Platinum-divinyltetramethyldisiloxane complex (200 μL) was added. Dry air was purged through the reaction mixture for a couple of minutes. The reaction was allowed to proceed for 6 h at boiling. The toluene was removed by means of a vacuum rotary evaporator and the product was stored at lower temperatures.

The structure is confirmed by ^{1}H NMR in TFA-d_1: δ (ppm) 0.45 (CH_3Si), 0.5 ($OCH_2CH_2C\underline{H}_2Si$),

2.25 ($OCH_2C\underline{H_2}CH_2Si$), 4.21 ($OC\underline{H_2}CH_2CH_2Si$). 4.56 (-$CH_2N$-, oxazoline), 5.41 (-$OCH_2$-, oxazoline).

SCA1 and SCA2 were prepared analogously. The hydrosilylation of the coupling agent 3 (Scheme1) with poly(methylhydro)siloxane resulted in a partially crosslinked product. 45 % of the isolated product SCA4 were insoluble in toluene (24 h at room temperature).

Blend Preparation

The blends were prepared in a DACA Micro Compounder with a capacity of 4.5 cm^3 (200°C, 100 rpm, 10 min.). The blend composition was PS / PA / SCA = 80 / 15 / 5 (v%). The mixing was carried out by two processing methods. In the one-step mixing method all blend components were fed together to the Micro Compounder. In the two-step method PA and the SCA were mixed first and in a second step the corresponding amount of PA/SCA was mixed with the PS. The blends containing both SCA1 and SCA2 were prepared by premixing of PA with SCA1 and PS with SCA2. Every mixing step proceeded for 10 min. To catalyse the crosslinking sometimes Pt solution (20 µl) were added to the mixture.

Analysis

The ^1H NMR spectra were recorded on a Bruker DRX 500 spectrometer operating at 500.13 MHz. Deuterated trifluoroacetic acid (TFA-d_1) was used as solvent. The morphology of the blends was investigated by scanning electron microscopy (SEM) using an SEM LEO 435 VP (Leo Elektronenmikroskopie) operating with an acceleration voltage of 5 to10 kV. The analysis was carried out on cryo cuts after chemical etching with trifluoroacetic acid for 4h at room temperature (TFA), which dissolves the PA phase and the non-crosslinked compatibilizer. All samples were sputtered with gold. Atomic force microscopy (AFM) was performed on smooth cryo-cuts by means of a NanoScope IV - Dimension 3100 (Veeco). The measurements were done in the tapping mode. The topography and phase images have been detected simultaneously. The scan conditions we choose according to Magonov[13] (free amplitude > 100 nm, set-point amplitude ratio 0.5) to get stiffness contrast in the phase image, that means bright features in the phase image are stiffer than dark areas.

Results and Discussion

Synthesis of SCA and Model Coupling Reactions

The hydrosilylation reaction is the most widely used method for preparing organofunctional polysiloxanes starting from poly(methylhydro)siloxanes.[14] The reaction is usually catalysed by platinum complexes. In our approach, two model polysiloxanes were used for the reaction of allyloxy derivatives of various coupling agents in order to synthesize polysiloxane-containing coupling agents containing organofunctional groups in the side chain (Scheme 2). One has a low Si-H content (25-29 % of the monomer unit) and the second one consists of 100 % Si-H units. The reaction scheme and the detailed description of the syntheses of SCA3 as example are given in the experimental section.

To proof the selective reactivity of the coupling agent SCA3 has been melt mixed with the individual blend components in about stoichiometric composition of PA/SCA3 = 75/25 (v%) and PS/SCA3 = 95/5 (v%), respectively. Fig. 1 shows the ^{1}H NMR spectra (onsets) of SCA3 (a), PA (b), and their reaction product (c) and the proposed chemical structure of the reaction product. The strong reduction of the signal at 3.2 ppm (methylene protons neighboured to the amino endgroup) and the appearance of strong signals of oxazoline groups bounded next to the reacted oxazinone group (a' and b') indicates that an almost complete reaction the amino group occurred with the complementary oxazinone group. The decrease of Si-H (5.0 ppm) group concentration indirectly proves the crosslinkability of the polysiloxane coupling agents during melt mixing. This may be caused by the presence of moisture and the high temperatures (Scheme 3). The oxazoline groups remain stable under these conditions. We also tried to prove directly the coupling reaction of the oxazoline functional groups and carboxylic acid groups of PS in melt. Due to the higher molecular weight of PS and, therefore, the small amount on COOH-groups, and the overlapping of the peaks of the virgin components and the reaction product a direct proof was not possible. However, in earlier works the selectivity of the oxazoline-carboxylic acid reaction (and also of the oxazinone-amine reaction) was proven.[10,15]

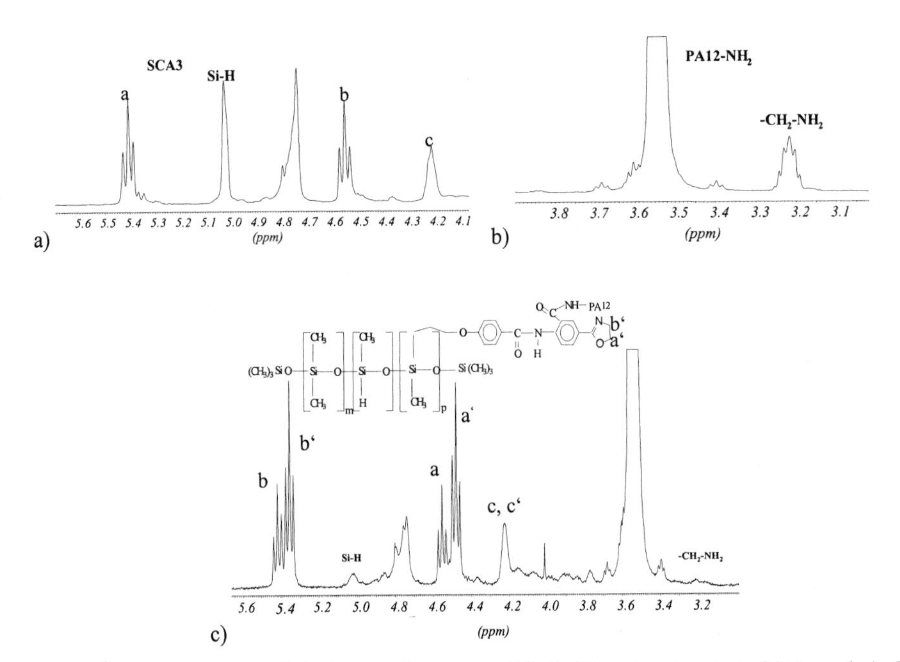

Fig 1. ^{1}H NMR spectra of a) the coupling agent SCA3, b) amino-terminated PA, and c) the coupling product between SCA3 and PA in melt. (measured in TFA-d_1).

Blend Morphology

AFM studies show that the SCA are immiscible with PA and PS and form an own phase. However, the size of the SCA3 is much smaller in PS (in the some 10 nm scale, not shown) than in PA where spheres in the size of 1 µm are visible (Fig. 2a). Such big SCA3 particles are not detectable in the PS/PA blend, neither after the one-step (Fig. 2b) nor the two step processing (not shown). The soft SCA3 (appearing dark in the phase contrast mode, right pictures) can be detected as small particles in the PS matrix (up to 200 nm size), within the PA particles in a some 10 nm scale (the detection is complicated by the overlapping with the partial crystalline structure of the PA), and also at the interface between PS and PA. In any case at least a partial location of the SCA (and therefore of the crosslinkable units) at the interface is possible.

SEM reveals that in the blends PA always forms the dispersed phase. Due to its rather low molecular weight (and therefore low viscosity) compared to the PS matrix and the favourable interactions between the amino and carboxylic groups of both blend components the particle size is very small (200 – 500 nm, Fig. 3a). The addition of SCA1, SCA2, or SCA3 (Fig. 3b) results in a significant increase of the particle size with very broad size distribution (0.3 – 3 μm). We assume that the bulky functional groups of the SCA hinder interactions between the complementary carboxyl and amino groups of the blended polymers and that the rheological conditions in the blend components are changed due to the addition of the SCA and their reactions with the blend components. The low viscous SCA3 may even hamper the load transmission from the matrix to the dispersed phase in the shear field during melt mixing. Studies on the influence of the SCA on the rheological properties of the blends and their components are in progress.

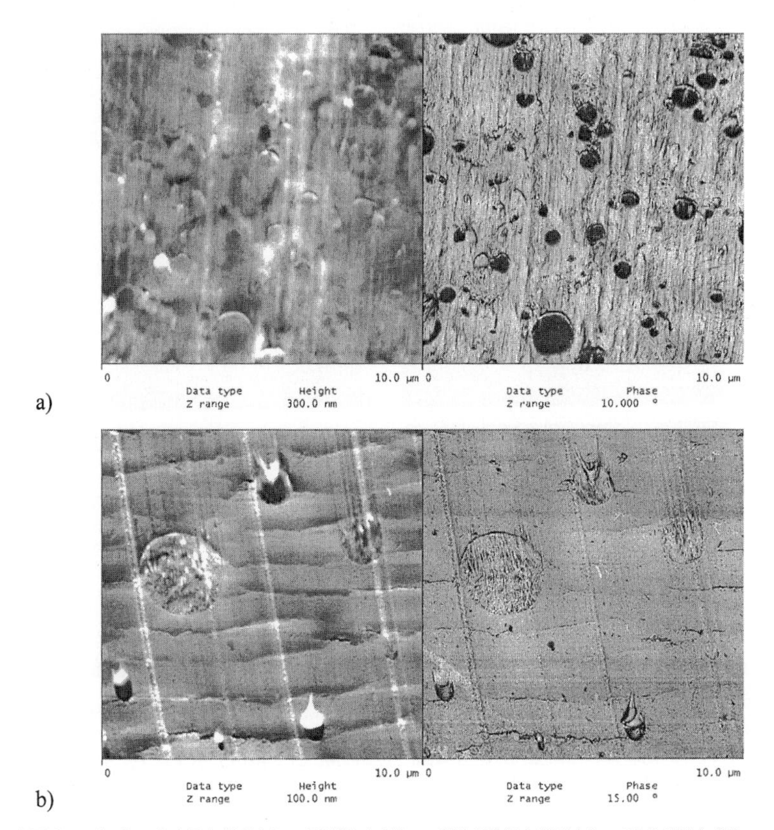

Fig. 2: AFM analysis of a) PA/SCA3 = 75/25 (v%) and b) PS/PA/SCA3 = 80/15/5 (v%)

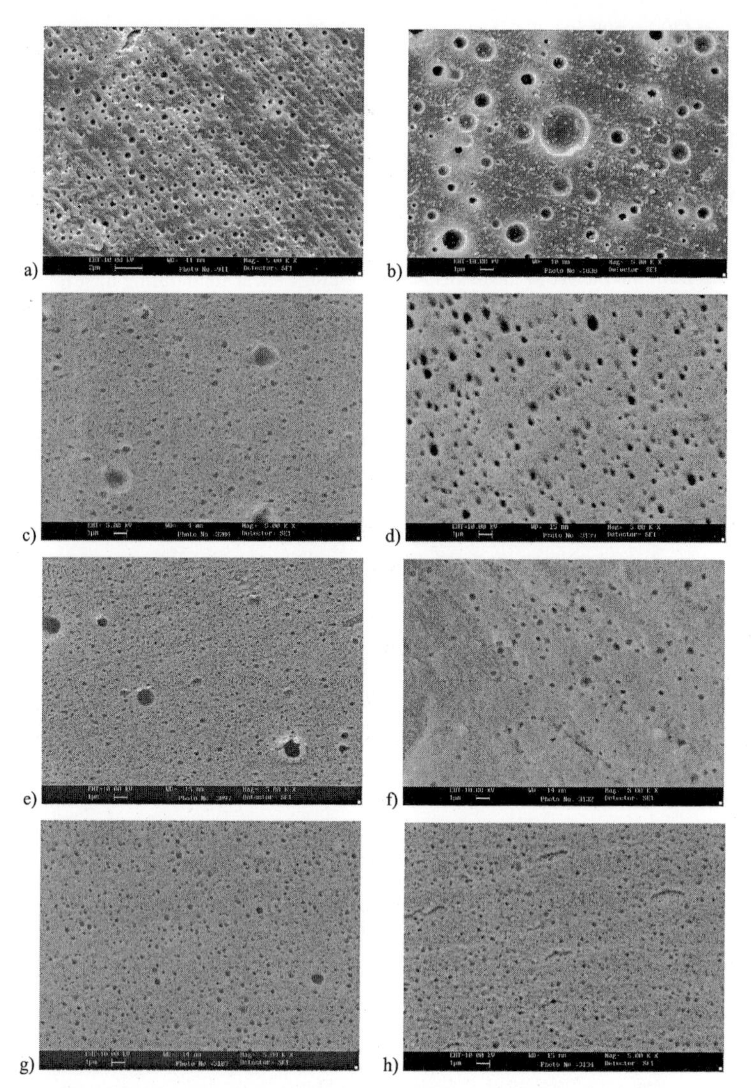

Fig. 3: SEM of PA/PS/SCA blends in dependence on the composition and processing conditions (frame size 24 × 17 μm; a) PA/PS = 84/16, etched cut; b) PA/PS/SCA3, one-step mixing; c) PA/PS/SCA3, two-step mixing; d) like b) with Pt; e) like c) with Pt; f) PA/PS/SCA1/SCA2 with Pt; g) PA/PS/SCA4, two-step mixing h) PA/PS/SCA4, two-step mixing with Pt).

When the SCA3 is premixed with the PA (which enables the reaction between components) and than PS is added the morphology is finer than in the one-step process (Fig. 3c), where the SCA3 locates favourably in the PS phase due to the better compatibility. The addition of Pt catalyst, which favours the crosslinking of the hydrosilane units according to Scheme 3, results in finer blend structures as well in the one-step process (Fig. 3d compared to Fig. 3b) as in the two-step process (Fig. 3e compared to Fig. 3c). The increased crosslinking density seems to favour a fine distribution of the PA phase. Even a mixture of PA/SCA1 with PS/SCA2 exhibits a rather fine morphology but only when Pt is added (Fig. 3f). The Pt-catalysed crosslinking enables the coupling between SCA1 and SCA2, which are chemically bound to the PA or PS, respectively. Thus the adhesion between both blend components is enhanced and a better dispersion during the melt mixing is possible. When the partially pre-crosslinked SCA4 was used a very fine distribution of the PA phase is observed (Fig. 3g), indicating similar good adhesion between both polymers. An increased crosslinking density due to the addition of Pt hardly influences the blend morphology (Fig. 3h). The pre-crosslinking and the additional crosslinking during processing seem to be sufficient for good interfacial adhesion.

$$-Si-H + H_2O \xrightarrow[\text{heat}]{\text{Pt catalyst}} -Si-OH + H_2$$

$$-Si-H + HO-Si- \xrightarrow[\text{heat}]{\text{Pt catalyst}} -Si-O-Si- + H_2$$

$$-Si-OH + HO-Si- \xrightarrow[\text{heat}]{\text{Pt catalyst}} -Si-O-Si- + H_2O$$

Scheme 3: Crosslinking reactions of polyhydrosilanes during processing

Summary and Conclusions

The use of polymeric hydrosilane containing coupling agents (SCA), which can prepared by hydrosilylation of allyloxy containing functional precursors with polyhydrosiloxanes, represents an attractive way for the modification of the interface in heterogeneous polymer blends, e.g. by crosslinking. We investigated the influence of the novel type of coupling agents on the morphology of reactive PS/PA blends. The processing conditions and the degree of polysiloxane crosslinking, which can be increased by the addition of Pt catalyst, strongly influence the dispersibility of the minor PA phase in the PS matrix. To achieve fine dispersions good adhesion between the phases is necessary. This can be reached by premixing the SCA with the complementary reactive polymers followed by blending the premixed components. Crosslinking of the SCA, which is in any cases at least partially located at the PS/PA interface, supports the load transfer from the matrix to the particles in the shear field during melt mixing, thus causing fine dispersions of the PA phase in the PS matrix.

The described mixing mechanism offers the opportunity to locate reactive sites at the interface of heterogeneous blends suitable for crosslinking and other chemical modification. To study the effects of such modifications will be a part of our future work.

Acknowledgements

The authors are grateful for the financial support by the Deutsche Forschungsgemeinschaft (Collaborative Research Centre 287). We would like to thank Degussa for providing the PA and Dr. H. Komber for performing the NMR-analysis.

[1] C. Z. Chuai, K. Almdal, J. Lyngaae-Jorgensen, *Polymer* **2003**, *44*, 481.
[2] N. C. Beck Tan, S. –K. Tai and R. M. Briber, *Polymer* **1996**, *37*, 3509.
[3] C. A. Orr, J. J. Cernohous, P. Guegan, A. Hirao, H. K. Jeon, C. W. Macosko, *Polymer* **2001**, *42*, 8171.
[4] Z. Yin, C. Koulic, C. Pagnoulle, R. Jerome, *Macromolecules* **2001**, *34*, 5132.
[5] B. Majumdar, H. Keskkula, D. R. Paul, G. Harvey, *Polymer* **1994**, *35*, 4263.
[6] P. Charoensirisomboon, T. Inoue, M. Weber, *Polymer* **2000**, *41*, 6907.
[7] P. Charoensirisomboon, T. Inoue, M. Weber, *Polymer* **2000**, *41*, 4483.
[8] K. Dedecker, G. Groeninckx, *Polymer* **1998**, *39*, 4993.
[9] J. Luston, F. Böhme, H. Komber, G. Pompe, *J. Macromol. Sci. Chem.*, **1998**, *A35*, 1045.
[10] L. Jakisch, H. Komber, F. Böhme, *J. Polym. Sci., Part A: Polym. Chem.* **2003**, *41*, 655.
[11] H. Baumert, R. Mülhaupt, *Macromol. Rapid. Commun.* **1997**, *18*, 787.
[12] H. Malz, H. Komber, D. Voigt, J. Pionteck, *Macromol. Chem. Phys.* **1998**, *199*, 583.
[13] S. N. Magonov, V. Elings, M.-H. Whangbo, *Surface Science* **1997**, *375*, L385.
[14] B. Marciniec, J. Gulinski, L. Kopylova, H. Maciejewski, M. Grundwald-Wyspianska, M. Lewandowski, *Appl. Organomet. Chem.* **1997**, *11*, 843.
[15] L. Jakisch, H. Komber, L. Häußler, F. Böhme, *Macromol. Symp.* **2000**, *149*, 237.

Sulfonated Multiblock Copoly(ether sulfone)s as Membrane Materials for Fuel Cell Applications

*Antje Taeger, Claus Vogel, Dieter Lehmann, Wolfgang Lenk, Kornelia Schlenstedt, Jochen Meier-Haack**

Institute of Polymer Research Dresden, P.O.Box 120411, 01005 Dresden, Germany
E-mail: mhaack@ipfdd.de

Summary: Arylene ether multiblock copolymers of the $(AB)_n$-type with various degrees of sulfonation have been prepared by a two-step polycondensation procedure. Multiblock copolymers in high yields and of high molecular weights were obtained. For comparison random copolymers with the same overall composition were synthesized. The theoretical ion-exchange capacities (IEC) of the materials were ranging from 0.50 mmol/g to 1.25 mmol/g. The water-uptake of the multiblock copolymers showed a linear dependency from the IEC and was increasing with increasing IEC. No differences were observed between random and block copolymers. Furthermore, the hydrolytic stability of aromatic sulfonic acid groups was investigated in this study. Aromatic sulfonic acids, having additional electron donating groups, especially in ortho- or para-position to the sulfonic acid group are sensitive to hydrolytic desulfonation. On the other hand electron-withdrawing groups in meta-position showed a stabilizing effect.

Keywords: ion-exchange membranes; multiblock copolymers; stability of aromatic sulfonic acids; sulfonated poly(ether sulfone)

Introduction

Recently, sulfonated polymers became of interest in processes requiring preferential transport of cations for example electrodialysis or in fuel cell applications. For the latter, perfluorosulfonic acid polymer membranes such as Nafion® (DuPont), Flemion® (Asahi Glass), DowMembrane® (Dow Chemical) are among the very few commercially available membranes for fuel cells due to their outstanding chemical properties and high proton conductivity. However, the Nafion®-type membranes have the disadvantages of (i) high costs of approx. 600 US$/m^2, (ii) proton conductivity reduction above 100°C and (iii) high water and methanol crossover, which is particularly disadvantageous for direct methanol fuel cells. These drawbacks have initiated worldwide research activities for developing new materials for proton exchange membranes, ranging from partially fluorinated to non-fluorinated fully aromatic materials mainly in order to reduce costs.

 DOI: 10.1002/masy.200450620

Overviews of recent developments in membrane research for fuel cells have been published for example by Savadogo [1], Rikukawa and Sanui [2], Costamagna and Srinivasan [3, 4] as well as in a special issue of *J. Membr. Sci.* [5 – 12].

Research activities have been focussed on sulfonated polyimides (PI), poly(ether ether ketone)s (PEEK), polysulfones (PSU), poly(ether sulfone)s (PES) and others. One approach to obtain sulfonated materials is the chemical modification of commercially available polymers by treatment with different types of sulfonation agents [13 – 17]. Less extensively explored is the preparation of an alternative material via the introduction of ionic groups onto the polymer backbone, using sulfonated monomers [18 - 24]. This method allows for a better control both of the sulfonation degree and the ionic group distribution. The synthesis of both random and block copoly(ether sulfone)s was reported by Wang et al.[18 - 20]. The preparation of polyimides using 4,4`-diamino-biphenyl-2,2`-disulfonic acid, 4,4`-oxydianiline and oxydiphthalic dianhydride or 1,4,5,8-naphthalene tetracarboxylic dianhydride was described for example by Faure et al. [21], Cornet et al. [22, 23] and Genies et al. [24].

Moreover, it is well known from organic chemistry textbooks [25] that the sulfonation of aromatic rings is a reversible process especially at low pH and at elevated temperature. However, this aspect has not been considered in the literature dealing with sulfonated polymers so far. Most often only the thermal stability is discussed.

Therefore, one aim of this study was to investigate the stability of sulfonated aromatic compounds. Model compounds, representing a repeating unit of the respective polymer chains were synthesized and treated at 135°C either in pure water or in 0.5N HCl. The samples were characterized by NMR spectroscopy before and after the treatment. The presence of sulfuric acid in the test samples was tested by titration with NaOH and precipitation of $BaSO_4$ by addition of $BaCl_2$ to the aqueous test solutions.

Secondly, $(AB)_n$ multiblock copolymers having non-sulfonated blocks (A-blocks) and highly sulfonated blocks (B-blocks) were prepared with the aim of providing polymeric materials with a phase-separated morphology. The phase of A-blocks should enable good mechanical stability and a reduction of the swelling of the highly sulfonated B-blocks whereas the B-blocks should provide a high proton conductivity. The properties of multiblock copolymers are compared with those of random copolymers of the same overall composition. All polymers are characterized in terms of water-uptake and ion exchange capacity.

Experimental

Materials

Bis-(4-fluorophenyl)sulfone (DFDPS) and chlorotrimethylsilane (TMSCl) were supplied by Aldrich. Hydroquinone sulfonic acid sodium salt (HQSA-Na) and 4,4`-dihydroxybiphenyl (DHBP) and N,N-dimethylacetamide (DMAc) were purchased from Acros. N-methyl-2-pyrrolidone (NMP) was purchased from Merck.

K_2CO_3 was dried in vacuum for 24 h at 150 °C. NMP was dried by subsequent distillation from P_4O_{10} and K_2CO_3 under reduced pressure. It was stored over molecular sieves (4 Å).

Silylation of Bisphenols

1 mol (186 g) of DHBP was suspended in a mixture of 500 ml dry toluene and 2.5 mol (244.5 g) of hexamethyl disilazane (HMDS). The reaction mixture was refluxed until the evolution of ammonia ceased (4 – 8 h). The toluene and excess of HMDS were removed using a rotary evaporator and the product was purified by distillation in vacuum (bis-TMS-DHBP: 140°C at $3x10^{-3}$ mbar, yield > 95%).

0.5 mol of HQSA-Na was suspended in 1.5 l dry THF. Per mol of functional groups 1.1 mol of an equimolar mixture of HMDS and chlorotrimethylsilane were added to the suspension. The mixture was refluxed for 96 h under exclusion of moisture. After cooling to room temperature precipitated ammonium hydrochloride and sodium chloride were filtered off and the THF was removed under reduced pressure using a rotary evaporator. The tris-silylated product was purified by distillation under reduced pressure (tris-TMS-HQSA: 146°C at $5.5x10^{-2}$ mbar). The yield was higher than 93%.

Synthesis of polymers

For the preparation of the poly(ether sulfone)s a three-necked round-bottom flask, equipped with a gas inlet tube, a magnetic stirrer, a still and a $CaCl_2$, was charged with equimolar amounts, typically 20 mmo,l of DFDPS and a mixture of silylated bisphenols with a predetermined ratio of tris-TMS-HQSA and bis-TMS-DHBP. The monomers were dissolved in dry NMP giving a 10 – 20 wt.-% polymer solution. After complete dissolution of the monomers, a double molar amout of K_2CO_3 was added to the solution and the temperature was raised to 150°C for 24 h. The reaction mixture was then filtered through a sintered-glass filter to remove unsoluble contents and the polymer was precipitated by pouring the reaction mixture into a 10-fold excess of cold ethanol. The precipitate was filtered off and washed thorougly with hot ethanol. The polymer was dried to constant weight at 80°C in vacuum.

In the case of block copolymers the A-block oligomers (hydroxy terminated) and B-block oligomers (fluoro terminated) were at first prepared separately. The multiblock copolymers were finally obtained by combining the reaction mixture of the A- and B-block and the reaction was completed by stirring the combined mixtures at 150°C overnight.

Membrane preparation

Membranes were prepared by casting a polymer solution containing approx. 20 wt.-% polymer in DMAc on a glass plate. The film thickness was set to 700 μm. After removal of the solvent at 80°C under normal pressure for 8 h and 16 h at 80°C in vacuum, the membranes were peeled off by immersing the glass plate in water. Finally the membranes were dried to constant weight at 80°C in vacuum. The resulting membranes had a thickness of approximately 80 μm.

Measurements

Prior to characterization all polymers were dried at 120°C in vacuum to constant weight. The viscosities were measured with an automated Ubbelohde viscosimeter thermostated at 25°C. The polymer concentration was 2 g/l in DMAc. ^{1}H- and ^{13}C-NMR experiments were carried out on a Bruker AMX 500 spectrometer (500 MHz for ^{1}H) in 5 mm o.d. sample tubes using deuterated DMSO as solvent and chemical shift reference. GPC measurements were performed on a Knauer GPC equipped with two Zorbax PSM Trimodal S columns and a RI detector. A mixture of DMAc with 2 vol.-% water and 3 g/l LiCl was used as eluent. Poly(vinylpyrrolidone) samples were used as standards for molecular weight calibration.

Water-uptake

A piece of dry membrane was soaked in water for 24 h at 80°C. The excess water is wiped off gently with a tissue and the membrane is weighed immediately. Subsequently, the membranes were dried at 80°C in vacuum to constant weight. The water-uptake is calculated using equation (1) and is given in percentage increase in weight

$$W = \frac{m_{wet} - m_{dry}}{m_{dry}} x100\% \quad (1)$$

where m_{wet} and m_{dry} are the weights of the water-swollen and dry membrane sample, respectively.

Ion-exchange capacity

The ion-exchange capacity (IEC, with units mmol/g of dry polymer) of the sulfonated membranes was measured using the standard experimental method of immersing the membrane in 1N HCl for 24 h, followed by soaking in distilled water to remove excess acid for another 24 h. Finally, the membrane samples are soaked for 1 day in 2 M NaCl solution (exchange of H^+ by Na^+ within the film) and then titrating the solution with 0.01N NaOH to determine the concentration of the exchanged protons.

Results and Discussion

Stability of aromatic sulfonic acids

Despite the fact, that the sulfonation of aromatic rings is reversible at elevated temperatures and at acidic conditions, most often only the thermal properties of proton-conducting membranes in an inert atmosphere are discussed in the literature. In this study sulfonated model compounds as well as polymer samples were used to test the hydrolytic stability in water and dilute acid (0.5N HCl) at 135°C for 168 h. The sulfuric acid formed during hydrolysis was determined quantitatively by titration of the test solution with NaOH and qualitatively by precipitation of $BaSO_4$ after the experiment.
As a result, nearly all compounds having electron-donating substituents at the sulfonic acid bearing ring are sensitive to desulfonation under the chosen test conditions. On the other hand, compounds having electron-withdrawing substituents were stable in most cases (Fig 1).
Furthermore, not only desulfonation was detected under test condition but also cleavage of imide and amide bonds leading to a reduction of the molecular weight of the polymers and the loss of mechanical stability. A more detailed discussion of the results will be published in a forthcoming paper.

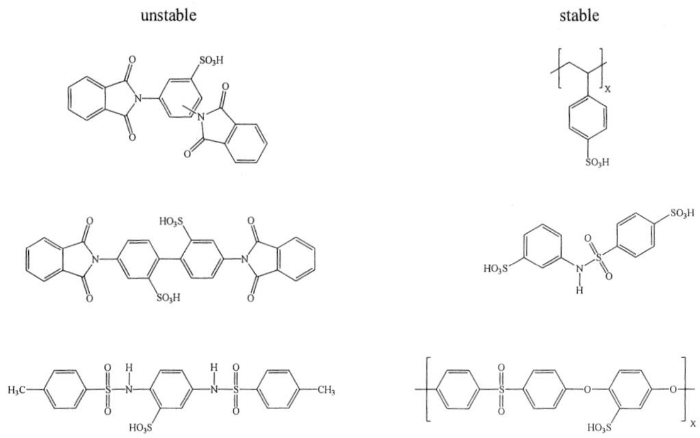

Figure 1. Examples for unstable and stable compounds under test conditions

Properties of sulfonated poly(ether sulfone)s

The so-called „silyl-method“ was employed to prepare poly(ether sulfone)s from difluoro compounds and trimethylsilyl derivatives of bisphenols. The general chemical structure of the synthesized poly(ether sulfone)s is given in Figure 2. The silyl-method has the advantage that the silylated bisphenols, especially hydroquinone sulfonic acid are much better soluble in dipolar aprotic solvents such as NMP than their non-silylated counterparts. Secondly, water and hydrohalic acid as reaction products are avoided, thus minimizing the risk of side reactions. Thirdly, the combination of silylation and distillation is an effective method for the purification of many monomers.

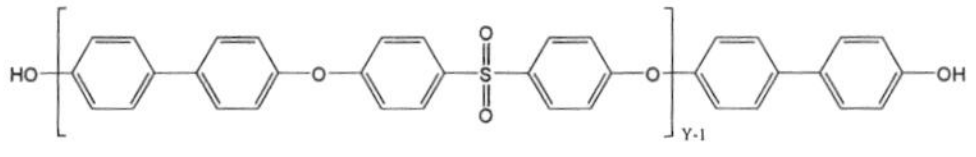

A)

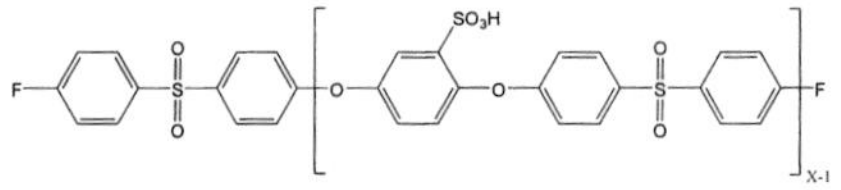

B)

Figure 2. Chemical structure of A) A-block and B) B-block with X and Y ranging between 1 and 21 in the multiblock copolymers

As outlined in Table 1, the poly(ether sulfone)s were obtained in high yields and with molecular weights sufficient for the preparation of membranes. The number of blocks in the multiblock copolymers varies from 2 to 8 depending on the molecular weight of the individual block building oligomers. In general it can be said that the higher the molecular weight of the oligomers the lower the number of blocks in the final block copolymer.

Table 1. Yields and molecular weights of poly(ether sulfone)s.

Sample[1)]	Yield (%)	$\eta_{inh.}$ [2)] (dl/g)	M_w [3)] (g/mol)	M_n (g/mol)	M_w/M_n	number of blocks
HPA	> 95	1.02	141850	36050	3.94	-
HPB	> 95	0.66	52750	14750	3.58	-
MBC 5/5	> 95	0.81	47450	12550	3.78	3/3
MBC 10/5	> 95	0.57	44000	20700	2.13	4/4
MBC 15/5	> 95	0.52	53700	24900	2.16	3/3
MBC 20/5	> 95	0.43	28650	9950	2.88	1/1
MBC 10/10	> 95	1.02	65360	18150	3.60	2/2
MBC 20/20	> 95	0.88	68250	16050	4.25	1/1
RC 1/1	> 95	1.18	63300	15900	3.98	-
RC 2/1	> 95	0.89	61300	18700	3.28	-
RC 3/1	> 95	0.46	35450	11550	3.07	-
RC 4/1	> 95	0.64	51350	16250	3.16	-

1) HPA = non-sulfonated homopolymer; HPB = sulfonated homopolymer; MBC = multiblock copolymer with numbers indicating the respective block length (A/B); RC = random copolymer with numbers indicating the ratio of non-sulfonated to sulfonated monomers

2) determined in DMAc at 25°C; c = 2g/l; capillary 0a

3) measured in DMAc containing 2 vol.-% water and 3 g/l LiCl, poly(vinylpyrrolidone) standards were used for calibration

Table 2 shows the water-uptake and ion-exchange capacities for poly(ether sulfone) membranes. The ion exchange capacities, determined by NMR spectroscopy are in good agreement with the values calculated from the initial monomer composition. However, using the titration method, the values are somewhat lower, which is explained by the fact that only a part of the sulfonic acids groups is accessible to the sodium ions.

On the other hand, with the NMR spectroscopy all sulfonic acid groups are detected, regardless of whether they play an active role in the ion exchange process or not. Therefore, the IEC determined by the titration method gives a more realistic value than the NMR spectroscopy concerning the behavior of the membrane in the fuel cells.

The water-uptake shows a linear dependency from the ion exchange capacity of the membranes. It is in the same range as observed for the Nafion membranes. Unexpectedly, no difference between the water-uptake of random and multiblock copolymers was observed in contrast to sulfonated polyaramide membranes [27]. From studies following the polymerization process of random sulfonated copoly(ether sulfone)s, it is known that during the preparation of random copolymers a block like structure is obtained due to differences in the reactivity of the monomers [28]. Therefore it is likely, that the morphology of the membranes prepared from random or block copolymers is very similar resulting in a similar behavior of the membranes.

Table 2. Ion exchange capacity and water uptake of poly(ether sulfone) membranes.

Sample[1)]	Water/uptake (%)	H_2O/SO_3H mol/mol	Ion exchange capacity (mmol/g) calculated from monomer comp.	titration	NMR
HPA	1.4	-	0	-	-
HPB	n.d.[1)]	-	2.48	-	-
MBC 5/5	34.4	19	1.24	0.83	0.99
MBC 10/5	21.7	14	0.83	0.67	0.86
MBC 15/5	12.6	12	0.62	0.46	0.59
MBC 20/5	9.5	9	0.51	0.35	0.59
MBC 10/10	28.3	14	1.24	0.84	1.14
MBC 20/20	34.8	16	1.24	0.90	1.19
RC 1/1	33.2	16	1.24	1.15	1.14
RC 2/1	16.9	12	0.83	0.85	0.81
RC 3/1	11.8	11	0.62	0.45	0.61
RC 4/1	8.4	10	0.50	0.08	0.48
Nafion 117	28.9	21	0.91[2)]	0.77	-

[1)] not determined due to solubility in water
[2)] reference [26]

The diffusion coefficients for methanol and protons in sulfonated poly(ether sulfone) membranes together with the values for a Nafion membrane are displayed in Figure 3. The procedures for the measurement and the calculation of the diffusion coefficients were adopted from the literature [29]. As expected, the diffusion coefficients are increasing with increasing IEC, due to higher hydrophilicity resulting in an enhanced swelling of the membranes (Table 2).

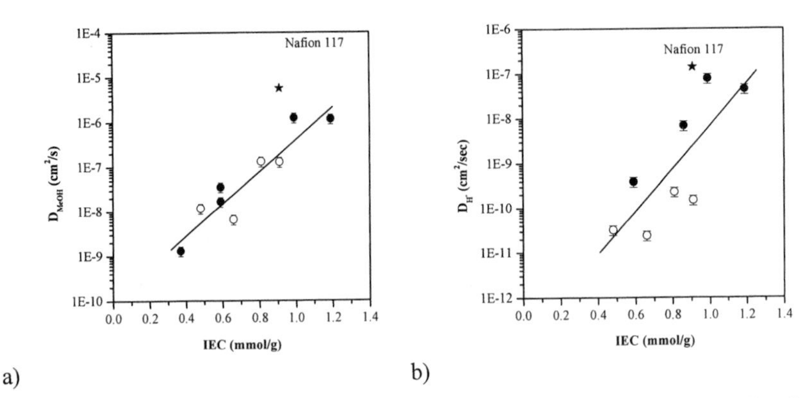

a) b)

Figure 3. Methanol (a) and proton (b) diffusion coefficients in sulfonated poly(ether sulfone) membranes; ● = MBC, O = RC; T = 50°C; c_{MeOH} = 20 wt.-%; c_{HCl} = 0,5 mol/l.

At comparable IEC the methanol diffusion coefficients for poly(ether sulfone) membranes are one to two orders of magnitude lower than that of the Nafion membrane, indicating improved barrier properties for methanol (lower methanol crossover) of poly(ether sulfone) membranes. Still for most membranes the proton diffusion coefficients are also lower than that of the Nafion membrane.

The findings for the proton diffusion coefficients are reflecting in the results of conductivity measurements (Fig. 4a). The conductivities of sulfonated poly(ether sulfone) membranes are approximately by a factor of 2 lower than that of a Nafion membrane. However, the fuel cell performance of the introduced poly(ether sulfone) membranes is quite similar to that of Nafion, except for the BC 20/20 membrane (Fig. 4b). For this membrane a high methanol crossover was observed, which was in the same range as for the Nafion sample. All other membranes showed a lower methanol permeation during the fuel cell tests, as was expected from the diffusion coefficients measurements.

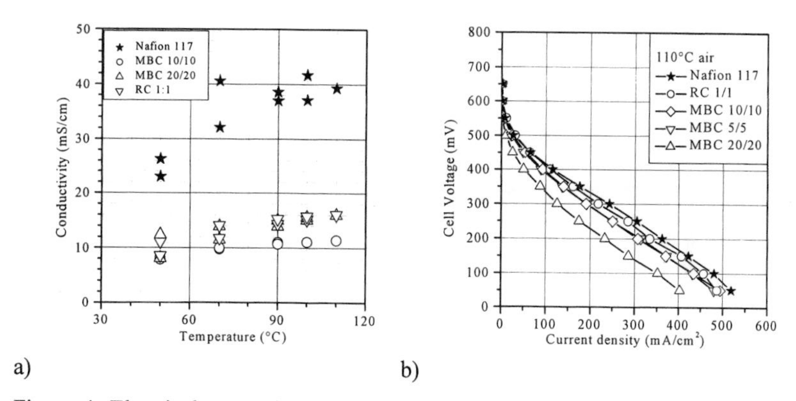

Figure 4. Electrical properties of poly(ether sulfone) membranes compared to Nafion 117 membrane; a) conductivity at 100% relative humidity; b) polarization curves from DMFC tests at 110°C (1.5 mol/l MeOH; 2.5 bar air)

Conclusions

Aromatic sulfonic acid groups are susceptible to hydrolytic degradation, especially at elevated temperatures, acidic pH and with additional electron donating substituents on the ring. Such substituents facilitate the electrophilic attack of the proton, leading to desulfonation.

Sulfonated arylene ether sulfone block copolymers as well as random copolymers with predetermined IECs were successfully synthesized by the "silyl-method". Membranes with an IEC ≥ 1 mmol/g showed a fuel cell performance slightly less than Nafion but with lower methanol crossover.

Acknowledgements

The authors gratefully acknowledge the financial support of this work by the HGF Strategiefond project "Membranes and Electrodes for Direct Methanol Fuel Cells". Furthermore we like to thank S. P. Nunes (GKSS), R. Reißner and E. Gülzow (DLR) for performing the membrane tests.

[1] O. Savadogo, J. New Mat. Electrochem. Systems **1998**, *1*, 47.
[2] M. Rikukawa K. Sanui, Progr. Polym. Sci. **2000**, *25*, 1463.
[3] P. Costamagna, S. Srinivasan, J. Power Sources **2001**, *102*, 242.
[4] P. Costamagna, S. Srinivasan, J. Power Sources **2001**, *102*, 253.
[5] G. Alberti, M. Casciola, L. Massinelli, B. Bauer, J. Membr. Sci. **2001**, *185*, 73.
[6] P. Genova-Dimitrova, B. Baradie, D. Foscallo, C. Poinsignon, J. Y. Sanchez, J. Membr. Sci. **2001**, *185*, 59.
[7] S. Haufe, U. Stimming, J. Membr. Sci. **2001**, *185*, 95.
[8] I. Honma, S. Nomura, H. Nakajima, J. Membr. Sci. **2001**, *185*, 83.
[9] D.J. Jones, J. Rozière, J. Membr. Sci. **2001**, *185*, 41.
[10] J. Kerres, J. Membr. Sci. **2001**, *185*, 3.
[11] K.D. Kreuer, J. Membr. Sci. **2001**, *185*, 29.
[12] F. Finsterwalder, G. Hambitzer, J. Membr. Sci. **2001**, *185*, 105.
[13] H. Tang, P. N. Pintauro, Q. Guo, S. O`Connor, J. Appl. Polym. Sci. **1999**, *71*, 387.
[14] H. Tang, P. N. Pintauro, J. Appl. Polym. Sci. **2001**, *77*, 49.
[15] B. Lafitte, L. E. Karlsson, P. Jannasch, Macromol. Rapid Commun. **2002**, *23*, 896.
[16] A. Dyck, D. Fritsch, S. P. Nunes, J. Appl. Polym. Sci. **2002**, *86*, 2820.
[17] M.-S. Kang, Y.-J. Choi, I.-J. Choi, T.-H. Yoon, S.-H. Moon, J. Membr. Sci. **2003**, *216*, 39.
[18] F. Wang, Y. Kim, M. Hickner, T. A. Zawodzinski, J. E. McGrath, Polymeric Materials: Science & Engineering **2001**, *85*, 517.
[19] F. Wang, Q. Ji, W. Harrison, J. Mecham, R. Formato, R. Kovar, P. Osenar J. E. McGrath, Polymer Preprints **2000**, *41*, 237.
[20] W. Harrison, F. Wang, J. Mecham, T. E. Glass, M. Hickner, J. E. McGrath, Polymeric Materials: Science & Engineering **2001**, *84*, 688.
[21] French Patent 96/05707 (1996) S. Faure, P. Aldebert, R. Mercier, B. Sillion.
[22] N. Cornet, O. Diat, G. Gebel, F. Jousse, D. Marsacq, R. Mercier, M. Pineri, J. New Mat. Electrochem. Systems **2000**, *3*, 33.
[23] N. Cornet, G. Beaudoing, G. Gebel, Sep. Pur. Techn. **2001**, *22-23*, 681.
[24] C. Genies, R. Mercier, B. Sillion, N. Cornet, G. Gebel, M. Pineri, Polymer **2001**, *42*, 359.
[25] Organikum, 21st Edition, Wiley VCH, Weinheim 2001.
[26] F. Lufrano, G. Squadrito, A. Patti, E. Passalacqua, J. App. Polym. Sci. **2000**, *77*, 1250.
[27] A. Taeger, C. Vogel, D. Lehmann, D. Jehnichen, H. Komber, J. Meier-Haack, N.A. Ochoa, S.P. Nunes, K.-V. Peinemann, Reactive and Functional Polymers **2003**, *57*, 77.
[28] A. Taeger unpublished results
[29] M. V. Fedkin, X. Zhou, M. A. Hofmann, E. Chalkova, J. A. Weston, H. R. Allcock, S. N. Lvov, Mat. Lett. **2002**, *52*, 192.

Macromol. Symp. **2004**, *210*, 185-192

Nonstoichiometric Interpolyelectrolyte Complexes as Colloidal Dispersions Based on NaPAMPS and Their Interaction with Colloidal Silica Particles

Simona Schwarz,[1] *Stela Dragan**[2]

[1]Institute of Polymer Research, Hohe Strasse 6, 0 1069 Dresden, Germany
[2]"Petru Poni" Institute of Macromolecular Chemistry, Aleea Grigore Ghica Voda 41 A, 700487 Iasi, Romania

Summary: Preparation and characterization of some nonstoichiometric interpolyelectrolyte complexes (NIPECs) as stable colloidal dispersions by the interaction between poly(sodium 2-acrylamido-2-methylpropanesulfonate) (NaPAMPS) and three strong polycations bearing quaternary ammonium salt centres in the backbone, poly(diallyldimethylammonium chloride) (PDADMAC) and two polycations containing N,N-dimethyl-2-hydroxypropyleneammonium chloride units (PCA_5 and PCA_5D_1), have been followed in this study as a function of the polycation structure and polyelectrolyte concentration. Complex characteristics were followed by polyelectrolyte titration, turbidity and quasi-ellastic light scattering. Almost monodisperse NIPECs nanoparticles with a good storage stability were prepared when total concentration of polyelectrolyte was varied in the range 0.85-6.35 mmol/L, at a ratio between charges (n^-/n^+) of 0.7. NIPECs as a new kind of flocculants were used to flocculate a stable monodisperse silica suspension. The main advantage of NIPECs as flocculants is the broad flocculation window, which is a very important aspect for industrial applications.

Keywords: dispersions; flocculation; nonstoichiometric interpolyelectrolyte complexes; particle size; polyelectrolytes; silica

Introduction

Nonstoichiometric interpolyelectrolyte complexes (NIPECs), either soluble[1-12] or stable colloidal systems[13-22] have been obtained up to date, by the interaction between oppositely charged polyelectrolytes, depending on different factors such as: nature of the ionic groups, polymerization degree, flexibility, charge density, and the concentration of the opposite polyelectrolytes, molar ratio between opposite charges, properties of the reaction medium (pH, temperature, ionic strength, dielectric constant). NIPECs as stable colloidal dispersions bearing positive or negative charges in excess proved to be of a great interest for some main practical areas such as flocculants for

 DOI: 10.1002/masy.200450621

cellulose and clay dispersions, and organic compounds (dyes and surfactants) from the waste waters as well as for surface modification of different solid substrates.[13-18,20,22] The surface charge and sizes of the IPECs nanoparticles could be influenced by the structure of opposite polyions, polyions mixing ratio, polyions concentration, to mention a few among different parameters which can be varied. Formation and properties of the NIPECs obtained by the interaction between strong oppositely charged polyelectrolytes have been less investigated up to date.[23] Therefore, the aim of this study was to obtain first NIPECs as stable colloidal dispersions employing poly(sodium 2-acrylamido-2-methylpropanesulfonate) NaPAMPS as a strong polyanion and three strong polycations containing quaternary ammonium salt centres in the backbone, poly(diallyldimethylammonium chloride) PDADMAC and two polycations containing N,N-dimethyl-2-hydroxypropyleneammonium chloride units and different hydrophobic substituents (PCA_5 and PCA_5D_1), and then to use these reactive colloidal dispersions in the separation of silica nanoparticles comparative with polycations alone.

Experimental

Materials

PDADMAC[27] with a molar mass of 240000 g/mol, purchased from Aldrich, has been used as received. Polycations PCA_5 and PCA_5D_1 were synthesized and purified as it was presented before.[28] The intrinsic viscosity in 1 M NaCl at 25 °C of the samples employed in this study were as follows: 0.46 dL/g and 0.41 dL/g, respectively.

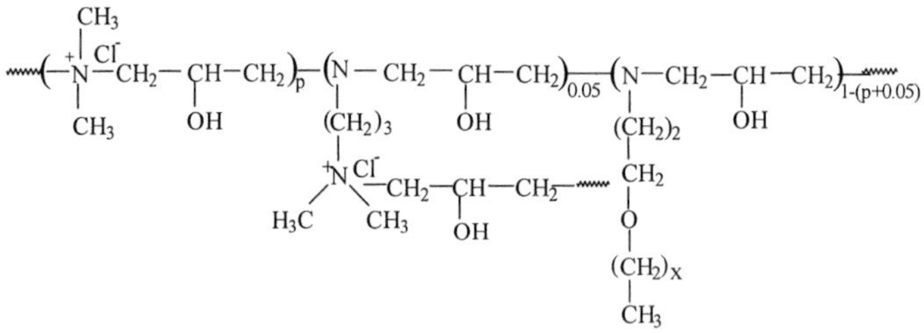

$$\mathrm{-\!\!\left(N^{+}(CH_3)_2\,Cl^- - CH_2 - CH(OH) - CH_2\right)_p\!\!-\!\left(N - CH_2 - CH(OH) - CH_2\right)_{0.05}\!-\!\left(N - CH_2 - CH(OH) - CH_2\right)_{1-(p+0.05)}\!-}$$

$$\mathrm{(CH_2)_3 - N^{+}Cl^-(CH_3)(H_3C) - CH_2 - CH(OH) - CH_2 -;\quad (CH_2)_2 - CH_2 - O - (CH_2)_x - CH_3}$$

p = 0.95, PCA_5; p = 0.94, x=9, PCA_5D_1

Scheme 1

NaPAMPS, synthesized and purified according to the method previously reported,[29] has been used as a strong polyanion, with a molar mass of 170000 g/mol, viscometrically determined.[30] Polyions characteristics were summarized in Table 1.

Table 1. Characteristics of opposite polyions employed in the IPECs preparation.

Sample	M_v, g/mol	M_u, g/charge	$B^{1)}$, nm
NaPAMPS	170000	229	0.25
PDADMAC	240000	162.7	0.5
PCA_5	-	140.35	0.57
PCA_5D_1	-	141.68	$0.57^{2)}$

$^{1)}$ distance between charges;
$^{2)}$ it was assumed to be the same like in PCA_5.

IPECs preparation

Aqueous solutions of the complementary polyelectrolytes with different concentrations were prepared by adequate dilution of the stock solutions (10 mM) with water deionized and purified by a Milli-Q system (Millipore, Eschborn, Germany). Variable volumes of the aqueous solution of the titrant (always NaPAMPS), were slowly dropped into the aqueous solution of polycation, under magnetic stirring, at room temperature (about 25 ^{o}C), until a certain ratio between opposite charges was achieved.[31] The IPECs dispersions were still stirred for 2 h, and were characterized after 24 h of storage, if other conditions were not mentioned.

Complex characterization

The NIPECs properties as a function of the polycation structure were followed first by the optical density of the IPECs nanoparticles at λ = 500 nm (OD_{500}). OD_{500} measurements were carried out with a Lambda 900 UV/Vis spectrometer (Perkin Elmer, UK). Quantitative determination of the polyelectrolyte in the starting solutions and the detection of free charges in the NIPECs dispersions were performed with a particle charge detector PCD 02, Müteck, Germany using either poly(sodium ethylenesulfonate) (PES) or PDADMAC with a concentration of 10^{-3} M, in dependence on the nature of the charges in excess. Quasi-elastic light scattering measurements were performed with a Zetasizer 3000 (Malvern Instruments Ltd, UK) at a scattering angle of 90°. The operating wavelength was 633 nm. Z-averaged particle sizes (R_h) and polydispersity index of the nanoparticles were evaluated by this method.

Flocculation test

Model suspensions of silica (Geltech Inc.) with a particle size of 200 nm were used for the flocculation tests. The flocculation studies were carried out with a conventional test equipment for the characterization of the flocculation and sedimentation process under static conditions. With a pivoted frame, the stirring conditions can be hold the same for the six graduated glasses of this equipment. After the addition of the appropriate amount of polycation solution or NIPEC-dispersions, followed by stirring and 20 minutes sedimentation, the supernatant fraction was separated and then, the efficiency of flocculation was determined by measuring the OD_{500} (Lambda 900, Perkin Elmer, UK) and the height of the precipitated solid fraction.

Results and Discussion

Information on the influence of polycation structure, polyanion molar mass and polyion concentration on the formation and characteristics of the complex nanoparticles based on NaPAMPS, at different ratios (n^-/n^+: 0.5 – 1.4) between opposite charges, have been already presented in detail elsewhere.[31] The main conclusions resulted from that study were used in the present work to prepare reactive NIPECs nanoparticles with positive charges in excess. Thus, OD_{500} values as a function of the ratio between opposite charges for the complex formation between NaPAMPS as a titrant and different polycations as starting solutions showed that the IPECs dispersions had a low turbidity before the complex stoichiometry, correlated with the abscisa values of the rising curves corresponding to the one-half of the turbidity maximum, and reflects a small influence of the polycation structure, OD_{500} values being a little lower for the complex formed with PDADMAC than those prepared with PCA_5 and PCA_5D_1 as polycations. Close to the isoelectric point (iEP), the OD_{500} values strongly increased, mainly because the sizes of the complex particles increased. Polyion concentrations strongly influenced nanoparticle sizes, R_h, both before and after the complex stoichiometry.[31] NaPAMPS with a molar mass of 170000 g/mol has been selected to prepare NIPECs dispersions with different concentrations in polyelectrolytes because, at this molar mass, polyanion concentration could be varied in a large range.

A very important aspect related to the possible applications of the NIPECs as reactive nanoparticles is their stability in dependence on the polyelectrolyte structure and the total concentration of the polyelectrolytes. NIPECs nanoparticle characteristics prepared with NaPAMPS as added polyion, at a constant ratio between charges, $n^-/n^+ = 0.7$, in dependence on the polycation structure and the total concentration of polylectrolyte, were measured after 24 h and 48 h, the values being collected in Table 2.

Table 2. Characteristics and stability of the nonstoichiometric polyelectrolyte complexes based on NaPAMPS.

Polycation	C_{PE}, mM	OD_{500} 24 h	OD_{500} 48 h	Rh, nm 24 h	Rh, nm 48 h
PDADMAC	1.59	0.0505	0.047	58.4	58.5
	3.18	0.186	0.191	76.9	76.25
	4.77	0.4494	0.449	89.2	88.55
	6.36	0.888	0.898	99.55	99.5
PCA_5D_1	3.18	0.1664	0.1705	75.45	73.5
	4.77	0.4363	0.4365	90.75	90.75
	6.36	0.6571	0.692	93.55	95.05
	1.59	0.048	0.049	61.45	60.07
PCA_5	3.18	0.1695	0.1676	76.35	76.05
	4.77	0.441	0.441	91.9	91.05
	6.36	0.759	0.759	98.45	99.25

It is well known that when polyelectrolytes are used as flocculants in waste water purification, the optimum flocculation concentration domain is usually very narrow and restabilization of the refractory pollutants can easily take place by the overdosing of the polyelectrolyte. The NIPECs colloidal dispersions synthesized up to date showed a larger optimum flocculation concentration, being considered as new reactive nanoparticles.[17,18]

Flocculation efficiency of the NIPECs dispersions synthesized in this paper has been tested, comparative with the starting polycations, in the flocculation of silica monodisperse stable nanoparticles (Figures 1, 2, 3). As one can observe from Figures 1, 2 and 3, at concentrations higher than those corresponding to the optimum concentration for flocculation, redispersion took place very quickly, when polycations where used alone. PDADMAC and PCA_5 alone show a very narrow range of flocculation and the residual turbidity corresponding to the optimum flocculation concentration remained about 0.4 g/mg for PDADMAC. PCA_5D_1, containing 1 mol % hydrophobic side chains, shows a broader flocculation range with a low residual turbidity. For industrial applications it is very important to have a broad flocculation window because the main parameters of the waste waters (solid content, ionic strength, pH, presence of surfactants, which act as stabilisators for the fine suspensions) could change in very large limits in one day.

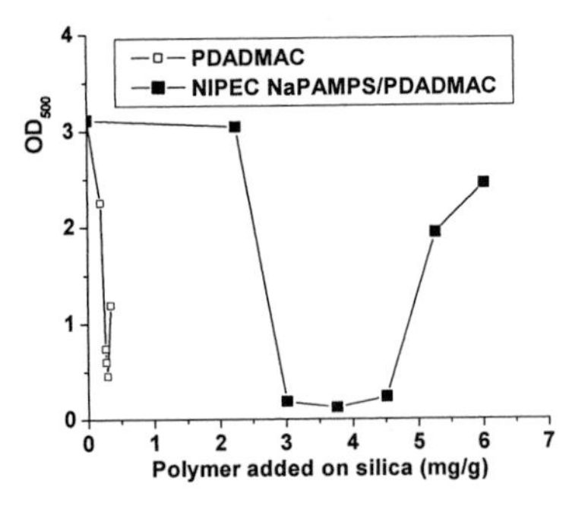

Fig. 1. Flocculation test of silica nanoparticles with PDADMAC comparative with NIPEC prepared with NaPAMPS added on PDADMAC.

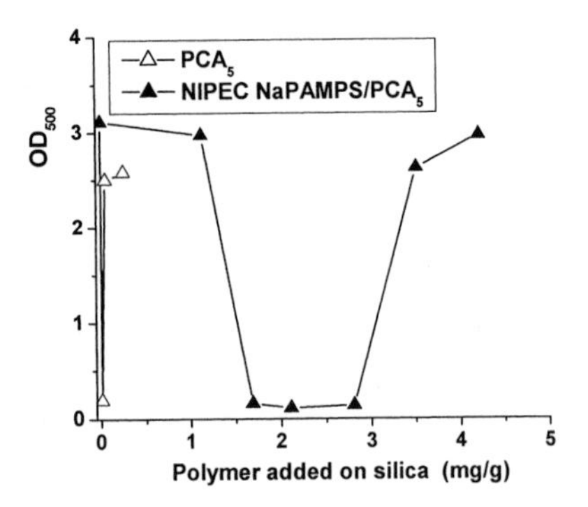

Fig. 2. Flocculation test of silica nanoparticles with PCA_5 comparative with NIPEC prepared with NaPAMPS added on PDADMAC.

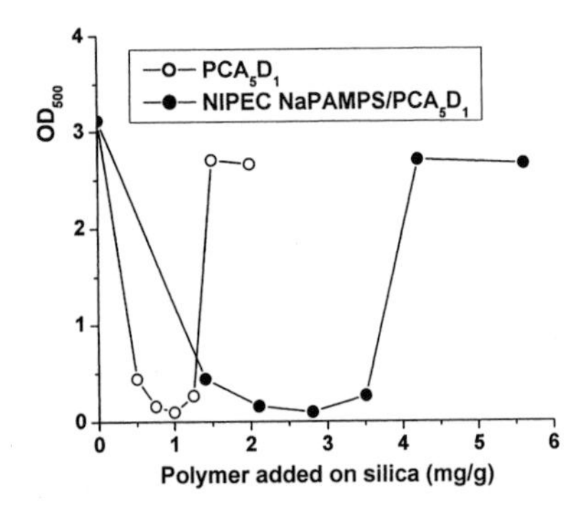

Fig. 3. Flocculation test of silica nanoparticles with PCA_5D_1 comparative with NIPEC prepared with NaPAMPS added on PCA_5D_1.

The main difference showed by the NIPECs dispersions used as flocculants compared with pure polycations was the much larger window of flocculation. The amount of polyelectrolytes corresponding to the optimum flocculation concentrations is higher because at a molar ratio of 0.7 the excess of positive charges is not so high. On the other hand, the residual turbidity is much lower than in the case of the polycation alone since the hydrophobic parts from the complex itself have also a contribution.

The influence of the polycation structure was clearly observed in the flocculation process of silica. Thus, the window of the optimum flocculation concentration was broader in the case of PCA_5D_1 because of the higher hydrophobicity induced by the presence of 1 mol % of decyloxypropylamine side chains. The largest range concentrations where the flocculation took place were found in the case of the polycation PCA_5D_1, both for pure polycation and for the NIPEC prepared on its basis. These results show that the presence of a small content of hydrophobic side chains has a positive influence on the flocculation window. Concerning the values of the optimum concentration, the higher values found in the case of NIPECs are explained by the lower amount of charges, which result from the difference between the ratio n^-/n^+ corresponding to the iEP and the ratio n^-/n^+ of 0.7 selected for the NIPECs preparation. But, even if the amount of flocculant is very important, NIPECs can be used as specialized flocculants for wastes containing nanoparticles with low charge such as disperse dyes from the textile factories or slurries from microelectronics, which could be considered as "refractory pollutants". Such solid-liquid separation processes are not well solved by the comercial polycations up to date.

Acknowledgements

Authors gratefully acknowledge the DAAD for funding this research and the contribution of Mr. Bernd Keßler to the flocculation measurements.

[1] E. Tsuchida, Y. Osada, K. Sanada, *J. Polym. Sci.: Polym. Chem. Ed.* **1972**,*10*, 3397.
[2] V.A. Kabanov, A.B. Zezin, *Pure Appl. Chem.* **1984**, *56*, 343.
[3] V.A. Kabanov, *Macromol. Chem. Macromol. Symp.* **1986**, *1*, 101.
[4] V.A. Kabanov, A.V. Kabanov, *Macromol. Symp.* **1995**, *98*, 601.
[5] J.F. Gohy, S.K. Varshney, S. Antoun, R. Jérôme, *Macromolecules* **2000**, *33*, 9298.
[6] H.Dautzenberg, , W. Jaeger, *Macromol. Chem. Phys.* **2002**, 203, 2095.
[7] A.N. Zelikin, D. Putman, P. Shastri, R. Langer, V.A. Izumrudov, *Bioconjugate Chem.* **2002**, *13*, 548.
[8] A.N. Zelikin, V.A. Izumrudov, *Macromol. Biosci.* **2002**, *2*, 78.
[9] H. Dautzenberg, *Macromol. Chem. Phys.* **2000**, *201*, 1765.
[10] H. Dautzenberg, A. Zintchenko, C. Konak, T. Reschel, V. Subr, K. Ulbrich, *Langmuir*, **2001**, *17*, 3096.
[11] A. Zintchenko, H. Dautzenberg, K. Tauer, V. Khrenov, *Langmuir* **2002**, *18*, 1386.
[12] A.N. Zelikin, N.I. Akritskaya, V.A. Izumrudov, *Macromol. Chem. Phys.* **2001**, *202*, 3018.
[15] G. Kramer, H.-M. Buchhammer, K. Lunkwitz, *Colloids Surfaces A: Physicochem. Eng. Aspects* **1997**, *122*, 1.
[16] G. Petzold, S. Schwarz, H.-M. Buchhammer, K. Lunkwitz, *Angew. Makromol. Chem.* **1997**, 253, 1.
[17] G. Petzold, A. Nebel, H.-M. Buchhammer, K. Lunkwitz, *Colloid Polym. Sci.* **1998**, *276*, 125.
[18] H.-M. Buchhammer, G. Petzold, K. Lunkwitz, *Langmuir* **1999**, *15*, 4306.
[19] S. Dragan, S. Schwarz, H.-M. Buchhammer, Polymeric Materials, Halle/Saale, September 25-27, **2002**, Proceedings, p. 338.
[20] L. Gärdlundm, L. Wågberg, R. Gernandt, *Colloids Surfaces A: Physicochem. Eng. Aspects* **2003**, *218*, 137.
[21] H.-M. Buchhammer, M. Mende, M. Oelmann, *Colloids Surfaces A: Physicochem. Eng. Aspects* **2003**, *218*, 151.
[22] R.J. Nyström, J.B. Rosenholm, K. Nurmi, *Langmuir* **2003**, *19*, 3981.
[23] J. Chen, J.A. Heitmann, M.A. Hubbe, *Colloids Surfaces A: Physicochem. Eng. Aspects* **2003**, *223*, 215.
[24] S. Dragan, D. Dragan, M. Cristea, A. Airinei, L. Ghimici, *J. Polym. Sci.:Part A:Polym. Chem.* **1999**, *37*, 409.
[25] S. Dragan, I. Dranca, I. Maftuleac, L. Ghimici, T. Lupascu, *J. Appl. Polym. Sci.* **2002**, *84*, 871.
[26] S. Dragan, M. Cristea, „Polyelectrolyte Complexes. Formation, Characterization and Applications" Review for *Recent Research Developments in Polymer Science*, (Gayathri A. Ed.) Transworld Research Network , Kerala, India, Nr. 7, 2003 in press.
[27] W. Jaeger, U. Gohlke, M. Hahn, Ch. Wandrey, K. Dietrich, *Acta Polym.* **1989**, *40*, 161.
[28] S. Dragan, L. Ghimici, *Angew. Makromol. Chem.* **1991**, *192*, 199.
[29] M. Cristea, S. Dragan, D. Dragan, B.C. Simionescu, *Macromol. Symp.* **1997**, *126*, 143.
[30] L.W. Fischer, A. Sochor, J.S. Tan, *Macromolecules* **1977**, *10*, 949.
[31] E. S. Dragan, S. Schwarz, *J. Polym. Sci.:Part A:Polym. Chem.* in press.

Inorganic, Polymeric and Hybrid Colloidal Carriers with Multi-Layer Reactive Shell

V. Novikov, A. Zaichenko, N. Mitina, O. Shevchuk, K. Rayevska, V. Lobaz, V. Lubenets, Yu. Lastukhin*

Lviv Polytechnic National University, St. Bandera Str., 12, Lviv, 79013, Ukraine
E-mail: zaichenk@polynet.lviv.ua

Summary: Main experimental approaches for obtaining polymer, inorganic and hybrid colloidal particles as well as the tailored functionalization of their surface by oligoperoxide surfactants (OPS) and metal complexes (OMC) on their basis are discussed in the paper. The methods proposed enable to combine the stage of the formation of colloidal polymer, siliceous, metal and metal-oxide particles with the stage of their surface modification by functional surface-active oligoperoxides, which are sorbed irreversibly. Novel functional particles are studied by chemical, colloidal-chemical, rheological methods and scanning electronic microscopy. The occurrence of metal and metal oxide particle formation in distinct zones correlates well with the particle size distribution. The availability of reactive ditertiary peroxidic fragments on the particle surface as a result of OPS or OMC sorption causes their reliable protection, hydrophobity and ability to form free radicals and participate in elementary stages of radical processes.

Keywords: "core-shell" polymerization; functional colloidal particles; kinetics; metal complexes; oligoperoxide surfactants; radical emulsion polymerization

Introduction

Last decade, investigation of the methods of functional particle synthesis with the reactive polymeric shell providing their sedimentation stability, compatibility with various matrixes, reactivity and variety of specifically targeted properties is of a significant interest for obtaining carriers of biomedical application and active fillers for microelectronics [1-4]. The creation of theoretical and experimental bases of the synthesis of novel surface-active oligoperoxides and metal complexes on their basis causes the possibility of obtaining new functional polymeric, inorganic and hybrid colloidal particles [5,6]. Modification of dispersed filler surface by the technique of homogeneous nucleation in the presence of such oligoperoxides provides increased adsorption value due to high activity of the newly formed surface as well as the control of filler particle size and size distribution. New functional surface-active water- soluble OPS and OMC are sorbed onto the surface of dispersed polymer and mineral fillers causing localization of the necessary quantity of hydrophobic or hydrophilic peroxide-containing fragments. Such radical-

 DOI: 10.1002/masy.200450622

forming sites cause the possibility of the initiation of polymerization from the surface of solid phase [7-9], which is a prospective technique for obtaining filled polymer composites. The study of the principal approaches for the formation of active functional particles is the aim of the work.

Experimental

Oligoperoxide surfactants (OPS) on the basis of vinyl acetate (VA), maleic anhydride (MA), and 5-tert.butylperoxy-5-methyl-1-hexene-3-yne (VEP) were synthesized by polymerization at 333K in ethyl acetate using azobisisobutyronitrile (AIBN) as initiator. Surface-active oligoperoxidic metal complexes (OMC) were obtained by interaction of corresponding OPS with copper cations in organic medium at room temperature.

The monomers – styrene (St), methylmethacrylate (MMA), N-vinylpyrrolidone (N-VP), butyl acrylate (BA), acrylic acid (AA), VA were purified by double vacuum distillation. MA was purified by vacuum sublimation and after purification, its melting point was 325K (literature datum[10]: 325.9K). The peroxidic monomer VEP was purified by vacuum distillation (active oxygen content = 8.79% (calcd 8.75%)). AIBN was purified by recrystallization from ethanol. Other monomers and solvents (Merck) as well as metal salts were used as received.

Discussion

The formation of primary reactive polymer particles with functional shell

Among the variety of the methods for obtaining functional polymer particle the methods using OPS and OMC as universal and convenient tool are, in our opinion, the most prospective for the formation of reactive polymer particles of tailored size and functionality.

1. Water dispersion polymerization initiated by OMC

Earlier [11,12] we have studied the formation of primary reactive polymer particles with functional shell by the technique of homogeneous nucleation during water dispersion polymerization initiated by OMC. The main regularities of controlled radical polymerization initiated by oligoperoxide Me^{n+}-containing surfactants witness about the possibility of obtaining polymer water dispersions with unimodal particle size distribution and reactive functional shell liable to radical, condensation and other reactions [12].

The special case of the formation of primary reactive colloidal polymer particles-carriers with antibacterial properties by this technique is water dispersion polymerization initiated by OMC - graft –poly (styrene—maleic acid —4-methacrylamid-1-methylsulfonate) (Scheme 1).

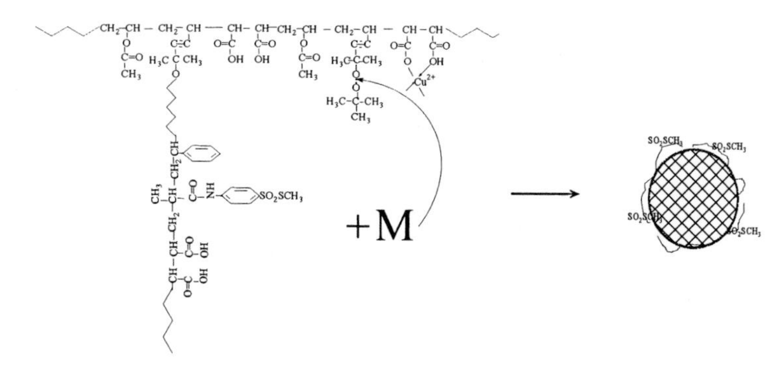

Scheme 1. The scheme of the synthesis of colloidal polymer particles with antibacterial properties by water dispersion polymerization initiated by OMC on the basis of peroxide-containing graft copolymer

2. Dispersion of condensation resin melts or solutions in water in the presence of OPS or OMC

Another prospective method for obtaining polymer colloidal particles with reactive functional shell is provided by mechanical dispersion of unsaturated polyester, alkyd or epoxy resin melts or solutions in water with OMC acting simultaneously as surfactant, stabilizer and modifier. As a result highly stable artificial water dispersions of the colloidal particles with the resin core and oligoperoxide shell are formed (Table 1).

Table 1. Characteristic of artificial water dispersions of polycondensation resins

Resin	%, OPS	additional surfactant	additional surfactant content per H_2O, %	pH	Water dispersion solid content, %	Latex particle size, μм
polyester PN-15	1.0 2.5 5.0	Rycinox-80 (oxyethylated rycinolyc acid)	10.0	6,5-7,5	45.0 44.5 46.0	0.22 0.14 0.1
Alkyd GF-188	1.0 2.5 5.0	Tinoram (N'-alkyl-N'N''-(3-aminopropyl)-amine)	7.5	3.5-4.0	40.0 39.0 39.5	0.16 0.13 0.1
Epoxy ED-16	1.0 2.5 5.0	Rycinox-80	7.5	6,5-7,5	33.0 33.0 32.0	0.20 0.15 0.12

The particle size is in the range of 0.1–0.4μm depending on the oligoperoxide surfactant concentration (Table 1). Water dispersion polymerization of the solutions of polycondensation resins in acrylic monomers (Fig. 1) enables to obtain particles with "core-shell" morphology and stable film forming colloidal systems on their basis combining the properties of these resins and polyacrylic polymers.

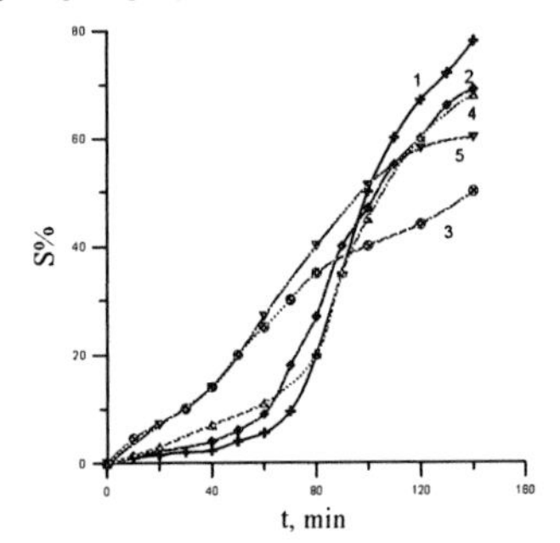

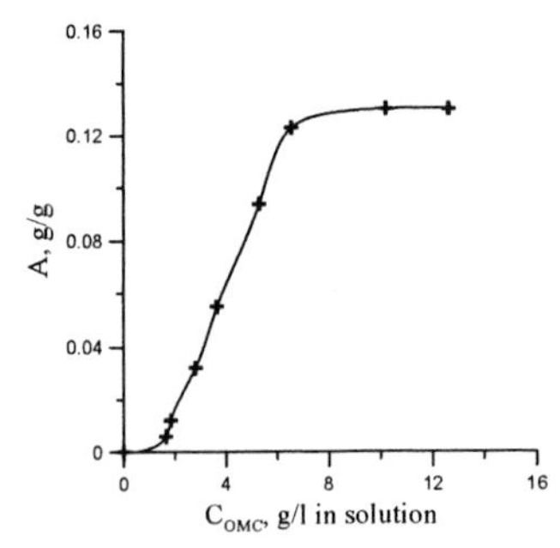

Fig. 1. BA—St—AA mixture conversion vs. emulsion polymerization time at the content of alkyd (**2, 3**) and epoxy (**4, 5**) resins: **1**- without resin, **2, 4** - 10% of resin per monomer mixture; **3, 5** - 40% of resin per monomer mixture

Fig. 2. Characteristics of OMC adsorption onto particles of Rhoplex acrylic latex

3. Sorption modification of industrial latexes by OPS or OMC

The sorption functionalization of the particles of industrial latexes by pre-determined amount of OMC is an interesting and practically feasible approach for obtaining primary reactive functional polymer particles. The investigation of OMC adsorption (Fig. 2) onto particle surface of industrial polyacrylic latex Rhoplex shows the achievement of the saturation of the surface at definite OMC concentration in solution as well as the decrease of the particle size, which is apparently due to the disaggregation of existing particles at the same time.

Adsorption saturation of the particle surface and the availability of radical-forming groups on them are evident from the results of chemical analysis and study of colloidal-chemical characteristics of polymer colloidal systems modified by OMC. This provides the possibility of carrying out graft copolymerization initiated from the particle surface leading to the formation of new tethered polymer chains with tailored length, functionality and reactivity.

4. Seeded polymerization initiated from the surface of the particles modified by OMC

The experimental results of seeded low temperature polymerization initiated from the surface of primary polymer particles represented by the scheme below (Scheme 2) and in Table 2 display the formation of composite particles with "core –shell" morphology and the possibility of obtaining multi-layer reactive shell as a result. It is evident (Table 2 , Fig. 3,4) that various monomers and monomer systems can be used for the formation of the second and third functional polymer shell on the particles resulting in the change of particle hydrophobic-hydrophilic properties, functionality and enhancement of their size.

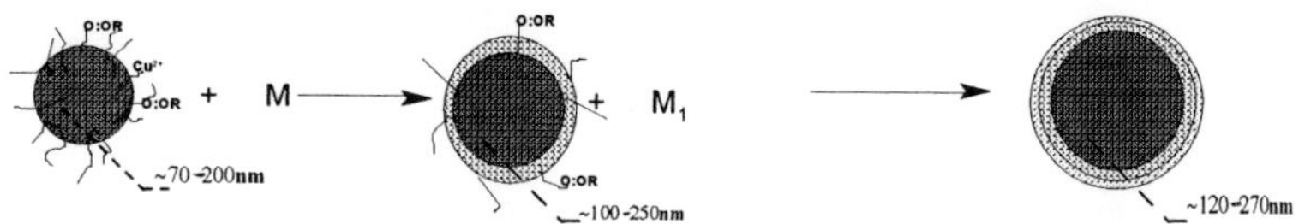

Scheme 2. The scheme of multi-stage seeded polymerization initiated from the particle surface

Table 2. Multi-stage seeded (co)polymerization initiated from the particle surface (293K)

First stage			Second stage*				Third stage **			
Latex particle structure	Dry residue, %	Dpart., μm	Monomer for the second shell***	Dry residue, %	Dpart., μm	W. %/h	Monomer for the third shell	Dry residue, %	Dpart. ,μm	W. %/h
Core St-BA-VEP 53:32:15 Shell OMC	22.0	0.015	F-MA	27.0	0.020	7.2	-	-	-	-
			Si-MA	25.5	0.018	9.0	-	-	-	-
			BA-GMA 90:10	28.0	0.020	10.2	-	-	-	-
Core St-BA-AA 70:25:5 Shell - OMC	23.0	0.010	VEP-BA 50:50	29.0	0.014	9.0	Si-MA	33.0	0.016	7.2
							F-MA	32.2	0.016	9.0

* - the formation of the second shell was initiated by residual OMC in the particle shell;
** - the formation of the third shell was initiated by additional OMC sorbed onto particle surface (0.5% per monomers);
*** - F-MA–2,2,3,3-tetrafluoropropyl-2-methacrylate; Si-MA – (3-trimethoxysilyl)propyl-2-methacrylate; GMA – (2,3-epoxy propyl)-methacrylate

One can see from the results of the investigation of the latex dynamic viscosity (Fig. 4) that the formation of new shell as a result of St seeded polymerization altered the particle surface functionality and weakened interaction between the particles. Moreover, the dependences of seeded polymerization rate on the concentration of sodium pentadecylsulphonate (E-30) and particle-initiator content in the systems (Fig. 5) proves the occurrence of the polymerization exceptionally on the primary particle surface leading to particle size enhancement. No new particles is formed during seeded polymerization initiated by particles modified by OMC.

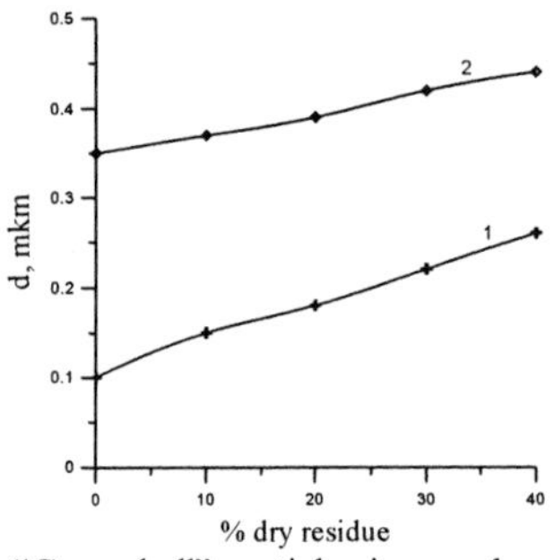

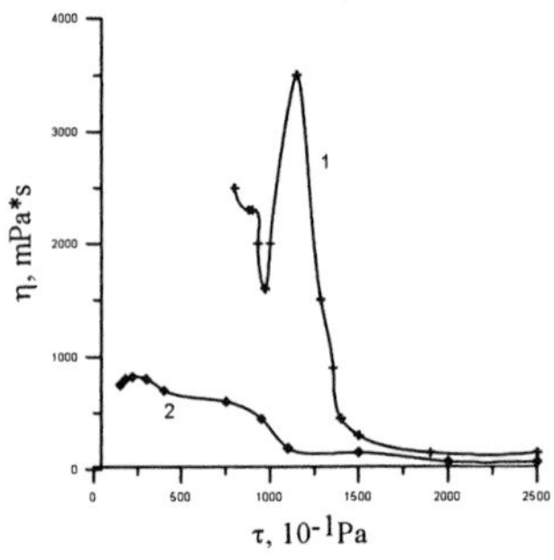

Fig. 3. "Core-shell" particle size vs. dry residue of initial dispersion (St seeded polymerization) 1 – unsaturated polyester dispersion, 2 – alkyd resin dispersion. (St conversion - 100%)

Fig. 4. Dynamic viscosity vs. shear stress for polyester dispersion modified by OMC (1) and after St polymerization initiated from surface (2) (dispersion solid content 27%)

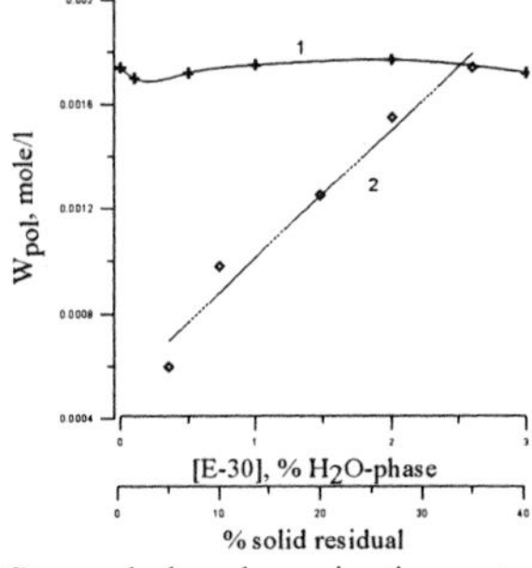

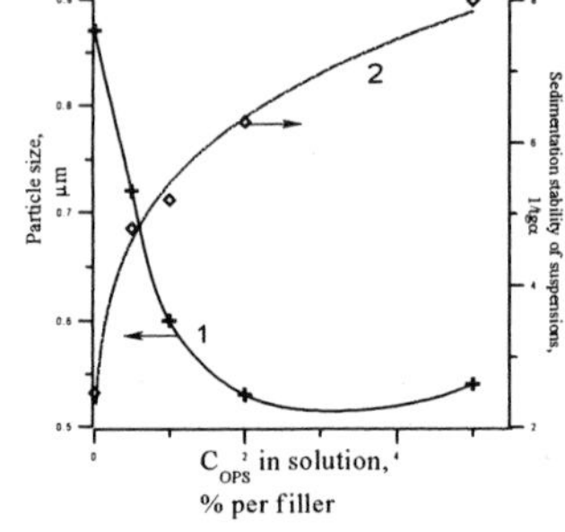

Fig. 5. St seeded polymerization rate vs. concentration of E-30 (1) and vs. dispersion solid residual (2) (initiator - OMC-3% with respect to monomer; 298K)

Fig. 6. Characteristics of Fe_3O_4 suspensions obtained via nucleation from the solution of the mixture of ferrous chlorides in the presence of OMC on the basis of N-VP-VEP-MA (the ratio [solid phase]:$[H_2O]$ = 1:10, T=333K)

The formation of primary reactive inorganic particles with functional shell

It is clear that the methods of obtaining inorganic and hybrid colloidal particles with reactive functional shell on their surface are similar (in many aspects) to the above-mentioned methods of the formation of functional polymer particles.

1. Homogeneous nucleation from salt solutions in the presence of OPS or OMC

Using this technique primary reactive inorganic particles with functional shell are formed by homogeneous nucleation from the solutions of corresponding metal salts in the presence of OMC as illustrated in scheme 3.

The characteristics of water suspensions of ferrous oxide (Fig. 6) show that the increase of OMC concentration in the reaction mixture causes the decrease of particle size and simultaneous enhancement of the amount of sorbed OMC.

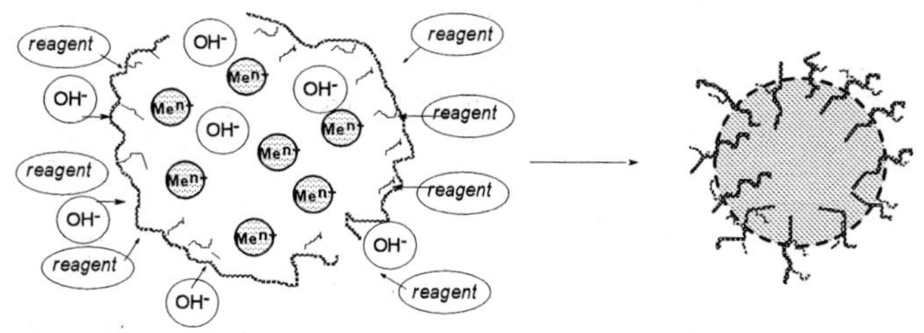

Scheme 3. The scheme of homogeneous nucleation of metal and metal oxide particles with functional OMC shell

The number average particle size distribution shows the tendency toward the formation of unimodal silver particles at their formation in the presence of OMC surfactant (Fig. 7). This can be explained as we have shown earlier [6] by the displacement of the reaction of particle nucleation into micelle-like structures formed by OMC, which are the exo-templates determining the particle size.

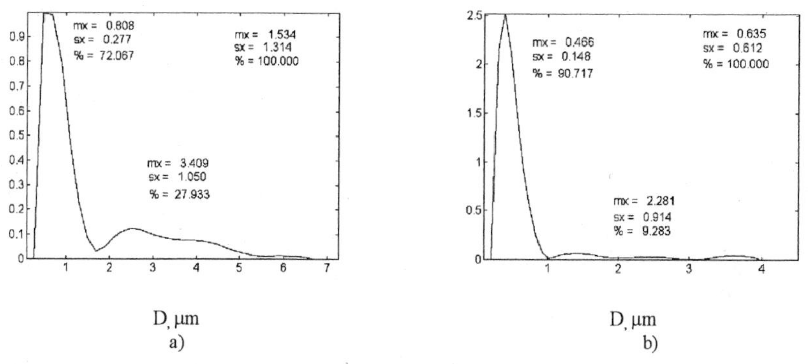

Fig. 7. Number-average distribution of silver particles obtained via reduction. A) synthesis without OMC; b) 5% of OMC with respect to Ag^{+} ions (0.09 % in solution)

2. The sorption modification of inorganic particle surface by OPS or OMC

The method of sorption modification of previously obtained inorganic particles by OMC also provides their tailored functionalization as a result of irreversible immobilization of ditertiary peroxide groups, hydroxyl, carboxy, epoxy and other functional fragments (Fig. 8).

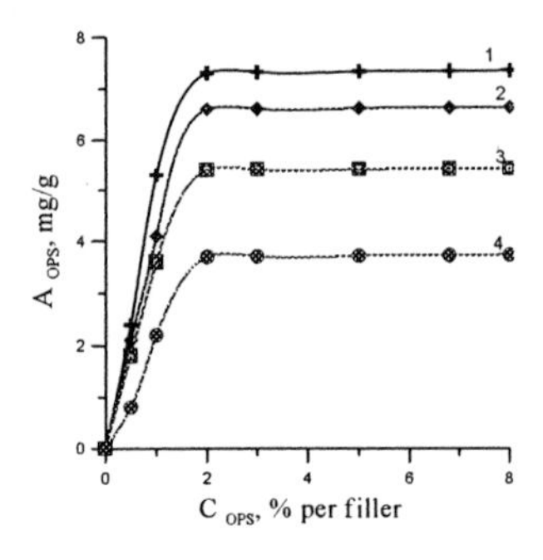

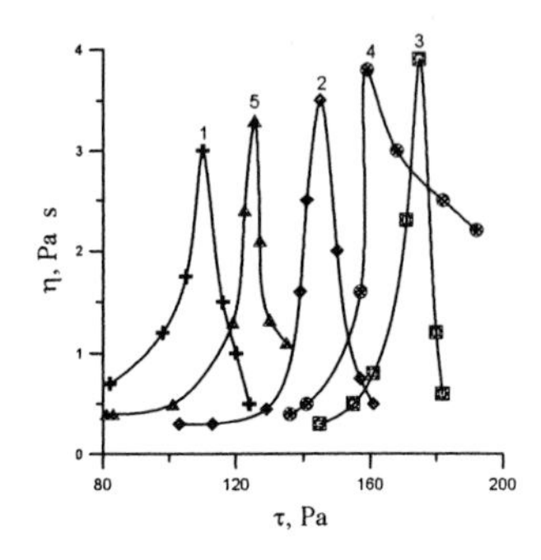

Fig. 8. Isotherms of OMC N-VP-VEP-MA sorption onto the surface of -Fe_2O_3 (1, 2) and Fe_3O_4 (3, 4). 1, 3 – before sample washing, 2,4 - after sample washing

Fig. 9. The dynamic viscosity of Fe_3O_4 suspensions modified by OMC N-VP-VEP-MA vs. shear stress. (C_{OMC}= 0%(1), 1% (2), 2% (3), 3% (4), 5% (5)

The dependence of dynamic viscosity of ferrous oxide suspension modified by various amount of OMC on shear stress (Fig. 9) witnesses, in our opinion, about the formation of multi-layer functional shell with different orientation of polar functional groups.

Table 3. Characteristics of the decomposition of OMC peroxide groups sorbed on colloidal Ni particle surface

Ditertiary peroxide fragments	Activation energy E, kJ/mole	lgA	$\Delta S^{\neq}$ J/mole K
OMC on Ni particle surface	37,0	2,0	-50,0
OMC in solution	42,0	2,5	-44,4

The experimental results of the investigation of the decay of ditertiary peroxide fragments immobilized on the particle surface (Table 3) show the significant difference between the activation parameters of the decomposition of peroxide groups on the surface and in solution, which is apparently due to the decrease of the degree of freedom of sorbed OMC molecules in comparison with the OMC molecules in solution.

3. Seeded polymerization initiated from inorganic particle surface modified by OMC

The presence of radical-forming sites on the particle surface causes the possibility of low temperature radical formation by the immobilized ditertiary peroxide groups as well as grafting of polymer chains to the surface with the formation of new functional shell at a given distance from the surface (Fig.10 and Table 4).

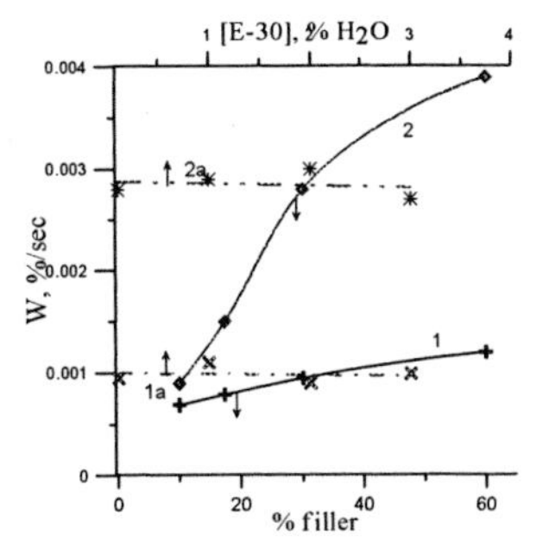

Fig.10. Water dispersion polymerization rate of VEP-GMA-St mixture vs. filler content (1,2) and concentration of emulsifier E-30 (1b, 2b). Initiation from the OMC modified filler surface: g-Fe_2O_3 ([OMC]=0,7%) (1, 1b) colloidal Ni particles ([OMC] =0,45%) (2, 2b); 291K

Table 4. Characteristics of copolymer GMA-VEP-St grafted to the surface of inorganic fillers (291K; monomers:H_2O=1:5; GMA-VEP-St 2:1:1)

Filler	Filler content, % wt	Content of copolymer grafted, %	Composition of copolymer grafted , %		
			GMA	VEP	St
γ-Fe_2O_3, [OMC]=0,7%	17,4	1,6	50,0	25,0	25,0
	30	2,5	60,0	19,0	21,0
	60	5,0	75,0	16,0	9,0
Colloidal Ni [OMC]=0,45%	17,4	0,8	65,0	5,5	29,5
	30	1,1	55,0	6,0	39,0
	60	2,5	50,0	10,0	40,0

One can see (Fig.10) that seeded polymerization initiated from the surface of inorganic particles obeys the same regularities peculiar to the polymerization initiated from the polymer particle surface, including the independence of the polymerization rate on the concentration of the additional emulsifier. This proves the occurrence of graft polymerization only on the particle surface and the impossibility of particle formation in the solution. The polymerization rate and conversion depend on the modified filler nature and content in the reaction system. The study of the particles after seeded polymerization witnesses about the increase of their size and tailored formation of grafted chains containing a definite amount of active epoxide and peroxide groups.

Conclusions

Various approaches of OMC and OPS use for the activation of colloidal particles providing polymer grafting onto their surface have been demonstrated in the work; This allows the formation of "core-shell" particle structure with fragments providing their tailored compatibility, functionality and reactivity.

Functional particles including magnetic ones contain spacers with functional groups, which are liable to radical and polymer-analogous transformations.

Acknowledgement

The authors express their thanks to Science and Technology Center in Ukraine for financial support of this work, Project # 1447.

[1] H. Kawaguchi. Progress in Polymer Science., ***2000***,25, p.1171-1210.
[2] M. Antonietti, E. Wenz, L. Bronstein, M. Seregina. Adv. Mater., ***1995***, 7, p. 1000.
[3] S. Rimmer. Designed Monomers and Polymers., ***1998,*** 1, p.89-96.
[4] M. Okubo, R. Takekoh, H. Sugano. Colloid. Polym. Sci., ***2000***, 278, p.559-564.
[5] A.Zaichenko, S.Voronov, A.Kuzayev, O.Shevchuk, V.Vasilyev. J. Appl. Polym. Sci. ***1998***, 70, p.2449-2455.
[6] A. Zaichenko, N. Mitina, M. Kovbuz, I. Artym, S. Voronov. J. Polym. Sci., ***2000,*** A38, p.516-527.
[7] A. Zaichenko , N. Mitina , O. Shevchuk , O. Hevus , T. Kurysko , N. Bukartyk , S. Voronov. Macromol. Symp.(React. Pol.), ***2001***, 164, p. 25-47.
[8] A. Zaichenko, S. Voronov, O. Shevchuk. Dopovidy Acad. Nayk Ukr., ***1999***, 5 , p.157-162, .
[9] A. Zaichenko, O. Shevchuk, S. Voronov, A. Sidorenko. Macromolecules ***1999***, 32, p. 5707-5711.
[10] . Moldavsky, B.; Kernos, Yu. *Maleic Anhydride and Maleic Acid*, Chemistry: Leningrad, **1977**.
[11] A. Zaichenko, N. Mitina, M. Kovbuz, I. Artym, S. Voronov. Macromol. Symp. (React. Pol.)., ***2001,***164, p.47-71
[12] A. Zaichenko, N. Mitina, K. Rayevska, M. Kovbuz, O. Hertsyk. Macromol. Symp.(***in press***)

Macromol. Symp. **2004**, *210*, 203-208

Monolayers of Reactive Cellulose Derivatives

Gerhard Wenz,[*1] Petra Liepold,[2] Nico Bordeanu[1]*

[1] Organic Macromolecular Chemistry, Saarland University, Stuhlsatzenhausweg 97, D-66123 Saarbrücken, Germany
E-mail: g.wenz@mx.uni-saarland.de
[2] FRIZ Biochem, München, Germany

Summary: Functional cellulose derivatives are very versatile materials for the creation of mono- and multilayer systems. Hydrophobic alkyl and trimethylsilyl celluloses form highly ordered Langmuir-Blodgett multilayers on hydrophobic substrates. Cellulose thiosulfates and methyl thio ethers were self-assembled on gold and silver surfaces to form hydrophilic monolayers. Cellulose layer systems are capable for chemical transformations under conservation of the structural order. They are suitable platforms for the investigation of molecular recognition at surfaces and the construction of sensor devices. Both biological ligands, e. g. biotin, and enzymes, e. g. horse radish peroxidase, could be attached to cellulose under conservation of their biological function.

Keywords: cellulose; enzymes; molecular recognition; nanolayers; self-assembly

Introduction

Cellulose derivatives are well suited for the formation of ultra-thin films on various substrates. This has several reasons: chains of cellulose derivatives are semi-rigid, their persistence lengths are in the range of 5-15 nm,[1, 2] cellulose can be functionalized by polmer analogous reactions[3] and cellulose is highly biocompatible. Lipophilic alkyl ethers, especially *i*-pentyl ethers, of cellulose form monolayers at the air water interface. These monolayers can be repetitively transferred onto planar hydrophobic substrates to form highly ordered Langmuir-Blodgett (LB) multilayers. The thickness per cellulose layer is 0.9 nm.[4] LB multilayers from trimethylsilyl-cellulose can be regenerated by gaseous HCl.[5] The resulting cellulose multilayers show a inter layer spacing of 0.4 nm which is similar to the one of native crystalline cellulose.[6] Biological receptor molecules, e. g. antibodies, were immobilized at regenerated cellulose LB multilayers functionalized by cyanur chloride for the construction of immunosensors using evanescent wave guide techniques.[7]

 DOI: 10.1002/masy.200450623

Beside the LB technique the so-called self-assembly method is used for the creation of nanoscopically defined monolayers. Alkane thiols and disulfides irreversibly form monolayers of two dimensional crystallinity onto gold and silver surfaces via Au-S bonds or Ag-S bonds, respectively.[8] These self-assembled monolayers (SAMs) offer several advantages to LB layers: there is no sophisticated instrumentation necessary, and the deposition of the monolayers can be controlled by *in situ* methods like surface plasmon resonance spectroscopy,[9] atomic force microscopy or impedance measurements,[10] and patterned SAMs can easily be produced by mico contact printing[11] or dip pen lithography.[12] SAMs are not only known from monomeric thiols but also from polymeric ones, such as thio derivatives of polystyrene and polyacrylates.[13] We synthesized water-soluble thiosulfate derivatives of cellulose and investigated the cellulose SAMs on gold and silver surfaces by ellipsometry, FT-IR and X-ray photon spectroscopy (XPS).[14-16] In the following we report on the synthesis of new highly water soluble cellulose thio derivatives and their formation of SAMs and describe coupling techniques for the immobilization of biomolecules to those cellulose SAMs.

Synthesis of thio derivatives of carboxymethyl cellulose

Commercial carboxymethyl cellulose (CMC) of a degree of substitution (DS) of 1.1 and a degree of polymerization (DP) of 925 was reacted with allyl glycidyl ether in NaOH solution. The partial DS values of the attached allyl-hydroxypropyl groups were in the range of 0.2 – 0.4 depending on the reaction conditions. Partial addition of tetrathionate to the allylic double bonds yielded the cellulose 2'',3''-bis-thiosulfate **TSHP-CMC**, while bromination in water and substitution by thiosulfate afforded the 2''-hydroxy-3''-thiosulfate **HTSHP-CMC** (s. eq. 1).

TSHP-CMC **HTSHP-CMC** (1)

Both compounds are highly water soluble. In addition, 6-*O*-tosylcellulose[17] was reacted with sodium methyl sulfide. The resulting methyl thioether was carboxymethylated to afford the water soluble cellulose derivative **MTh-CMC** (s. eq. 2).

(2)

R = H, CH_2COONa
MTh-CMC

Formation self-assembled monolayers (SAMs) on gold

The formation of SAMs on gold was detected *in situ* by surface plasmon resonance (SPR) spectroscopy using a Bio-Suplar 2 instrument from Analytical μ-Systems, Regensburg, Germany (http://www.micro-systems.de) equipped with a continuous flow cell.[18] Both the cellulose thiosulfates, **TSHP-CMC**, **HTSHP-CMC**, and the methyl thioether **MTh-CMC** spontaneously form SAMs on gold. A typical film thickness of 3 nm was derived from the change of the surface plasmon resonance angle. This thickness is in accordance with a multiple attachment of the cellulose chain. As a straight conformation of the chain would lead to much thinner monolayers with an estimated thickness of about 1 nm, we assumed that the polymer chains are bound in a curved conformation. SAM formation was completed within a few minutes for the cellulose thiosulfates, **TSHP-CMC** and **HTHP-CMC** (s. fig. 1). On the other hand, it took some more time (1 h) for the methyl thioether **Mth-CMC**.

The cellulose SAMs were hydrophilic, the contact angles vs. water were in the range of 15-30°. The cellulose SAMs showed little unspecific interactions with blood proteins, e. g. bovine serum albumin. Therefore they are very promising platforms for biosensor devices.

Molecular recognition at cellulose SAMs

As a first example of a detection of a molecular recognition event at a cellulose SAM, we choose the ligand/receptor system biotin/streptavidin. Streptavidin is a protein consisting of 123 amino acids which has a very high and selective affinity to biotin.[19] An amino-terminated biotin derivative was linked by standard N,N'-dicyclohexylcarbodiimide (DCC) coupling to **HTSHP-CMC**. The resulting conjugate formed monolayers on gold. Subsequent addition of a streptavidin solution leads to the built-up of a streptavidin layer on top of the cellulose layer. The streptavidin layer had a thickness of 6 nm, in good agreement with the thickness of a streptavidin molecule.[20] We concluded from this finding, that a dense streptavidin monolayer had been formed. Molecular recognition of antibodies by immobilized antigens will be investigated at cellulose SAMs in the future. The immobilization of ligands at functional cellulose monolayers appears advantageous to the well-known Biacore system,[21] as fewer reaction steps have to be performed on the surface for the creation of a sensor chip.

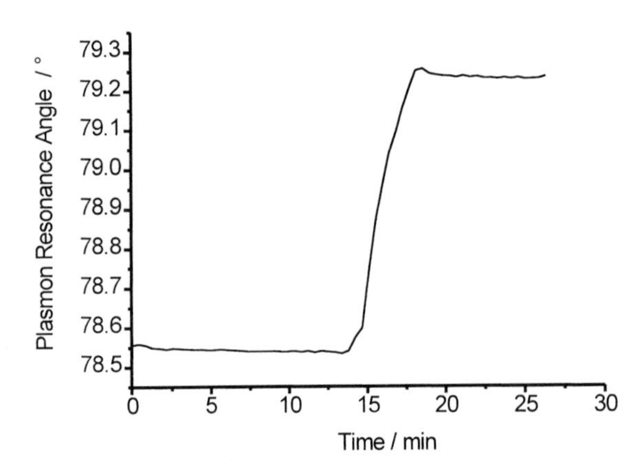

Fig. 1. Kinetics of the SAM formation from a solution of 0.1 % **TSHP-CMC** in 0.1 N HCl on a 50 nm gold layer on glass monitored by SPR at a wavelength of 670 nm.

Enzyme activity at cellulose SAMs

For the attachment of enzymes the cellulose thiosulfate **HTSHP-CMC** was reacted with cyanur chloride to the chlorotriazinyl derivative **HTSHP-Tz-CMC** (s. eq. 3). This compound is stable in aqueous solution and reacts selectively with amines at pH 8. It spontaneously forms monolayers on gold as shown by SPR. Horseradish peroxidase was immobilized to these reactive monolayers by incubation of the monolayer for 20 minutes in an aqueous solution of the enzyme at room temperature. Presumably lysine residues of the enzyme couple to the chlorotriazinyl groups of the monolayer.

As no further coupling reagents are necessary in the aqueous phase ("reagent free" coupling), adverse side reactions of the enzyme are minimized. Consequently, the enzyme remains still active after the immobilization, as proven by activity tests using the ABTS assay.[22] This example shows in principle, that enzymes can be tested in the nanoscopic environment of a cellulose monolayer. Interesting applications such as highly efficient screening tests for enzyme inhibitors are conceivable.

HTSHP-CMC → HTSHP-Tz-CMC (3)

HTSHP-CMC: R = H, CH_2COO^-

HTSHP-Tz-CMC: R = H, chlorotriazinyl-O-CH_2COO^-

Conclusion

Cellulose monolayers offer several advantages to classical SAMs. They are rapidly formed with uniform quality. They can be functionalized for the coupling of ligands and proteins. Biomolecules can be immobilized under gentle "reagent free" coupling conditions. The coupling density can be controlled with high reproducibility. As unspecific interactions are low, cellulose SAMs are suitable platforms for the construction of biosensors.

Acknowledgements

Financial support was provided as part of the focus program on "Cellulose and cellulose derivatives - molecular and supramolecular structural design" by the Deutsche Forschungsgemeinschaft (DFG) and as part of a German Israelian Project Cooperation "Building nanostructured devices by controlled assembly of monomers, polymers and nanoparticles" by Deutsches Zentrum für Luft- und Raumfahrt (DLR), International Bureau of BMBF.

[1] K. Kamide, M. Saito, *Macromol. Rapid Commun.* **1983**, *4*, 33.
[2] C. W. Hoogendam, A. De Keizer, M. A. C. Stuart, B. H. Bijsterbosch, J. A. M. Smit, J. A. P. P. Van Dijk, P. M. Van der Horst, J. G. Batelaan, *Macromolecules* **1998**, *31*, 6297.
[3] D. Klemm, B. Philipp, T. Heinze, U. Heinze, W. Wagenknecht, *Comprehensive Cellulose Chemistry II: Functionalization of Cellulose, Vol. 2*, Wiley-VCH, Weinheim, 1998.
[4] M. Schaub, C. Fakirov, A. Schmidt, G. Lieser, G. Wenz, G. Wegner, P. A. Albouy, H. Wu, M. D. Foster, C. Majrkzak, S. Satija, *Macromolecules* **1995**, *28*, 1221.
[5] M. Schaub, G. Wenz, G. Wegner, A. Stein, D. Klemm, *Adv. Mater.* **1993**, *12*, 919.
[6] A. A. Baker, W. Helbert, J. Sugiyama, M. J. Miles, *Biophys. J.* **2000**, *79*, 1139.
[7] F. Löscher, T. Ruckstuhl, T. Jaworek, G. Wegner, S. Seeger, *Langmuir* **1998**, *14*, 2786.
[8] C. D. Bain, E. B. Troughton, Y. T. Tao, J. Evall, G. M. Whitesides, R. G. Nuzzo, *J. Am. Chem. Soc.* **1989**, *111*, 321.
[9] D. K. Kambhampati, W. Knoll, *Curr. Opin. Coll. Interf. Sci.* **1999**, *4*, 273.
[10] M. W. J. Beulen, J. Bugler, M. R. De Jong, B. Lammerink, J. Huskens, H. Schonherr, G. J. Vancso, B. A. Boukamp, H. Wieder, A. Offenhauser, W. Knoll, F. C. J. M. Van Veggel, D. N. Reinhoudt, *Chem. Eur. J.* **2000**, *6*, 1176.
[11] B. D. Martin, S. L. Brandow, W. J. Dressick, T. L. Schull, *Langmuir* **2000**, *16*, 9944.
[12] J.-H. Lim, D. S. Ginger, K.-B. Lee, J. Heo, J.-M. Nam, C. A. Mirkin, *Angew. Chem. Int. Ed.* **2003**, *42*, 2309.
[13] C. Erdelen, L. Häussling, R. Naumann, H. Ringsdorf, H. Wolf, J. Yang, M. Liley, J. Spinke, W. Knoll, *Langmuir* **1994**, *10*, 1246.
[14] US 6,245,579 B1 (2001), Universität Karlsruhe, invs.: G. Wenz, D. F. Petri, S. W. Choi.
[15] S. Choi, H. Lauer, G. Wenz, M. Bruns, D. F. S. Petri, *J. Braz. Chem. Soc.* **2000**, *11*, 11.
[16] D. S. Petri, S. W. Choi, H. Beyer, T. Schimmel, M. Bruns, G. Wenz, *Polymer* **1999**, *40*, 1593.
[17] K. Rahn, M. Diamantoglou, D. Klemm, H. Berghmans, T. Heinze, *Angew. Makromol. Chem.* **1996**, *238*, 143.
[18] S. A. Zynio, A. V. Samoylov, E. R. Surovtseva, V. M. Mirsky, Y. M. Shirshov, *Sensors* **2002**, *2*, 62.
[19] P. C. Weber, D. H. Ohlendorf, J. J. Wendoloski, F. R. Salemme, *Science* **1989**, *243*, 85.
[20] B. A. Katz, *J. Mol. Biol.* **1997**, *274*, 776.
[21] W. Jager, *Carbohydr. Chem. Biol.* **2000**, *2*, 1045.
[22] K. Monde, H. Satoh, M. Nakamura, M. Tamura, M. Takasugi, *J. Nat. Prod.* **1998**, *61*, 913.

Peroxide-Containing Compatibilizer for Polypropylene Blends with Other Polymers

Yuri Roiter,[1] *Volodymyr Samaryk,*[1] *Sergiy Varvarenko,*[1] *Natalya Nosova,*[1] *Igor Tarnavchyk,*[1] *Jürgen Pionteck,*[2] *Petra Pötschke,*[2] *Stanislav Voronov**[1]

[1] National University "Lvivska Polytechnica", 12 S. Bandera Str., Lviv 79013, Ukraine
[2] Institute of Polymer Research Dresden, Hohe Strasse 6, D-01069 Dresden, Germany
E-mail: stanislav.voronov@polynet.lviv.ua

Summary: An approach to synthesize interfacial active peroxide graft copolymers, so called precompatibilizers, which are suitable for the universal compatibilization of one special polymer with a number of other polymers, has been presented. As example, this approach is illustrated by the reactive fusion of a random peroxide copolymer (VO) with polypropylene (PP) resulting in a VOgPP precompatibilizer. A mathematical model of the process of the VOgPP synthesis and the conduction of a full-factorial experiment have allowed both the optimization of the synthesis conditions and the prediction of its proceeding.
During blending PP with other polymers VOgPP localizes across the blend interphases and initiates radical processes leading to the *in situ* formation of final compatibilizer macromolecules, which are efficient just for the blends where they are formed. The universality of the precompatibilizer concept is demonstrated in PP blends of the thermoplastic/thermoplastic type (with polystyrene and polyethylene) and of the thermoplastic/thermoset type (with unsaturated polyester resin).

Keywords: compatibilization; peroxide graft copolymer; polymer blends; simulations; synthesis

Introduction

The creation of polymer blends allows to combine the properties of individual polymers providing the possibility of essential widening of an assortment of new valuable polymer materials. A general problem for the creation of new blends is the thermodynamically immiscibility of the majority of polymer combinations leading to phase separation and worse blend characteristics if the compatibility between the separated phases is poor. Today, this problem is overcome by the creation of compatibilizing systems, decreasing the interfacial tension and increasing the adhesion between the blend components [1]. The utilization of diblock, triblock, and graft copolymers as compatibilizers is efficient from this point of view [2], but their

 DOI: 10.1002/masy.200450624

separate synthesis is rather expensive. Reactive blending of polymers, where the compatibilizing systems are formed *in situ* by the reaction of both blend components with each other, is considered to be more promising [2,3]. However, still the main problem of the contemporary methods remains, their non-universality and the necessity of the elaboration of expensive compatibilizing systems for almost every set of polymers to be blended.

Since a universal property of almost all polymers blended today is their ability to participate in free radical reactions of chain transfer and recombination, a more universal concept for the *in situ* compatibilization is the formation of compatibilizing structures by radical processes. The utilization of low molecular weight substances as radical sources is accompanied with a number of drawbacks such as their volatility at blending temperatures, their lack of interfacial activity, and, consequently, their effects on the bulk properties of the blended polymers. Thus, in our opinion, the most promising route for a universal *in situ* creation of compatibilizing systems is the elaboration and application of interfacial active peroxide-containing high molecular weight precompatibilizers, capable to locate across the interphases in different blends and generating there free radicals which results finally in the formation of the compatibilizer molecules efficient just for the blends where they are formed.

Experimental

The peroxide random copolymer VO on the basis of 2-tert-butylperoxy-2-methyl-5-hexene-3-yne (VEP) and octyl methacrylate, which was synthesized using a known technique [4], has been utilized for the synthesis of the interfacial active peroxide-containing precompatibilizer VOgPP (Scheme 1). Due to VO interactions with PP in the melt (a process that is studied in this work) at the expense of partial decomposition of peroxide groups, an alloy is formed containing ≈35 %wt VOgPP, which has been utilized directly for the compatibilization of blends of PP with other polymers.

PP (Aldrich 428183, fraction extractable with heptane) has been utilized for the elaboration of the VOgPP synthesis. The same PP was used for the preparation of blends with polystyrene (PS, PS143E, BASF AG, Germany) and unsaturated polyester resin (UPR, PN-15, Armoplast Co., Ukraine). PP Novolen 1106 H (Targor, France) was used for the preparation of blends with polyethylene (PE, Hostalen GC 7260, Elenac GmbH, Germany).

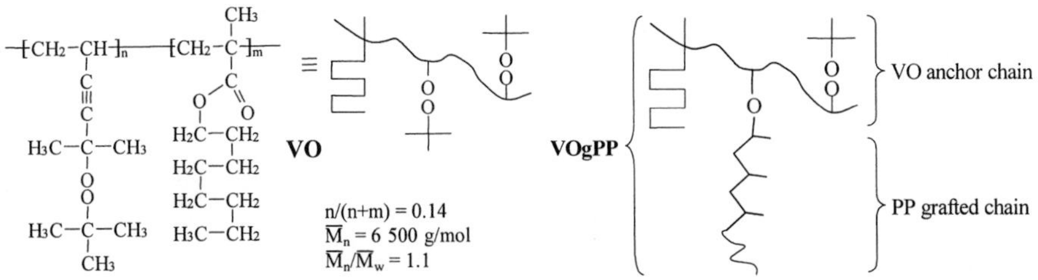

VO

VOgPP

VO anchor chain

PP grafted chain

$n/(n+m) = 0.14$
$\overline{M}_n = 6\ 500$ g/mol
$\overline{M}_n/\overline{M}_w = 1.1$

Scheme 1. Structure of the peroxide containing random copolymer VO (left) and of its fusion product with PP, the precompatibilizer VOgPP (right)

The melt blending of PP/PS and PP/PE was performed using a DACA Micro-Compounder (volume 4.5 cm^3, DACA-Instruments, USA) at 190 °C for 10 min at 100 rpm. Prepared blends were extruded as strands into water. Micrographs were obtained from cryocuts by means of a scanning electron microscope LEO VP435 (Zeiss, Germany) at 10 kV.

The thermoplastic/thermoset blends (PP/UPR) were prepared at 190 °C by mixing PP and UPR for 15 min. Then styrene preheated to 50 °C was added to the mixing chamber (with heating turned off) reaching an UPR/styrene ratio of 60/40 by weight. The mixture cooled down to 20 °C within 20 min under stirring. 5 %wt benzoyl peroxide (40 %wt blend in dibutyl phthalate, Aldrich) and 0.15 %wt 2,4-dimethylaniline (Aldrich), both upon the total quantity of UPR and styrene; were added to the blend before molding. Then, blends were cured at ambient temperature during 24 hours in form of specimens sized 0.5×1.0×8.0 cm^3. Charpy unnotched impact strength of cured PP/UPR-styrene samples was measured using a pendulum striker TiraTest 2000 (TIRA, Germany). Results were averaged over 12 measurements.

Results and Discussion

Elaboration of the synthesis conditions for the peroxide precompatibilizer

We have established earlier the ability of VO to bind with PP using ellipsometry and atomic force microscopy, and also to form crosslinked products [5]. During the performance of fusion of VO with PP, the formation of several fractions of polymer materials possessing different molecular weight (MW) was observed. Corresponding results were obtained by a consequent soxhlet extraction with hexane and heptane of obtained specimens (Figure 1).

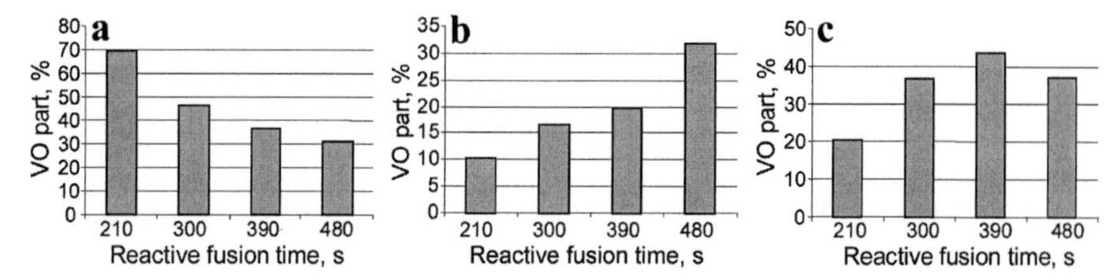

Figure 1. Dependence of VO part (determined by quantitative IR-spectroscopy) after the interaction with PP on the time of fusion at 190 °C (initial VO content: 2.5 %wt) in: **a** - hexane fraction (lower MW); **b:** heptane fraction (higher MW); **c:** heptane gel-fraction (highest MW).

One can see from Figure 1 that VO transforms steadily to higher molecular weight compounds at the proceeding of free radical processes in its melt with PP. It is worth to note, that if recombination proceeds between the macroradicals of VO and PP (which was at the solubility limit in heptane) then the resulting product will be part of the gel-fraction insoluble in the solvents utilized. The latter fraction includes also the crosslinked products.

The target process of the interaction studied is a recombination of VO and PP macroradicals with the formation of VOgPP. This part of the work is devoted to the optimization of this process. A mathematical model of the grafting has been built on the basis of the detailed scheme of possible reactions proceeding during the fusion interaction of VO and PP. The final integral equation of the current VO concentration can be presented as follows:

$$[VO] = \frac{[VO]_0}{e^{k_{C1} \cdot \tau}(1 + k_{C2}) - k_{C2}}, \qquad (1)$$

with [VO] and $[VO]_0$ are the current and initial VO concentration, respectively; and k_{C1} and k_{C2} are process effective constants.

Integral equations describing the formation of the target product VOgPP, C_{TP}, and the formation of non-target products of VO interaction (including the crosslinked products), C_{NP}, are:

$$C_{TP} = S_{TP} \cdot \frac{100}{\rho} \cdot [VO]_0 \cdot \left(1 - \frac{1}{e^{k_{C1} \cdot \tau}(1 + k_{C2}) - k_{C2}}\right), \qquad (2)$$

$$C_{NP} = S_{NP} \cdot \frac{100}{\rho} \cdot [VO]_0 \cdot \left(1 - \frac{1}{e^{k_{C1} \cdot \tau}(1 + k_{C2}) - k_{C2}}\right), \qquad (3)$$

where S_{TP} and S_{NP} are coefficients of the formation of target and non-target products, respectively;

and ρ is the reaction mass density. Values of constants and coefficients of the equations shown have been listed in Table 1.

Table 1. Values of the equation (**1**) constants and coefficients S_{TP} and S_{NP} from equations (**2**) and (**3**) obtained by optimization of experimental data in dependence on the conditions of proceeding the reactive fusion of VO and PP.

Temperature	*[VO]0*	$k_{C1}\times10^4$	k_{C2}	S_{TP}	S_{NP}
K	mol/dm³	s⁻¹		g/mol	g/mol
443	0.0325	3.3±0.7	7.5±0.6	4370±70	1320±80
	0.0185	5.6±0.3	5.1±0.2	4640±120	1000±60
453	0.0250	5.9±0.6	4.8±0.5	4610±40	1100±40
	0.0325	5.7±0.4	5.0±0.4	4500±70	1220±70
463	0.0325	9.6±0.3	3.1±0.4	3880±90	1780±70

The model built and its parameters optimized using the experimental data allow to find an optimal temperature of target VOgPP formation (Figure 2) and to predict the proceeding of the process of precompatibilizer formation during reactive fusion of VO with PP (Figure 3).

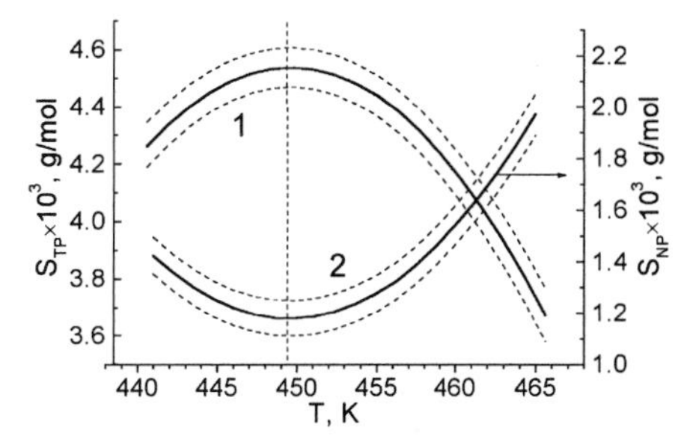

Figure 2. S_{TP} (**1**) and S_{NP} (**2**) coefficients vs. process proceeding temperature.

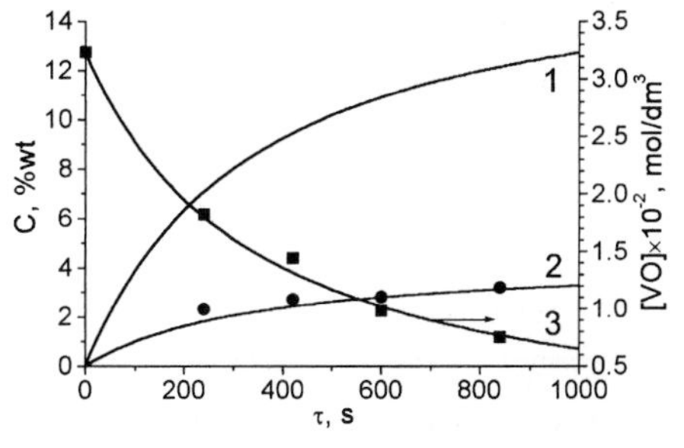

Figure 3. Predicted curves of VOgPP accumulation (**1**), non-productive VO interaction (**2**), and VO consumption (**3**). Points are experimental. T = 449 K, $[VO]_0$=0.0325 mol/dm³.

One can see from Figure 2 that the optimal temperature for yielding the maximum quantity of the target precompatibilizer VOgPP and minimum of byproducts of VO and PP interactions is 449 K. At this temperature a control experiment has been performed for the estimation of obtained model adequacy. Figure 3 represents the results of that experiment. It can be seen that calculated curves describe the experimental data satisfactorily. About 80 % VO is consumed for the formation of target VOgPP at optimal temperature. At this, grafting is performed statistically on the residue of one peroxide group of VO and, because of the dominant β-decomposition of PP radicals under the fusion conditions, by the end of polypropylene chain.

Full-factorial experiment performance

A study in accordance with a full-factorial orthogonal second-order design with three variable parameters (VO concentration, time, temperature) has been performed for the optimization of the reactive fusion of VO and PP. Graphic representation (Figure 4) of the responses allowed to obtain optimal values for the fusion conditions.

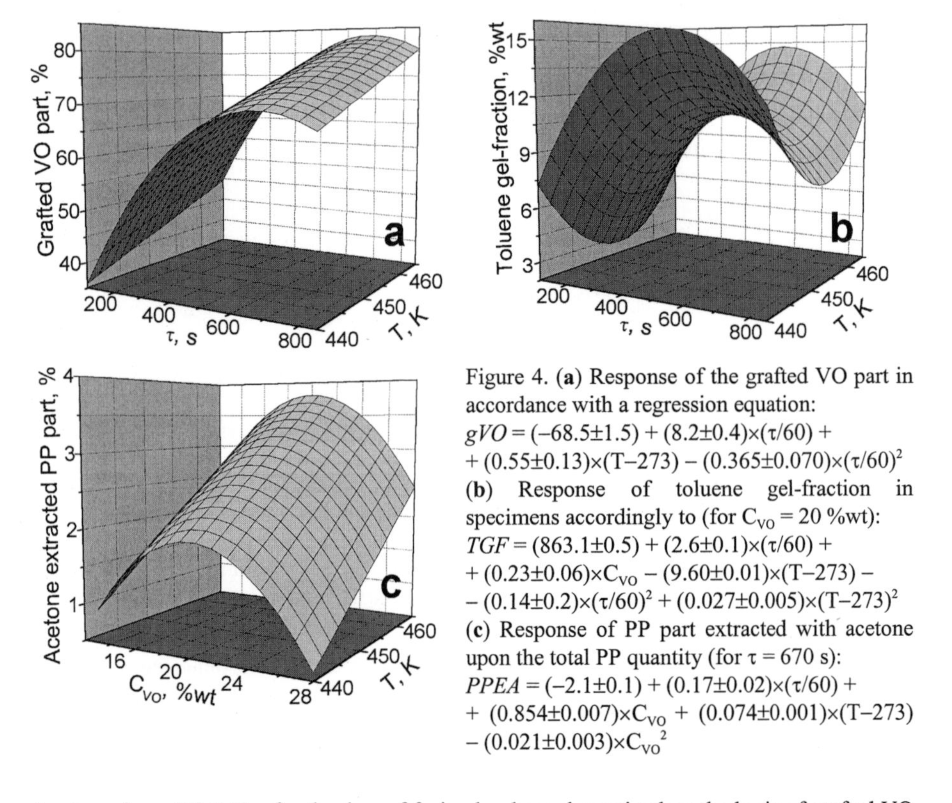

Figure 4. (**a**) Response of the grafted VO part in accordance with a regression equation:
$gVO = (-68.5\pm1.5) + (8.2\pm0.4)\times(\tau/60) + (0.55\pm0.13)\times(T-273) - (0.365\pm0.070)\times(\tau/60)^2$
(**b**) Response of toluene gel-fraction in specimens accordingly to (for $C_{VO} = 20$ %wt):
$TGF = (863.1\pm0.5) + (2.6\pm0.1)\times(\tau/60) + (0.23\pm0.06)\times C_{VO} - (9.60\pm0.01)\times(T-273) - (0.14\pm0.2)\times(\tau/60)^2 + (0.027\pm0.005)\times(T-273)^2$
(**c**) Response of PP part extracted with acetone upon the total PP quantity (for $\tau = 670$ s):
$PPEA = (-2.1\pm0.1) + (0.17\pm0.02)\times(\tau/60) + (0.854\pm0.007)\times C_{VO} + (0.074\pm0.001)\times(T-273) - (0.021\pm0.003)\times C_{VO}^2$

A value of $\tau = 670\pm160$ s for the time of fusion has been determined on the basis of grafted VO part response (Figure 4a) by differentiation of its regression equation. The optimal process temperature is T = 450±8 K determined from the toluene gel fraction (TGF) response (Figure 4b) based on minimizing the crosslinked product quantity. This agrees well with the temperature mentioned above (449±5 K), which was optimized independently from the coefficients of the target and non-target product formation.

At last, an optimal value of 20±3 %wt VO has been determined from the response of the amount of PP extractable with acetone (PPEA, Figure 4c). According to the activation parameters of the VEP peroxide group decomposition [6] 51 % of the peroxide groups decompose during 670 s at 449 K. This provides an active proceeding of PP radical β-decomposition and its grafting by one end. The latter fact is important to form the desired structure (Figure 5). This structure facilitates the diffusion of VOgPP in PP blends and its location across the blend interphases due to the PP fragments miscible with the PP-phase and the VO fragments incompatible with PP. The quantity of peroxide groups remained is satisfactory for the further VOgPP utilization for the creation of compatibilizing systems in PP blends with other polymers.

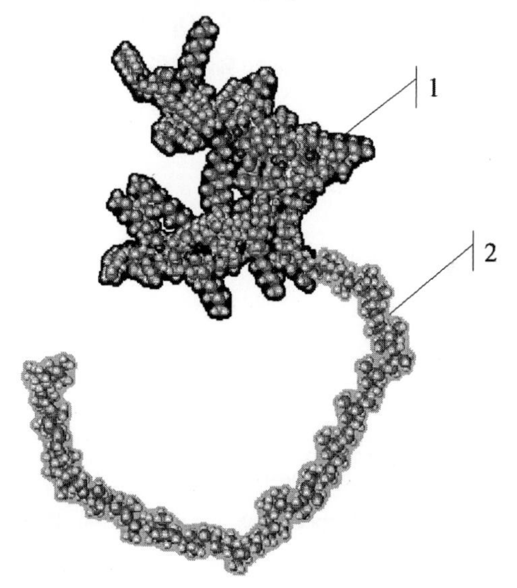

Figure 5. Model of a VOgPP molecule (9143 g/mol), optimized with HyperChem™ software using a MM+ method. **1** – anchor VO fragment; **2** – grafted PP fragment.

Application of peroxide precompatibilizer in blends

The VOgPP quantity necessary for the compatibilization of PP blends with other polymers can be calculated on the basis of the surface area of the dispersed PP phase in the blends and the area occupied by the anchor fragment of VOgPP molecule accounting for the Van der Waals radii (Figure 5, region dark circumscribed). The precompatibilizer synthesized under optimal conditions (an alloy of 35 %wt VOgPP in PP) has been applied for the compatibilization of both blend types: thermoplastic/thermoplastic and thermoplastic/thermoset blends.

In blends of PP with unsaturated polyester resin (UPR dissolved in styrene, see experimental part) the compatibilization with VOgPP resulted in dispersions with high stability (more than the observed 6 months). Noteworthy, the non-compatibilized composition was unstable and separated at once (one day as maximum) after preparation; which is in accordance with patent literature [7]. 2 months stability for such compositions is considered as applicable term.

Besides, an improvement of flexibility and impact strength of the specimens cured from the compatibilized compositions has been established. The maximal impact strength has been observed for the specimens obtained at a ratio of VOgPP to PP dispersed phase of 7 %wt, independent on the PP content in the blend (3 to 10 %wt) (Figure 6).

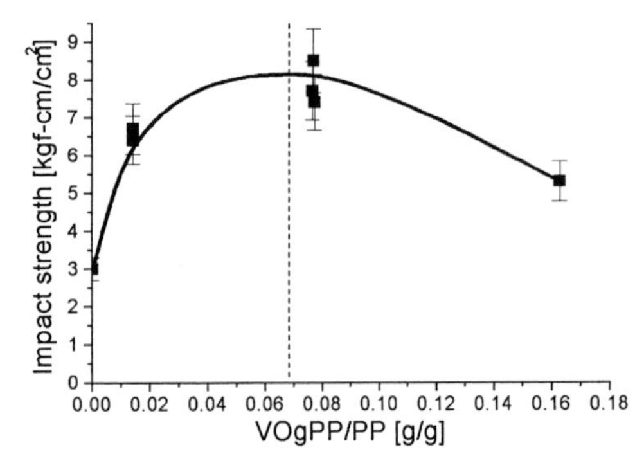

Figure 6. Impact strength of cured specimens obtained from PP/UPR compositions vs. VOgPP/PP ratio.

In addition, the precompatibilizer VOgPP has been successfully utilized for the compatibilization of thermoplastic/thermoplastic PP/PE and PP/PS blends (Figure 7), as reported earlier [6, 8].

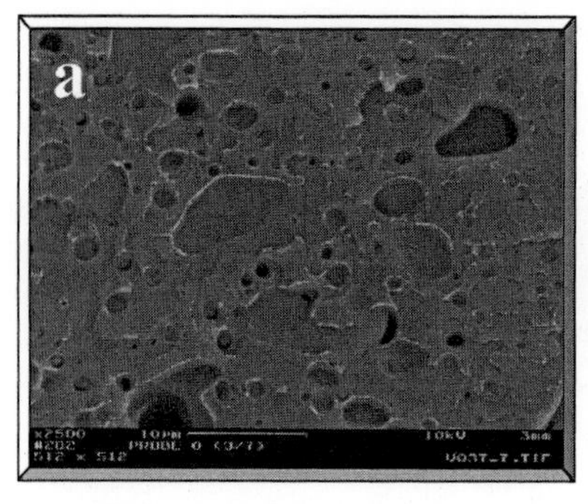

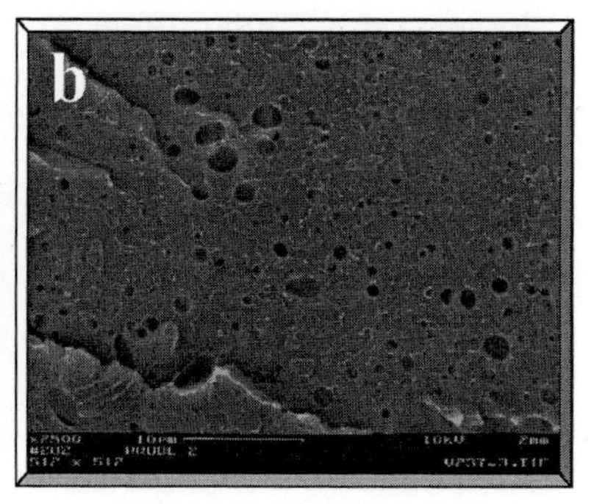

Figure 7. SEM-micrographs of cryofractured surfaces of PP/PS (30/70) blends (**a**) without VOgPP and (**b**) with 0.9 %wt VOgPP.

Conclusions

An approach has been elaborated for the synthesis of precompatibilizers universal for the compatibilization of polymer blends. In this approach a fragment of one polymer to be blended has to be grafted to an anchor copolymer. Fragments of the second blend copolymer will graft to the precompatibilizer during blending, forming the final compatibilizing structures.

Methods for the mathematical description and optimization of the synthesis conditions have been presented using as example the creation of an interfacial active precompatibilizer based on peroxide copolymer and polypropylene. These methods may be applied for the modeling of precompatibilizer formation for other polymer blends, which do not include polypropylene.

The universality of the precompatibilizer synthesized has been shown for the compatibilization of polypropylene with other thermoplastics and also with an UPR thermoset.

[1] R. P. Wool "*Polymer Interphaces. Structure and Strength*", Hanser Publishers, New York, 1995.

[2] M. J. Folkes , P. S. Hope, "*Polymer Blends and Alloys*", Blackie Academic and Professional, London, 1993.

[3] I. Piirma, "*Polymeric Surfactants*", Marcel Dekker, New York, Basel, Hong Kong, 1992, ch. 8.

[4] S. A. Voronov, V. Ya. Samaryk, S. M. Varvarenko, N. H. Nosova, Yu. V. Roiter, *Dopovidi NAN Ukrayiny (Reports of National Academy of Sciences of Ukraine)* **2002**, *6*, 147.

[5] N. Nosova, V. Samaryk, S. Varvarenko, Yu. Roiter, S. Voronov, Polymer surface activation with peroxide-containing copolymers, Tagung "Funktionspolymere für Systemlösungen", Darmstadt, 2002, abstract, p. 168.

[6] Yu. Roiter, V. Samaryk, N. Nosova, S. Varvarenko, P. Pötschke, S. Voronov, *Macromol. Symp.* **2001**, *164*, 377.

[7] US Patent 4,299,927 (1981), Eastman Kodak Company, inv.: J. R. Dombroski.

[8] S. Voronov, F. Böhme, J. Pionteck, P. Pötschke, V. Samaryk, Yu. Roiter, S. Varvarenko, N. Nosova, Polymer peroxide compatibilizer for reactive blending of carbon-chain polymers, Theses of EuroPolymer Foundation Congress, Eindhoven, 2001, S.3.N.214.

Adsorption of Poly(vinyl formamide-*co*-vinyl amine) onto Silica Particle Surfaces and Stability of the Formed Hybrid Materials

Florin Bucatariu,[1,3] *Frank Simon,*[1] *Stefan Spange,*[2] *Simona Schwarz,*[1] *Stela Dragan**[3]

[1] Institute of Polymer Research Dresden, Hohe Str. 6, D-01069 Dresden, Germany
[2] Chemnitz University, Department of Polymer Chemistry, Strasse der Nationen 62, D-09107 Chemnitz, Germany
[3] *Petru Poni* Institute of Macromolecular Chemistry, Aleea Grigore Ghica Voda 41 A, Ro-700487 Iasi, Romania

Summary: In order to produce silica/polyelectrolyte hybrid materials the adsorption of the polyelectrolyte poly(vinyl formamide-*co*-vinyl amine), P(VFA-*co*-VAm) was investigated. The adsorption of the P(VFA-*co*-VAm) from an aqueous solution onto silica surface is strongly influenced by the pH value and ionic strength of the aqueous solution, as well as the concentration of polyelectrolyte.
The adsorption of the positively charged P(VFA-*co*-VAm) molecules on the negatively charged silica particles offers a way to control the surface charge properties of the formed hybrid material. Changes in surface charges during the polyelectrolyte adsorption were studied by potentiometric titration and electrokinetic measurements.
X-ray photoelectron spectroscopy (XPS) was employed to obtain information about the amount of the adsorbed polyelectrolyte and its chemical structure.
The stability of the adsorbed P(VFA-*co*-VAm) was investigated by extraction experiments and streaming potential measurements. It was shown, that polyelectrolyte layer is instable in an acidic environment. At a low pH value a high number of amino groups are protonated that increases the solubility of the polyelectrolyte chains. The solvatation process is able to overcompensate the attractive electrostatic forces fixing the polyelectrolyte molecules on the substrate material surface. Hence, the polyelectrolyte layer partially undergoes dissolving process.

Keywords: dissolving mechanism; hybrid material; polyelectrolyte adsorption; poly(vinyl formamide-*co*-vinyl amine); silica; stability of polyelectrolyte layer

 DOI: 10.1002/masy.200450625

Introduction

The adsorption of polyelectrolytes onto various kinds of solid materials, for example inorganic oxide particles, resins, flat surfaces or fibres, is an intensively studied field of material research, due to the practical relevance of PE in adhesion, flocculation or wetting [1-4]. Decher *et al.* developed a new technology for synthesizing stable polyelectrolyte multilayers [5]. The authors used oppositely charged polyelectrolyte that have been adsorbed consecutively from the diluted aqueous solutions onto flat surfaces. Möhwald *et al.* [6-7] used similar techniques for synthesis of novel polymeric hollow spheres. The polyelectrolytes used by the mentioned authors, poly(sodium styrenesulfonate) or quaternary polyammonium salts are strong polyelectrolytes with a fixed number of charges along the polymer chain that do not form an equilibrium controlled by the surrounding media. The electrostatic interactions between the oppositely charged polyelectrolyte layers are also not influenced by the pH value of an aqueous solution. Hence, those multilayer systems are very stable in aqueous media. But these polyelectrolytes seem not suitable for further functionalization reactions under mild conditions.

P(VFA-*co*-VAm) is a weak polyelectrolyte containing a controllable number of reactive and accessible primary amino groups on the polymer backbone. It is an interesting candidate for silica surface functionalization in aqueous solutions [8-10], because it has a high chemical potential for subsequent derivatization reactions with electrophilic agents.

The present study reports the coating of silica particles with well-defined P(VFA-*co*-VAm) in order to control the surface properties of the formed hybrid material, like its surface charge density, acid-base characteristics and stability. Beside the entropy, the adsorption of P(VFA-*co*-VAm) and its adhesion is driven by electrostatic forces between the oppositely charged silica surface and the P(VFA-*co*-VAm)'s amino groups. However, the charge degree of P(VFA-*co*-VAm) depends on the protonation/deprotonation equilibrium of the amino groups. The protonation/deprotonation equilibrium can be easily influenced by the pH value of a surrounding aqueous media. Hence, the stability of the adsorbed polyelectrolyte layer can be also affected by the environmental conditions. Investigations of the polyelectrolyte layer stability show that weakly bounded polyelectrolyte molecules can be easily removed. But it seems impossible to remove the complete P(VFA-*co*-VAm) layer from the silica particles. Nevertheless, for practical applications and subsequent functionalization reactions it is recommended that the P(VFA-*co*-VAm) layer to be stabilized by cross-linking reactions [8].

To study the adsorption process of P(VFA-*co*-VAm) onto silica surfaces and the hybrid's stability, polyelectrolyte and potentiometric titration as well as electrokinetic measurements were employed. In addition, X-ray photoelectron spectroscopy (XPS) was applied to study the molecular structure of the produced hybrid materials.

Experimental

Materials

Kieselgel 60 (Merck, Darmstadt, Germany) a commercially available spherical silica was used as the inorganic substrate material. The diameter of the silica particles ranged between 15 and 40 μm. Kieselgel 60 particles are micro-porous (pore diameters ca. 5 nm).

The P(VFA-*co*-VAm) sample with molar mass of M_n = 15.000 g mol^{-1} was provided by BASF (Ludwigshafen, Germany). The degree of hydrolysis was 96 mol-%. This means that 96 mol-% of the former formamide groups of the PVFA chains were converted into amino groups.

Adsorption procedure

For the adsorption experiments, an aqueous stock solution containing 11.4 wt.-% P(VFA-*co*-VAm) was diluted to different concentrations. Then, 0.5 g silica was suspended in 50 ml of the diluted P(VFA-*co*-VAm) solutions. During the adsorption process, the suspension was gently shaken at room temperature. The modified particles were filtered off by slight suction. The hybrid materials were carefully washed with distilled water and dried in vacuum at 40 °C for 24 h.

X-ray photoelectron spectroscopy

XPS studies were carried out by means of an Axis Ultra photoelectron spectrometer (Kratos Analytical, Manchester, UK). The spectrometer was equipped with a monochromatic Al Kα ($h \cdot \nu$ = 1486.6 eV) X-ray source of 300 W at 15 kV. The kinetic energy of the photoelectrons was determined with a hemispheric analyzer set to pass energy of 160 eV for wide scan spectra. During all measurements electrostatic charging of the sample was over-compensated by means of a low-energy electron source working in combination with a magnetic immersion lens. Later, all recorded peaks were shifted by the same amount that was necessary to set the C 1s peak to 285.00 eV for saturated hydrocarbons.

Quantitative elemental compositions were determined from peak areas using experimentally determined sensitivity factors and the spectrometer transmission function.

Polyelectrolyte and potentiometric titration

Polyelectrolyte and potentiometric titration were performed with the particle charge detector PCD 02 (Mütek, Germany). For polyelectrolyte titration a 10^{-3} mol·l^{-1} solution of poly(sodium ethylenesulphonate) was used. Potentiometric titration was carried out between pH 3.5 and 10, using 0.1 mol·l^{-1} KOH and HCl, respectively.

Electrokinetic measurements

The electrokinetic measurements were performed as streaming potential experiments employing an EKA (Anton Paar, Austria). In a specially designed powder-measuring [11] cell the hybrid materials were packed as diaphragm, which was flown through by an aqueous KCl solution ($c = 0.001$ mol·l^{-1}). The pH dependent measurements start from pH = 6.8. The addition of 0.1 mol·l^{-1} HCl lowered the pH values, while the addition of 0.1 mol·l^{-1} KOH increased the pH values. After recording the streaming potential values of the acidic pH range the sample was exchanged for measuring the streaming potential values of the basic pH range. The values of the electrokinetic potential (zeta-potential ζ) were calculated from the measured streaming potential values according to the Smoluchowski's equation [11].

Results and discussion

Bare silica surface and P(VFA-co-VAm)

Silica particles suspended in water or in aqueous electrolyte solutions may be considered as a polyelectrolyte. The surface charging of the silica particles is either the result of dissociation processes of Brønsted-acidic silanol groups (Si–OH) forming negatively charged silanolate ions (Si–O$^-$) or proton adsorption yielding Si–OH_2^+ species. In a wide pH range the two charge generation mechanisms may take place simultaneously on the silica surface because the acidity of surface silanol groups can be quite different.

P(VFA-*co*-VAm) chains also have discrete charges; therefore, an important component of the driving force of P(VFA-*co*-VAm) adsorption onto silica surfaces should be the Coulomb force between the charge centres of the solid surfaces and the polycation.

Potentiometric titration in water shows that the point of zero charge (pzc = $pH|_{\Psi=0}$, where Ψ is the potential) is reached at pH = 2.5 for bare silica and at pH =10.2 for P(VFA-*co*-VAm) (Figure 1). As mentioned above, the dissociation reactions form negatively charged silanolate ions on the silica surface. The formation of ammonium groups lead to positively charged P(VFA-*co*-VAm) over a wide pH range (pH < pzc).

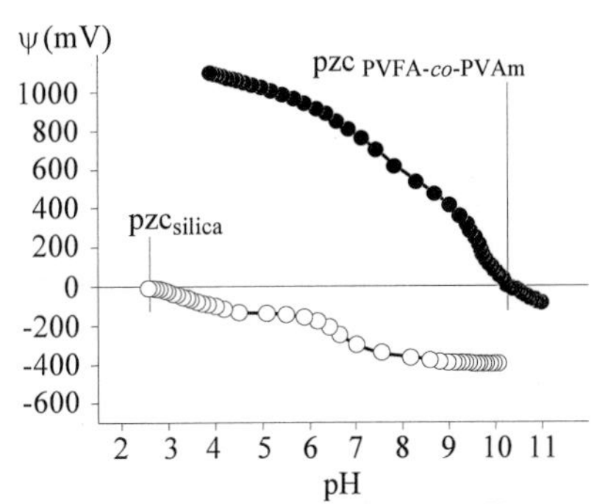

Figure 1. Potentiometric titration of bare silica (O) and P(VFA-*co*-VAm) (●)

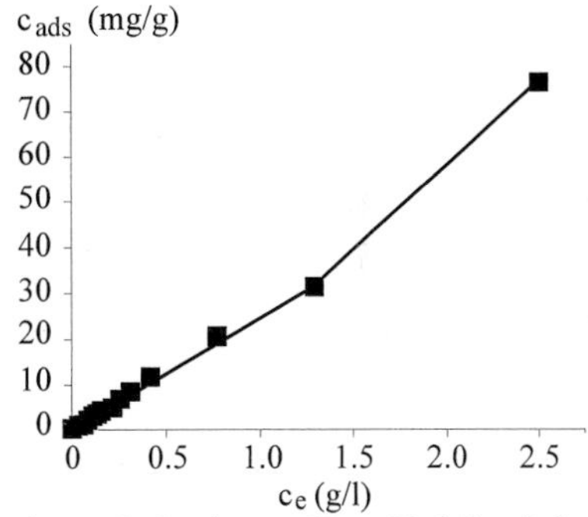

Figure 2. Isotherm (T = 22 °C) of the adsorption of P(VFA-*co*-VAm) onto silica from aqueous solution (c_e = equilibrium concentration of P(VFA-*co*-VAm) in the solution, c_{ads} = adsorbed amount of P(VFA-*co*-VAm)

Adsorption behaviour of P(VFA-co-VAm) onto silica surface

The adsorption behaviour of P(VFA-*co*-VAm) is shown in Figure 2. Surprisingly, results of polyelectrolyte titration of supernatant P(VFA-*co*-VAm) solution illustrate that no plateau region is observed (Figure 2). It is assumed that beside the adsorption of P(VFA-*co*-VAm) onto the silica particle surface an additional polyelectrolyte aggregation is initiated by increasing the polyelectrolyte concentration in the stock solution. Hence, the remained P(VFA-*co*-VAm) content in the supernatant solution appears smaller.

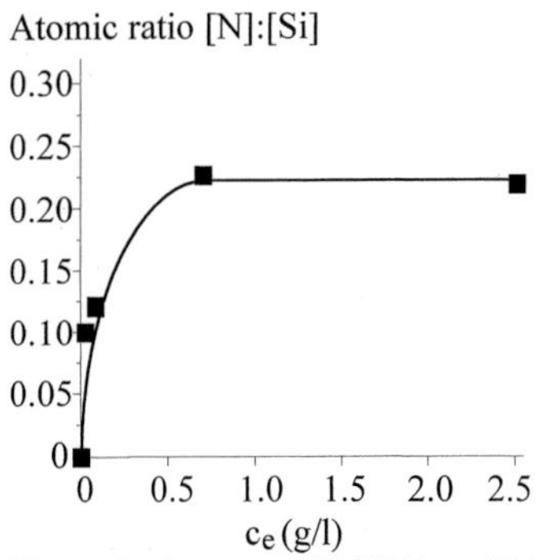

Figure 3. Amount of P(VFA-*co*-VAm) adsorbed onto silica surfaces determined by means of XPS. The amount of the P(VFA-*co*-VAm) is indicated by the nitrogen content [N], the number of unoccupied silica surface sites is indicated by the silicon value [Si] (c_e = equilibrium concentration of P(VFA-*co*-VAm) in the solution)

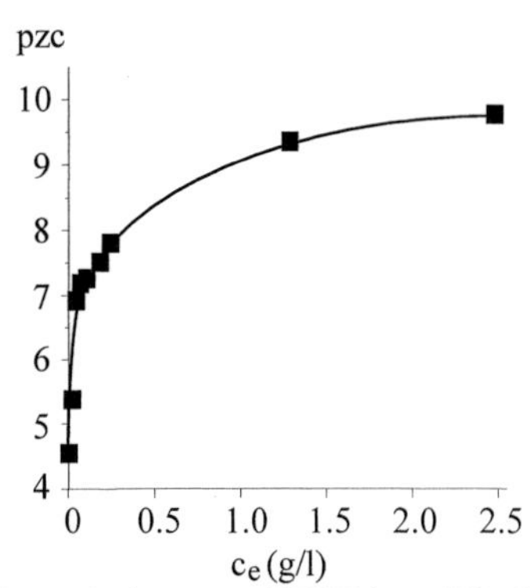

Figure 4. Amount of P(VFA-*co*-VAm) adsorbed onto silica surfaces measured by potentiometric titrations (c_e = equilibrium concentration of P(VFA-*co*-VAm) in the solution, the adsorbed amount of P(VFA-*co*-VAm) is expressed by the shift of the point of zero charge, pzc)

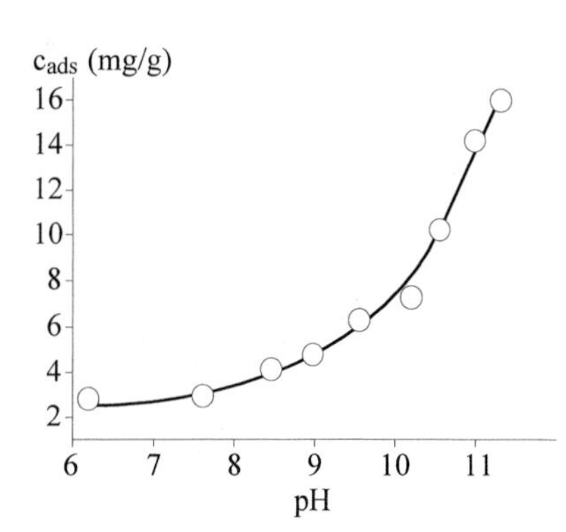

Figure 5. Influence of the stock solution's pH value on the P(VFA-*co*-VAm) amount adsorbed onto silica ($c_{P(VFA\text{-}co\text{-}Vam)}$ = 10^{-2} mol·l^{-1}, c_{ads} = adsorbed amount of P(VFA-*co*-VAm))

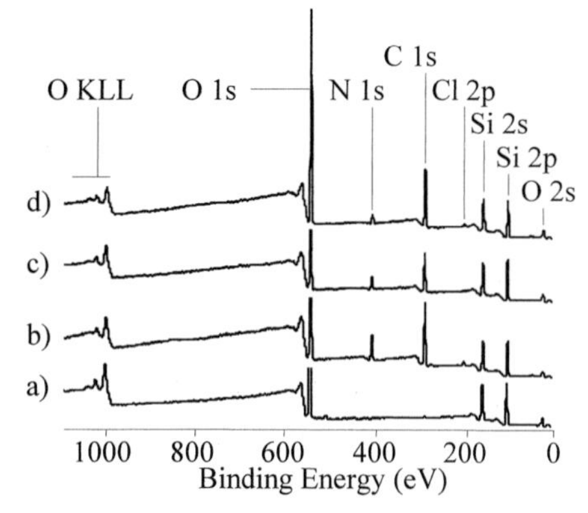

Figure 6. Wide-scan XPS spectra of bare silica (a), P(VFA-*co*-VAm) adsorbed onto silica (b), P(VFA-*co*-VAm) adsorbed onto silica after extraction (c), and P(VFA-*co*-VAm) adsorbed onto silica pre-treated with 2 mol·l^{-1} HCl (d)

During washing the freshly prepared P(VFA-*co*-VAm)/silica hybrid materials, all weakly bounded macromolecules will be removed. Only the polyelectrolyte molecules that are bound via strong interaction forces remain on the particle surface. Adsorption isotherms determined from XPS measurements (Figure 3) and potentiometric titration (Figure 4) of the P(VFA-*co*-VAm)/silica hybrid particles confirm this assumption by showing a clear plateau region. In the case of XPS it can be assumed that the N 1s peak appears only from the polyelectrolyte layer and Si 2p is a specific signal of the substrate material. Then, the [N]:[Si] atomic ratio should be an accurate measure of the amount of adsorbed polyelectrolyte. Potentiometric titration shows an increased shift of the pzc with increasing the polyelectrolyte concentration in the stock solution, which indicates the progress of compensation of the originally negative charge carriers on the silica surface by the adsorbed polycation (Figure 4).

Figure 5 shows the influence of the solution pH value on the adsorbed amount of P(VFA-*co*-VAm). At low pH values, the macromolecules have a high charge density and as a result of this the chains are stiff and stretched. Hence, the P(VFA-*co*-VAm) amount adsorbed onto silica surface is low. As can be seen in Figure 1 an increased pH value decreases the charge density of the P(VFA-*co*-VAm) molecules. The macromolecules will be more flexible and no electrostatic repulsion between the P(VFA-*co*-VAm) molecules could prevent the adsorption process. A higher amount of P(VFA-*co*-VAm) is adsorbed onto the silica surface (Figure 5).

Stability of the P(VFA-co-VAm) layer

The stability of the formed P(VFA-*co*-VAm) layers was tested by extraction experiments. Samples were put in a Soxhlet apparatus and extracted with distilled water over 72 h. As can be seen in Figure 6 and Table 1 during the extraction procedure the relative amount of nitrogen expressed as ratio [N]:[Si] is significantly decreased.

From this result it can be concluded that the adsorbed polyelectrolyte layer is not stable. The formation of ionic species on the polyelectrolyte chains may enhance the electrostatic interaction to the negatively charged silica surface, but with increasing the charging degree the solubility of the polyelectrolyte molecule is also increased. New amounts of pure water disturbs the primary adjusted adsorption equilibration and dissolves part of the weekly adsorbed macromolecules.

The partially removal of the polyelectrolyte layer was also observed in streaming potential experiments. It is expected that during the experiment the streaming liquid generates an additional mechanical stress, which should enhance the abrasion of the polyelectrolyte layer.

Table 1. XPS quantification report of P(VFA-*co*-VAm)/silica hybrid material before and after a Soxhlet extraction, and after the pre-treatment with 2 mol·l^{-1} HCl. The atomic ratio [N]:[Si] indicates the adsorbed amount of P(VFA-*co*-VAm).

	Atomic ratio [N]:[Si]		
	Before Soxhlet extraction	After Soxhlet extraction	After pre-treatment with 2 mol·l^{-1} HCl
Bare silica	0.000	0.000	0.000
P(VFA-*co*-VAm) adsorbed onto silica	0.231	0.204	0.178

Figure 7 shows the zeta-potential of a P(VFA-*co*-VAm)/silica hybrid material in dependence on the pH value of the streaming aqueous KCl solution. In the pH range 5 ≥ pH ≥ 9, the positive zeta-potential remains nearly constant (plateau, region *A* in Figure 7). Here, all amino groups that are able to form positively charged ammonium salt species are protonated. At pH > 9 the ammonium species are gradually deprotonated by the excess of OH^- ions in the aqueous solution and as a consequence the zeta-potential value decreased. The isoelectric point (iep = $pH|_{\zeta=0}$) of the P(VFA-*co*-VAm)/silica hybrid excellently agrees with the pzc of the P(VFA-*co*-VAm) solution (Figure 1). This indicates that the P(VFA-*co*-VAm) fully covers the silica surface and determine the charging behavior of the hybrid material. In the acidic range (pH < 5, region *B* in Figure 7), a high content of protonated amino groups is expected and the measured zeta-potential should agree with the potential of the plateau phase. However, at pH ≈ 3.4 the zeta potential shows a minimum. That minimum can be explained by the instability of the adsorbed polyelectrolyte layer in an acidic environment. The stepwise addition of HCl results in an increased solvatation of macromolecules, which is the first step of their dissolution. The streaming liquid removes the dissolved polymers, and support so the dissolving process by changing the adsorption equilibrium to the side of the solved species. In order to proof the assumption mentioned above, the hybrid material was treaded with 2 mol·l^{-1} HCl. As can be seen in Figure 7, the HCl treatment does not change the shape of the function $\zeta = \zeta(\mathrm{pH})$ in the region *A*. The minimum observed for the non-treated sample was not found. The plateau phase is on a lower level and the iep is shifted to a lower value of pH. The iep corresponds with the number of Brønsted base sites on the sample surface. The shift of the iep to a lower value indicates that the weakly bound P(VFA-*co*-VAm) layer is removed by the HCl treatment. Hence, a decrease of the zeta-potential in the region *B* cannot be observed for the HCl pre-treated sample.

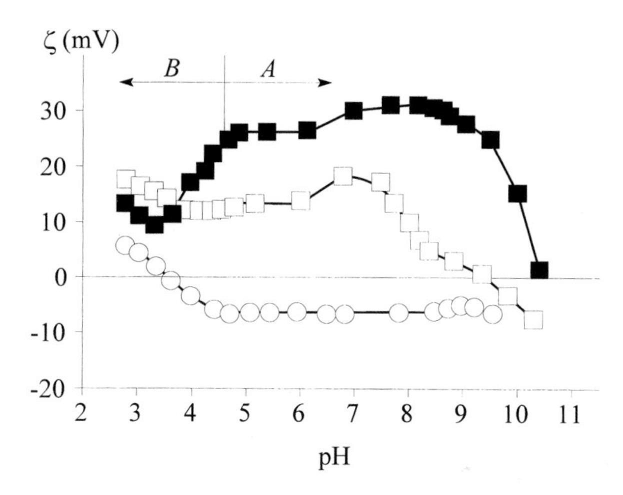

Figure 7. Zeta-potential values determined from streaming potential measurements in dependence on pH of an aqueous 0.001 mol·l^{-1} KCl solution. The measurements were carried out with bare silica (○), P(VFA-*co*-VAm)/silica hybrid material ([N]:[Si] = 0.231, ■), and the same P(VFA-*co*-VAm)/silica hybrid material pre-treated with 2 mol·l^{-1} HCl for 24 h ([N]:[Si] = 0.178, □)

The curves of the zeta-potential-pH dependence of the P(VFA-*co*-VAm)/silica hybrid materials show that the polyelectrolyte layer is not stable in strong acidic aqueous solutions. They show also that it is not possible to completely wash-off the polyelectrolyte layer. For subsequent functionalization reactions and the hybrid's applications it is recommended to enhance the polyelectrolyte layer stability by cross-linking reactions. The free and reactive amino groups of the P(VFA-*co*-VAm) chains offer the possibility to carry out such crosslinking reactions employing bi-functional low-molecular weight agents, metal clusters or fullerenes [8].

Conclusions

The adsorption of P(VFA-*co*-VAm) onto silica particle surfaces was studied. During the adsorption process, the excess of the polyelectrolyte forms aggregates that were weakly adsorbed on the silica surface. Those polyelectrolyte aggregates can be easily washed-off. On the silica particle surface remains a certain amount of P(VFA-*co*-VAm) that can be controlled by the adsorption conditions. The adsorbed polyelectrolyte layer is partly instable in acidic aqueous media. The protonation of the amino groups increases the solubility of the adsorbed P(VFA-*co*-VAm) molecules, attractive electrostatic interactions can be overcompensated and part of the polyelectrolyte chains can be desorbed, while the major part of the adsorbed polyelectrolyte molecules remain stable. For the subsequent derivatization reactions, which can be carried out using the reactive and accessible primary amino groups of the P(VFA-co-VAm)/silica hybrid particles, cross-linking reactions are recommended to stabilize the adsorbed P(VFA-co-VAm) layer.

[1] H. Dautzenberg, W. Jaeger, J. Kötz, B. Philipp, C. Seidel, D. Stscherbina *Polyelectrolytes: formation, characterization and application* **1994**, Carl Hanser, Munich.

[2] D. Bauer, E. Killmann, W. Jaeger, *Prog. Colloid Polym. Sci.* **1998**, *109*, 161.

[3] C. Huguenard, J. Widmaier, A. Elaissari, E. Pefferkorn *Macromolecules* **1997**, *30*, 1434.

[4] E. Gailliez-Degremont, M. Bacquet, J. Laureyns, M. Morcellet *J. Appl. Polym. Sci.* **1997**, *65*, 871.

[5] G. Decher *Science* **1997**, *227*, 1232 and references therein.

[6] G. B. Sukhorukov, E. Donath, H. Lichtenfels, E. Knippel, M. Knippel, H. Möhwald *Coll. Surf.* **1998**, *A137*, 253.

[7] E. Donath, G. B. Sukhorukov, F. Caruso, S.A. Davis, H. Möhwald *Angew. Chem.* **1998**, *110*, 2324 and references therein.

[8] S. Spange, T. Meyer, I. Voigt, M. Eschner, K. Estel, D. Pleul, F. Simon *Adv. Polym. Sci.* **2003**, *165*, Springer Verlag, Heidelberg (2003), ISBN 3-540-00528-5.

[9] E. Poptoshev, M.W. Rutland, , P.M. Claesson *Langmuir* **1999**, *15*, 7789.

[10] T. Serizawa, , K. Yamamoto, M. Akashi *Langmuir* **1999**, *15*, 4682.

[11] H.J. Jacobasch, F. Simon, C. Werner, C. Bellmann *Technisches Messen: Sensoren, Geräte, Systeme* **1996**, *63*, 447.

Macromol. Symp. **2004**, *210*, 229-235

Switching and Structuring of Binary Reactive Polymer Brush Layers

Leonid Ionov,[1] *Manfred Stamm,*[1] *Sergiy Minko,*[2] *Frank Hoffmann,*[3] *Thomas Wolff**[3]
[1] Institut für Polymerforschung Dresden, Hohe Strasse 6,Dresden, 01069, Germany
[2] Department of Chemistry, Clarkson University, Potsdam, New York 13699-5614, USA
[3] Institut für Physikalische Chemie, Technische Universität Dresden, 01062 Dresden, Germany
E-mail: thomas.wolff@chemie.tu-dresden.de

Summary: Switchable binary polymer brushes grafted to Si-wafers were prepared from hydrophilic and hydrophobic polymer components. When exposed to solvents, either the hydrophobic or the hydrophilic component extends in to the liquid phase, depending on the polarity of the solvent. The hydrophilic component was poly-2-vinylpyridine; the hydrophobic component was made photocrosslinkable in that a polystyrene copolymer containing a photodimerizing chromophore was used. In this system surfaces differing in water contact angle between 60° and 100° can be produced by variation of the solvent. The chromophore was phenylindene, which forms crosslinks upon direct UV-irradiation. Therefore, the polystyrene component can be fixed in the extended or collapsed state. It will be shown that by irradiation through an appropriate mask, surfaces can be structured and the structures fixed.

In both the systems structural patterns differing in surface properties were produced and fixed photochemically.

Keywords: networks; photochemistry; stimuli-sensitive polymers surfaces; thin films;

Introduction

Tethered polymer layers[1] were shown to be effective for the colloidal stabilization[2], size exclusion chromatography[3], control of adhesion[4], lubrication[5], liquid-crystal displays[6], biomaterials[7], etc. Chemical grafting of polymers ensures the stable polymer-solid interaction via covalent bonds that is very important for the most applications. Via tuning the parameters, which control the brush, properties (grafting density, chain length, chemical composition of the chains) one can approach a variety of nano-scale structures and thin film properties.

 DOI: 10.1002/masy.200450626

A highly interesting class of brushes is the mixed polymer brushes consisting of two incompatible polymers tethered to the substrate. Anchoring prevents the macroscopic segregation of polymers[8]. The theoretical analysis of nano-scale phase segregation in binary brushes results in a complicated phase diagram and plenty of thin film morphologies[9]. Depending on the solvent quality, layered and rippled phases (or their mixture) were observed experimentally. A transition between different morphologies upon external stimuli (solvent, temperature, etc.) results in switching of the surface properties of the film, e.g., switching from hydrophilic to hydrophobic[10] or from smooth to rough.

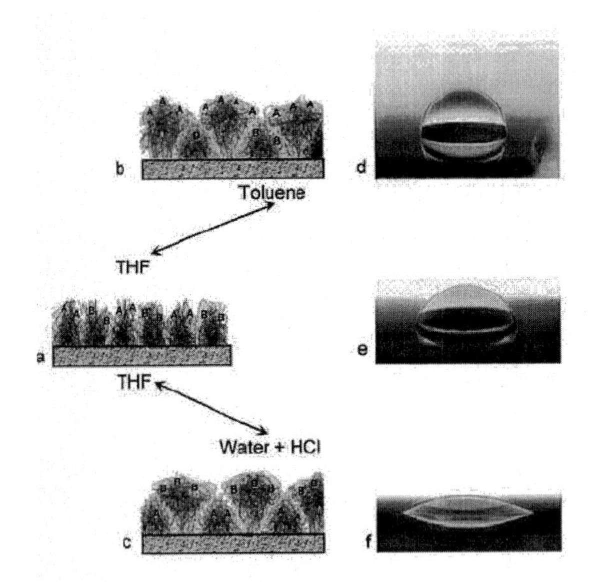

Figure 1. Wetting of the heterogeneous polymer brush composed of polystyrene and poly(2-vinylpyridine) chains on a Si-wafer

The concept of switching of surface properties is illustrated on Fig. 1. At least two different polymers A and B containing functional groups F1 and F2, respectively, may be grafted to a solid substrate to form a heterogeneous polymer brush (Figure 1a). The ratio between chain length of the polymers A and B, the composition of the brush and the nature of the functional groups F1 and F2 affect layer properties such as roughness, wettability, reactivity, adhesion to other materials, biocompatibility, etc. It is clear that interplay of the above-mentioned parameters of the heterogeneous brush allows approaching a wide variety of diverse surface properties. Besides, the heterogeneous brush has the unique capability to change the properties responding on a change of surroundings. Let us assume that the initial stage (Figure1a) was obtained in a nonselective solvent with respect to both polymers A and B. Then a change of the surroundings by addition of a selective solvent (e.g. for polymer A), or a change of pH (if one or both A and B are sensitive to pH) brings about selective swelling of the polymer A, and collapse of chains of the polymer B (Figure 1b). If we use a selective solvent for the polymer B the inverse behavior of the layer is expected (Figure 1c). A degree of coverage of the top of the brush by one of the polymers depends on the composition of surrounding and layer and can be adjusted very carefully to any desirable value. Consequently, one may obtain the desirable composition of the top layer with respect to F1 and F2 as a respond to the composition of the surrounding (liquid or a gas phase).

Crosslinking is a well-known method of modifying bulk properties of polymers. For the thin polymer layers the field of photocrosslinking of monomers, oligomers and polymers has grown into an important branch of polymer science. The photocrosslinkable and photofunctional polymers find wide applications in the field of optical photolithography, for printing plates, photocurable coatings, photorecorders, photoconductors and photosensitizers for organic synthesis. The widely used photoresists are based on various photoactive groups like cinnamate esters, acetylene, stilbene, azide, arylidene, etc. The most common method to produce crosslinks is the use of low molecular crosslinking agents. However, mixing and distribution of the crosslinking agent adds additional problems. This can be avoided if one of the comonomers is capable of crosslinking. Pendant photoreactive groups are introduced during homo- and copolymerization. Examples of utilizations of this approach are polyvinylcinnamates, acetylenes, and polymers with bound anthryl, stilbazolium, tetrazole, benzylidenephthalimidine, and stilbene substitutents. Photocrosslinking is widely used for the modification of thin polymer films. Nevertheless, this process was not applied to the polymer brush-like layers. Here we report on two distinct attempts in this field.

Polyisoprene system

We have chosen step - by - step "grafting to"[8] of the polyisoprene (PI) – poly-2-vinylpyridine (P2VP) pair of polymers onto a Si-wafer. The choice of PI as a hydrophobic component is based on the very high concentration of double bounds. Crosslinking of them can ensure a dense polymer network, enough to suppress switching.

Data of water contact angles are reported in Table 1 and show that the mixed brush can be switched from a hydrophobic surface (upon exposure to toluene) to a hydrophilic surface when water at pH = 2) is substituted for toluene. The hydrophobicity of the brush-like layer is easily accounted for by the selective interactions of toluene with PI. In contrast, PI is insoluble in low pH water, which is a protonating agent for P2VP. The top of the brush is then hydrophilic and ionized. In contact with ethanol, a selective solvent for P2VP, the top layer is non-ionized P2VP, and the surface is less hydrophilic (Table 1). After exposure to toluene, the brush has been irradiated by UV light through a mask to cross-link polyisoprene selectively in the illuminated areas. Table 1 shows that the contact angle of water for the film previously exposed to toluene changes from 80° before local cross-linking to 69° after crosslinking. This can be attributed to the modification of the polymers by the photoinitiator. To minimize this effect and to extend the range of switching, it would be effective to select appropriate photoinitiators (the selection should be based on polarity/ solubility of the photoinitiator molecules). The most important fact is that switching is suppressed in the irradiated area. The non-irradiated area, however, retains its capability of switching.

Table 1. Water Contact Angle Data on non-UV-irradiated and on UV-irradiated areas upon exposure to different solvents

solvent	mixed PI/P2VP brush	
	Θ_A on non – UV irradiated area	Θ_A on UV irradiated area
toluene	80°	69°
ethanol	69°	69°
acidic water (pH=2)	42°	69°

The different behavior of irradiated and non-irradiated areas in different media can be used for sensors. After photoprinting at the micrometer scale, the patterned brush is washed with ethanol for several minutes, dried, and then exposed to water vapor. No pattern is developed on the surface (Figure 2b). The same treatment is repeated with water (pH = 2) instead of ethanol. The local condensation of water droplets reveals quite clearly the image printed on the brush (Figure

2a). A nicely contrasted image is formed because the light reflection changes with the size and shape of the water droplets. The image can be erased (Figure 2b) merely by washing with ethanol or neutral water (pH = 6.5). This general observation can be repeated at will, which is the evidence that the film is sensitive to acidic water and that this specific interaction can be repeatedly visualized.

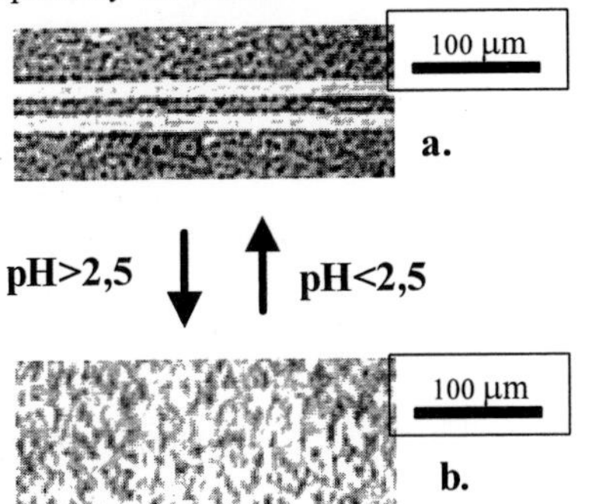

Figure 2. Adsorption of water drops (optical microscopy, bar – 100 μm) on the polymer brush with developed patterns by exposure to water with pH<2,5 (a.) and hidden by water with pH>2,5 (b.)

Phenylindene system

Another switchable system different from the one above was also investigated. Instead of polyisoprene a photosensitive styrene/2-(4-styryl)-indene copolymer was used as a photocrosslinkable hydrophobic component, cf. [11]. The hydrophilic part of the binary brush was again taken by poly-2-vinylpyridine. Photochemical cross-linking occurs when a ground state phenylindene moiety encounters an excited phenylindene moiety during the lifetime of the latter. Thereby defined crosslinks are brought about via photodimerization, i.e. the formation of cyclobutane derivatives (Figure 3). Compared to systems that need a photoinitiator such a system is advantageous in that the degree of cross-linking can be adjusted by switching on and off the light and in that no photoinitiator affects the surface properties.

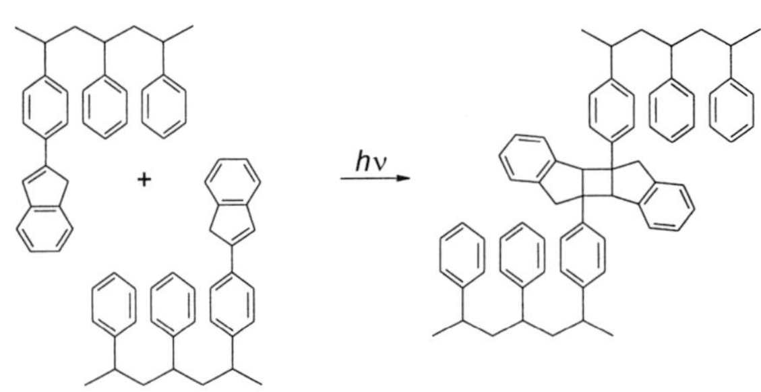

Figure 3. Mechanism of copolymer cross-linking via phenylindene photodimerization

In order to synthesize the polymer containing the phenylindene chromophore it was necessary to use a monomer with a protected double bond at the five membered indene ring to avoid network formation during polymerization. This special monomer, i.e. 2-trimethylsiloxy-2-(4-styryl)-indane, is formed in a two-step-synthesis: 1. coupling of 2-indanone to 4-bromostyrene, 2. addition of trimethylsilylchloride.

The copolymer used here contained 9 % styrylindene.

P2VP and the styrene/2-trimethylsiloxy-2-(4-styryl)-indane-copolymer were attached simultaneously to the Si-wafer by a photochemical grafting method[12], which makes use of the photoreductive addition of hydrocarbons to benzophenone. This photochemical grafting method has several advantages: special anchor groups at the polymer chain ends are not required, solid supports with benzophenone-layers can be stored easily, and the grafting conditions are mild.

Table 2. Water contact angle data of photochemically attached layers upon exposure to different solvents

system	solvent	
	toluene	hydrochloric acid pH=2
P2VP/PS (1:1) brush with 9 % cross-linker (protected)	92°	57°
PS brush with 9 % cross-linker (protected)	90°	90°
P2VP brush	70°	50°

After the removal of the trimethylsilyl- protection groups by acidic hydrolysis and the according formation of the phenylindene double bond, the grafted polymer brush is sensitive to light with wavelengths below 400 nm. After exposure to hydrochloric acid at pH = 2, the brush has been irradiated by UV light through a mask to cross-link phenylindene groups selectively in the illuminated areas. Then the brush was switched to the other state by treatment with toluene and the whole surface was irradiated to fix the pattern. Figure 4 shows the pattern produced as the result of this procedure. The visibility of the pattern here does not rest on water vapor condensation as in the polyisoprene system.

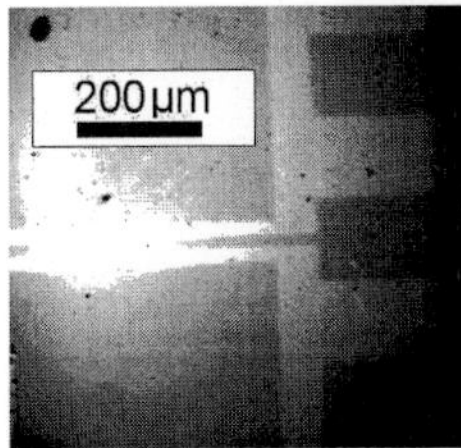

Figure 4: Photographical image of photochemically patterned Si-wafer. Light areas have been initially illuminated with UV light and show different reflectivity as compared to the dark areas irradiated after switching the state of the brush

Conclusions

Structural patterns (differing in surface properties) in switchable binary polymer brushes can be generated and fixed via selective photochemical cross-linking.

Acknowledgement

We are grateful for financial support of this project by DFG (SFB 287 "Reactive Polymere").

[1] A Halperin, M. Tirel, T. P. Lodge, *Adv. Polym. Sci.* **1992**, *100*, 31.
[2] L. Quali, J. François, E. Pefferkorn, *J. Colloid. Interface Sci.* **1999**, *214*, to appear. 215 (1): 36-42
[3] Y. Cohen, R. S. Faibish, M. Rovira-Bru, in *„Interfacial Phenomena in Chromatography"*, Pefferkorn, E. Ed.; Marcel Dekker, Inc: New York **1999,** Chapter 7.
[4] H. R. Brown, V. R. Deline, P.F. Green, *Nature* **1989**, *341*, 221.
[5] J. Klein, *Annu. Rev. Mater. Sci.* **1996**, *26*, 581.
[6] B. Peng, D. Johannsmann, J. Rühe, *Macromolecules* **1999**, *32*, 6759.
[7] J. Rühe, R. Yano, J. S. Lee, P. Köberler, W. Knoll, A. Offenhäusser, *J.Biomater. Sci.Polymer Edn* **1999**, *10,* 859.
[8] S. Minko, S. Patil, V. Datsyuk, F. Simon, K.-J. Eichhorn, M. Motornov, D. Usov, I Tokarev, M. Stamm, *Langmuir* **2002**, *18*, 289-2969
[9] S. Minko, M. Müller, D. Usov, A. Scholl, C. Froeck, M. Stamm, *Phys. Rev. Lett.* **2002**, *88*, 035502-1
[10] S. Minko, M. Müller, M. Motornov, M. Nitschke, K. Grundke, M. Stamm, *J. Am. Chem. Soc.* **2003**, 125 (13), 3896-3900.
[11] R. Schinner, T. Wolff, *Colloid Polym. Sci.* **2001** (279) 1225-1230.
[12] O. Prucker, C. A. Naumann, J. Rühe, W. Knoll, C. W. Frank, *J. Am. Chem. Soc.* **1999**, *121,* 8766-8770.

Reactive Methacrylate Systems for Dental Bioadhesives

Mirosław Gibas,[*1] Tomasz Kupka,[2] Marta Tanasiewicz,[2] Witold Malec[3]*

[1] Department of Organic Chemistry, Bioorganic Chemistry and Biotechnology, Silesian University of Technology, Gliwice, Poland
E-mail: gibas@polsl.gliwice.pl
[2] Department of Preclinical Dentistry, Silesian Medical Academy, Katowice, Poland
[3] Institute of Non-Ferrous Metals, Gliwice, Poland

Summary: The procedure of restoring missing crown part of a human tooth with use of experimental materials, including new functional dimethacrylate monomers, is presented. The essential steps are precise evaluation of root canal cavities and providing a good adhesion of restorative materials to tooth structures. Magnetic resonance microscopy is considered to be a valuable supporting technique in restorative dentistry.

Keywords: adhesives; dental polymers; dimethacrylates; magnetic resonance microscopy; shear bond strength

Introduction

Dental bioadhesives are a sort of restorative materials destined to provide sealing and mechanical stabilization of a restoration of a hard tooth tissue made from an artificial filling material. The prefix *bio* suggests that a material is at least biocompatible, some healing action towards neighbouring tissue is desirable as well.

The dental adhesive systems are composed of various methacrylate monomers and an initiating system; the latter may be either a photoinitiator in light-cured system or peroxide/tertiary amine in chemically-cured one. The most important ingredients of dental adhesives are hydrophilic methacrylates containing reactive functional groups, e.g. carboxylic, anhydride, amine and phosphate ones. Those provide an affinity either to collagen or hydroxyapatite contained in dentin. A reactive group should be joined via some spacer group to the polymerizable methacrylate double bond. The latter copolymerizes with other monomers contained in an adhesive resin and in a binder of a composite filling material [1, 2].

Some more complex restorations require reinforcing by use of devices made of metals or

 DOI: 10.1002/masy.200450627

metal alloys. Adhesives destined for such a purpose should be capable of bonding metal surface as well.

We have developed a series of new multifunctional methacrylate monomers, having the structure below, intended to be employed in dental adhesives of an universal type, i.e. capable to bond:

- composite restoration to hard tooth tissue;
- composite to amalgam and other metal alloys
- amalgam restoration and other metal alloys to hard tooth tissue;

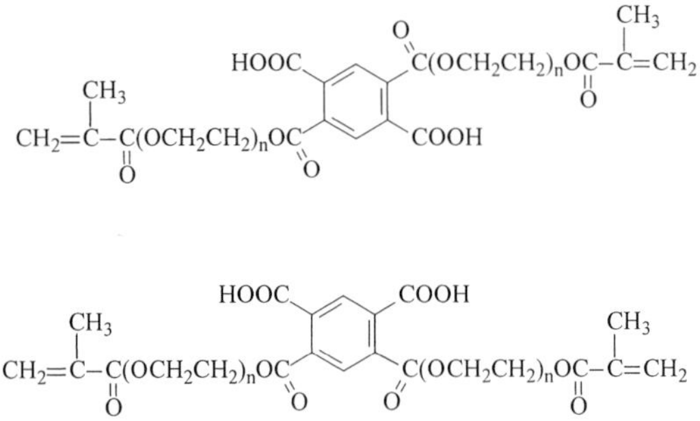

$CH_2{=}C(CH_3){-}C(O)(OCH_2CH_2)_nOC(O){-}C_6H_2(COOH)_2{-}C(O)(OCH_2CH_2)_nOC(O){-}C(CH_3){=}CH_2$

para isomer

n=2 PM2EDM
n=3 PM3EDM
n=4 PM4EDM

meta isomer

The PM2EDM, PM3EDM and PM4EDM monomers may be thought to be homologues of PMDM, well-known HEMA-based dental monomer (n=1 in the scheme) [3]. However, structural features of the new series are advantageous, since flexible oxyethylene spacers provide both functionalities more unconstraint in action as well as miscibility with other methacrylate resins.

This work is aimed to evaluate a performance of the new methacrylate-based adhesive systems employed in a complex, total restoration of a missing tooth crown, involving an individual intra-root canal fixation (so-called “individual post’) made of metal alloys. Since a precise knowledge on root canal cavities shape and volume is essential to cast the post precisely, the procedure has been supported by MRM (Magnetic Resonance Microscopy) technique.

Materials and methods

The monomers were synthesized and the adhesive systems were formulated according to the recent report [4]. Both light-cured composite and cobalt-based alloy used to the restoration are experimental materials elaborated by the authors [5]. Experiments were performed onto extracted human teeth. Magnetic resonance microscope working at 4.7 T equipped with DRX console was used to visualize tooth root canals interior, more details are given in [6, 7]. Shear bond strength of composite material to cobalt-based alloy was measured with the aid of Instron apparatus according to the procedure previously reported [5].

Results and discussion

When rebuilding missing crown parts of teeth by use of restorative materials and intra-root fixation, the dentists need to perform precise mapping of the shape and the volume of root cavities. Few imaging techniques are known, among which magnetic resonance seems to be promising as a support to the impression mapping methods [6, 7]. Figure 1 presents exemplary images obtained *in vitro* for the extracted human tooth. A series of images enables creation of three-dimensional picture of a cavity.

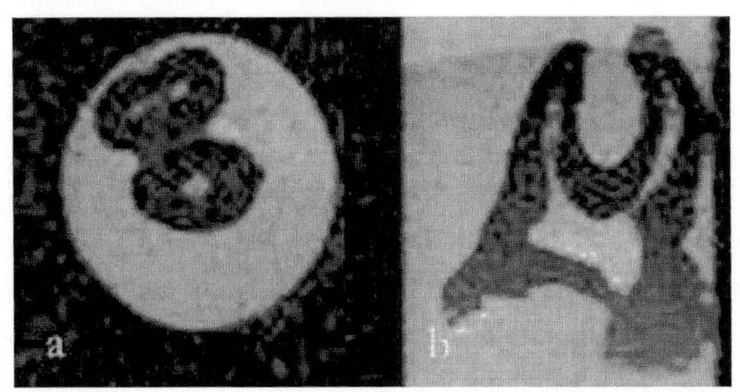

Figure 1. The MRM images of a human tooth root (a) transverse section, (b) longitudinal section.

The data obtained from MRM have been used to verify an impression negative of root canals before subsequent preparation of an appropriate mould to cast the individual post using an experimental cobalt-based alloy (Figure 2a). The tooth root canals were coated with the self-cured adhesive resin based on the PM3EDM monomer to provide bonding of the alloy to the dentin and then the post was introduced therein. After that, the crown part of a tooth was rebuilt gradually, layer by layer, with light-cured composite material. Prior to applying the first layer, the light-cured version of adhesive resin as above was introduced onto the upper surface of the post to provide bonding of the alloy to the composite. The final restoration is shown in Figure 2b.

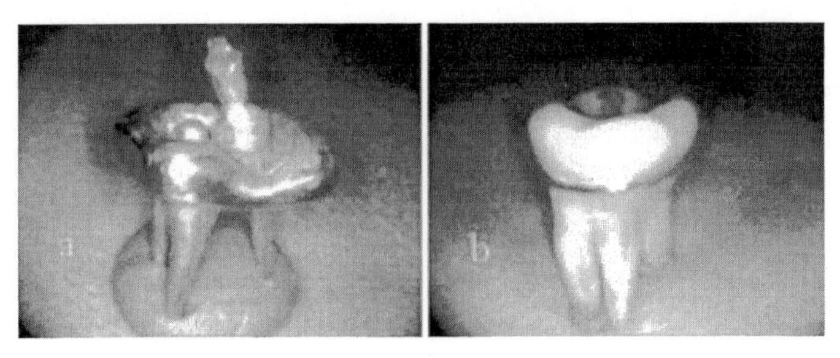

Figure 2. (a) An individual intra-root post, (b) a crown restored with composite material onto the post.

The performance of a restoration made as above can not be evaluated directly without long-term observation period. However, mechanical strength of composite/metal and metal/dentin bonds provided by the adhesive material and measured *in vitro* in standardized way, may be helpful in predicting future properties of the restoration. Thus, shear bond strength measured on samples of the composite bonded to the cobalt-based alloy with the PM3EDM-based adhesive amounted to 5.35 ± 3.82 MPa for samples of polished metal surface and 17.92 ± 3.22 MPa for non-polished ones [5]. The results are satisfactory, however, thermocycling experiments are necessary to verify the results obtained in terms of long-term performance.

[1] B. E. Causton, "Medical and Dental Adhesives", in *Materials Science and Technology*, vol.14: *Medical and Dental Materials*, D. F. Williams, vol. ed., VCH 1992, p.285

[2] International Symposium on Adhesives in Dentistry, *Operative Dentistry Suppl.5*, Omaha 1992

[3] R. L. Bowen, E. N. Cobb, J. E. Rapson, *J.Dent.Res.* **1982,** *61*, 1070

[4] M. Gibas, B. Gruszka, I. Szarzec, T. Kupka, M. Tanasiewicz, W. Malec, *International Polymer Seminar Gliwice 2003, Proceedings,* **2003,** in press

[5] T. Kupka, *Magazyn Stomat.* **2003,** *13,* 68

[6] M. M. Tanasiewicz, T. W. Kupka, W. P. Węglarz, A. Jasiński, M. Gibas, *J.Dent.Res.* **2003,** *82*, Special Issue B, 3047

[7] M. Tanasiewicz, W. P. Węglarz, T. Kupka, Z. Sułek, M. Gibas, A. Jasiński, *Stomat. Współczesna*, **2002,** *9,* 9

Crystallization Behavior and Thermal Properties of Blends of Poly(3-hydroxybutyate-co-3-valerate) and Poly(1,2-propandiolcarbonate)

Long Chen,[1] *Meifang Zhu,**[1] *Liyuan Song,*[1] *Hao Yu,*[1]
Yu Zhang,[1] *Yanmo Chen,*[1] *H. J. Adler*[2]

[1] College of Material Science & Engineering, State Key Laboratory for Modification of Chemical Fiber & Polymeric Materials, Dong Hua University, Shanghai 200051, China
E-mail: zmf@dhu.edu.cn
[2] Institute of Macromolecular Chemistry & Textile Chemistry, Dresden University of Technology, Mommsenstr. 13, D-01069 Dresden, Germany

Summary: Crystallization behavior of blends of poly(3-hydroxybutyrate-co-3-hydroxyvalerate) (PHBV) and poly(1,2-propandiolcarbonate) ($PR(CO_2)$) has been investigated by polarized light microscopy (PLM). The spherulite growth rates (SGR) of all blends were faster than that of pure PHBV, and the spherulite growth rates of PHBV in the PHBV/$PR(CO_2)$ blends reduced with increasing $PR(CO_2)$ weight fraction. There are two melting peaks in both the pure PHBV and the PHBV/$PR(CO_2)$ blends. The melting peak of PHBV/$PR(CO_2)$ blends was reduced by lower temperature about 20K as compared to PHBV and the higher temperature melting peak was increased by about 10K in the blends.
Keywords: blends; crystallization behavior; thermal properties

1. Introduction

Among environmentally friend polymeric materials, the bacterial organ poly(hydroxyalkanoate)s (PHAs) is a kind of high-molecular-weight polyesters produced by a wide range of microorganisms as intracellular carbon and energy reserve materials[1]. Poly(hydroxybutyrate) (PHB) are principal type of PHAs. PHB is relatively abundant in the environment and may be found in soil bacteria[2], estuarine micro-flora[3], blue-green algae[4] and micro-bially treated sewage[5]. The percentage of PHB in these cells is normally low, between 1 and 30%, but under controlled fermentation condition of corn

 DOI: 10.1002/masy.200450628

starch and nitrogen limitation and overproduction of PHB can be encouraged directly and yields increased to about 70% of dry cell weight[6]. In all cases the polymer occurs as discrete granules within the living organisms and may be extracted from them using a variety of organic solvents such as chloroform[7], ethylene dichloride[8], methylene chloride[9], pyridine[10] and propylene carbonate[11].

Recently PHAs have attracted much attention as an excellent biocompatible, bioabsorbable and biodegradable thermoplastic[12]. However, there are several shortcomings to commercial use of the PHB. Because it is produced from microorganisms, it is relatively expensive compared with other biodegradable polymers. Also it has a very narrow processing window due to their poor thermal stability[13] and it is rather brittle for use below the glass transition temperature (*Tg*). To reduce its brittle character, various copolymers that have different types of aliphatic polyester units, for example poly(hydroxybutyrate-*co*-hydroxyvalerate)(PHBV) presents over PHB homopolymer the advantages of improved thermostability and toughness[14-18].

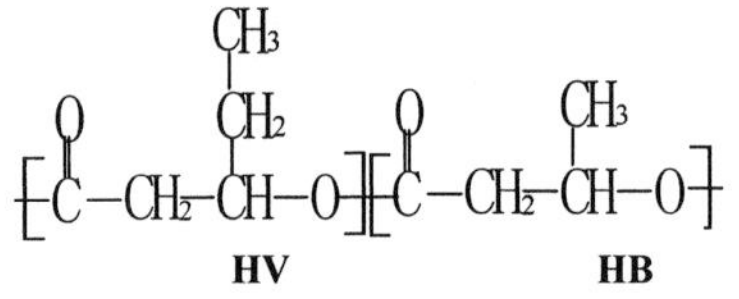

$$\left[-\overset{O}{\overset{\|}{C}}-CH_2-\underset{}{CH}(CH_2CH_3)-O- \right]\left[-\overset{O}{\overset{\|}{C}}-CH_2-CH(CH_3)-O- \right]$$

HV HB

Unit structure of PHBV

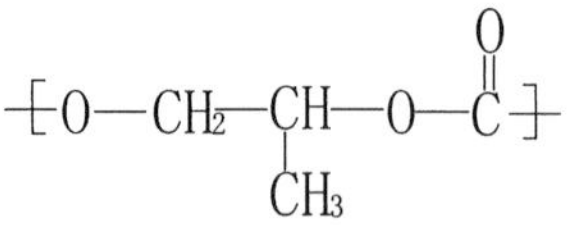

$$\left[-O-CH_2-CH(CH_3)-O-\overset{O}{\overset{\|}{C}}- \right]$$

Unit structure of $PR(CO_2)$

Fig. 1. Chemical structures of PHBV and $PR(CO_2)$.

Another approach to modify the properties of PHB and PHBV is to form polymer blends with synthetic or biodegradable polymers. Through blending with other polymers, the cost of the final materials can be reduced and the mechanical properties of PHB and PHBV can be improved. Many investigators have studied thermal behavior, crystallization, and mechanical properties of blends of the PHB and PHBV with synthetic and other biodegradable polymers such as poly(ethylene oxide)[19,20], poly(vinyl acetate)[21], poly(vinyl alcohol)[22], cellulose ester[23], poly(ε-caprolactone)(PCL)[24], and poly(styrene-*co*-acrylonitrile)(SAN)[12]. In this paper, the effect of carbon dioxide copolymer resin (PR(CO_2)) on the thermal properties and crystallization behavior of PHBV/PR(CO_2) blends will be discussed in detail. Chemical structures of PHBV and PR(CO_2) are shown in Figure 1.

2. Experimental

2.1 Materials

Poly(hydroxybutyrate-co-hydroxyvalerate)(PHBV) (15 mol% HV content) used in this study was obtained from Hongzhou Tian'an Bioproducts Ltd. Co., P.R.C.. The characteristics of PHBV are shown in Table 1.

Table 1. Characteristics of PHBV sample.

Sample	HV Content(%)	$\overline{Mw}$ [1]	$\overline{Mn}$ [1]
PHBV	15	4680000	268000

[1]Supplied by the source factory.

PR(CO_2) used in this study was supplied by Changchun Institute of Applied Chemistry Chinese Academy of Sciences.

2.2 Preparation of Samples

PHBV was dried in vacuum at 80° C for 24 h. Then PHBV and PR(CO_2) were melt blended in DACA microcompounder at 175° C for 15 min, which is a discontinuous twin-screw compounder in a 4.5ccm scale. The screw speed was 75rpm. The composition and the code of the blends investigated are given in Table 2.

Table 2. Composition and code of PHBV/PR(CO_2) blends.

Code	PHBV	P90	P80	P70	P60	P40	P30	P20	P10	PR(CO_2)
Blend composition PHBV/PR(CO_2) (wt % ratio)	100/0	90/10	80/20	70/30	60/40	40/60	30/70	20/80	10/90	0/100

2.3 Polarized Light Microscopy (PLM)

The growth of PHBV spherulites in the blends were observed with an Olympus BX51 polarized light microscope, equipped with a hot stage and an Olympus exposure control unit. Sample sandwiched between two thin glass slides were melted for 1 min on a hot-plate preheated at 190°C. It was then quickly transferred onto the hot stage of the microscopy, which was maintained at a desired temperature (T_c). The sample was allowed to crystallize isothermally.

2.4 Differential Scanning Calorimetry (DSC)

The thermal properties of all samples were analyzed by using a Mettler differential scanning calorimeter, Toledo Star System. Temperature calibration was performed using indium [melting temperature (Tm)=156.6°C, ΔHf=28.5J/g]. To measure the melting temperature and crystallization temperature of PHBV/RE blends, blend samples of 5 to 10 mg were heated in a nitrogen atmosphere from -50°C to 200°C at heating rate of 10°C/min and then cooled to 0°C at a rate of 10K/min.

3. Results and Discussion

3.1 Crystallization Behavior of PHBV/PR(CO_2) Blends

Figure 2 shows a series of optical micrographs of the PHBV/PR(CO_2) blend system. In Figure 2, PHBV shows the typical black cross under PLM. In PHBV/PR(CO_2) blends there are several black areas within the spherulites, which could be caused by PR(CO_2) penetrating to the spherulites of PHBV or light barriers by PR(CO_2) and PHBV distributed in different layers. The influence of impurity on the growth of spherulites of the pure PHBV is shown in Figure 3. In Figure 3(a), it can be seen that the spherulite of PHBV does not grow in a beeline when it comes into contact with impurities. As shown by Figure 3(a), spherulite of PHBV grows along the edge of impurity. The complete spherulite of PHBV contained impurity is shown in Figure 3(b). In Figure 3(b), no concentric ring like patterns appeared. In contrast to this, the concentric ring like patterns appeared on the spherulites of P80, P40 as shown in Figure 1.. So the black areas on the appearance of spherulites in PHBV/PR(CO_2) blends should be caused by light barriers by PR(CO_2) and PHBV distributed in different layers, i.e., PR(CO_2) did not enter the spherulite in PHBV/PR(CO_2) blends.

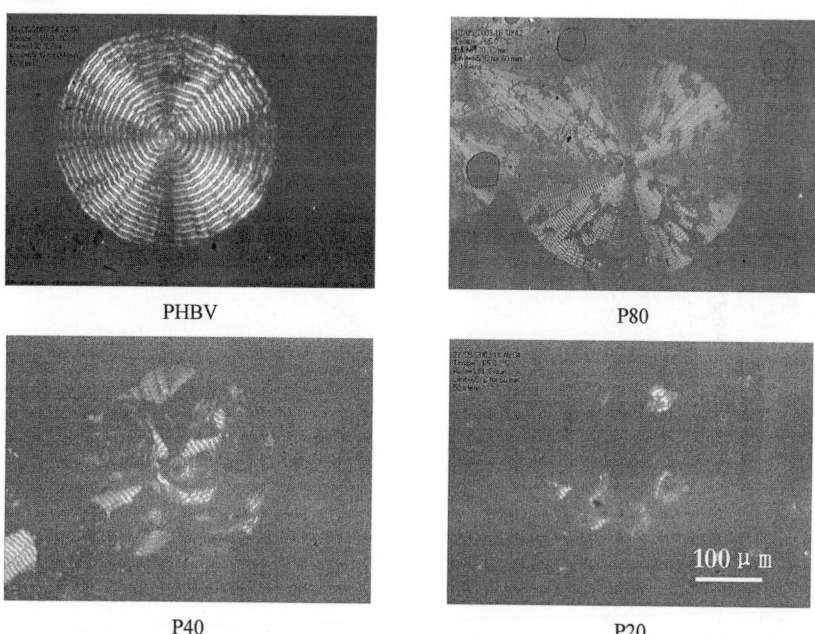

Fig. 2. A series of optical micrographs showing the spherulitic morphology of PHBV/PR(CO_2) blends with PR(CO_2) content.

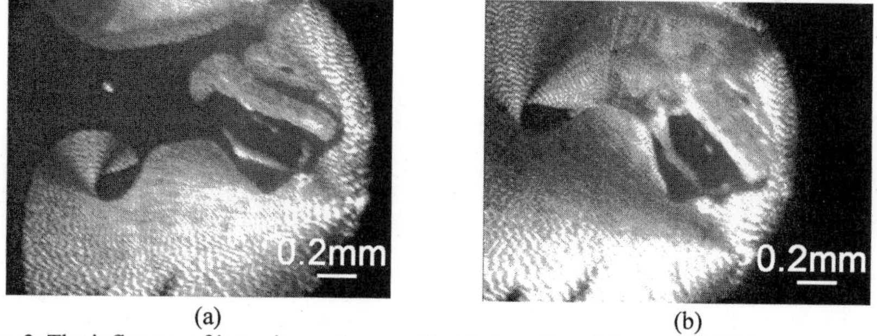

Fig. 3. The influence of impurity on the growth of spherulite of the pure PHBV.

The spherulite diameter of PHBV in the PHBV/PR(CO_2) blends was measured. It can be seen the change of the spherulite diameter of PHBV with time is linear (illustrated as Fig. 4). Thus the spherulite growth rate of PHBV can be calculated from the slope. Table 3 shows the spherulite growth rates of PHBV in the PHBV/PR(CO_2) blends. The spherulite growth rate of PHBV in the blends is higher than that of the pure PHBV. And the spherulite growth rates (SGR) of all blends are faster than that of pure PHBV, and decreases slightly with increasing PR(CO_2) weight fraction.

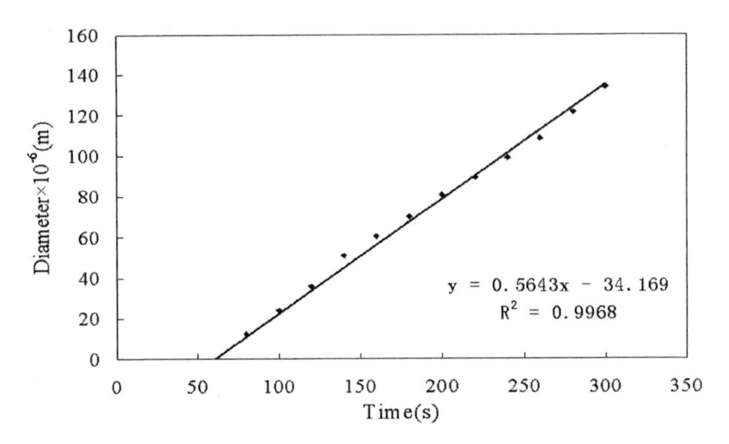

Fig.4. PHBV spherulite growth linear fit.

Table 3. The spherulite growth rate (SGR) of different samples.

Samples	PHBV	P90	P80	P70	P60	P40
SGR (m/s)	5.6×10^{-7}	9.7×10^{-7}	9.6×10^{-7}	8.6×10^{-7}	8.1×10^{-7}	8.0×10^{-7}

3.2 Thermal Behavior of PHBV/PR(CO_2) Blends

Thermal behavior of PHBV/PR(CO_2) blends was studied using DSC. Figure 5 shows the DSC heating thermograms of different PHBV/PR(CO_2) blends. In Figure 5(a), it can be seen that the T_g of the PR(CO_2) is almost unchanged at about 45° C in the blends. This result indicates that the blends of PHBV and PR(CO_2) are immiscible under the mixing condition. Y. S. Chun and W. N. Kim have studied the thermal properties of PHBV/poly(ε-caprolactone) (PCL) blends[12]. They have reported that the $\Delta C_{P}s$ of PHBV in the PHBV/PCL blends was too small to be detected clearly since the magnitude of this transition was too low. Therefore, the T_g of PHBV in the PHBV/PR(CO_2) blends which is expected to appear at about 7° C could not be measured by DSC.

In Figure 5(b), there are two melting peaks on both the pure PHBV and the PHBV/PR(CO_2) blends curves. Table 4 summarizes the thermal parameters of PHBV/PR(CO_2) blends as obtained by DSC. The melting temperature of a semicrystalline polymer is primarily a function of the crystal size. So the two melting peaks of the pure PHBV and the PHBV/PR(CO_2) blends resulted from two crystal sizes formed during the quick cooling in the samples preparation. The melting peak at about 150° C resulted from a crystal size that is similar to poly(hydroxyvalerate)(PHV) crystal size. The melting peak at about 170° C resulted from a crystal size that approaches PHB crystal size. The amorphism of PR(CO_2) resulted in no melting peak on the pure PR(CO_2) curve. In addition, it can be seen that the two melting peaks of the P80, P70, P30 and P20 PHBV/PR(CO_2) blends are shifted. That one at 145° C is reduced y about 20K and that one at 170° C is increased by about 10K. These results indicate that the crystal size that resulted in the melting peak at 145° C in the PHBV/PR(CO_2) blends much approaches that of poly(hydroxyvalerate)(PHV). Similarly, the crystal size that resulted in the melting peak at 170° C in the PHBV/PR(CO_2) blends much approaches that of PHB.

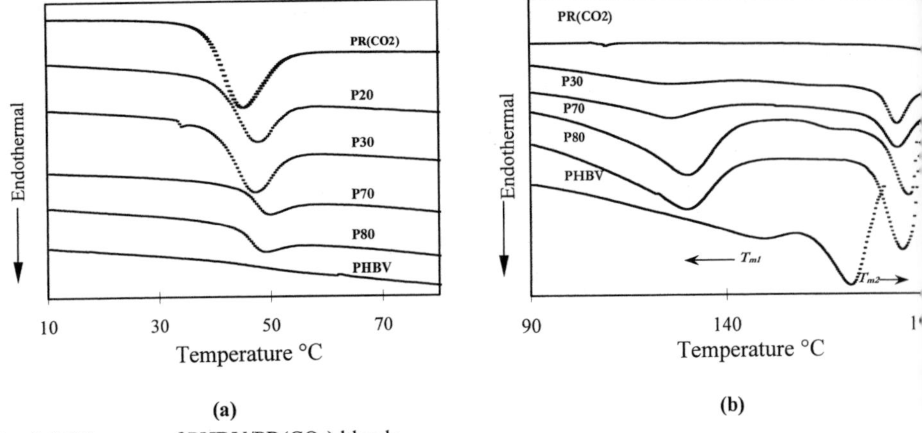

Fig. 5. DSC curves of PHBV/PR(CO_2) blends

Table 4. Thermal parameters of PHBV/PR(CO_2) blends.

Codes	T_{m1}(°C)	T_{m2}(°C)	ΔH_m(J/g)
PHBV	147	172	23
P80	129	184	26
P70	129	186	24
P30	125	183	25
P20	123	183	24
PR(CO_2)	-[1]	-[1]	-[1]

T_{m1}, T_{m2} and ΔH_m are the onsets of lower temperature and higher temperature peaks and the melting enthalpy in Figure 5(b), respectively.

[1] Not detected.

4. Conclusion

As detected by polarized light microscopy and DSC, the blends of PHBV and PR(CO_2) are immiscible under the mixing condition. There are two melting peaks of both the pure PHBV and the PHBV/PR(CO_2) blends and no melting peak of the pure PR(CO_2). The higher temperature melting peak resulted from a crystal size that is similar to PHB crystal size. The lower temperature melting peak resulted from a crystal size that approaches poly(hydroxyvalerate)(PHV) crystal size. Amorphism of PR(CO_2) resulted in no melting peak of the pure PR(CO_2). In addition, the two melting peaks of the P80, P70, P30 and P20 PHBV/PR(CO_2) blends moved toward the lower and higher temperatures compared with that of the pure PHBV, respectively. It indicates that the crystal size that resulted in the lower temperature melting peak in the PHBV/PR(CO_2) blends much approaches that of poly(hydroxyvalerate)(PHV). Similarly, the crystal size that resulted in the higher temperature melting peak in the PHBV/PR(CO_2) blends much approaches that of PHB. Optical micrographs indicate that PHBV shows typical black cross under PLM, but there are several black areas on the appearance of spherulite in PHBV/PR(CO_2) blends. The black areas on the appearance of spherulite in PHBV/PR(CO_2) blends should be caused by light barriers by PR(CO_2) and PHBV distributed in different layers, i.e., PR(CO_2) did not enter the spherulite in PHBV/PR(CO_2) blends. Furthermore, the spherulite growth rates (SGR) of all blends are similar and higher than that of pure PHBV.

Acknowledgement

This work is financially supported by Shanghai Dawn Project (2000SG27) and Ministry of Education Key Project (No. 00059).

The author wishes to thank “211 Project” of “China Tenth Five Years plan”.

[1] F. Gassner, A. J. Owen, *Polymer International*, 1996, 215
[2] E. A. Dawes and P. J. Senior, *Adv. Microbial*, 1973, 10, 138
[3] J. S. Herron, J. D. King and D. C. White, *Appl. Environ. Microbiol.*, 1978, 35, 251
[4] N. G. Carr, *Biochem. Biophys. Acta*, 1966, 120, 308
[5] L. L. Wallen and W. K. Rohwedder, *Enviro. Sci. Technol.*, 1974, 8, 576
[6] A. C. Ward, B. I. Rowley and E. A. Dawes, *J. Gen. Microbiol.*, 1977, 102, 61
[7] D. G. Lundgren, R. Alper, C. Schnaitm and R. H. Marchessault, *J. Bacteriol.*, 1965, 245
[8] P. A. Holmes, L. F. Wright and B. Aldersc, *European patent application,* 1979,15123
[9] J. N. Baptist, *US Patent*, 1962, *3044 942*
[10] Idem, *US Patent*,1962, *3036 959*
[11] R. M. Lafferty, *Chem. Rundsch.,* 1977, *30,* 15
[12] Yong Sung Chun, Woo Nyon Kim, *J. Appl. Polym. Sci.*, 2000, 77(13), 673
[13] Naoyuki Koyama, Yoshiharu Doi, *Macromolecules*, 1996, 29, 5843
[14] Griffin, G. J. L. *Chemistry and Technology of Biodegradable Polymers*, Chapman & Hall: London, 1994
[15] Organ, S. J., Barham P. J., *J. Mater Sci.*, 1991, 28, 1368
[16] Gross R. A., De Mello C., Lenz R. W., Brandl H., Fuller R. C., *Macromolecules*, 1989, 22, 1106
[17] Preusting H., Nijenhuis A., Witholt B., *Macromolecules*, 1990, 23, 4220
[18] Morin F. G., Marchessault R. H., *Macromolecules*, 1992, 25, 576
[19] Avella M., Martuscelli E., *Polymer*, 1988, 29, 1731
[20] Yoon J. S., Choi C. S., Maing S. J., Choi H. J., Lee H. S., Choi S. J., *Eur. Polym. J.*, 1993, 29, 1359
[21] Greco P., Martuscelli E., *Polymer*, 1989, 30, 1475
[22] Azuma Y., Yoshie N., Sakurai M., Inoue Y., Chujo R., *Polymer*, 1992, 33, 4763
[23] Lotti N., Scandola M., *Polymer Bull*, 1992, 29, 407
[24] Gassner F., Owen A. J., *Polymer*, 1994, 35, 2233

Preparation of PA6/Nano Titanium Dioxide (TiO_2) Composites and Their Spinnability

Meifang Zhu,[1] Qiang Xing,[1] Houkang He,[1] Yu Zhang,[1] Yanmo Chen,[1] Petra Pötschke,[2] Hans-Jürgen Adler[3]*

[1] College of Material Science & Engineering, State Key Laboratory for Modification of Chemical Fiber & Polymeric Materials, Dong Hua University, 200051 Shanghai, China
[2] Institute of Polymer Research Dresden, Hohe Str. 6, D-01069 Dresden, Germany
[3] Institute of Macromolecular Chemistry & Textile Chemistry, Dresden University of Technology, Mommsenstr. 13, D-01069 Dresden, Germany
E-mail:zmf@dhu.edu.cn

Summary: In this work, surface modification technique with coupling agents and anchoring polymerization was adopted to tailor the surface properties of nanoscaled titanium dioxide (TiO_2). Ethyl glycol sols with TiO_2 were prepared in order to simulate the dispersibility of differently modified TiO_2 in a molten polyamide 6 (PA6) matrix. The modified TiO_2 were melt compounded with PA6 and composites and fibers were prepared. The average filler diameter of 47 nm (in composites) and 44 nm (in fibers) indicated homogeneous dispersion of TiO_2 in the matrix, whereas unmodified TiO_2 showed agglomerated structures in the PA6 matrix. The mechanical properties of the composite fibers were improved as compared to pure PA6 fibers and composite fibers with unmodified TiO_2.

Keywords: composites; mechanical properties; nano-TiO_2; PA6; surface modification; spinnability

Introduction

As one of novel high-tech materials, composite materials and functional fibers of polymer/ inorganic nanosized particles are attracting more and more attention and interest these years[1,2]. In these kinds of materials, the inorganic phase is dispersed at a nanoscale level in the polymer matrix phase. Due to the special structural characteristics of polymer/inorganic nanocomposites, they can exhibit high-order characteristics such as optical transparency, special dielectric properties, electrical conductivity (in case of conductive fillers), nonlinear optical effects, quantum confinement effects, biological compatibility, and biological

 DOI: 10.1002/masy.200450629

activity[3]. The properties of such nanocomposites can be widely controlled by the state of nanofiller dispersion. In this context, the dispersibility of nanofillers in organic solvents or polymeric matrices is one of the hotspot and puzzle of research and application[4, 5].

Among inorganic nanoparticles, nanosized titanium dioxide (TiO_2) is one of the most promising materials in research and application fields because of its versatile functions such as photocatalytic activity, far-infrared radiation, anti-bacterium properties, UV-resistance, antistatic behaviour et al.[6]. A lot of novel composites and fibers with good functions were prepared based on its special properties[7-8].

Many reports were published about polyolefin or polyester based nano-TiO_2 composites, but the work is rare on polyamide based composites. This work concerns to improve the dispersibility of modified nano-TiO_2 in polyamide 6 (PA6) matrix. Silane coupling agents and anchoring polymerization were used to modify the surface properties of TiO_2. An ethyl glycol sol with nanoparticles was prepared in order to simulate the dispersibility of modified TiO_2 in a molten PA6 matrix. The structure and properties, such as dispersion of the nanofillers, spinnability, and mechanical properties of the obtained polyamide/TiO_2 nanocomposite and fibers prepared theirfrom were characterized.

Experimental Section

Materials

The PA6 (T_m=226°C) was supplied by Shanghai Yu Hang Special Chemical Fiber Factory, China and nanoscale TiO_2 (NT) particles were commercially producted by Jiangsu Hehai Nano Science & Technology Ltd., China. A silane coupling agent (CA) was used to modify the surface properties of TiO_2 with structures as RO-Si-R′-C=C. The solvents ethanol, acetone, and glycol and monomers such as vinyl dianhydride (VDA) as well as initiator benzoyl peroxide (BPO) were analytical class.

Surface modification of TiO_2

Coupling agent modification

The dried TiO_2, silane coupling agent, and ethanol as solvent were premixed and added into the flask. Stirred for 5 min and dispersed for 30 min with an ultrasonic disintegrator (Model CQX 25-06) the mixture completed reaction after 5 hr at 80°C in circumfluence condition. The dosages of coupling agents are given in Table 1.

Table 1.Dosages of coupling agents

Samples No.	Coupling agent (g /10g TiO_2)
NT0	0
NT3	0.3
NT5	0.5
NT8	0.8
NT10	1.0

Anchoring polymerization modification

The anchoring polymerization modification is based on the modified TiO_2- NT10. The dried NT10, monomer VDA, and initiator BPO were mixed and added into the acetone and dispersed for 30 min with an ultrasonic disintegrator. The reaction took place at 60°C for 5 hr in order to prepare the anchoring polymerization modified TiO_2 (APNT10).

Characterization

Preparation of ethyl glycol/ TiO_2 sol and sedimentation properties test

The ethyl glycol (EG) sol with TiO_2 was prepared with 0.005g modified TiO_2 added in 50 ml EG dispersing for 30min with an ultrasonic disintegrator. The sedimentation of TiO_2-EG sol was proceeded by a Centrifuger (Model 800, Shanghai Ultrasonic Instrument Factory, China), using a centrifugation speed of 3250 rpm at 0, 0.5, 1, 2, 4, 8, 15, 30 min. The absorbance of the sol was detected by an UV-Vis spectrometer (Model 7520, Shanghai Analyse Instrument Factory, China, wavelength= 536 nm). The sedimentation characteristics of nano-TiO_2 in EG as measured by the absorbance of the sol at different centrifugation times, was used to estimate the dispersibility of TiO_2 in molten PA6.

Chemical surface analysis of TiO_2

The surface of the modified TiO_2 was analysed using FTIR spectroscopy. A FTS-185 (Bio-Rad, USA) was used. The dried TiO_2 and KBr were mixed and pressed into flakes.

Preparation of the nanocomposites and fibers

The mixing of the composites (PA6/TiO_2: 97/3 wt%) was proceeded in a DACA Micro Compounder (DACA Instruments, USA) with 4.5 ccm volume applying a rotation speed of 100 rpm at 250°C for 8 min.

As-spun PA6 and PA6/ TiO_2 composites fibers were prepared on an ABE Spinner Instrument (ABE Corp., Japan) at 270°C and a take-up velocity of 800 m/min was used. The as-spun fibers were drawn using a Barmag 3010 Drawer (BARMAG, Germany) set to 70°C for the heat roller and to 120°C for the winding roller.

Morphology observation

The morphology of nano-TiO_2 particles was investigated using TEM (Model: JEM-100CX II, JEOL, Japan). Samples for scanning electron microscopy (SEM) of composites and fiber were prepared by cryo-fracturing in liquid nitrogen. The fractured sections were sputtered with gold and observed at 10 kV acceleration voltage on a JSM-5600LV (JEOL, Japan).

Results and Discussion

TiO_2 surface modification

Coupling agent modification

Most nano inorganic dioxides contain abundant hydroxyl groups[9]. As shown in Figure 1, the methoxy- or ethoxygroups of the coupling agent can react with the hydroxyl groups on the inorganic particles surface. The created long alkyl tails have a good compatibility with the polymer matrix[10]. Figure 2 and 3 show the Ti-O-Si transmittance peak at 1042 cm^{-1} evidently, implying that the coupling agent was grafted onto the surface by chemical bonds.[10] The peaks at 1378cm^{-1} and 1639 cm^{-1} predicated the C=C bond which is suitable for the compatibility with the polymer matrix since it can act as the active point for the further surface modification[11]. The grafted coupling agents formed an organic layer on the inorganic TiO_2 particle surface which can serve as the compatibilizer for an organic medium, solvent, or polymer matrix.

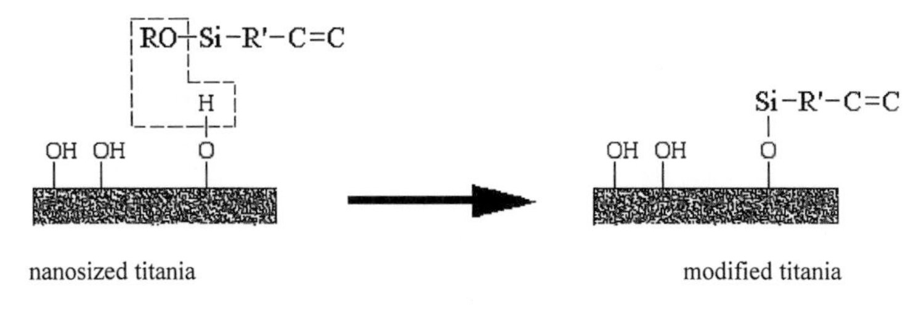

Figure 1. Reaction scheme for modification of nanosized titania with coupling agent

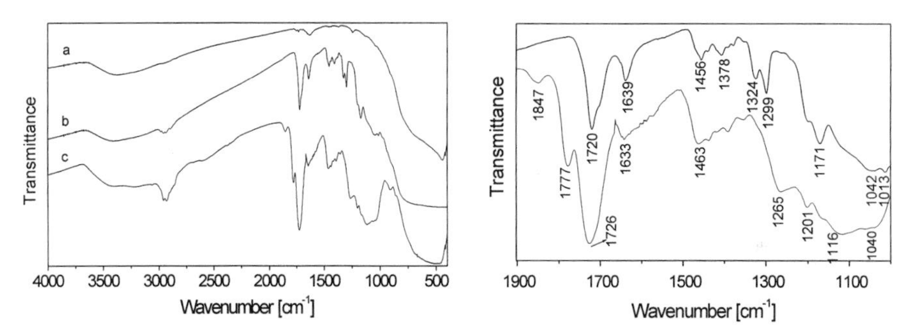

Figure 2. FTIR spectra of NT0 (a), NT10 (b), and APNT10 (c), curves horizontally shifted

Figure 3. FTIR finger mark spectra of NT10 (upper curve) and APNT10 (lower curve)

Anchoring polymerization modification

The surface graft polymerization is imported to improve the coverage ratio and to obtain long organic tails onto a substrate in order to get better compatibility and dispersibility in a medium. The anchoring polymerization surface modification is one of the free-radical surface graft polymerizations emerged as a simple method for obtaining a high surface chain coverage[12-13]. In this case, the active sites, vinyl groups on surface graft polymerization of vinyl monomers, were introduced by the formerly coupling agent modification as shown in Figure 4. The FTIR spectrum (Figure 2,3) of the PVDA-grafted TiO_2 (APNT10) particles displays a series of new peaks as compared to the NT10. The spectrum reveals two peaks at 1777 cm^{-1} and 1847 cm^{-1} which are characteristics of the C=O bond stretching introduced by the VDA, and a group of partially overlapped peaks between 1116 cm^{-1} and 1265 cm^{-1}, owning to the O-C bond stretching[12].

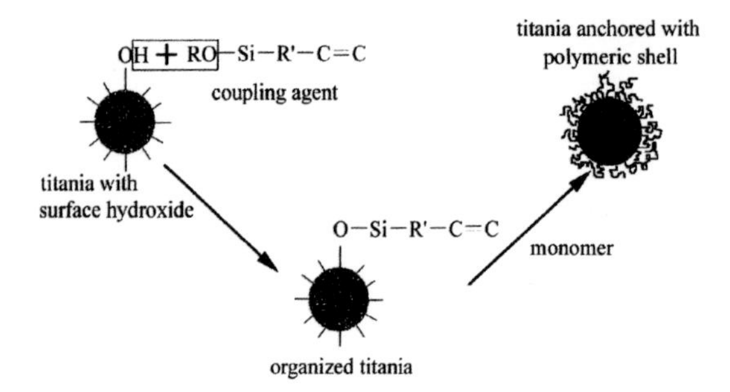

Figure 4. Schematic representation of the modification of TiO_2 through anchoring polymerization

In addition, there is a broad peak around 2930 cm^{-1} corresponding to the H-C bond stretching of the hydrocarbon backbone of the grafted polymer chains[14]. The Ti-O-Si peak (1042 cm^{-1}) was remained in FTIR spectrum of APNT10 implying the silane coupling agents were not removed during the polymerization. However, the peak at 1633 cm^{-1} becomes weaker than in the NT10. This peak corresponds to the C=C bonds on the TiO_2 surface which are reacting as the active sites with the VDA monomer, thus, reducing their content. From the FTIR results we can conclude that the VDA was polymerized from the silane-coupling agents to get a thick polymer shell around the TiO_2 surface as indicated in Figure 4. This modification should be favourable for the TiO_2 dispersibility.

Morphology of modified TiO_2

Figure 5 shows the morphology as obtained by TEM of differently modified TiO_2. The unmodified TiO_2 particles agglomerated as shown in Figure 5(a) and separate particles can not be distinguished. It is difficult to break these agglomerates under common process condition, which is also named 'hard agglomeration'. On the contrary, the modified TiO_2 particles, shown in Figure 5(b,c), can be observed as divided particles. Within the agglomerated areas, clear contours are visible between the TiO_2 particles. This indicates that the modified TiO_2 particles are easier to disperse during the composite processing[15], named 'soft agglomeration'. The APNT10 particles have a looser agglomeration structure as compared to NT10 which is indicative for their thick and complete organic coverage.

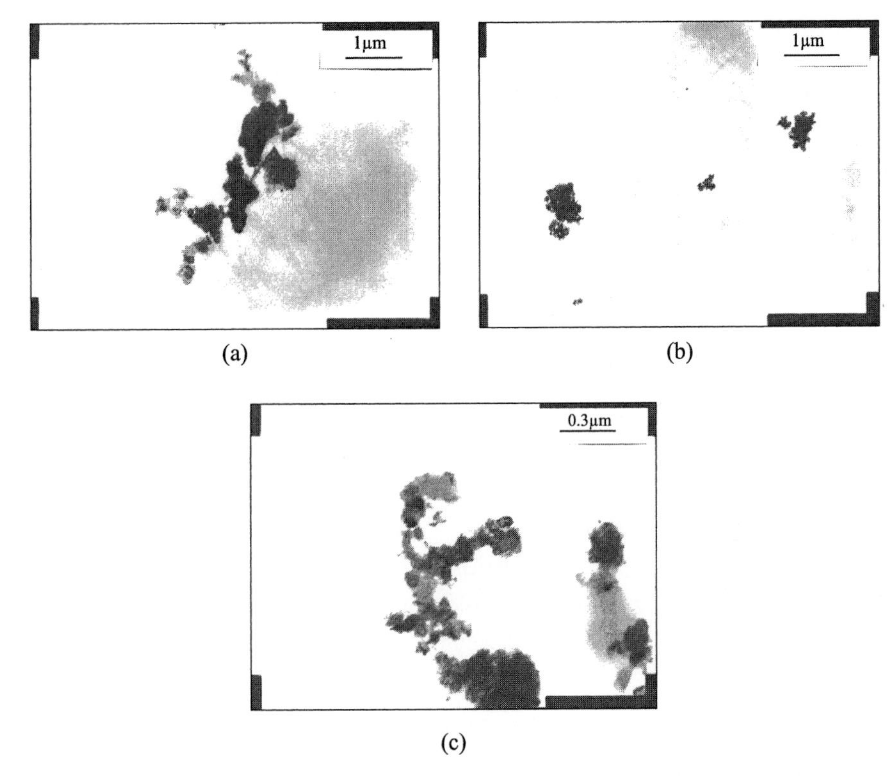

Figure 5. TEM micrographs of NT0 (a), NT10 (b) and APNT10 (c)

Sedimentation properties of modified TiO_2

The dispersion process of nanoparticles in a polymer matrix is difficult to observe and appraise, especially when forming the composites by melt mixing. In order to simulate the dispersibility of TiO_2 in a matrix, the sedimentation of TiO_2 in EG was introduced by using the EG as the substitute of molten PA6. Properties of EG and PA6 are shown in Table 2.

Table 2. The basic properties of ethylene glycol (EG) and PA6

Property	Ethylene glycol	PA6
Polarity /$[cal \cdot cm^{-3}]^{1/2}$	15.7	13.6
Density /$g \cdot cm^{-3}$	1.1135	1.23/1.08 (ρ_c/ρ_a)
Rheological behavior	Mucous (liquid)	Viscous (melt)

Figure 6 shows the absorbance of EG/TiO_2 sol with centrifugation time. As lower the sedimentation speed (as slower the absorbance reduction), the better the TiO_2 particles are dispersed in the EG. With increasing dosages of coupling agent the sedimentation speed is decreasing and the balance adsorbance is increasing. NT10 has the highest balance absorbance implying the best dispersibility and processablity in the PA6 matrix.

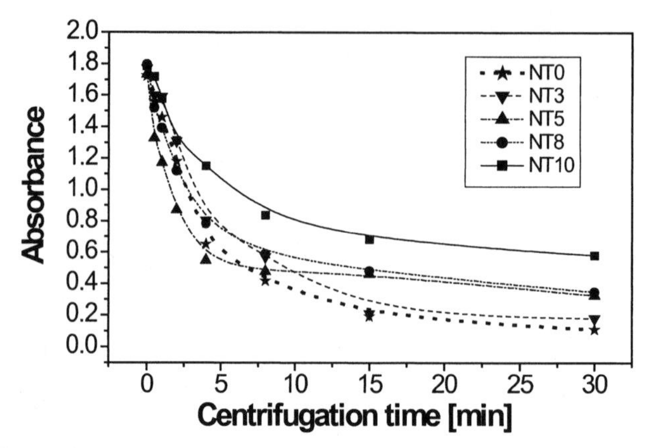

Figure 6. Sedimentation curves of TiO_2 modified with 0, 3, 5, 8, 10 wt% silane coupling agent (relative to the mass of titania)

Morphology of TiO_2 in PA6 composites and fibers

The properties of nanocomposites can be normally controlled by the dispersion of the nanoparticles in the polymer matrix. The dispersibility of the filler becomes the bottleneck problem for the performance of nanocomposites. Therefore, a lot of research is focused on this field. In this study, the grafted polymer layer consisting of the coupling agent and PVDA on the TiO_2 surface can be completely miscible with the surrounding PA6 medium. By this way, polymer detachment is prevented due to the covalent attachment of polymer chains to the substrate which also leads to a better stability of the dispersion.

Figure 7 shows the morphology of differently modified TiO_2 in PA6 composites. Some big agglomerates and black holes are visible in the PA6 matrix in Figure 7(a) where 3 wt% unmodified TiO_2 were added. These agglomerates are caused by the already introduced 'hard agglomeration' of unmodified TiO_2 which could not be broken during the blending process. The holes were produced by TiO_2 breaking off from the PA6 fractured section during the SEM sample preparation because of the poor binding with the PA6 matrix. The average diameter of the TiO_2 agglomerates is about 100 nm. The TiO_2 particles after surface modifications were dispersed in the matrix homogeneously and their average sizes are less than 50 nm. No holes could be detected on the sections surface. Therefore, the surface modification is effective to improve the dispersibility and the interface adhesion of TiO_2 to the PA6 matrix. This result is in accordance with the sedimentation result of TiO_2-EG sol. Comparing Fig. 7(b) to 7(c), the anchoring polymerization modification showed an further increase in adhesion.

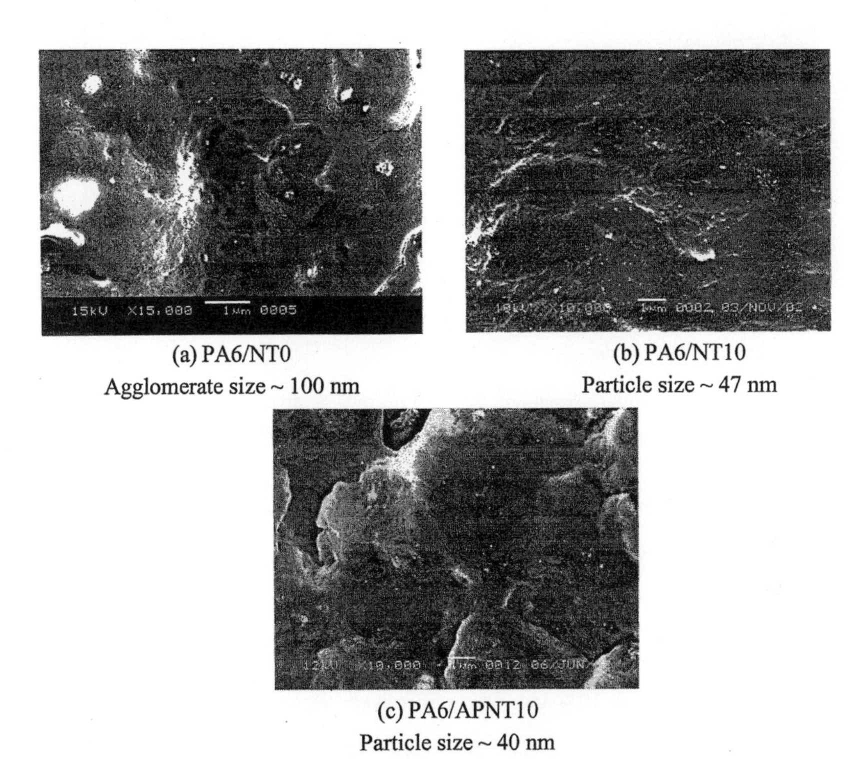

(a) PA6/NT0
Agglomerate size ~ 100 nm

(b) PA6/NT10
Particle size ~ 47 nm

(c) PA6/APNT10
Particle size ~ 40 nm

Figure 7. SEM micrographs of fractured sections of PA6/TiO_2= 97/3 wt% composites

The good dispersibility of the modified TiO_2 in composites supplied the precondition for the fiber formation of these composites. The dispersion morphology and statistics of particles size of NT10 in PA6 composite fiber is shown in Figure 8 and 9. The homogeneous dispersion of the modified TiO_2 particles already discussed in the composite is maintained after the spinning procedure. The amount of particles with sizes below 70 nm reached 95%, while the amount of agglomerates with sizes larger than 100 nm is less than 2%, as shown in Figure 9. During the fiber formation, the shear stress and strain tend to break the agglomerates of TiO_2 and, on the other hand, the high process temperature can destroy the organic layer on TiO_2 surface and can induce agglomeration. These two factors affect the final dispersion of TiO_2 in the PA6 fiber simultaneously. As the results show the particle size in the fibers is comparable to that of the composite indicating no agglomeration of TiO_2 during spinning.

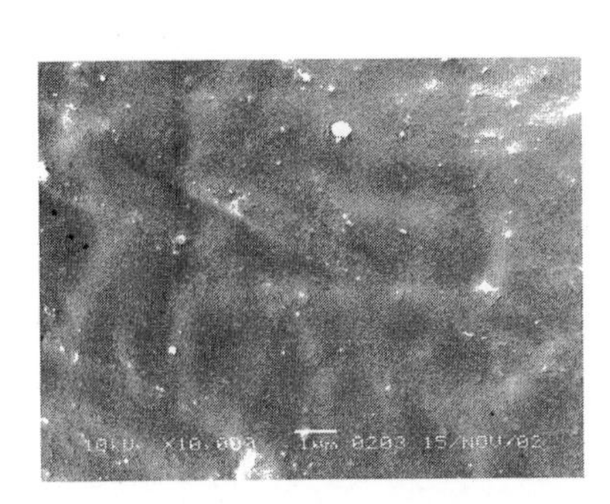

Particle size ~ 44 nm

Figure 8. SEM micrograph of fractured section of PA6/NT10 composite fiber

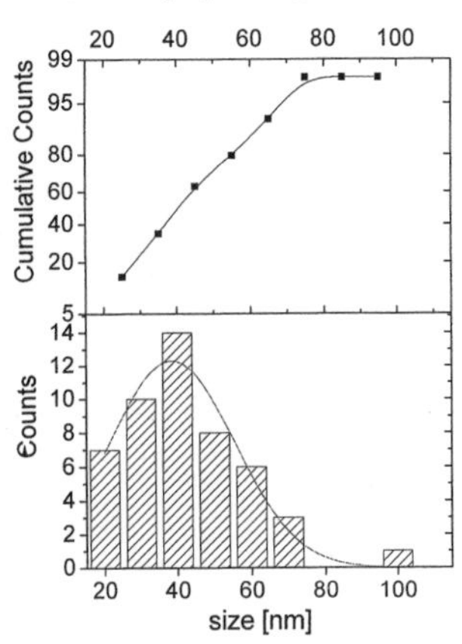

Figure 9. Statistics of TiO_2 particles size distribution in PA6/NT10 composite fiber

Spinnability of nanocomposites and properties of drawn fibers

It is known that lower particles size and more homogeneous dispersion lead to improved composite spinnability and result in better mechanical and functional properties of the fibers. This is the reason that the tenacity at break and the initial modulus of PA6/NT10 composites fiber were improved by about 10% and 20%, respectively, as compared to pure PA6 (Table 3). This indicates that TiO_2 acts as a reinforcing phase in the PA6 matrix [9]. The modification reagent on the TiO_2 surface also strengthens the interface adhesion to the polymer matrix.

Table 3. Fineness and mechanical properties of drawn fibers

Sample	Fineness /dtex·f^{-1}	Tenacity at break /cN·dtex^{-1}	Initial modulus /cN·dtex^{-1}	Elongation at break /%
100 PA6	143.2/60	3.47	45.3	18.4
97/3 PA6/NT0	135.5/60	3.47	40.8	10.2
97/3 PA6/NT10	135.6/60	3.80	59.9	16.0

Conclusions

The silane coupling agent and anchoring polymerization methods are effective to modify the surface properties of nano TiO_2 with improved dispersibility and interface adhesion in a PA6 matrix. The 'hard agglomeration' state of compactly assembled unmodified TiO_2 particles was replaced by a 'soft agglomeration' of modified TiO_2 particles. The composite with modified TiO_2 showed better spinnability than the composites with unmodified TiO_2 filler as well as improved mechanical properties.

Acknowledgement

The authors thank the National 863 Project (No.2002AA302616), the Shanghai Nano Special Project (No.0219nm039), and the Shanghai Nanotechnology Application Plateform (No.0359nm008) for the financial support within the program "211 Engineering" of the "Tenth Five Plan".

[1] Maiti S N, Ghosh K. *J. Appl. Polym. Sci.* **1994**, *52*(8), 1091-103.
[2] Zhang M, Gao Y-F, Fu W, Jiang Y,. Zhou Q-X, Liu D-S, *Acta Materiae Compositae Sinica* **1998**, *15*(3), 71-4.
[3] Novak,B.M. *Adv. Mater.* **1993**, 5, 422-33.
[4] Nakamuru Y, Okabe S, Yoshimoto N, Iida T, *Polymer Composite.* **1998**, *6*, 407-14.
[5] Bories M, Huneault MA, Lafleur PG, *Int. Polym. Proces.s* **1999**, *14*(3):34-42
[6] Hagfelt A, Gratzel M., *Chem. Rev.* **1995**,*95*(1), 49-56.
[7] Takeda Chemical Inc Ltd. *Medical Textiles* **1998**, *10*, 2.
[8] Japan Patent: 10-259521, 1998.
[9] Norio T, Akira K, Kazue M, Keiko T, *Polymer Journal* **1990**, *22*(9), 827-33.
[10] Yu S Lipatov. *Polym. Sci. Technol.; Part B: Adhes. Adsorp. Polym.* **1980**, *12*, 601-7.
[11] Tripp C P, Hair M L. *Langmuir* **1992**, *8*, 1120-8.
[12] Brown T, Chaimberg M, Cohen, Y. *J. Appl. Polym. Sci.* **1992**, *44*, 671-7.
[13] Nguyen V, Yoshida W, Jou J. D, Cohen Y. *J. Polym. Sci., Part A: Polym. Chem.* **2002**, *40*(1), 26-31.
[14] Siberzan P, Leger L, Ausserre, D, Benatta J J. *Langmuir* **1991**, *7*, 1647-53.
[15] Pampach R, Haberkc K. *Ceramic Powders*. Amsterdam: Elsevier Scientific Pub. Company, 1983, 623.

Cooperative Hydration Effect on the Binding of Organic Vapors by a Cross-Linked Polymer and Beta-Cyclodextrin

Valery V. Gorbatchuk,[*1] Marat A. Ziganshin,[1] Ludmila S. Savelyeva,[1] Nikolay A. Mironov,[1] Wolf D. Habicher*[2]

[1]Department of Chemistry, Kazan State University, Kremlevskaya 18, Kazan, 420008, Russia
E-mail: valery.gorbatchuk@ksu.ru
[2]Technological University of Dresden, Institute of Organic Chemistry, Mommsenstr. 13, D-01062, Dresden, Germany

Summary: A cooperative hydration effect being favorable for the binding of organic vapors by cross-linked poly(N-6-aminohexylacrylamide) and beta-cyclodextrin was observed in ternary systems in the absence of liquid phase. For these systems the vapor sorption isotherms were determined by the static method of headspace gas chromatographic analysis at 298 K. The obtained isotherms show an increase of binding affinity for vapor of hydrophobic sorbates above a threshold value of receptor hydration. Further hydration gives a saturation of this affinity for the studied hydrophilic polyacrylamide derivative, while the affinity of beta-cyclodextrin for the hydrophilic sorbate ethanol even decreases. A similar behavior of this polymer and beta-cyclodextrin at the change of their hydration helps to explain the observed cooperative hydration effect in terms of clathrate formation.

Keywords: clathrate formation; cooperative hydration effects; headspace GC analysis; macromolecular receptor; vapor sorption isotherms

Introduction

An important feature of an efficient and selective macromolecular receptor is a substrate binding cooperativity that mimics the cooperative properties of proteins. The molecular design of such a receptor is often confined to the synthesis of molecularly imprinted polymers with the binding sites similar to those of antibodies and enzymes.[1] Thermotropic polymers with their specific hydrophobic-hydrophilic balance mimic the cooperative stimulus-responsive behavior of proteins[2] including protein folding.[3] This property linked to the change of a receptor hydration is used in the encapsulation/release of drugs by polymers in biomedical applications.[2] The change of protein hydration gives also a number of other cooperative effects such as the

 DOI: 10.1002/masy.200450630

cooperative hydration influence on the rates of enzymatic reactions,[4] on the binding of hydrophobic and relatively large neutral monofunctional compounds[5] and on the protein specific heat capacity.[6] This hydration cooperativity can be seen in a hydration history effect for enzymes.[7] Besides, some proteins have more specific cooperative behavior as in the binding of oxygen by hemoglobin in water solution.[8]

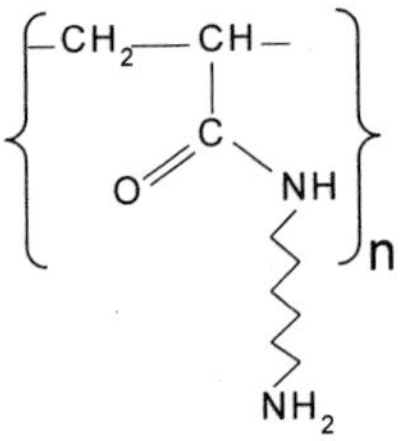

cross-linked poly(N-6-aminohexylacrylamide)
(PAA)

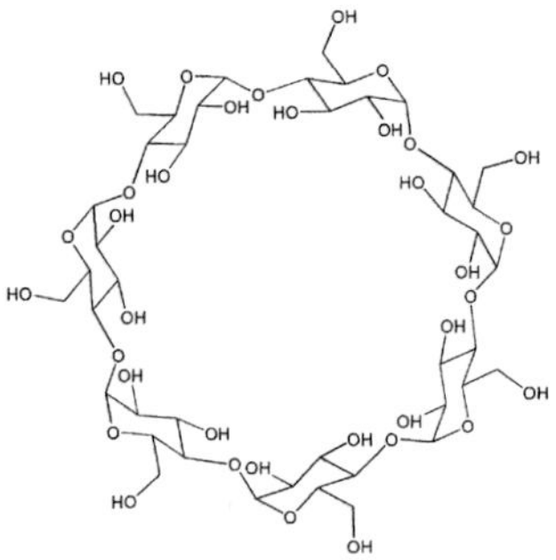

beta-cyclodextrin
(BCD)

To reveal the structural requirements for a receptor to have the biomimetic cooperative properties the structure-property relationship was studied in this work for hydration effect on the binding properties of very different types of hydrophilic receptors: cross-linked poly(N-6-aminohexylacrylamide) (PAA) and beta-cyclodextrin (BCD). The vapor sorption isotherms obtained for these receptors in ternary systems with water-organic vapors in the absence of liquid phase were compared with the data on the cooperative behavior of human serum albumin in the same conditions[5] and hydrophobic calixarene, for which the cooperative binding of guest vapor is observed in binary host-guest systems.[9] This is the first study showing a cooperative hydration effect that helps a hydrophilic polymer and macrocyclic receptor to bind hydrophobic substrates instead of interfering with this process.

Experimental Part

The cross-linked poly(N-6-aminohexylacrylamide) (PAA) and beta-cyclodextrin (BCD) were purchased in Aldrich and ICN, respectively. The PAA studied was in the shape of beads with the average diameter of 100 μm. It was dried for 5 months over P_2O_5 at room temperature before the sorption experiment to remove hydration memory effects. BCD was dried for 8 hours at 100°C and 1 mm Hg. To prepare the prehydrated BCD its dried 200-mg samples in open 15-ml glass vials were saturated by water vapor in hermetically closed vessel at 298 K. The residual hydration of dried PAA and BCD and the water contents in the prehydrated BCD were determined using the method of thermogravimetry as described elsewhere.[10]

The vapor sorption isotherms were determined by static method of headspace GC analysis as described earlier.[5,10] For this experiment the samples of dried PAA (100 mg), dried (200 mg) or prehydrated (234 mg) BCD were equilibrated for 72 hours in the hermetically closed 15-ml vials with organic or water-organic vapors at 298 K. The equilibration time was chosen using the estimations of the sorption kinetics and kinetics of sorbate evaporation. The liquid sorbate or water-organic mixture with 6:94 and 1:4 organic component/water volume ratios for PAA and BCD, respectively, was dosed in the open small internal vials to avoid a direct contact between a liquid sorbates and solid receptor. The absence of liquid phase after equilibration was checked visually. The error of sorbate thermodynamic activity P/P_0 determination is in the range from 10% at P/P_0 <0.10 to 5% at P/P_0 >0.5. The error of sorbate uptake A determination is 5%. Each error is mostly systematic and has low influence on the shape of obtained sorption isotherms. The total hydration value h of receptor was calculated using its initial hydration value and the amount of water added with the sorbate. The correction on the water contents in the vapor phase of the system was made. The estimated error of the receptor hydration value is ±0.002 h (= 0.002 g H_2O per g of dry receptor).

Results and Discussion

The vapor sorption isotherms obtained for ternary systems with PAA and BCD hydrated *in situ*, together with the saturation by the vapor of organic component, are given on Figure 1 and 2, respectively, showing two different 2D cross sections of the three-dimensional isotherms. On the Figures 1A and 2A 'pure liquid sorbate – receptor' partition coefficient $A/(P/P_0)$ is plotted as a function of receptor hydration h. On Figures 1B and 2B the sorbate uptake A is plotted as function of sorbate activity for the same isotherms. Besides, on Figures 1 and 2 the vapor sorption isotherms for ternary systems of human serum albumin[5] prepared in the same way are shown for comparison. The contents of the water/organic mixtures added to the studied receptors were chosen to reduce the contribution of PAA plasticization by organic component and to create optimal conditions for the clathrate formation with BCD. A simultaneous saturation of dried hydrophilic macromolecular receptors by water-organic vapors provides more reproducible sorption data because it is easier to remove the hydration memory effects from this initial state of sorption process.

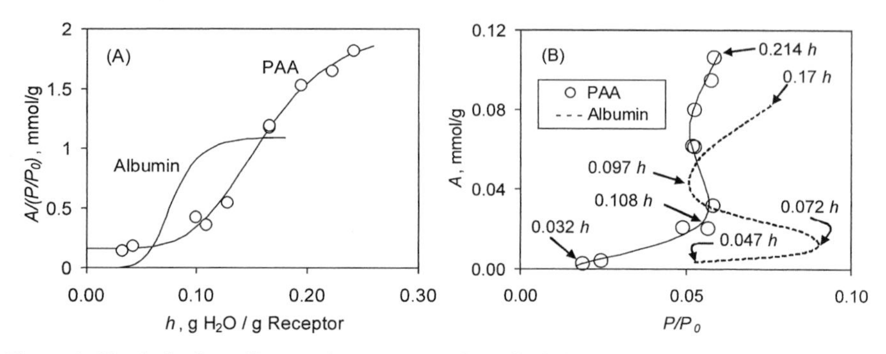

Figure 1. The hydration effect on the vapor sorption of ethyl acetate by cross-linked poly(N-6-aminohexylacrylamide) (PAA) with initial hydration 0.004 *h* at 298 K in two different presentations of the same sorption isotherm: (A) as a function of 'pure liquid sorbate – polymer' partition coefficient [$A/(P/P_0)$ *vs.* polymer hydration h] and (B) sorbate uptake A *vs.* sorbate activity P/P_0. The contents of organic component in liquid water-organic mixture added to the initially dried polymer is 6 vol. %. The vapor sorption isotherm of ethyl acetate by human serum albumin (initial hydration 0.01 *h*) determined in the same conditions is from Ref.5. The lines are drawn to guide the eye.

The sigmoidal shape of the vapor sorption isotherm of ethyl acetate by the studied cross-linked polyacrylamide derivative (Figure 1A, B) reveals rather high cooperativity of the hydration effect on the binding properties of this polymer. The hydration up to 0.11 *h* does not increase much the free volume of PAA accessible for organic component. Above hydration level of 0.11 *h* the sorption affinity of PAA for ethyl acetate cooperatively increases (Figure 1A) indicating a small decrease of sorbate activity P/P_0 at the increase of sorbate uptake A (Figure 1B). This cooperativity is similar to that of albumin[5] in the same conditions (Figure 1). The observed inflections of sorption isotherm for PAA on Figure 1B are less deep than for albumin. A separate liquid phase is not observed after equilibration in all studied range of PAA hydration.

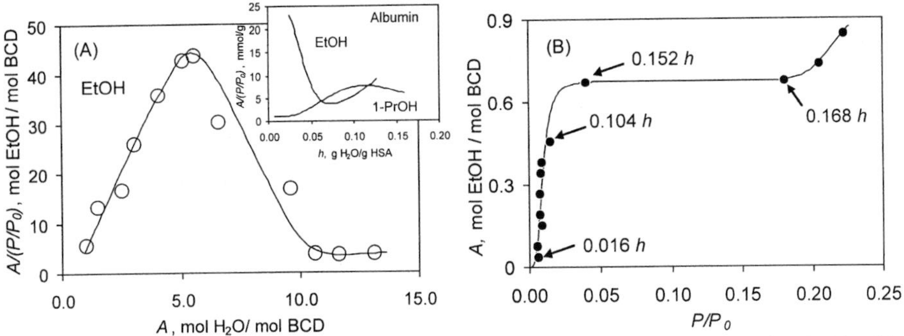

Figure 2. The hydration effect on the vapor sorption of ethanol by beta-cyclodextrin (BCD) with the initial hydration of 0.008 h at 298 K in two different presentations of the same sorption isotherm: (A) as a function of 'pure liquid sorbate – polymer' partition coefficient $A/(P/P_0)$ *vs.* BCD hydration and (B) sorbate uptake A *vs.* sorbate activity P/P_0. The contents of organic component in liquid water-organic mixture added to the initially dried polymer is 20 vol. %. The inset shows the vapor sorption isotherms of ethanol and 1-propanol by human serum albumin (initial hydration 0.01 h, contents of the organic component in the water-organic mixture added is 6 vol. %, $T = 298$K) from Ref.5. The lines are drawn to guide the eye.

The sigmoidal shape of sorption isotherms both of PAA and albumin can be explained supposing formation of clathrate 'water + ethyl acetate + receptor'. The formation of similar clathrates for γ-chymotrypsin was confirmed by X-ray data.[11] The same may be supposed to make a significant contribution to the sorption of ethyl acetate by PAA.

To make more definite conclusion on the nature of the molecular interactions causing the sorption of organic component in this system, we studied the binding properties of another hydrophilic receptor, BCD, which binds organic compounds in solid phase only through clathrate formation.[12] For BCD the isotherm of ethanol vapor sorption in the presence of the simultaneously added water show a definite saturation, Figure 2. In the '$A/(P/P_0)$ *vs.* h' presentation this ethanol sorption isotherm resembles the isotherm of 1-propanol vapor sorption by albumin in comparable conditions[5] (Figure 2A, inset), which also has a maximum in these coordinates, but less expressed. The isotherm of ethanol vapor sorption by albumin has a very different shape,[5] Figure 2A, inset. For albumin no saturation by ethanol at high hydration is observed, but this isotherm has a minimum at 0.07 h, which reflect a Langmuir shape of ethanol sorption isotherm (A *vs.* P/P_0) at low ethanol activity, P/P_0, for dried albumin[13] (Figure 3, inset).

The experimental point on the saturation part of the sorption isotherm on Figure 2B at the BCD hydration 0.152 h corresponds to the molar ratio of solid phase components 1 : 0.7 : 9 (BCD : ethanol : water), which is not much far from the composition of BCD-ethanol-water clathrate 1 : 1 : 8 studied by neutron diffraction and X-ray methods.[14] The isotherm inflection above BCD hydration 0.168 h is caused by the existence of separate liquid phase in the system after its equilibration. In binary BCD-water system a separate liquid phase forms above hydration 0.175 h at water activity P/P_0>0.93.[15] So, according to the slope of sorption isotherms at high receptor hydration, the ability of three hydrophilic macromolecular receptors to perform a saturation limit by organic component at the simultaneous hydration increase decreases in the order: BCD>albumin>PAA. This may be caused by the decreasing contribution of clathrate formation for the sorbate binding by these receptors in the same order. The influence of the receptor homogeneity on the shape of hydration effect on the ethanol binding by BCD and its cooperativity can be seen in the comparison of the binding properties of dried and hydrated BCD. The obtained vapor sorption isotherms of ethanol by dried BCD with residual hydration 0.008 h and of benzene by BCD dried and prehydrated to 0.172 h are shown on Figures 3 and 4, respectively. For comparison the ethanol sorption isotherms for albumin[13] and clathrate forming host *tert*-butylcalix[4]arene (TBC4)[16] earlier obtained in the same conditions are shown on Figure 3. Besides, the isotherm of benzene vapor sorption by TBC4[16] is given on Figure 4. No sorption of propanols, ethyl acetate and benzene from vapor phase exceeding 0.2 mol/mol BCD was observed for dried BCD (0.008 h) at sorbate activity P/P_0 < 0.8. The observed sigmoidal shape of the ethanol sorption isotherm for dried BCD is the same as

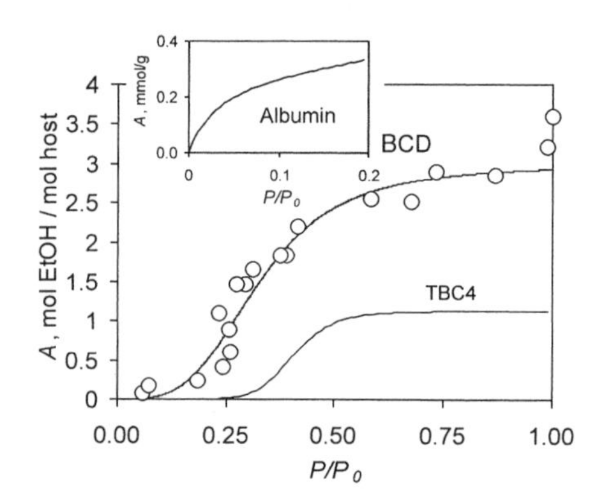

Figure 3. The vapor sorption isotherm of ethanol by the dried beta-cyclodextrin (BCD) with the residual hydration of 0.008 h at 298 K. The inset shows the isotherm of ethanol vapor sorption by dried human serum albumin (hydration 0.01 h) determined in the same conditions from Ref.13. The vapor sorption isotherm of ethanol by solid *tert*-butylcalix[4]arene (TBC4) at 298 K is from Ref.16. The sigmoidal lines were calculated using Hill equation (1).

for TBC4 (Figure 3) and is typical for clathrate formation at the binding of guest vapor by homogeneous solid hosts.[9,17] Any inhomogeneity caused, for example, by the presence of second guest in the solid host may reduce the threshold value of guest activity P/P_0 almost to zero removing the apparent cooperativity.[16] This may be a cause of Langmuir-like isotherm shape of ethanol vapor sorption by dried albumin at low ethanol activity P/P_0 (Figure 3, inset). Dry albumin has practically no surface of protein-air interface that can give a significant Langmuir contribution to the sorption.[13]

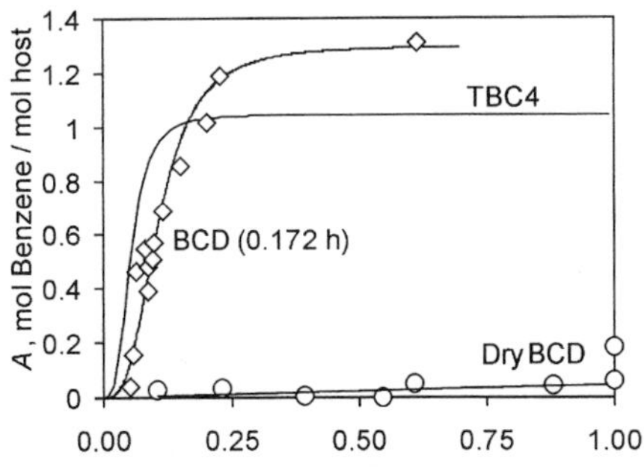

Figure 4. The vapor sorption isotherms o benzene by the dry and prehydrated beta-cyclodextrin (BCD) with the constant hydration of 0.008 *h* and 0.172 *h*, respectively, at 298 K. The vapor sorption isotherm of benzene by *tert*-butylcalix[4]arene (TBC4) at 298 K is from Ref. 16 . The sigmoidal lines were calculated using Hill equation (1).

The isotherms obtained for BCD were fitted by Hill equation[9]:

$$A = SC(P/P_0)^N / (1 + C(P/P_0)^N) \qquad (1)$$

where inclusion stoichiometry S, a co-operativity parameter N and a sorption constant C are the fitting parameters. Their values for ethanol isotherm are S = 3.0 (mol EtOH/mol BCD), N=3.5 and C = 49.3. The ethanol threshold activity at the 50% saturation is equal to $a_{0.5S} = \exp(-(\ln C)/N) = 0.33$. The lower is the value of $a_{0.5S}$, the higher is host-guest affinity.[9] The comparison of isotherms given on Figures 2-4 shows how much the hydration changes the binding properties of BCD. Hydrated BCD has much lower threshold value of ethanol activity, $a_{0.5S}$ <0.007, and lower inclusion stoichiometry, S = 0.7, than the dried receptor. These changes cannot be non-cooperative, but a measure of this cooperativity is not easily seen. The more profound hydration effect is observed for benzene sorption by BCD. Dry BCD with residual hydration 0.008 *h* does not bind benzene vapors (Figure 4). The same was observed at least up to BCD hydration of 0.06 *h* and benzene activity P/P_0 = 0.8. But BCD with high hydration of 0.172 *h* binds benzene with almost the same efficiency (S=1.3, N=3.0, C=800, $a_{0.5S}$ = 0.11) as TBC4 (Figure 4), which does not need water for it. So, as well as albumin and PAA in systems with ethyl acetate, BCD has a hydration threshold for the binding of benzene. This threshold is a cause of S-like shape of sorption isotherms on Figure 1A.

Conclusions

Obtained results are the first data describing a favorable cooperative hydration effect on the binding of organic vapors by a hydrophilic cross-linked polymer and beta-cyclodextrin. Despite these two hydrophilic receptors and earlier studied protein (albumin) have very different structure, they have important common features. Without hydration all of them have rather tight packing preventing from the binding of relatively large or/and hydrophobic compounds like ethyl acetate and benzene. Still the structure of the studied receptors is relatively flexible, which can be seen from the cooperative increase of their binding ability above a threshold hydration value. The shape of this hydration effect is similar for these three types of receptors. Albumin occupies an intermediate position between the studied cross-linked polyacrylamide derivative and beta-cyclodextrin by its ability to perform saturation by organic compound at the simultaneous hydration increase. This ability, decreasing in the order BCD>albumin>PAA, may be linked to the capacity of hydrated receptor to form clathrates with organic component.

This work was supported by the programs RFBR-tat. (No.03-03-96188), URBR (No. UR.05.01.035) and BRHE (REC007).

[1] K. Mosbach, K. Haupt, *Chem. Rev.* **2000**, *100*, 2495.
[2] I. Y. Galaev, B. Mattiasson, *Trends Biotech.* **1999,** *17*, 335.
[3] D. W. Urry, *J. Phys. Chem. B* **1997**, *101*, 11007.
[4] G. Bell, A. E. M. Janssen, P. J. Halling, *Enzyme Microb. Technol.* **1997**, *20*, 471.
[5] V. V. Gorbatchuk, M. A. Ziganshin, B. N. Solomonov, *Biophys. Chem.* **1999**, *81*, 107.
[6] R. B. Gregory, in: "*Protein-solvent interactions*", R.B. Gregory, Ed., Marcel Dekker, New York 1995, p.191.
[7] T. Ke, A. M. Klibanov, *Biotechnol. Bioeng.* **1998**, *57*, 746.
[8] M. F. Perutz, A. J. Wilkinson, M. Paoli, G. G. Dodson. *Annu. Rev. Biophys. Biomol. Struct.* **1998**, *27*, 1.
[9] V. V. Gorbatchuk, A. G. Tsifarkin, I. S. Antipin, B. N. Solomonov, A. I. Konovalov, P. Lhotak, I. Stibor, *J. Phys. Chem. B* **2002**, *106*, 5845.
[10] N. A. Mironov, V. V. Breus, V. V. Gorbatchuk, B. N. Solomonov, T. Haertle, *J. Agric. Food Chem.* **2003**, *51*, 2665.
[11] N. H. Yennawar, H. P. Yennawar, G. K. Farber, *Biochemistry* **1994**, *33*, 7326.
[12] W. Saenger, J. Jacob, K. Gessler, T. Steiner, D. Hoffmann, H. Sanbe, K. Koizumi, S. M. Smith, T. Takaha, *Chem. Rev.* **1998**, *98*, 1787.
[13] V. V. Gorbatchuk, M. A. Ziganshin, B. N. Solomonov, M. D. Borisover, *J. Phys. Org. Chem.* **1997**, *10*, 901.
[14] T. Steiner, S. A. Mason, W. Saenger, *J. Am. Chem. Soc.* **1991**, *113*, 5676.
[15] G. S. Pande, R. F. Shangraw, *Int. J. Pharm.* **1995**, *124* , 231.
[16] V. V.Gorbatchuk, A. G.Tsifarkin, I. S.Antipin, B. N.Solomonov, A. I.Konovalov, *Mendeleev Commun.* **1997**, 215
[17] T. Dewa, K. Endo, Y. Aoyama, *J. Am. Chem. Soc.* **1998**, *120*, 8933.

Surface Properties and Swelling Behaviour of Hyperbranched Polyester Films in Aqueous Media

Y. Mikhaylova, E. Pigorsch , K. Grundke, K.-J. Eichhorn, B. Voit*

Institut für Polymerforschung Dresden e. V., Hohe Str. 6, 01069 Dresden, Germany

Summary: Thin films of hydroxyl (POH) and carboxyl (PCOOH) terminated aromatic hyperbranched polyesters (HBPs) were prepared by spin coating on silicon wafers and subsequently annealed above their glass transition temperature (T_g). The surface properties and the swelling behaviour of these films in aqueous buffer solutions were studied as a function of annealing time using contact angle measurements and ellipsometry. Non-annealed films were hydrophilic with surface free energies of 51 mJ/m^2 for POH and 49 mJ/m^2 for PCOOH, respectively. The swelling behaviour of the polymer films in buffer solution with pH 7.4 was described in terms of changes of the thickness and effective refractive index of the swollen layer. Under identical conditions a lower water uptake was found for hydroxyl terminated HBPs (POH) which were annealed more then 2 h. The lower water uptake correlates with the surface properties of the films. The annealed films were less hydrophilic. Their surface free energy was 38 mJ/m^2 independent of the annealing. Films of carboxyl terminated HBPs (PCOOH) showed similar surface properties after annealing. However, these films were unstable under the same conditions in aqueous solutions. Stable PCOOH films were obtained by additional covalent binding to the substrate using an epoxy silane as a coupling agent.

Keywords: grafting to; hyperbranched polymers; surface free energy; swelling behaviour

Introduction

The increased attention to thin films of HBP on solid substrates is connected with their potentially interesting properties as chemical sensors, diagnostic tools and functional coatings due to their high number of functional groups in a highly branched architecture.[1,2] In a previous paper[3], we studied the surface properties of thin films of hyperbranched aromatic polyesters terminated with hydroxyl and carboxyl groups and found that these polymers had hydrophilic surfaces caused by acidic functional groups at the outermost surface. Using spectroscopic ellipsometry and reflectometric interference spectroscopy, the swelling behaviour of the HBP films was studied at different atmospheric humidity. From the results, it could be concluded that HBP films can be used potentially as sensoric materials. In this work, we are interested in the

 DOI: 10.1002/masy.200450631

application of HBP films as biofunctional materials. Since the media in which biomolecules, such as proteins, usually interact are aqueous solutions, we studied the behaviour of HBP films in buffer solutions (pH=7.4) and used hyperbranched aromatic polyesters having hydroxyl and carboxyl end groups.

The films were characterized with regard to their chemical composition, thickness, optical constants and morphology using FTIR, NMR, ESCA, ellipsometry methods and scanning force microscopy. Their surface properties were determined by contact angle and zeta potential measurements. The swelling behavior of the films in contact with electrolyte solutions was investigated in situ by spectroscopic ellipsometry.

Experimental section

Materials. Hydroxyl (POH) and carboxyl (PCOOH) terminated hyperbranched polyesters with the same backbone structure were synthesized by Schmaljohann[4] using melt polycondensation of AB_2 monomers. Before the investigations the polymers were purified and structures were checked by NMR and FTIR measurements. The information concerning structure and properties are shown in Figure 1 and Table 1.

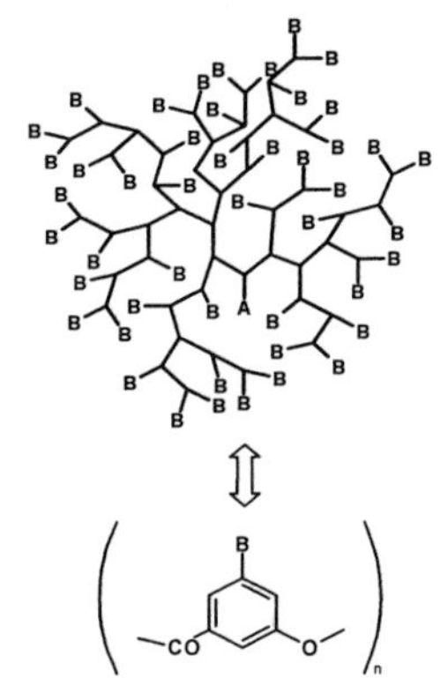

B = OH or COOH

Figure 1. Schematic structure of HBPs

Table 1. Chemical characterisation of HBP

material	M_w [g/mol]	T_g [°C]	$T_{annealing}$ [°C]	branching degree
POH	16.000	227	240	0.60
PCOOH	15.500	214	240	n.d.

n.d. – not determined

Substrates. The substrates were polished silicon wafers with native (~ 2 nm, for Null-ellipsometry) and thermally oxidized (~ 55 nm, for spectroscopic ellipsometry) silicon dioxide layers. For cleaning of the silicon wafers dichloromethane of 99.5% purity (Acros) in an ultrasonic bath (2 times for 5 min) and a hydroperoxide bath (a mixture of water, ammonia solution (25%) and hydrogen peroxide (30%) in volume ratio 1:1:1) at 60 °C for 1 hour were used. Then the samples were washed with the reagent–grade water produced by Milli-Q filtration system and dried with argon. Using the hydroperoxide bath, it was possible to transform the surface (Si-O-Si) of the natural SiO_2 layer into a surface dominated by Si-OH groups without destruction of the surface.

Preparation of the non–grafted polymer layers. A 1 wt% solution of HBP in distilled tetrahydrofuran (THF, Merck) was used for spin coating. Annealing of HBP films above T_g leads to removing of voids and inhomogeneities in the polymer layer due to release of stress within the films and better segregation between fragments.[5] The resulting roughness (rms roughness value) of all polymer films was in the range of 2 nm (determined by AFM for a scan size of 20x20 μm^2).

Preparation of the grafted polymer layers. The grafted PCOOH layers were prepared via a "grafting to" approach using 3-glycidoxy-propyltrimethoxy silane (GPS)[6,7] as coupling agent (Figure 2). For the chemisorption the GPS was used as 1 wt% solution in dry toluene and contacted with the wafer surface for 16 h in a special box with an argon atmosphere. Then the wafers were washed two times in dry toluene in the argon box and three times in ethanol for removing non-grafted GPS. Afterward the wafers were dried under argon stream. The thickness of the GPS layer obtained by Null-ellipsometry was in the range of 0.6 nm. PCOOH were prepared by dip coating from 2 wt% solution in THF. Then the samples were annealed above T_g at 240 °C in vacuum oven (Transparent Drying Oven Buchi model T0-51) for 6 h. The non-grafted polymer was removed by a THF soxhlet extraction for 4 h.

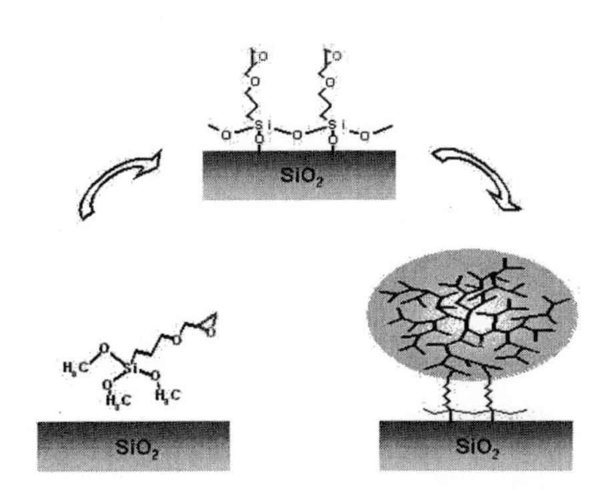

Figure 2. Grafting of PCOOH to the silicon substrate

Sample Characterisation

Ellipsometry. The thickness and optical constants of the HBP films were measured by a multiwavelength ellipsometer M-44 (VASE, J.A. Woollam Co., Inc). The samples were positioned in the central area of a quartz cell and held by a special Teflon holder. The light beam was directed to the middle of the sample surface at an incidence angle of 68 °C. The ellipsometry measurement was started with determination of the thickness and refractive index of the dry film in air at 22°C and 40-50% relative humidity. Then buffer solution (phosphate buffer, pH 7.4, Aldrich) was introduced in to the cell and subsequent time dependent ellipsometry autoscans were started.

The different steps of PCOOH grafting were characterized by Null-ellipsometry (SE402, Sentech) with a He–Ne laser (λ = 633 nm) and an angle of incidence 70°. The layer thickness values were calculated using the corresponding refractive indices of the polymers determined by spectroscopic ellipsometry and Cauchy layer optical models (Table 2).

Table 2. Refractive indices (n) and extinction coefficients (k) used for thickness calculations (λ = 630 nm)

material	Si[1]	SiO_2[1]	GPS[2]	POH[2]	PCOOH[2]
n	3.875	1.4571	1.4290	1.6500	1.6170
k	0.016	0	0	0	0

where [1] and [2] – optical constants are taken from literature and determined by spectroscopic ellipsometr, respectively.

The swelling degree (SD) of the layers was defined by following equation:

$$SD\ (\%) = [(D_\infty - D_0) / D_0] \times 100 \tag{1}$$

where D_∞ and D_0 are the equilibrium thickness values of the HBP film in swollen and dry state, respectively.

Contact angle measurements. Surface wettability and hydrophilicity of HBP films on smooth silicon wafers were investigated by advancing and receding contact angle measurements (DSA-

10 Krüss, Hamburg, Germany) using fresh ultrapure water (MilliQ, Millipore) in laboratory atmosphere. Each reported contact angle measurement represents an average value of 2 or 3 separate drops on different areas. The solid surface tension (γ_{SV}) was calculated by combining the Young equation:

$$\gamma_{LV} \cos\Theta = \gamma_{SV} - \gamma_{SL} \tag{2}$$

and the equation of state for solid-liquid interfacial tensions:[8]

$$\gamma_{SL} = \gamma_{LV} + \gamma_{SV} - 2\sqrt{\gamma_{LV} \cdot \gamma_{SV}} \cdot e^{-\beta(\gamma_{LV} - \gamma_{SV})^2} \tag{3}$$

where γ_{LV} and γ_{SL} are the liquid–vapour and solid–liquid interfacial tension, respectively. β is an empirical constant (0.0001247). Combining eq. (2) and eq. (3) one yields the following relation:

$$\cos\Theta = -1 + 2\sqrt{\frac{\gamma_{SV}}{\gamma_{LV}}} e^{-\beta(\gamma_{LV} - \gamma_{SV})^2} \tag{4}$$

According to eq. (4) the solid surface tension γ_{SV} can be calculated from the experimentally measured contact angle (Θ) and the known γ_{LV} of water (72.5 mN/m).

For the investigation of the contact angles as function of time the dynamic measurements of the Axisymmetric Drop Shape Analysis – profile (ADSA) was used.[9,10] The same samples were used as for the ellipsometric measurement.

Results and Discussion

Swelling measurements of POH

The initial measurements of the dry film gave a thickness in the range from 34 to 37 nm with the refractive index of 1.650 at λ = 630 nm (Figure 3). These values were taken as starting points for the following measurements. Time-dependent data of the thickness of POH layers in the buffer solution were obtained using a simple optical model (Si/SiO_2/swollen polymer/water)[11]. Figure 3 clearly shows the strong effect of the film annealing on the swelling behaviour of POH films. An increase of the annealing time correlates with a decrease of swelling. This behaviour can be a result from a combined effect of the improved structural reorganization of hydrophobic and hydrophilic segments press towards to become better segregated within the film [5,12]. The swelling kinetics can be described by a two steps mechanism: During the first step 90 % of water is sorbed in the film within seconds. This can be classified by Fickian diffusion law. Water molecules penetrate into microvoids and free volume of the HPB and interact via hydrogen bonds. As result of the corresponding out-of-plane expansion a thickness increase is observed. At the same time the uptake of the "optical thinner" water (n=1.33) gives a decrease of the effective refractive index of the swollen layer. The second, much slower, step is due to relaxation process within the layer. The changes in conformation of the polymer molecules allow the layer to absorb the additional water molecules.

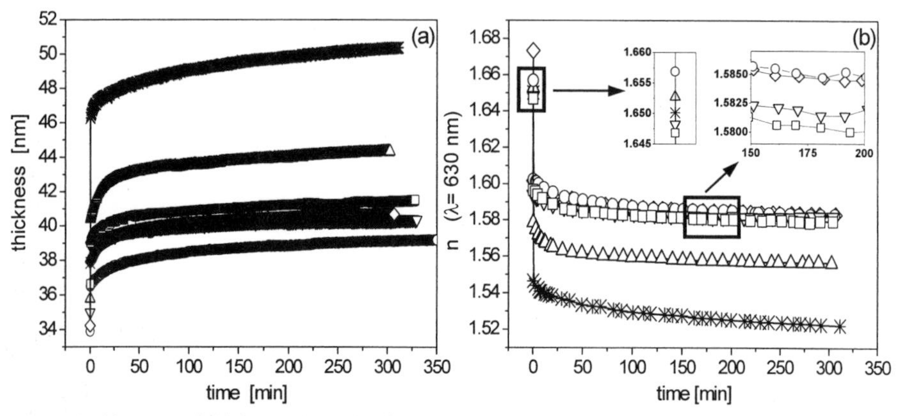

Figure 3. Changes of thickness (a) and refractive index (b) of POH films annealed for different times: 0h (*), 1h (Δ), 2h (◊), 3h (○), 4h (∇), 5h (□)

The refractive indices for the swollen films of POH annealed from 3 till 5 hours are similar and are around 1.58. The swelling degrees at different annealing time are shown in Figure 4. The similar swelling degrees of polymers annealed from 3 to 5 hour demonstrate that a time of 3 h is enough for full segment reorganisation within the polymer film.

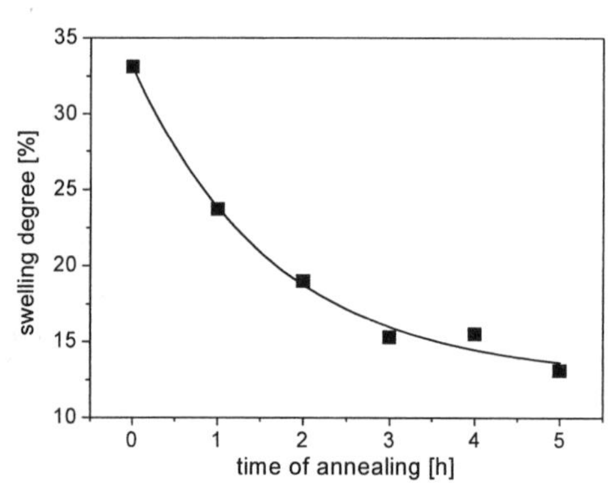

Figure 4. Equilibrium swelling degree of POH films as a function of annealing time

To prove the reversibility of the swelling process, the POH films were removed - from buffer solutions after experiment, washed with water, dried under an argon stream and stored under constant conditions for three days.[10] Then the measurement of the polymer layer thickness was repeated. It gave a thickness having a variation of about 4 % from initial state. Hence, a completely reversible swelling process occurs.

The results of the water contact angle measurements for POH films with different time of annealing are presented in Table 3. It can be clearly seen, that changes in the surface-energetic properties occurred after 1 hour of annealing and that after longer annealing time further changes did not take place. The non-annealed surface has a surface free energy of 51 mJ/m^2 while the annealed surfaces yield surface free energies of about 38 mJ/m^2 independent of the annealing time.

Table 3. Effect of annealing time on advancing and receding contact angles and solid–liquid surface tension of POH surface

time of annealing [h]	advancing contact angle (Θ_a) [deg]	receding contact angle (Θ_r) [deg]	$\Delta\Theta = \Theta_a - \Theta_r$ [deg]	γ_{sv} [mN/m]
0	54.0 ± 0.4	20.1 ±1.3	55.1	51.2
1	75.2 ± 0.8	34.6 ±1.3	40.6	38.3
2	75.1 ±1.0	36.4 ± 1.2	38.7	38.3
3	75.8 ±0.6	33.7 ±1.4	42.1	37.9
4	75.6 ± 0.9	35.1 ± 0.6	40.5	38.0
5	76.6 ± 1.4	37.9 ± 1.3	38.7	37.4

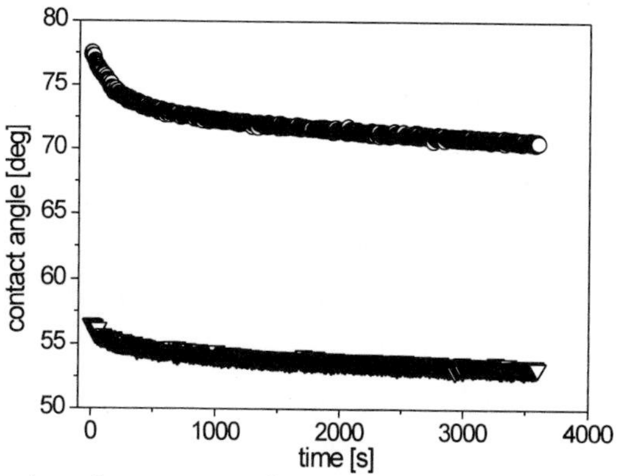

Figure 5. Time-dependent contact angle measurements of water on annealed (○) an non-annealed (▽) POH films using ADSA

If a water droplet is placed on the POH film and the time-dependence of the contact angle is measured by ADSA a small decrease of the water contact angle is observed after a few minutes on both, the non-annealed and the annealed POH surface (Figure 5). These changes in the water contact angle can be related to the swelling of the film.

Grafting reaction and surface characterisation of PCOOH

The contact angles and surface free energies of simply spin-coated, non–attached PCOOH layers as a function of the annealing time are shown in Table 4.

Table 4. Results of the contact angle measurement for spin-coated, non-grafted PCOOH annealed at different times

time of annealing [h]	advancing contact angle (Θ_a) [deg]	receding contact angle (Θ_r) [deg]	$\Delta\Theta = \Theta_a - \Theta_r$ [deg]	γ_{sv} [mN/m]
0	57.7 ± 1.6	19.6 ±1.0	38.1	48.6
2	69.9 ± 0.3	25.3 ±0.4	44.6	41.2
4	73.8 ± 1.5	28.4 ±0.1	45.4	38.8
6	73.8 ± 1.1	29.4 ± 0.3	44.4	38.8

Different annealing times were used to find the optimal conditions for preparation of the polymer films. The surface-energetic properties of the PCOOH films were changed similarly to the POH films by annealing: The originally hydrophilic PCOOH surface was changed into a surface with a lower surface free energy. After an annealing time of 4 h, no further change in the surface properties was observed. In the case of PCOOH, it is assumed that intermolecular and intramolecular reactions between carboxylic groups of PCOOH occur during annealing. As a result anhydride groups could be formed in dependence on temperature and humidity. It is known from contact angle measurements on dry hydrolyzed (i.e. exhibiting carboxylic groups) and annealed (i.e. having anhydride surface groups) polymer films that anhydride groups are less hydrophilic.[13]

Figure 6 shows the "dissolution behaviour" of PCOOH/GPS reactive layers after different times of annealing to initiate the thermal grafting process. The initial film thickness on the GPS modified wafer after spin-coating was ~ 35 nm. The films with 2 and 4 hours of annealing disappeared from the surface of the wafer. The six hours annealed film was not removed from the substrate. It swells strongly, as expected, up to 55 nm.

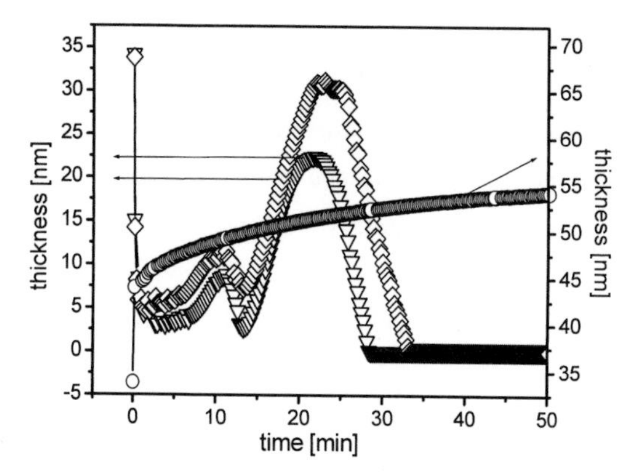

Figure 8. Time dependent swelling of non-extracted PCOOH/GPS samples with different times of annealing to proceed the grafting reaction: 2 h (▽), 4 h (◇) and 6 h (○)

These results show that non-attached PCOOH layers are not stable under the used conditions (pH ≥7). Therefore, it is necessary to form PCOOH layers which are covalently bound to the substrate. The surface properties of the GPS and PCOOH grafted layers after extraction are presented in Table 5. The advancing contact angle of the grafted PCOOH layer was found in the range between contact angles of non-annealed and 2h annealed non-grafted PCOOH (Tab.4).

Table 5. Results of the contact angle measurements on GPS and grafted PCOOH (non-grafted polymer was removed by soxhlet extraction)

material	Thickness [nm]	advancing contact angle (Θ_a) [deg]	receding contact angle (Θ_r) [deg]	$\Delta\Theta = \Theta_a - \Theta_r$ [deg]	γ_{sv} [mN/m]
GPS	0.6	47.1 ± 1.3	29.2 ± 3.0	17.9	55.5
graft–PCOOH	12.0	63.7 ± 1.5	17.7 ± 1.3	46.0	45.4

Conclusions

In summary, we have investigated the surface properties and the swelling behaviour of different annealed HBPs. It was shown that the stability of non–covalently attached HBP layers in aqueous solutions was strongly dependent on the type of functional groups. Stable films of PCOOH were obtained only by covalent grafting onto the modified substrate. POH layers swell clearly, reversibly and with similar kinetics (very fast initial phase, then reaching a plateau), and a corresponding change in thickness and refractive index was measured. The swelling behaviour strongly depends on: initial layer thickness, type of functional groups, buffer type, annealing procedure. The lower water uptake of the annealed films correlated with the less hydrophilic surface properties. Therefore, for the application of HBP films as biofunctional materials, a control of the swollen HBP layer is necessary if subsequent protein adsorption experiments will be carried out which are in the scope of our interest at present.

Acknowledgment

This work is supported by the DFG-SFB 287-02/EG (B6).

(1) B. Voit, *J. Polym. Sci. Part A: Polym. Chem* **2000**, *38*, 2506.

(2) G. Belge, D. Beyerlein, C. Betsch, K.-J. Eichhorn, G. Gauglitz, K. Grundke, B. Voit, *J. Anal. Bioanal. Chem.* **2002**. *374*. 403.

(3) D. Beyerlein, G. Belge, K.-J. Eichhorn, G. Gauglitz, K. Grundke, B. Voit, *Macromol. Symp.* **2001**,*164*, 117.

(4) D. Schmaljohann "Funktionalisierung von hochverzweigten Polyestern für den Einsatz als Beschichtungs- und Blendmaterial" Herbert Utz Verlag, Wissesnchaft, München **1998**

(5) Y.Tang, J.R. Lu, A.L. Lewis, T.A. Vick, P.W. Stratford, *Macromolecules*, **2001**, *34*, 8768.

(6) S. Minko, S. Patil, V. Datsyuk, F. Simon, K.-J. Eichhorn, M. Motornov, D. Usov, I. Tokarev, M. Stamm, *Langmuir* **2002**, *18*, 289.

(7) A. Sidorenko, X.W. Zhai, A. Greco, V.V. Tsukruk, *Langmuir* **2002**, *18*, 3408.

(8) D.Y. Kwok, A.W. Neumann, *Adv. Colloid Interface Sci.* **1999**, *81*, 167.

(9) D.Y. Kwok, T. Gietzelt, K. Grundke, H.-J. Jacobasch, A.W. Neumann, *Langmuir* **1997**, *13*, 2880.

(10) K. Grundke, C. Werner, K. Pöschel, H.-J. Jacobasch, *Colloids and Surfaces Part A* **1999**, *156*, 19.

(11) R. F. Peez; D. L. Dermody; J. G. Franchina; S. J. Jones; M. L. Bruening; D. E. Bergbreiter; R. M. Crooks, *Langmuir* **1998**, *14*, 4232.

(12) Y. Tang, J.R. Lu, A.L. Lewis, T.A. Vick, P.W. Stratford, *Macromolecules*, **2002**, *35*, 3955.

(13) P. Uhlmann, S. Skorupa, C. Werner, K. Grundke, submitted for publication in Langmuir.

Heteropolyacids Dispersed within a Polymer Matrix as a New Catalytic Systems with Controlled Oxidative-Reductive and Acid-Base Active Centers

Wincenty Turek,[1] *Mieczyslaw Lapkowski,*[1,2] *Adam Pron,*[3] *Joanna Debiec,**[1] *Agnieszka Wolna,*[1] *Wojciech Domagala*[1]

[1]Silesia Universty of Technology, Department of Chemistry, Institute of Physical Chemistry and Technology of Polymers, ul. Strzody 9, 44 - 100 Gliwice, Poland
[2]Institute of Coal Chemistry, Polish Academy of Sciences, ul. Sowinskiego 5, 44 - 121 Gliwice, Poland
[3]Laboratoire de Physique des Metaux Synthetiques, UMR5819 (CEA-CNRS-Univ.J.Fourier-Grenoble), DRFMC, CEA-Grenoble, 17 Rue des Martyrs, 38054 Grenoble Cedex 9, France

Summary: Polymer support such as polypyrrole was selected as a matrix for heteropolyacid $H_5PMo_{10}V_2O_{40}$ in an attempt to prepare heterogeneous catalysts containing two different active centers: protons and transition metal ions. Exchanging protons from heteropolyanions dispersed in polymer matrix into ferric or ferrous ions cause the modifications of their catalytic properties. It is manifested by decrease of activity of acid-base centers and increase of activity of oxidative–reductive centers. Oxidation state of iron in all samples before and after catalytic reaction is the same (Fe(III)), but their structure is not similar. For catalysts doped with ferric ions the structural order is much more pronounced than for these doped with ferrous ions.

Keywords: catalysts; conducting polymers; heterogeneous catalysis; heteropolyacids; polypyrrole

Introduction

Polypyrrole (PPy), poly(N-methylpyrrole) (P*N*MPy) and polyaniline (PANI) hold a special position among conjugated polymers. This ows to the ease of their preparation and good environmental stability, which make these polymers exceptionally attractive from application point of view. Possible applications of the properly doped PANI and PPy include heterogeneous catalysis. A number of studies in which catalytically active centers have been incorporated into these polymers can be found in the literature [1,2,3]. They have proven to be efficient catalysts of several industrially important reactions such as alcohol conversion, olefins oxidation, to name a few [4,5,6]. Hetereopolyacids such as $H_4SiW_{12}O_{40}$ or $H_5PMo_{10}V_2O_{40}$ are well known for their interesting catalytic properties. They have two separated active sites i.e. acid-base (protons) and

 DOI: 10.1002/masy.200450632

oxidative-reductive (transition metal ions) [7]. Depositing heteropoliacids in polymer matrix causes molecular dispersion of active species in the whole volume of polymer support and gives new kind of heterogeneous catalysts susceptible for diverse modifications [8]. In our search to diversify the redox properties of these new heterpolyanion – conjugated polymer systems we have recently prepared new catalysts by doping a well known conjugated polymer – polypyrrole – with complex heteropolyanions of the following type: $HFePMo_{10}V_2O_{40}{}^{-}$. These heteropolyanions contain four redox centers of different type – three associated with three different transition metals and the fourth one being a Brønsted center. In this communication we present the studies of the catalytic behavior our new catalyst using i-propanol conversion as a test reaction. We also present a detailed characterization of the catalysts by EPR, X-ray diffraction, TG, DTA and ^{57}Fe Mössbauer spectroscopy before and after the catalytic test.

Experimental

In an attempt to prepare catalysts with versatile but controllable properties we prepared $H_5PMo_{10}V_2O_{40}$ acid which was then used as an oxidizing-polymerizing agent for pyrrole. Such one-step reaction leads to the doped form of the polymer in which $H_4PMo_{10}V_2O_{40}^-$ or $H_3PMo_{10}V_2O_{40}^{2-}$ anions serve as dopants. In the next step of the catalyst preparation the remaining protons, present in the dopant, were exchanged for transition metal cations of variable oxidation state, for example Fe(III)/Fe(II). After the synthesis the presence of heteropolyanions as dopants in the prepared polymer was confirmed by C,H,N,Mo elemental and XRD analysis . The calculation of the doping level from analytically determined C/Mo and N/Mo ratios gave, within the experimental error, the same value of $PPy(HPA)_{0.17}$ where PPy denotes polypyrrole mer involving one ring and HPA denotes heteropolyanion dopant ($H_4PMo_{10}V_2O_{40}$). In the next step H^+ ion was exchanged either with Fe^{2+} or Fe^{3+}. To verify the presence and the oxidation state of iron Mössbauer and EPR spectra of resulting $PPy(HPA)_{0.17}/Fe^{2+}$ and $PPy(HPA)_{0.17}/Fe^{3+}$ catalysts were measured.

EPR measurements were performed using an X-band (9.3 GHz) spectrometer with modulation of magnetic field of 100 kHz. The microwave frequency was recorded. EPR spectra were measured with attenuation of 20 dB (~0.7 mW) to avoid microwave saturation of resonance absorption curves. Isopropanol conversion was studied as a test reaction. The conversion reaction was carried out in an oxygen-free atmosphere. The concentration of isopropyl alcohol in nitrogen was 1.79 mol%. Conversion levels ranged from 5 to 20 %.

Mössbauer spectra of the catalysts tested were recorded at 4.2 K. A typical thickness of the absorber was 5 mg/cm^2 of neutral iron. Co-(RH) was used as a Mössbauer source. An α-Fe absorber operating at room temperature was used for the velocity calibration. The spectra were recorded in a constant acceleration mode and analyzed in a least-squares procedure.

Powder X-ray diffractograms were obtained using Cu Kα radiation (λ = 1.54184 Å) on a Simens diffractometer D5005 (AXS-Bruker) in the 2θ range from 3 to 60°.

Results and discussion

X-ray diffractograms of pure $H_5PMo_{10}V_2O_{40}$ heteropolyacid acid as well as PPy+$H_5PMo_{10}V_2O_{40}$ are shown in Fig. 1 and 2, respectively.

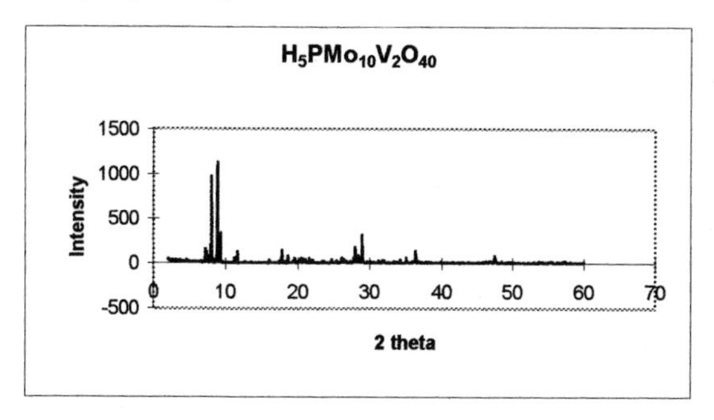

Fig 1. X-ray diffractogram of $H_5PMo_{10}V_2O_{40}$

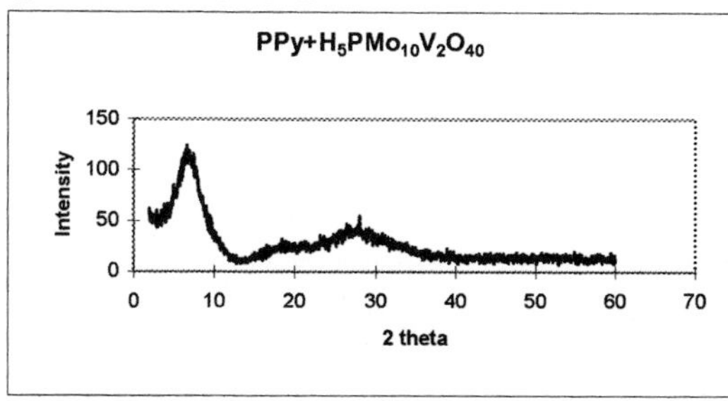

Fig 2. X-ray diffractogram of PPy+$H_5PMo_{10}V_2O_{40}$

The former presents a set of clearly defined Bragg reflections typical of crystalline solids. The diffractogram of the catalyst is different. In this case we notice the total absence of reflections characteristic of the crystalline heteropolyacid. This can be considered as an indication of the molecular dispersion of the dopant within the polymer matrix. The diffractogram of the catalyst is somehow similar to the analogous X-ray patterns reported for other conjugated polymers doped with heteropolyacids [9]. In particular, a strong reflection corresponding to d = 13.1 Å (1) should be noticed which usually is interpreted as originating from polymer chain/dopant/polymer chain stacking. Two broad halos with maxima corresponding to d = 5.3 Å (2) and d = 3.2 Å (3) can also be distinguished. They are usually ascribed to the repeat distances along the polymer chain. Evidently the crystallographic order in this direction is much more poor than the polymer/dopant/polymer stacking order. The exchange of protons for iron ions has minimal effect on the structural order in the doped polymer. The results of this exchange were checked by EPR analysis.

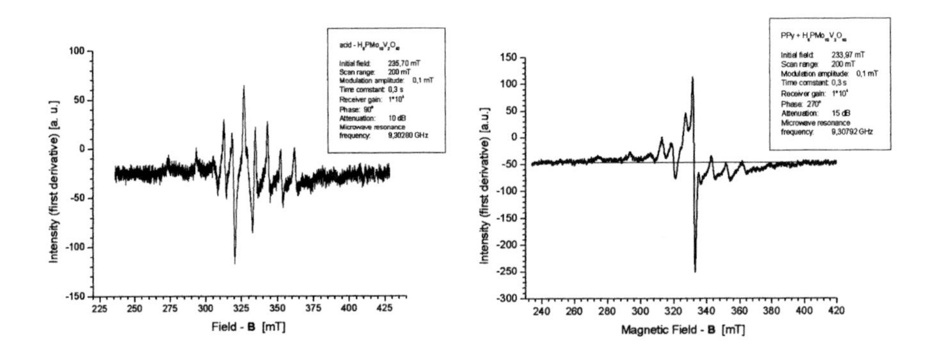

Fig 3. EPR spectra of samples a) acid $H_5PMo_{10}V_2O_{40}$ b) PPy + $H_5PMo_{10}V_2O_{40}$

The presented EPR spectra unequivocally show that in the prepared catalysts, the dopants are molecularly dispersed via the formation of dopant-polymer chain association characteristic for doped conjugated polymers (Fig. 3.). The EPR spectrum of $H_5PMo_{10}V_2O_{40}$ acid doped polypyrrole modified by addition of Fe^{3+} ions proves that it is possible to dope transition metal ions into this type of catalysts. Fe^{3+} ions quenched the EPR lines of nonorganic paramagnetic centers (from heteropolyanions) in the studied polymers. Total quenching of the heteropolyacid EPR signal is observed (Fig. 4.).

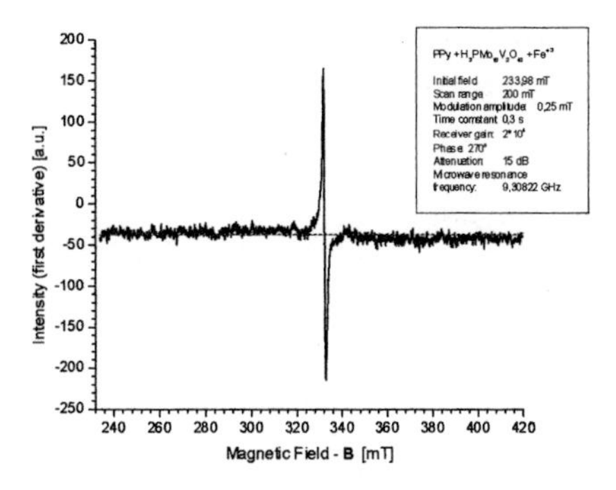

Fig 4. EPR spectrum of PPy + $H_2Fe(III)PMo_{10}V_2O_{40}$

To verify the oxidation state of iron inserted into the polymer matrix as well as to characterize its nearest coordination sphere we have used ^{57}Fe Mössbauer effect measurements. The Mössbauer parameters of the investigated catalysts measured before and after the catalytic tests are collected in Table 1 whereas the representative Mössbauer spectra are presented in Fig. 5, Fig. 6.

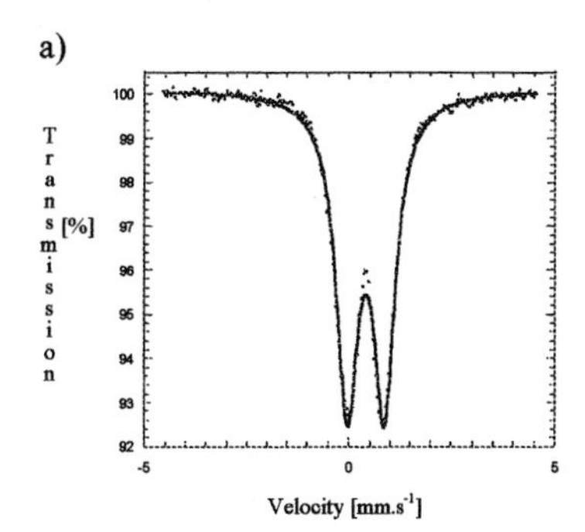

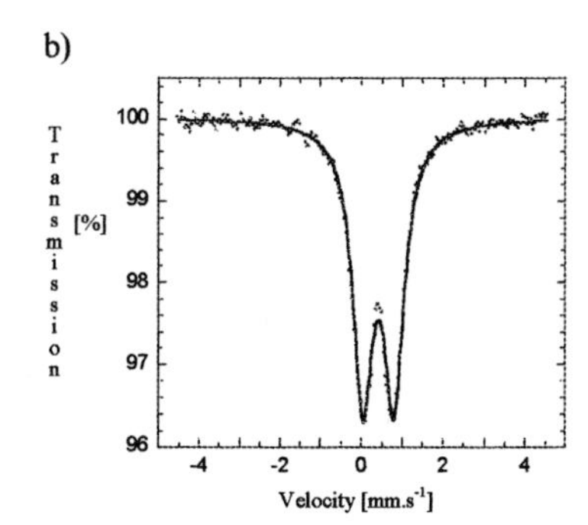

Fig 5. Mössbauer spectra of PPy+$H_2Fe(III)PMo_{10}V_2O_{40}$, a) before reaction, b) after reaction

Interestingly, the Mössbauer parameters of all samples studied are characteristic of high spin Fe(III), independently of the type of iron ion used for the exchange of protons Fe(II) or Fe(III). This means that in the case of PPy+$H_3Fe(II)PMo_{10}V_2O_{40}$, Fe(II) is converted into Fe(III) in the course of the catalysts preparation. Moreover Mössbauer parameters of $PPy+H_3Fe(II)PMo_{10}V_2O_{40}$ and $PPy+H_2Fe(III)PMo_{10}V_2O_{40}$ are very similar.

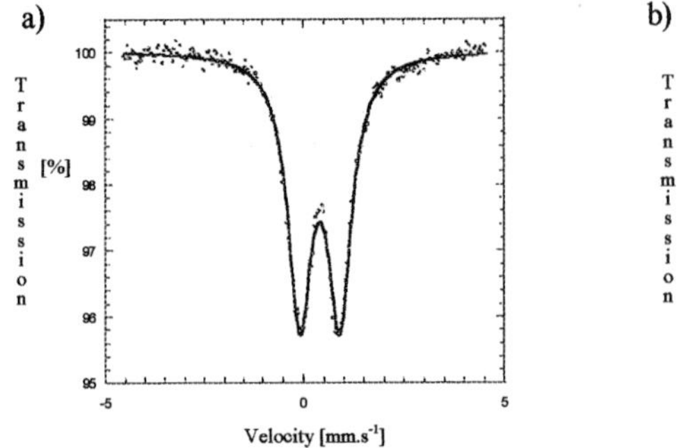

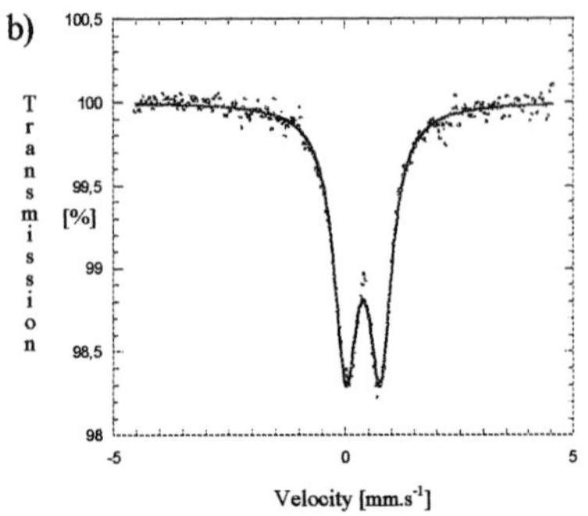

Fig 6. Mössbauer spectra of $PPy+H_3Fe(II)PMo_{10}V_2O_{40}$, a) before reaction, b) after reaction

Table 1. The Mössbauer parameters of $PPy+H_3Fe(II)PMo_{10}V_2O_{40}$ and $PPy+H_2Fe(III)PMo_{10}V_2O_{40}$ measured at 4.2 K

Catalyst	Isomer shift* [mm/s]	Quadrupole splitting [mm/s]	Line width [mm/s]
$PPy+H_3Fe(II)PMo_{10}V_2O_{40}$ before reaction	0.48	0.98	0.68
$PPy+H_3Fe(II)PMo_{10}V_2O_{40}$ after reaction	0.49	0.74	0.58
$PPy+H_2Fe(III)PMo_{10}V_2O_{40}$ before reaction	0.49	0.89	0.62
$PPy+H_2Fe(III)PMo_{10}V_2O_{40}$ after reaction	0.49	0.76	0.57

*With respect to α-Fe at room temperature

The spectra can be fitted with only one doublet which means that only one type of structurally equivalent iron sites are present in the polymer matrix [10]. Despite the fact that both types of catalysts contain iron ions whose spectroscopic features are almost identical, their catalytic behavior is different.

The investigated catalysts are thermally stable in the temperature range of catalytic tests (from 384 to 459 K).

The thermogravimetric measurements show in this temperature range only a small weight decrease of 12 % which can be correlated to a weak endothermic DTA peak with a maximum at 393 K.

In order to verify the influence of proton exchange for iron ions, we have compared the catalytic results of three types of samples: $PPy+H_3Fe(II)PMo_{10}V_2O_{40}$, $PPy+H_2Fe(III)PMo_{10}V_2O_{40}$ and the sample containing no iron $PPy+H_5PMo_{10}V_2O_{40}$.

Catalytic conversion of isopropanol can lead to two reaction products, namely acetone and propene. Acetone being the product of dehydrogenation is formed on redox centers whereas propylene is formed on acid-base centers as the product of dehydration. The results of catalytic tests carried out $PPy+H_3Fe(II)PMo_{10}V_2O_{40}$ and $PPy+H_2Fe(III)PMo_{10}V_2O_{40}$ are shown in Fig. 7a and Fig. 7b, respectively. For comparative reasons the same Arrhenius plots for $PPy+H_5PMo_{10}V_2O_{40}$, i.e. the catalyst in which protons were not exchanged with Fe^{2+} or Fe^{3+}, are shown in Fig. 7c.

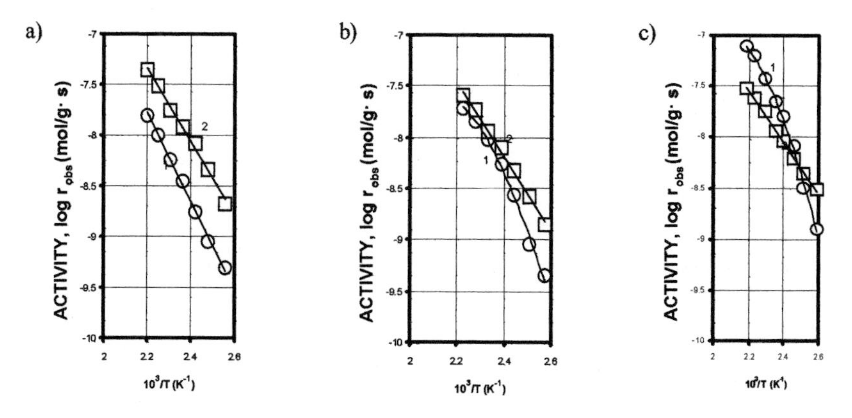

Fig 7. Arrhenius plots of isopropyl alcohol decomposition over a) $PPy+H_3Fe(II)PMo_{10}V_2O_{40}$, b) $PPy+H_2Fe(III)PMo_{10}V_2O_{40}$, **c) $PPy+H_5PMo_{10}V_2O_{40}$**, 1 – Propene, 2 - Acetone

The most striking effect of the exchange of protons with iron ions in the dopant inserted to the polypyrrole matrix is a pronounced change in the catalysts selectivity. The determined selectivities are collected in Table 2 whereas activation energies derived from the Arrhenius plots are listed in

Table 3. Both $PPy+H_3Fe(II)PMo_{10}V_2O_{40}$ and $PPy+H_2Fe(III)PMo_{10}V_2O_{40}$ exhibit much more pronounced redox activity as compared to $PPy+H_5PMo_{10}V_2O_{40}$. This is not unexpected since dehydration of isopropanol involves, in the first step, the protonation of the molecule followed by the abstraction of water molecule with simultaneous creation of a carbocation. Re-arrangement of this carbocation with simultaneous transfer of proton to the catalyst results in the formation of propylene. Since the exchange of H^+ with iron ions efficiently removes active protons from the system the redox activity of the catalyst must increase on the expense of its acid-base activity. One should also note that in $PPy+H_3Fe(II)PMo_{10}V_2O_{40}$ and $PPy+H_2Fe(III)PMo_{10}V_2O_{40}$ the selectivity towards the product formed on redox centers (acetone) is much less temperature dependent than in the case of $PPy+H_5PMo_{10}V_2O_{40}$. The latter shows strongly decreasing redox activity with increasing temperature. The results of Mössbauer spectroscopy studies of both iron containing catalysts after the catalytic tests unequivocally shows that in the course of the test reaction no change in the oxidation state of iron takes place (see Table 1 and Fig.5, 6.

Table 2. The selectivity of isopropyl alcohol conversion

Catalyst	Selectivity [%]					
	393 [K]		418 [K]		443 [K]	
	Propene	Acetone	Propene	Acetone	Propene	Acetone
$PPy+H_5PMo_{10}V_2O_{40}$	40.0	60.0	63.5	36.5	71.5	28.5
$PPy+H_3Fe(II)PMo_{10}V_2O_{40}$	26.6	73.4	41.5	58.5	43.2	56.8
$PPy+H_2Fe(III)PMo_{10}V_2O_{40}$	16.8	83.2	20.8	79.2	24.5	75.5

Table 3. The activation energy of isopropyl alcohol conversion to propene (E_{a1}) and to acetone (E_{a2})

Catalyst	Activation energy, E_a [kJ/mol]	
	E_{a1} (Propene)	E_{a2} (Acetone)
$PPy+H_5PMo_{10}V_2O_{40}$	107.2	49.2
$PPy+H_3Fe(II)PMo_{10}V_2O_{40}$	114.9	72.8
$PPy+H_2Fe(III)PMo_{10}V_2O_{40}$	88.2	67.9

It should be however noted that the value of quadrupole splitting recorded for iron ions in both types of catalysts studied is surprisingly high for high spin Fe(III). This means that some type of distortion occurs upon insertion of heterpolyanions into the polymer matrix which increases the lattice term of the quadrupole splitting. After the catalytic test the quadrupole splitting value decreases (Table 1). Evidently the distortion of the inserted heteropolyanions is partially removed via some relaxation phenomena induced by extended heat treatment. The decrease of the quadrupole splitting value is accompanied by a decrease in the spectral linewidth consistent with better structural order in the samples which underwent thermal treatment.

Conclusions

To summarize, we have demonstrated that using a simple reaction of proton – iron ions exchange it is possible to tune the selectivity of polypyrole supported heteropolyanions catalysts. Since this ion exchange involves the removal of Brønsted centers, the selectivity towards the redox reaction products significantly increases on the expense of the products of the acid-base catalysis.

[1] Huang S.W., Neoh K.G., Kang E.T., Han S.H., Tan K.L., *J Mater Chem* **1998**, *8*, 1743
[2] Kowalski G., Pielichowski J., *Synlett* **2002,** *12,* 2107
[3] M. Misono, Chem. Commun., **2001**, 1141
[4] M. Hasik, W. Turek, E. Stochmal, M. Lapkowski, A. Pron, J. Catal., **1994**, 147 ,544,
[5] E. Stochmal-Pomarzanska, M. Hasik, W. Turek, A. Pron, J. Mol. Catal., **1996**, 114, 267
[6] W. Turek, E. Stochmal-Pomarzanska, A. Pron, J. Haber, J. Catal., **2000**, 189, 297
[7] Makoto M., *Chem. Commun.*, **2001**, 1141-1152
[8] Hasik M., Pron A, Pozniczek J., Bielanski A., Piwowarska Z., Kruczała K. and Dziembaj R., *J. Chem. Soc., Faraday Trans.* **1994**, *90*, 2099
[9] M. Hasik, J.B. Raynor, W. Luzny, A. Pron, *New J. Chem.* **1995**, 19, 1155
[10] N. N. Greenwood, T. Gibbs, *Mossbauer Spectroscopy*, Chapman and Hall, London 1971

Reactive Compatibilization in Phase Separated Interpenetrating Polymer Networks

Tatiana Alekseeva, Yuri Lipatov, Sergey Grihchuk, Natali Babkina*

Institute of Macromolecular Chemistry of National Academy of Sciences of Ukraine, Kharkovskoye Shausse 48, 02160 Kiev, Ukraine
E-mail: todos@ukrpack.net

Summary: The effects of compatibilizing additives (monomethacrylic ester of ethylene glycol (MEG) and oligo-urethane-dimethacrylate (OUDM)) on the kinetics of interpenetrating polymer network (IPN) formation based on cross-linked polyurethane and linear polystyrene and its influence on the microphase separation, viscoelastic and thermophysical properties have been investigated. It was established, that various amounts (3-10 mass%) of the additive MEG and 20 mass% OUDM introduced into the initial reaction system prevent microphase separation in the IPN. In the course of the reaction the system undergoes no phase separation up to the end of reaction, as follows from the light scattering data. The viscoelastic properties of modified IPN are changed in such a way that instead of two relaxation maxima characteristic of phase-separated system, only one relaxation maximum is observed, what is result of the formation of compatible IPN system. The position of this relaxation transition depends on the system composition and on the reaction conditions.

Keywords: compatibilization; interpenetrating polymer network (IPN); microphase separation; relaxation transitions

Introduction

The problem of improvement of compatibility of polymers by introduction of various compatibilizers attracts a great attention. [1-5] The analysis of the data available allows to distinguish two kinds of compatibilization. The essence of the first consists in the reinforcement of the interface in incompatible blends of linear polymers. The interaction between the two constituent phases may be improved by introduction of some bifunctional additives. The most popular type of such additives are block-copolymers, which are localized at the interface and can physically interact with both phases through the two constituent blocks. Usually, the compatibilizer lowers the interfacial tension between phases which leads to the decrease of the average dimensions of a dispersed phase. Such compatibilizers do not affect the thermodynamic stability of the system, but it changes the morphology leading to a finer dispersion degree of components in the course of blending.

 DOI: 10.1002/masy.200450633

The second definition, based on thermodynamic principles, assumes that a compatibilizer is distributed throughout the whole volume of the system. In this context, the compatibility means the formation of a single-phase stable system characterized by the decrease of the mixing free energy. Two incompatible (immiscible) polymers become compatible (miscible) in the presence of a third component. Thermodynamic compatibility is achieved when the free energy of mixing becomes negative.

Interpenetrating polymer networks are one of the most distributed types of polymer blends. Mainly, these systems are formed in the course of chemical reactions of formation of two non-interacting components (simultaneous IPNs). The thermodynamic incompatibility arises in these systems during reaction, accompanied by the incomplete microphase separation. Such systems are characterized by the existence of two glass transition temperatures T_g corresponding to the evolved phases.

Improving compatibility of IPNs may be achieved by the parallel reaction of grafting [1], by introduction of compatibilizers, [2] or by changing the reaction kinetics (the ratio of the rate of formation of constituent networks).[3] Better compatibility may be also the result of the sequential curing, changing the sequence of the networks formation, [4] and due to specific interaction between opposite charged groups or due to formation of hydrogen bonds. [5] The grafting of one network onto another is realized by the introduction of some heterofunctional monomers or oligomers into the initial reaction mixture, which are capable to interact with both networks in the course of their formation.

It is known that the formation of IPN is characterized by superposition of chemical kinetics of network formation and physical kinetics of phase separation, both processes proceeding under nonequilibrium conditions. The reaction kinetics is one of the most important factors determining the onset of the phase separation in the system. It can expect that the presence of compatibilizer in the reaction system will affect both the kinetics of reactions and conditions of phase separation in the reaction system.

In the present work the effects of compatibilizing additives (monomethacrylic ester of ethylene glycol (MEG) and oligourethanedimethacrylate (OUDM)) on the kinetics of the formation of IPN based on cross-linked polyurethane and linear polystyrene and their influence on the microphase separation, viscoelastic and thermophysical properties of the IPN have been investigated.

Experimental

Semi-IPNs were obtained by simultaneous curing of crosslinked polyurethane (PU) in the presence of styrene at 333 K. PU was based on the macrodiisocyanate (MDI) consisting of 2,4-2,6 toluene diisocyanate (TDI), poly(oxypropylene glycol) (PPO, MM 1000) and trimethylolpropane as crosslinking agent. The PU/PS ratio was taken 70/30, 50/50 and 30/70 by mass. The concentration of the initiator [I] for the radical polymerization of styrene (2,2'-azo-bis-isobutyronitrile) was $1 \cdot 10^{-2}$ mol/l. The concentration of the catalyst [kt] for PU formation (dibutyl tin laurate) was $0.3 \cdot 10^{-5}$ mol/l.

As a compatibilizing agents monomethacrylic ester of ethylene glycol (MEG) and oligourethanedimethacrylate (OUDM) were used. OUDM was synthesized from macrodiisocyanate (TDI, PPO, MM 1000) with MEG at a ratio of 1:2 at 313 K with the catalyst-dibutyl tin dilaurate (0.01 mass%) according to

$$\underset{\text{MDI}}{\text{OCN–R–NCO}} + 2\ \underset{\text{MEG}}{\text{HO–CH}_2\text{–CH}_2\text{–O–C(O)–C(CH}_3\text{)=CH}_2} \rightarrow$$

$$\underset{\text{oligourethanedimethacrylate (OUDM)}}{\text{CH}_2\text{=C(CH}_3\text{)–C(O)–O–CH}_2\text{–CH}_2\text{–O–C(O)–NH–R–NH–C(O)–O–CH}_2\text{–CH}_2\text{–O–C(O)–C(CH}_3\text{)=CH}_2}$$

where R is $\text{–C}_6\text{H}_3\text{(CH}_3\text{)–NH–C(O)O–[–CH}_2\text{–CH(CH}_3\text{)–O–]}_{17}\text{O(O)C–NH–C}_6\text{H}_3\text{(CH}_3\text{)–}$.

The reaction kinetics of PU and polystyrene (PS) formation without and in the presence of compatibilizer was investigated at 333 K using a differential calorimeter DAK-1-1A by methods described elsewhere. [6] The kinetic parameters of the reaction of PU formation have been calculated from the kinetic equation for a second-order reaction: $K = \alpha/(1-\alpha)\ C_0\tau$ [6, 7].

The onset of microphase separation (MPS) during curing (τ_{MPS}) was established by the light scattering from the cloud points, using the photocalorimeter FPS-3. From the time dependence of the intensity of light scattering we have calculated the coefficient of the rate of growth of fluctuations in the system R_q as the ratio of ln $\boldsymbol{I}/\tau$, which characterizes the rate of MPS. [8] The compositions of the systems under consideration, the kinetic parameters of reactions, and parameters of MPS are given in Table 1.

The viscoelastic properties were studied by dynamic mechanical analysis (DMA) using a frequency relaxometer. The measurements were done on the frequency of forced sinusoidal vibration of 100 Hz in the temperature interval 220-450 K. The IPNs were studied also by differential scanning calorimetry using the calorimeter DSK-D of Perkin-Elmer type in the temperature interval 150-400 K. The parameters of relaxation transitions (increments of heat capacities ΔC_p at T_g) and data of DMA are presented in Table 2.

Results and Discussion

Kinetic investigations

Introduction of MEG containing two functional groups into the semi-IPN PU/PS for simultaneously proceeding reaction lead to the decrease of the rate of PU formation and increase of the beginning rate of PS formation respecting the initial one free of additive (Table 1, Figure 1). It is connected with chemical interaction between functional groups of MEG and component of PU-phase, as well as component of PS-phase, forming grafted IPNs.

Figure 1 displays the kinetic curves of the PU and PS formation for semi-IPNs with 5 mass % of MEG at various ratio PU/PS and for the system without MEG. Increasing amount of PU components in the reaction system in both cases increases the rate of urethane and of PS formation (Table 1). This effect for PS formation is probably connected to the increasing the viscosity of the reaction system due to higher amount of PU. It lead to sharp reducing the termination constant (K_t) as against the growth constant (K_{gr}) and accordingly increasing the K_{gr} / K_t ratio and the chain growth rate V_{gr} according to eguation: $V_{gr} = K_{gr} [M] V_{in}^{1/2} / K_t^{1/2}$ [7]. The maximum values of reduced rate ($W_{red} = V/[M]$, where $V = dM/dt$ is reaction rate and M is the amount of unreacted monomer) of PS formation in initial IPNs change in the following series: 30/70 < 50/50 < 70/30 as a result of the growth in viscosity. Introduction of MEG in IPNs decreases the maximum values W_{red} of PS formation in the series 50/50 > 70/30 > 30/70 (Figure 2).

Table 1. Parameters of microphase separation (MPS) and reaction rate of formation initial semi-IPN and with different contents of MEG and OUDM.

PU/PS mass %	MEG, mass %	OUDM mass %	$K^{1)} \cdot 10^5$, $kg \cdot mol^{-1} \cdot s^{-1}$	$max W_{red} \cdot 10^2$, min^{-1}	τ_{MPS}, min	$R(q) \cdot 10^2$, min^{-1}	$\alpha^{2)}$	
							PU	PS
70/30	–	–	10.2	1.08	18.0	1.84	0.10	0.003
70/30	3.5	–	8.7	1.01	–	–	–	–
70/30	5.0	–	7.1	0.88	–	–	–	–
70/30	10.0	–	5.9	0.88	–	–	–	–
70/30	–	2.0	10.9	1.06	23.0	1.70	0.13	0.001
70/30	–	5.0	12.2	0.86	30.0	1.38	0.12	0.005
70/30	–	10.0	13.3	0.75	–	–	–	–
70/30	–	20.0	14.3	0.57	–	–	–	–
50/50	–	–	8.7	0.83	35.0	1.62	0.13	0.004
50/50	5.0	–	5.7	0.97	–	–	–	–
50/50	–	20.0	12.2	0.91	–	–	–	–
30/70	–	–	7.2	0.37	50.0	1.43	0.08	0.004
30/70	5.0	–	3.9	0.58	–	–	–	–
30/70	–	20.0	15.2	0.73	–	–	–	–

1) –constant of the reaction rate of urethane formation;
2) – conversion degree at the onset of MPS

Introduction of various amounts of MEG into the reaction system changes the kinetics parameters of IPN formation respecting initial one (Table 1). MPS does not proceed during all the time of reaction (light scattering data). The film obtained remain transparent.

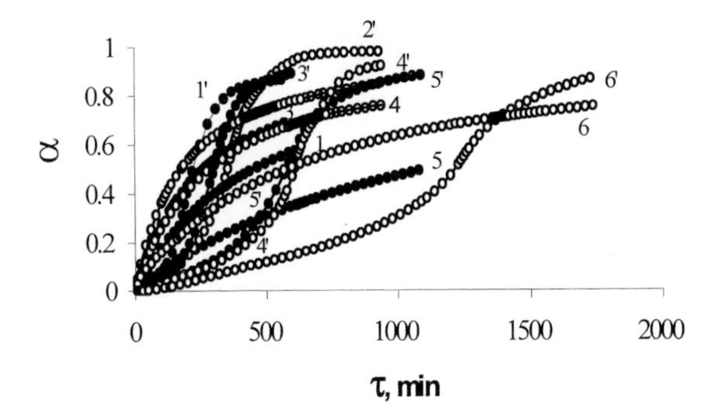

Figure 1. Kinetic curves for PU (1-6) and PS (1'-6') formation in initial semi-IPN (2,2', 4,4', 6,6') and with 5 mass % of MEG (1,1', 3,3', 5,5') at different PU/PS ratio, mass %: 1,1', 2,2' – 70/30; 3,3', 4,4' – 50/50; 5,5', 6,6' – 30/70.

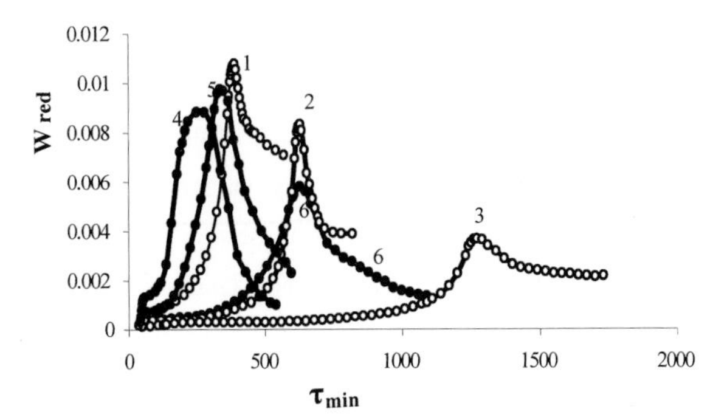

Figure 2. Dependence of reduced rate (W_{red}) of PS formation in IPNs (initial IPN – open symbol; IPN with 5 mass % MEG – close symbol) on time at different PU/PS ratio, mass % : 1, 4 – 70/30; 2, 5 – 50/50; 3, 6 – 30/70.

Introduction of OUDM (2-20 mass %) into semi-IPN PU/PS increases the PU formation rate and reduces the rate of PS formation at different PU/PS ratio respecting initial one (Table 1). Increasing amount of PU components in the reaction system with 20 mass % of OUDM decreases the rate of PU formation in series: 30/70 >70/30 > 50/50 and maximum values of reduced rate of PS formation change in the following series: 50/50 > 30/70 > 70/30 (Figure 3, Table 1). Introduction of 2-5 mass % of OUDM into the reaction system delays the microphase separation, but introduction of 10 and 20 mass % of OUDM prevents microphase separation during IPN formation (Table 1).

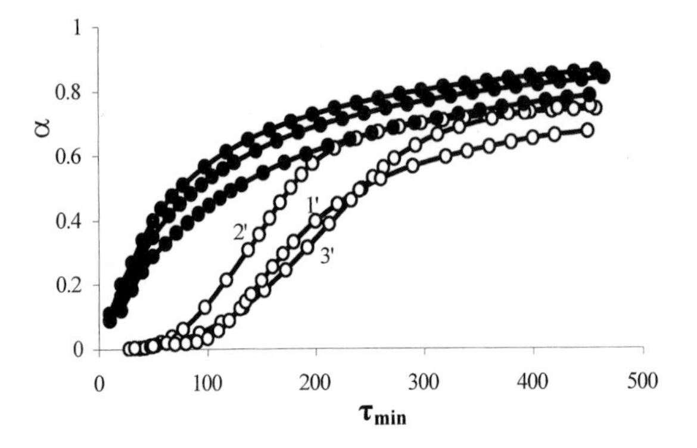

Figure 3. Kinetic curves for PU (1-3) and PS (1'-3') formation in semi-IPN with 20 mass % of OUDM at different PU/PS ratio, mass %: 1,1' – 70/30; 2,2' – 50/50; 3,3' – 30/70.

Dynamic Mechanical Analysis (DMA)

From DMA data of Table 2 and Figure 4a follows that initial IPNs are typical two-phase systems at all component ratios. Introduction of 20 mass % of OUDM in semi-IPN of various compositions essential changes the relaxation properties of the system. Experimental data of Table 2 and Figure 4b shows that for semi-IPNs only one relaxation maximum is observed. Introduction of 5 mass % of MEG in semi-IPNs of various compositions shows only one relaxation transition too (Figure 4c). Its position and height depend on the ratio of the components in the IPN. Thus the increasing amount of the rigid PS component in the IPN containing the compatibilizer show the increase of the single T_g by almost 50 K for the 70/30 compared for the 30/70 IPN. That means that the glass transition in compatibilized IPN depends strongly on the component ratio, being an additional proof for the formation of compatible systems. The reduction in the broadness of the relaxation peak and corresponding increasing the height of maximum also confirm the increasing compatibility and diminishing sizes of phase domains.

By increasing the amount of MEG the glass transition temperature practically does not change (T_g = 323-328 K), although height of relaxation peak (tg δ_{max}) increases (Table 2). One may suppose that increasing amount of the compatibilizer provides the formation of more homogeneous system. In all cases during the reaction no turbidity was observed. For IPNs containing 5 and 10 mass % of OUDM two glass transition temperatures were observed.

Table 2. Parameters of relaxation transition on DSC and DMA data for semi-IPN and grafted IPN based on PU and PS

PU/PS, mass%	OUDM mass%	MEG, mass%	DSC				DMA			
			T_g, K		ΔC_p, J/g·K		T_g, K		tg δ_{max}	
			PU	PS	PU	PS	PU	PS	PU	PS
100/0	–	–	251	–	0.68	–	283	–	0.88	–
0/100	–	–	–	368	–	0.42	–	388	–	3.26
70/30	–	–	255	373	0.55	0.35	288	408	0.68	0.46
70/30	–	3.5	268*		0.65*		323*		0.66*	
70/30	–	5.0	278*		0.68*		328*		0.70*	
70/30	–	10.0	283*		0.85*		323*		0.96*	
70/30	2.0	–	255	368	0.50	0,25	–	–	–	–
70/30	5.0	–	255	368	0.50	0,20	293	398	0.44	0.18
70/30	10.0	–	255*		0.60*		300	375	0.39	0.39
70/30	20.0	–	255*		0.60*		358*		0.52*	
50/50	–	–	258	370	0.40	0.36	293	418	0.59	0.65
50/50	–	5.0	295*		0.48*		368*		0.84*	
50/50	20.0	–	275*		0.50*		358*		0.89*	
30/70	–	–	258	368	0.35	0.35	303	413	0.43	0.88
30/70	–	5.0	330*		0.40*		375*		1.35*	
30/70	20.0	–	315*		0.30*		373*		1.00*	

* – one relaxation transition is observed

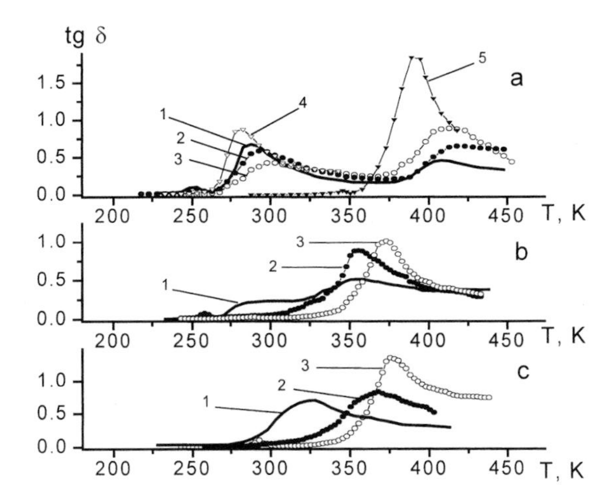

Figure 4. Temperature dependence of mechanical losses for a – initial semi-IPN (1-3), PU (4), PS (5); b – semi-IPN with 20.0 mass % of OUDM (1-3); c – semi-IPN with 5.0 mass % of MEG (1-3) at various ratio PU/PS , mass %: 1 - 70/30, 2 – 50/50, 3 – 30/70.

With increasing the amount of OUDM the glass transition temperature of PU-phase increases from 288 K (without OUDM) to 300 K (with 10 mass % of OUDM) and the T_g of PS-phase decreases from 408 K to 375 K. Introduction of 20 mass % of OUDM into IPNs results in only one glass transition temperature at various ratio PU/PS. With increasing the content of the rigid PS component into the system T_g and height of relaxation peak increase (Table 2).

Differential Scanning Calorimetry (DSC)

In Table 2 the DSC data for the initial specimens of pure components, PU and PS and for the same components in the presence of OUDM and MEG are given. The temperature dependences of heat capacity for IPNs show two jumps: a low- temperature jump, which is related to the glass transition of the oligoether component of PU network at 255-258 K, and a high-temperature jump at 368-370 K, the PS glass transition. The introduction of 2-5 mass % of OUDM into IPNs does not change the glass transition temperature of the PU and PS phase but causes a decrease in the heat capacity jump at T_g respecting initial one. IPNs of various composition obtained in presence of the 10-20 mass % of OUDM show only one jump the heat capacity in the glass temperature region with different values of T_g and ΔC_p. With increasing the content of the rigid PS component in the system ΔC_p decreases and T_g increases. An decrease in ΔC_p is related to the increase of the degree of polymer ordering at temperatures above T_g. It may be supposed that OUDM chemically interact with the PS component and physically interact with PU-component, therefore increasing the content of the rigid PS in the system leads to an increase of their T_g.

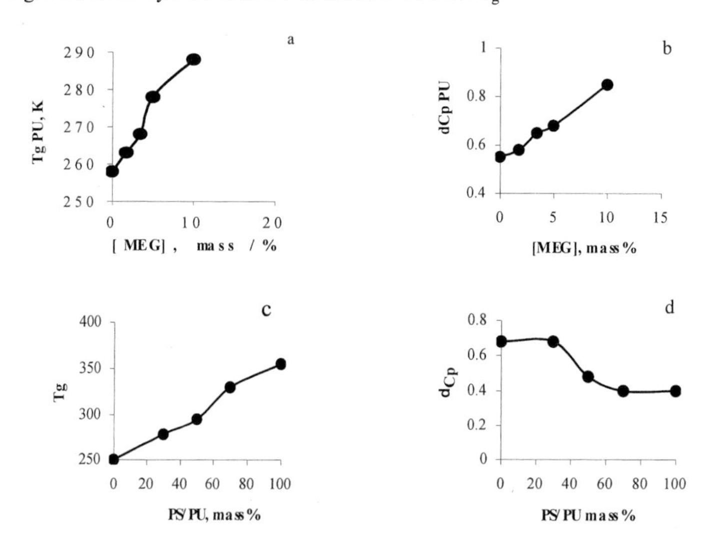

Figure 5. Dependence of glass transition temperature T_g (a, c) and the value of heat capacity jump ΔC_p (b, d) on contents of MEG (a, b) in semi-IPN (PU/PS = 70/30 mass %) and on ratio components PU/PS in semi-IPN with 5 mass % of MEG (c, d) (DSC-data).

The effect of various amounts of MEG was studied for IPN of the 70/30 composition. It was observed only one heat capacity jump in the temperature interval 263-283 K, depending on MEG amount. Values of T_g and ΔC_p increases with increasing MEG concentration (Table 2, Figure 5a, b). One can suppose that due to chemical grafting of PU onto PS via MEG the flexibility of PU component lowers as it can be judged from the increasing glass transition temperature of the system in relation to the T_g of the PU-component. The growth of ΔC_p by introducing MEG may be connected with the formation of an irregular and more defective network structure. With increasing the content of the PS component in the system ΔC_p decrease and T_g increases respecting initial one (Figure 5c, d).

Conclusions

The possibility of compatibilization of IPN's by addition of monomethacrylic ester of ethylene glycol (MEG) and oligo-urethane-dimethacrylate (OUDM) is established. Due to their chemical nature these compatibilizers are able to chemical interaction with components of IPNs (OH-groups and double bounds). As result the formation kinetics of IPNs, the onset of microphase separation, the viscoelastic and the thermophysical properties are essentially changed.

It was established, that various amount (3.5-10.0 mass %) of additives MEG and 20.0 mass % OUDM introduced into the initial reaction system prevent microphase separation of the system. In the course of the reaction the system undergoes no phase separation up to the end of reaction, as follows from the light scattering data. The viscoelastic properties of modified IPNs are changed in such a way that instead of two relaxation maxima characteristic for phase-separated system, only one relaxation maximum is observed. The position of this relaxation transition depends on the system composition and on the reaction conditions.

[1] T.J. Hsu, L.J. Lee, *J. Appl. Polym. Sci.***1988**, *36*, 1157.
[2] Y.S. Lipatov, T.T. Alekseeva, N.V. Babkina, *J. Polym. Mater.* **2001**, *18*, 201.
[3] P. Zhou, Q. Xu, H.L. Frisch, *Macromolecules*. **1994**, *27*, 938.
[4] M.S. Sanchez., G.G. Ferrer, C.T. Cabanilles, *Polymer*, **2001,** *42*, 10071.
[5] E.F. Cassidy, H.X. Xiao, K.C. Frisch, H.L. Frisch, *J. Pol. Sci., Polym. Chem. Ed.,* **1984,** *22*, 1851.
[6] Y.S. Lipatov, T.T. Alekseeva, V.F. Rosovitsky, *Rep. Acad. Sci. USSR.* **1989,** *307*, 883.
[7] Y.S. Lipatov, T.T. Alekseeva, V.F. Rosovitsky, N.V. Babkina, *Polymer.* **1992**, *33*, 610.
[8] V.M. Andreeva, A.A. Tager, I.S. Tukova, L.F. Golenkov, *Polym. Sci: Part A,* **1977,** *19*, 2604 (Russia).

Dynamics of Hyperbranched Polymers and Dendrimers: Theoretical Models

Alexander Blumen,[*1] *Aurel Jurjiu,*[1] *Thorsten Koslowski*[2]

[1]Theoretical Polymer Physics, Hermann-Herder-Str. 3, University of Freiburg, D-79104 Freiburg, Germany
[2]Institute for Physical Chemistry, Albertstr. 23 a, University of Freiburg, D-79104 Freiburg, Germany

Summary: Chemical reactions depend in many ways on the dynamics of the underlying reactants, and an important aspect is the distance covered by the reactants before the reaction act occurs. Hence, even diffusion-limited reactions between *point* particles in confined geometries and in low-dimensional systems display decay forms which are very different from those obtained from simple chemical kinetics. Clearly, even more complex decay forms hold for macromolecules, given their internal degrees of freedom. Here we discuss how the dynamics of macromolecules in solution relates to their topological structure and focus on the motion of macromolecular segments (monomers) under the influence of external fields. After a general survey of the method of generalized Gaussian structures (GGS) we recall the wealth of forms which are observed, depending on the topology and on the microscopic dynamics involved. Paradigmatic are the findings for the class of hyperbranched macromolecules; to these belong the dendrimers. While the dendrimers do not show a typical scaling behavior, as found, say, for linear chains, the situation is different for particular classes of regular hyperbranched polymers which are fractal. We end by discussing their pattern of motion in the GGS-Rouse-Zimm picture.

Keywords: dendrimers; fractals; hyperbranched macromolecules; reaction kinetics; scaling

Introduction

In chemical reactions an important aspect is the distance travelled by the reacting units before the reaction act occurs. In the absence of external fields, under isotropic and homogeneous conditions, an important quantity to this respect is the mean square displacement. In general the Einstein-relation holds, so that $\langle \mathbf{R}^2(t) \rangle$, the mean square displacement of a particle, is related in a

 DOI: 10.1002/masy.200450634

straightforward way to its mean drift $\langle \mathbf{R}(t)\cdot\mathbf{F}\rangle$ in a constant external field **F**. For diffusion-limited reactions at first encounter, an even more important quantity is $S(t)$, the mean number of distinct sites visited by the reacting particle during the time *t* [1]. Now, many investigations have shown that the decay forms of even the simplest chemical reactions, say of the type

$$A + B \to B \tag{1}$$

may strongly differ from the simple exponential decay predicted by the classical kinetic reaction scheme of Eq. (1), see Ref. [1] for a review. Most of the decay forms turn out to be stretched exponentials [1]. With $\phi \neq 1$ and C, $\tilde{C}$ being constants, one has, say the concentration $\Phi(t)$ of A:

$$\Phi(t) = C\exp\left(-\tilde{C}t^{\phi}\right). \tag{2}$$

The reasons why Eq. (2) holds are manifold, and ϕ depends on the space in which the particles move, on the restrictions which they find on their way and on the local fields. Related to this is the fact that even for point-like particles, one finds [1] that $\langle \mathbf{R}(t)\cdot\mathbf{F}\rangle$ and $S(t)$ are not linear functions of *t*, obeying in general:

$$S(t) \sim t^{\beta} \text{ and } \langle \mathbf{R}(t)\cdot\mathbf{F}\rangle \sim t^{\gamma}\text{, with } \beta \neq 1 \text{ and } \gamma \neq 1. \tag{3}$$

In general, the relation between $S(t)$ and $\Phi(t)$ is indirect [1]. The situation gets even more complex when one considers polymers, which have internal degrees of freedom. Here exact solutions are very scarce, an exception being, for instance, the trapping of the end monomer of a polymer diffusing along a linear chain [2].

We will hence restrict our task here to evaluate (as far as possible, analytically) the displacement of macromolecular segments under the influence of external fields. We note that recent observations on the microscopic level allow to monitor such motions: Thus parts of the polymer can be moved by optical tweezers or by attached magnetic beads; furthermore one can follow the local motions through fluorescence and magnetic resonance techniques [3-5].

The displacement of segments (monomers) under the influence of external fields provides valuable information on the macromolecule to which the monomers belong. In general, the motion is anomalous, in the sense of Eq. (3). Such a behavior will clearly influence the course of chemical reactions involving such macromolecules. Here we study this motion using Rouse-Zimm-type models and concentrate on hyperbranched structures [6,7]. In very general fashion, the dynamics of macromolecules is connected in approaches based on linear response to the eigenmodes and eigenfunctions of the system. We find that while for dendrimers the response functions do not scale [6,8,9], i.e. do not even depend algebraically, as a power law, on time or frequency [10,11], a different behavior arises for special classes of hyperbranched structures. As we will show, regular hyperbranched fractals (RHFs) do scale [7], in a way quite similar to the subdiffusive laws of Eq. (3), a feature well-known from the study of linear macromolecules.

Generalized Gaussian Structures and Relaxation

To focus our ideas on particular polymer topologies we choose to perform our study based on the Rouse-Zimm approaches [7-10,12-15] and on their extension to generalized Gaussian structures GGS [16-18]. A GGS consists of beads subject to friction, connected to each other by springs. The configuration of GGS is given by the set of position vectors $\{\mathbf{R}_k\}$, where $\mathbf{R}_k(t) = (R_{xk}(t), R_{yk}(t), R_{zk}(t)) = (X_k(t), Y_k(t), Z_k(t))$ is the position vector of the kth bead at time t. The potential energy $U(\{\mathbf{R}_k\})$ reads

$$U(\{\mathbf{R}_k\}) = \frac{K}{2} \sum_{\beta,m,n} R_{\beta m} A_{mn} R_{\beta n} - \sum_{\beta,n} F_{\beta n} R_{\beta n} \,. \tag{4}$$

Here K is the spring constant, β runs over the components x,y, and z, and the GGS is taken into account through the $N \times N$ matrix $\mathbf{A} = (A_{ij})$. The matrix $\mathbf{A}$ is the so-called connectivity, adjacency or Laplace matrix [16,19,20], and is symmetric: its diagonal element A_{ii} equals the number of bonds emanating from the ith bead, and its off-diagonal elements A_{ij} are either equal to –1 if i and j are connected by a bond, or zero otherwise.

The interactions mediated by the solvent such as the hydrodynamic interactions (HI) are taken into account through the HI-tensor (mobility matrix) [13-15,21,22] between the *i*th and *j*th beads. In a simplified picture $\mathbf{H}_{ij}$ reads:

$$\mathbf{H}_{ij} = \left(\delta_{ij} + \alpha < R_{ij}^2 >^{-1/2} \left(1 - \delta_{ij}\right)\right)\mathbf{I} \equiv H_{ij}\mathbf{I}, \tag{5}$$

with $\alpha = \zeta_r l \sqrt{6/\pi}$ and $\zeta_r = \zeta / 6\pi\eta_0 l$, where l is the average bond length, ζ the friction constant and η_0 the solvent's viscosity. As is usual in the Rouse-Zimm picture, the components of each bead experience the influence of random forces, here denoted by $f_i(t)$; these are taken to be zero-centered, i.e., $\langle f_i(t) \rangle = 0$ and Gaussian distributed.

We observe now that the motion of the different components X_i, Y_i, and Z_i decouples. Hence one may restrict oneself to either one of the components, say Y_i. Setting $\sigma = K/\zeta$ and $\mathbf{Y} = (Y_1, Y_2, ..., Y_N)^T$, $\mathbf{f} \equiv (f_1, f_2, ..., f_N)^T$ and $\mathbf{F} \equiv (F_1, F_2, ..., F_N)^T$, where T denotes the transposed vector, leads to the Langevin equation [8,9,18,23-25];

$$\frac{d\mathbf{Y}(t)}{dt} + \sigma\mathbf{HAY}(t) = \frac{1}{\zeta}\mathbf{H}[\mathbf{f}(t) + \mathbf{F}(t)], \tag{6}$$

where the matrix A is defined as after Eq. (4). Equation (6) has the following formal solution:

$$\mathbf{Y}(t) = \frac{1}{\zeta}\int_{-\infty}^{t} dt' \exp[-\sigma(t-t')]\mathbf{HA}]\mathbf{H}[\mathbf{f}(t') + \mathbf{F}(t')]. \tag{7}$$

To bring Eq. (7) to a more manageable form, one proceeds by diagonalizing the product **HA**, i.e. by determining N linearly independent normalized eigenvectors $\mathbf{Q}_i$ of **HA**, so that $\mathbf{HAQ}_i = \lambda_i \mathbf{Q}_i$. **HA** has only one vanishing eigenvalue, which we denote by λ_1. A further simplification of Eq. (7) arises when a constant external force **F** acts on a single monomer. Assuming now that this monomer is picked randomly, but once chosen *fixed* (quenched disorder), one obtains [8,9,16-18, 26,27] for the doubly averaged *Y(t)* (averaged both over the thermal fluctuations and over the positions of all the monomers in the GGS):

$$\langle\langle Y(t)\rangle\rangle = \frac{F\widetilde{H}_{1,1}t}{N\zeta} + \frac{F}{\sigma N\zeta}\sum_{i=2}^{N}\frac{1-\exp(-\sigma\lambda_i t)}{\lambda_i}\widetilde{H}_{ii}. \tag{8}$$

Here $\widetilde{H}_{ii} = \sum_{k,l} Q^{-1}_{ik} H_{kl} Q_{li}$ and $\widetilde{H}_{1,1}$ simplifies to $\widetilde{H}_{1,1} = \sqrt{N}\sum_k Q^{-1}_{1k} H_{k1}$. In the Rouse case, where **H=I**, one has even that $\widetilde{H}_{ii} = 1$.

Hyperbranched Structures: Dendrimers and Regular Hyperbranched Fractals

In previous work we have studied topologically distinct structures, ranging from star molecules to Sierpinski-type fractals and to networks composed of subunits [8,9,18,26-33]; here we restrict ourselves to hyperbranched macromolecules. Their study is, in general, much simplified, since they are (topologically-speaking) trees, i.e. they are devoid of loops. A most prominent subclass of hyperbranched macromolecules form the dendrimers; one dendrimer is depicted on the left side of Fig. 1. Dendrimers start from a central core from which f arms emerge. In Fig. 1 this central part is indicated by a dashed circle. Now, at each new generation the ends of the arms get $(f-1)$ new arms attached to them. In the case of an ideal dendrimer structure the growth ends at the g-th generation, and it is customary to consider the central core to be the zeroth generation. Viewed topologically, such dendrimers are chemical realisations of finite Cayley-trees. For $f=3$ the dendrimer consists of $N = 3\cdot(2^g-1)+1$ beads, and its number of bonds is $3\cdot(2^g-1)$.

We now turn to another class of hyperbranched polymers, namely to regular hyperbranched fractals (RHF), which we also present in Fig. 1. Our study is motivated by the search of scaling [10,11]; we recall that such RHFs (generalized Vicsek fractals) [7] obey scaling. For the construction we take again $f=3$. Fig. 1 shows schematically the structure at generation $g=3$. Iteratively, one starts from the RHF at $g=1$, indicated in Fig. 1 by a dashed circle. To this object one attaches through 3 bonds, in star-wise fashion, 3 identical copies of itself, obtaining the RHF at $g=2$, which consists of 16 beads. The iterative procedure is now obvious; Fig. 1 presents on its right side the finite RHF for $f=3$ and $g=3$.

The regular pattern of Fig. 1 can be readily generalized to arbitrary f. It turns out that all these objects are fractal [7]. Embedded in the 2*d*-Euclidean space, such a RHF has a fractal dimension $\bar{d}_r$ of $\bar{d}_r = \ln(f+1)/\ln 3$. This differs from the fractal dimension corresponding to the mass of the RHF in solution, $\bar{d}_f$, by a factor of two, so that $\bar{d}_f = 2\bar{d}_r$. Furthermore, in many dynamical features, such as $\langle\langle Y(t)\rangle\rangle$, the spectral dimension $\tilde{d}$ enters; for general RHF one has [7] $\tilde{d} = 2\ln(f+1)/\ln(3f+3)$.

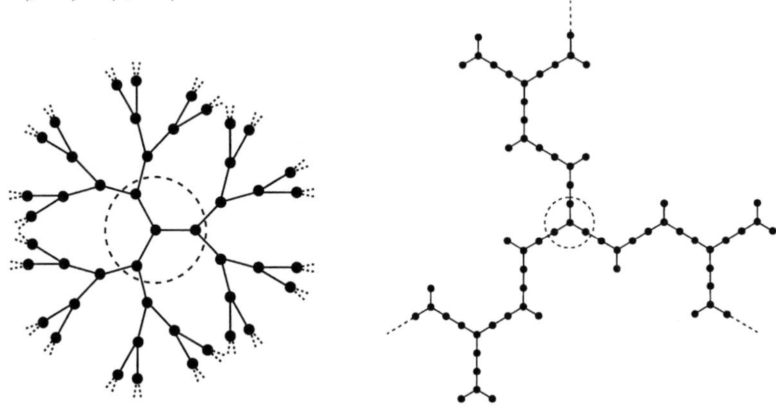

Figure 1. Left: Dendrimer with functionality $f = 3$ at the fourth generation, $g = 4$. Right: Regular hyperbranched fractal (RHF) with $f = 3$ at generation $g = 3$.

For the possible chemical realization of RHFs we note from Fig. 1 that RHFs are built from structural entities whose valence *c* equals 1,2 and *f*, entities which we denote by M_1, M_2 and M_f, respectively. One has a large variety of M_f entities at one's disposal, such as the building blocks of polycarbosilanes [34] or copolyesters [35]. For $f = 3$, condensed triarylamines are interesting candidates; they have been synthesized as bridged molecules [36] and exist in a polymer phase [37]. A fact to be remarked, see Fig. 1, is that chemical compounds corresponding to RHFs with small *f*, say $f = 3$, do not suffer for large *g* from the extreme overcrowding found in dendrimers.

Drift Patterns

As shown above, the evaluation of $\langle\langle Y(t)\rangle\rangle$, Eq. (8), requires the knowledge of the eigenvectors and eigenvalues of the matrices **A**, **HA** and $\mathbf{Q}^{-1}\mathbf{HQ}$. In the presence of HI one needs first to diagonalize **A** in order to obtain the $\langle R_{ij}^2\rangle$, needed in Eq. (5) for **H**; the diagonalization of **HA** leads then to the eigenvalues required in order to compute $\langle\langle Y(t)\rangle\rangle$ based on Eq. (8). In the absence of HI (Rouse case), it is sufficient to know all the eigenvalues of **A**. For dendrimers and for RHFs the evaluation of the eigenvalues of **A** is a simple matter, since one can avoid the direct numerical diagonalization of **A** by making use of analytical recurrence relations, see Refs. [7,24,33] for a review. For **HA**, on the other hand, direct diagonalization and inversion procedures [38-40] are necessary, which imply (because of restricted computer time and memory) an upper limit of roughly 5.000 segments for the polymers to be treated.

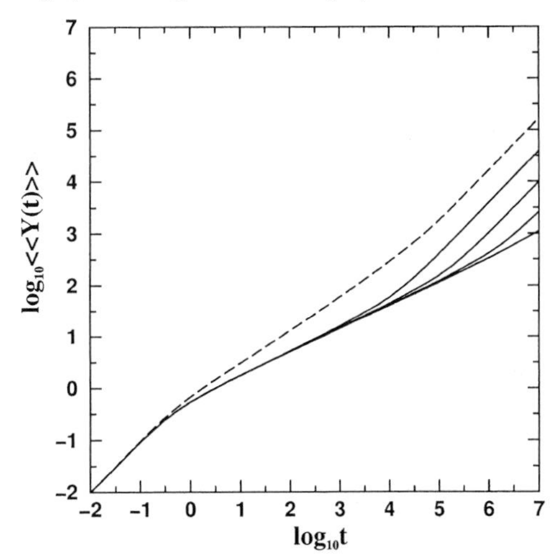

Figure 2. The averaged monomer displacement $\langle\langle Y(t)\rangle\rangle$ under an external force. Displayed are in doubly-logarithmic scales to basis 10 and in dimensionless units both t and the normalized $\langle\langle Y(t)\rangle\rangle$, see text for details

The results of our numerical evaluations for $\langle\langle Y(t)\rangle\rangle$ are presented in Fig. 2 for RHFs with $f=3$. We set $\sigma=1$ and $F/\zeta=1$. The calculations in the Rouse case (given by continuous curves) are for the systems $N = 4^4, 4^5, 4^6$, and 4^7 from above. For the system with $N = 4^6$ we have also performed calculations under HI-conditions, taking $\zeta_r = 0.25$; the results are given through a dashed curve. Note that in Fig. 2 the scales are doubly logarithmic. Clearly evident from the figure is that at long times one reaches a linear domain, $\langle\langle Y(t)\rangle\rangle \simeq F\tilde{H}_{11}t/(N\zeta)$, see Eq. (8). Because of the N dependence of $\langle\langle Y(t)\rangle\rangle$, in Fig. 2 the curves belonging to RHFs of different sizes are shifted with respect to each other. On the other hand, for very short times all of the curves merge; this is the domain where $\langle\langle Y(t)\rangle\rangle = Ft/\zeta$. In the logarithmic scales of Fig. 2 these two domains appear as straight lines with slope 1. Now, as often stressed [6-9,17,18,29-31], the intermediate domain is that which is typical for the polymer considered. In Fig. 2 the intermediate domain appears as a straight line, whose slope γ is less than unity. This demonstrates nicely that $\langle\langle Y(t)\rangle\rangle$ scales, i.e. obeys Eq. (3), with $\gamma \simeq 0.45 \simeq 1 - \tilde{d}/2$ in the absence and $\gamma \simeq 0.64$ in the presence of HI. This finding is remarkable, especially when compared to other fractal structures, such as the Sierpinski-gaskets. There one finds that scaling is obeyed in the Rouse case (no HI), but not in the Zimm case (with HI) [30]. We hence expect that diffusion-limited reactions involving RHFs might lead to decay laws obeying Eq. (2).

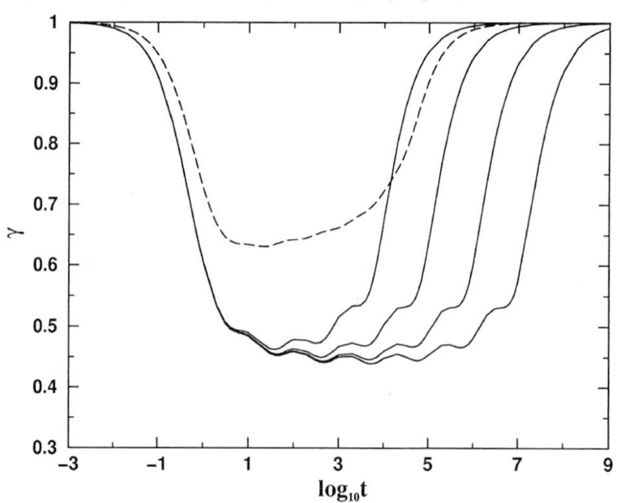

Figure 3. The slope γ (derivative) of Fig. 2, see text for details.

In order to render the scaling analysis more quantitive, we plot in Fig. 3 the derivative of the curves of Fig. 2; for forms obeying Eq. (3) the derivative would equal γ. In both cases (with and without HI) the intermediate domains show very broad minima; superimposed on these is a waviness due to the hierarchical structure of the particular RHF.

We stop to note that dendrimers behave differently in the intermediate range; neither in the presence nor in the absence of HI do dendrimers show scaling. What one observes (see Refs. [8,9,18] for details) is a somewhat logarithmic dependency, when one plots $\langle\langle Y(t)\rangle\rangle$ in double logarithmic scales. The reason for this is to be found in the exponential growth of the dendrimer in each generation, an aspect which gets mirrored in its dynamics. It follows that the observables discussed in this paper, but also more macroscopically-minded quantities, such as the mechanical moduli, can differentiate well between dendrimers and RHFs.

Summary

In this paper we have concentrated on regular hyperbranched fractals (RHFs), a class of polymers whose dynamical properties obey scaling; these objects have well-determined fractal and spectral dimensions. Using the GGS-formalism we have calculated the displacement of RHF-type macromolecules in solution, where we have also taken the influence of hydrodynamic interactions into account. The general picture that emerges is that RHF-type macromolecules do obey scaling, fact with differentiates them from the dendrimers. RHFs are also prime candidates for future studies of diffusion-limited reactions involving macromolecules with complex topology. We do hope that substances with RHF-architecture will be soon synthesized, thus allowing for their experimental investigation.

Acknowledgements

We acknowledge gratefully the help of the Deutsche Forschungsgemeinschaft. A.B. thanks the Fonds der Chemischen Industrie and BMBF for their support.

[1] A. Blumen, J. Klafter, and G. Zumofen, in *Optical Spectroscopy of Glasses*, I. Zschokke ed., D. Reidel Publ. Co., Dordrecht, 1986, p. 199
[2] S. Nechaev, G. Oshanin, and A. Blumen, J.Stat.Phys. 98, 281 (2000)
[3] D. Wirtz, Phys. Rev. Lett. 75, 2436 (1995)
[4] S.R. Quake, H. Babcock, and S. Chu, Nature (London) 388, 151 (1997)
[5] J.W. Hatfield and S.R. Quake, Phys. Rev. Lett. 82, 3548 (1999).
[6] C. von Ferber, A. Blumen, J. Chem. Phys. 116, 8616 (2002)
[7] A. Blumen, A. Jurjiu, Th. Koslowski, and Ch. von Ferber, Phys. Rev. E 67, 061103 (2003)
[8] R. Kant, P. Biswas, and A. Blumen, Macromol. Theory Simul. 9, 608 (2000)
[9] P. Biswas, R. Kant, and A. Blumen, J.Chem.Phys. 114, 2430 (2001)
[10] A. Blumen, Philos. Mag. B 81, 1021 (2001)
[11] I.M. Sokolov, J. Klafter, and A. Blumen, Physics Today 55 (11), 48 (2002)
[12] P.E. Rouse, J. Chem. Phys. 21, 1272 (1953)
[13] B.H. Zimm, J. Chem. Phys. 24, 269 (1956)
[14] M. Doi and S.F. Edwards, *The Theory of Polymer Dynamics* (Clarendon, Oxford, 1986)
[15] A.Yu Grosberg and A. R. Khokhlov, *Statistical Physics of Macromolecules* (AIP Press, New York, 1994)
[16] J.-U. Sommer and A. Blumen, J. Phys. A 28, 6669 (1995)
[17] H. Schiessel, Phys. Rev. E 57, R5775 (1998)
[18] P. Biswas, R. Kant, and A. Blumen, Macromol. Theory Simul. 9, 56 (2000)
[19] A. Kloczkowski, J.E. Mark, and H.L. Frisch, Macromolecules 23, 3481 (1990)
[20] G. Allegra and F. Ganazzoli, Prog. Polym. Sci. 16, 463 (1991)
[21] J.G. Kirkwood and J. Riseman, J. Chem. Phys. 16, 565 (1948)
[22] J. Rotne and S. Prager, J. Chem. Phys. 50, 4831 (1969)
[23] R. La Ferla, J. Chem. Phys. 106, 688 (1997)
[24] C. Cai and Z.Y. Chen, Macromolecules 30, 5104 (1997).
[25] Z.Y. Chen and C. Cai, Macromolecules 32, 5423 (1999)
[26] H. Schiessel, Ch. Friedrich, and A. Blumen, in *Applications of Fractional Calculus in Physics*, edited by R. Hilfer (World Scientific, Singapore 2000) p. 331
[27] Ch. Friedrich, H. Schiessel, and A. Blumen, in *Advances in the Flow and Rheology of Non-Newtonian Fluids*, edited by D.A. Siginer, D. McKee, and R.P. Chhabra (Elsevier, Amsterdam, 1999).
[28] A. Blumen and A. Jurjiu, J. Chem. Phys. 116, 2636 (2002)
[29] A. Jurjiu, Ch. Friedrich, and A. Blumen, Chem. Phys. 284, 221 (2002)
[30] A. Jurjiu, Th. Koslowski, and A. Blumen, J. Chem. Phys. 118, 2398 (2003)
[31] A. Blumen, A. Jurjiu, and Th. Koslowski, Macromol. Symp. 191, 141 (2003)
[32] A.A. Gurtovenko, and A. Blumen, J. Chem. Phys. 115, 4924 (2001)
[33] A.A. Gurtovenko, Yu.Ya. Gotlib, and A. Blumen, Macromolecules 35, 7481 (2002)
[34] C. Lach, P. Müller, H. Frey, and R. Mülhaupt, Macromol. Rapid Commun. 18, 253 (1997)
[35] A. Möck, A. Burgath, R. Hanselmann, and H. Frey, Macromolecules 34, 7692 (2001); A. Möck, A. Burgath, R. Hanselmann, and H. Frey, Polym. Mater. Sci. Eng. 80, 173 (1999); A. Burgath, A. Möck, R. Hanselmann, and H. Frey, Polym. Mater. Sci. Eng. 80, 126 (1999)
[36] C. Lambert and G. Nöll, JACS 121, 8434 (1999); C. Lambert, W. Gachler, E. Schmäzlin, K. Meerholz, and C.Bräuchle, J. Chem. Soc. Perkin Trans. 2, 577 (1999).
[37] C. Lambert, private communication.
[38] B.T. Smith et al., *Matrix Eigensystem Routines-EISPACK Guide, Lecture Notes in Computer Science*, Vol. 6 (Springer, Berlin, 1976)
[39] B.S. Garbow, J.M. Boyle, J.J. Dongarra, and C.B. Moler, *Matrix Eigensystem Routines-EISPACK Guide Extension, Lecture Notes in Computer Science*, Vol. 51 (Springer, Berlin, 1977)
[40] J.J. Dongarra, J.R. Bunch, C.B. Moler, and G.W. Stewart, *LINPACK User's Guide* (SIAM, Philadelphia, PA 1979).

A Polyelectrolyte Bearing Metal Ion Receptors and Electrostatic Functionality for Layer-by-Layer Self-Assembly

*Henning Krass, Georg Papastavrou, Dirk G. Kurth**

Max Planck Institute of Colloids and Interfaces, Research Campus Golm, 14424 Potsdam, Germany
Tel: +49 331-567 9211; Fax: +49 331-567 9202
E-mail: kurth@mpikg-golm.mpg.de

Summary: A polyelectrolyte (BiPE) containing bipyridine ligands as metal ion receptors and quaternary ammonium groups is described, which can be assembled via electrostatic interactions or metal ion coordination. Electrostatic layer-by-layer self-assembly of BiPE with sodium poly(styrene sulfonate) (PSS) as oppositely charged component results in striated multilayers. The BiPE/PSS multilayers can reversibly bind and release transition metal ions including Fe(II), Ni(II), and Zn(II). Formation of 2-D arrays of metallo-units is achieved by μ-contact stamping transition metal salts onto the BiPE/PSS interface. Also, multilayers of BiPE are readily assembled through metal ion coordination. Due to the reversible nature of metal ion coordination, exposure of the multilayers to EDTA causes instant disassembly of the layer, a property needed to implement stimulus triggered release functions. The importance of metal ion coordination for multilayer formation is demonstrated by force-distance curves measured with AFM.

Keywords: atomic force microscopy; metal ion coordination; multilayers; μ-contact printing; thin films

Introduction

The sequential deposition of oppositely charged polyelectrolytes has become a buoyant procedure to fabricate thin films.[1] The wide spread interest in the so-called electrostatic layer-by-layer self-assembly (ELSA) method rests on the simplicity of layer formation, the excellent thickness control, and it's broad applicability. Metal ions have properties that are of special interest for functional devices.[2,3] They provide a range of binding strengths, and ligand exchange kinetics, which are needed for reversible and switchable interaction sites. The ability to interact with photons and electrons alike makes metal ion coordination compounds or metallo-supramolecular

 DOI: 10.1002/masy.200450635

modules (MEMOs) of particular interest as functional components in devices and materials.[4] In addition, metal ions possess interesting magneto-optical properties that are relevant for the construction of advanced materials.[5] Using metal ion ligand interactions in film fabrication, therefore, offers interesting perspectives including reversible interaction sites for controlled assembly and release as well as sensing.[6] In this first account, some of the perspectives and potential features of metal ion coordination as binding motive in thin films are investigated. First, we will discuss electrostatic layer-by-layer self-assembly of BiPE with PSS and metal ion binding of the resulting multilayers, including μ-contact printing. Then, we describe a procedure to assemble multilayers through metal ion coordination and investigate the influence of metal ion–ligand interactions on the adhesion in these multilayers as determined by force-distance curves measured with AFM (Scheme 1).

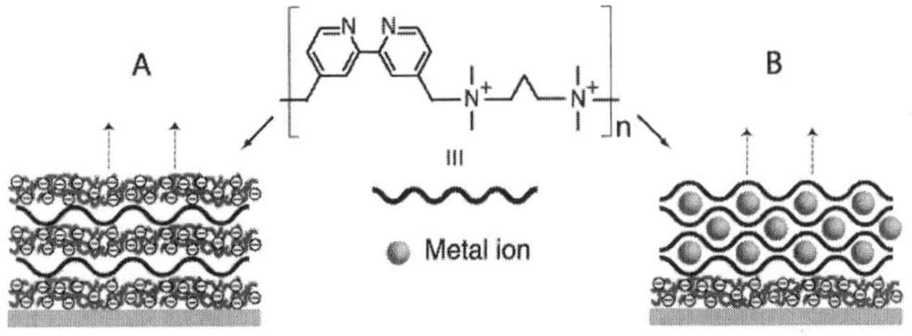

A

B

Metal ion

Scheme 1. Structure of the bi-functional polyelectrolyte (BiPE) with bipyridine metal ion receptors and positively charged ammonium groups. The polyelectrolyte can be assembled through electrostatic interactions with negatively charged polymers, such as polystyrene sulfonate (A) as well as through metal ion coordination (B).

Experimental Section

The synthesis of BiPE and thin film assembly was performed according to literature procedures.[7]

Results and discussion

Multilayers through electrostatic interactions. The quaternary ammonium groups in BiPE provide permanent charges that open a route to fabricate multilayers through electrostatic interactions with oppositely charged polyelectrolytes, such as PSS. Multilayers are deposited onto the PEI (polyethylenimine)/PSS cushion by alternating immersion in solutions containing BiPE and PSS and intermittent washing steps. To elucidate the internal structure the interfacial roughness, the multilayers were subjected to X-ray reflectance (XRR) measurements. Figure 1 shows a representative XRR curve of a $PEI(PSS/BiPE)_3PSS$ film and the corresponding fit.

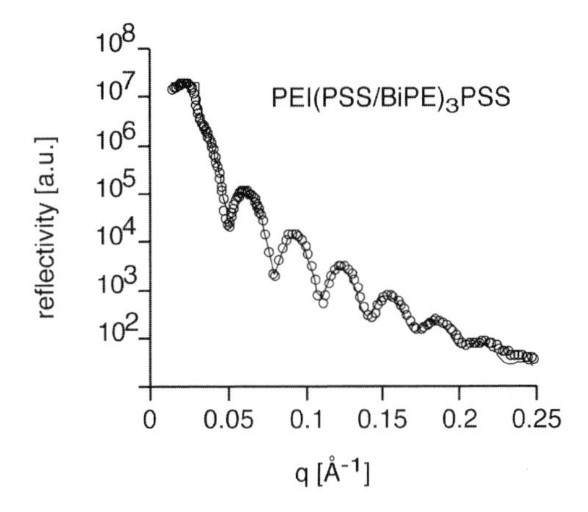

Figure 1. Representative experimental (dots) and calculated (lines) XRR curves for a $PEI(PSS/BiPE)_3PSS$ multilayer (Fit parameters: electron density δ Si: 7.44 10^{-6}; SiO: 7.33 10^{-6}; film: 3.2 10^{-6}; thickness SiO: 0.6 nm; film: 20.5 nm; total interfacial roughness: 1 nm).

As much as seven Kiessig interference fringes are discernible, which demonstrate that the multilayer is homogeneous in thickness and composition.[8] The detailed analysis of the reflectance curve reveals a film thickness of 20.5 nm and a roughness of approx. 1 nm. Similarly, the multilayer $PEI(PSS/BiPE)_2PSS$ has a thickness of 13.4 nm. The average thickness per PSS/BiPE layer pair, therefore, amounts to 5.6 nm, or approx. 2.8 nm per layer, which is in agreement with the surface coverage determined by UV/Vis spectroscopy (not shown). Independent measurements of the thickness by optical ellipsometry confirm these results.

Metal ion coordination and exchange in PSS/BiPE multilayers. The $(PSS/BiPE)_n$ multilayers can coordinate transition metal cations through the bipyridine groups. With many transition metal ions bipyridine forms octahedral tris-bipyridine complexes with the general formula $M(bipy)_3$. With Fe(II) complex formation is particularly strong and gives rise to characteristic and strong metal-to-ligand-charge-transfer (MLCT) bands in the UV/Vis spectrum (Fig. 2).[9]

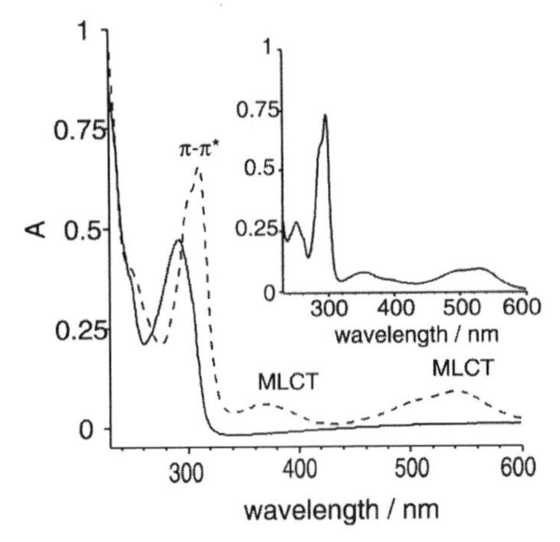

Figure 2. UV/Vis-spectra of a PEI(PSS/BiPE)$_8$ multilayer before (solid line) and after (dashed line) immersion in Fe(II) solution and of Fe(bipy)$_3$(NH$_4$)$_2$(SO$_4$)$_2$ in aqueous solution (insert).

To verify metal ion binding, the (PSS/BiPE)$_n$ multilayer is treated with an aqueous solution containing Fe(II) ions. Figure 2 shows the UV/Vis-transmission spectrum of a PEI(PSS/BiPE)$_8$ multilayer before and after metal ion insertion. Initially, we observe the characteristic bipyridine transitions at 247 nm and 290 nm. After metal ion coordination, the absorption band at 290 nm shifts to 310 nm and increases in intensity. Furthermore, new bands at 371 nm and 540 nm appear, which are assigned to MLCT bands of the Fe(bipy)$_3^{2+}$ complex. The similarity of the spectra of (PSS/Fe(II)-BiPE)$_8$ and Fe(bipy)$_3$(NH$_4$)$_2$(SO$_4$)$_2$ in solution (insert) in terms of band positions and relative intensities confirms formation of the tris-bipyridine metal ion complex in the multilayer.[10]

In order to demonstrate the reversibility of metal ion coordination within the multilayer, the Fe(II) containing multilayer is exposed to ethylenediaminetetraacetic acid (EDTA). In basic solution, EDTA has a strong propensity to coordinate Fe(II) (pk = 14 at pH 11).[11] Removal of the Fe(II) ions from the multilayer is readily confirmed by UV/Vis spectroscopy, in particular by the disappearance of the MLCT bands (not shown).

Due to the versatile coordination chemistry of bipyridine, incorporation of other metal ions, such as Ni(II) or Zn(II), in these layers give similar results. In case of Zn(II), metal ion coordination is accompanied with multilayer fluorescence (Figure3). The similarity of the fluorescence spectra of

(PSS/Zn(II)-BiPE) and $Zn(bipy)_3^{2+}$ in solution confirms as in the previous case formation of the tris-bipyridine complex.

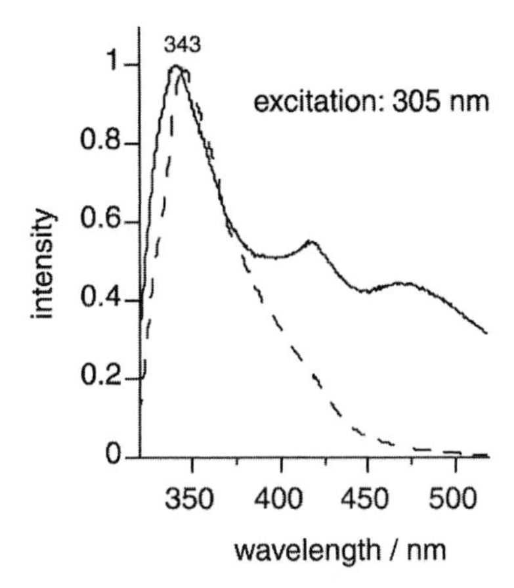

Figure 3. Normalized fluorescence spectra of a $PEI/(PSS/Zn(II)\text{-}BiPE)_{11}$ multilayer (solid line) and $Zn(bipy)_3^{2+}$ in acetonitrile (dotted line).

Lateral patterning with metal ion coordination. The immobilization of metal ion receptors at the interface offers the possibility to confine metal ions to pre-defined areas. Lateral patterning of the surface is achieved by μ-contact stamping the transition metal salt on the (PSS/BiPE) multilayer followed by rinsing with water.

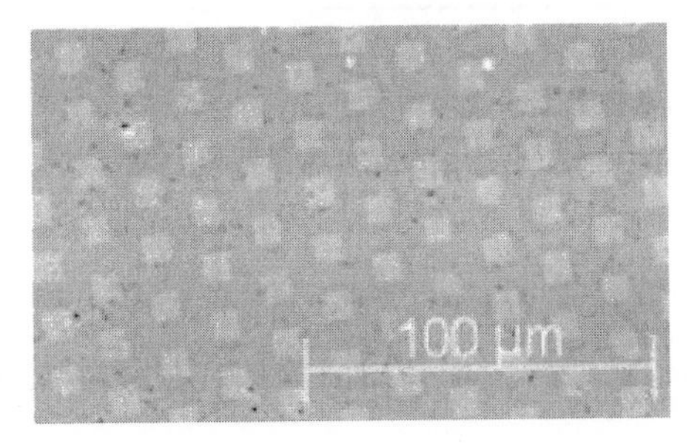

Figure 4. Phase-contrast microscopy image of a patterned $PEI(PSS/BiPE)_{10}$ surface. Lateral patterning with Fe(II)-ions is achieved by μ-contact stamping the metal salt. The continuous dark area represents the imprinted (PSS/Fe(II)-BiPE) region. (The contrast was enhanced using digital editing.)

Figure 4 shows a representative microscopy image of a quartz slide coated with 10 BiPE/PSS layers after μ-contact printing of $(NH_4)_2Fe(SO_4)_2$. The darker area in the image corresponds to regions where salt is imprinted. UV/Vis spectroscopy of the sample confirms metal-ion coordination and complex formation. We, therefore, conclude that μ-contact stamping of the transition metal salt results in spatial confinement of metallo-units. The fact that the metallo-pattern is preserved after rinsing demonstrates the high fidelity of metal ion–ligand interactions and the thereby resulting spatial confinement of the metal ions in the multilayer.

Multilayer formation with coordinative interactions. In a second approach, we explore the metal ion ligand interaction to form coordination bonds between adjacent BiPE layers. First, we generate a PEI/PSS/BiPE cushion, onto which subsequent BiPE layers are deposited through metal ion coordination. The PEI/PSS/BiPE film shows the characteristic π-π* transition at 290 nm of the (uncoordinated) bipyridine groups.

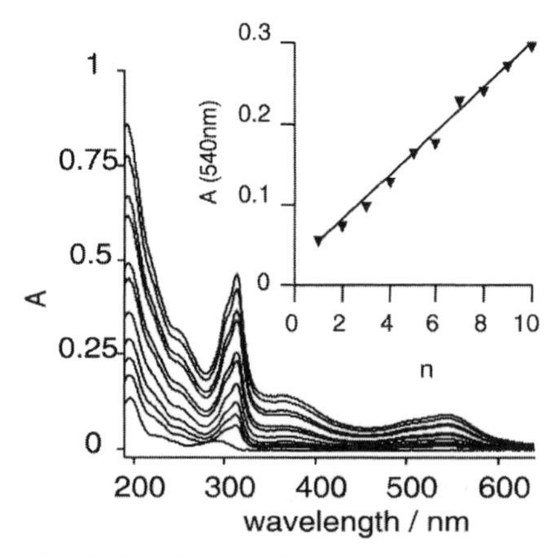

Figure 5. UV/Vis spectra of a PEI/PSS(Fe(II)-BiPE)$_n$ multilayer (n=1-10) on quartz substrate deposited by metal ion coordination. The bottom trace shows the π-π* transition of uncoordinated bipyridine at 290 nm. Interlayer linkage by metal ion coordination shifts the band to 314 nm. Metal ion coordination is also confirmed by the characteristic MLCT transition at 542 nm. The insert shows the increase of absorbance at 541 nm as a function of the number of layers.

Alternating immersion of the substrate in solutions containing (a) $(NH_4)_2Fe(SO_4)_2$ and (b) BiPE results in multilayer formation. Film growth is indicated by an increase of the MLCT bands and the π-π* transition (Figure 5). The appearance of the MLCT band and the shift of the π-π* transition proves metal ion coordination (*vide supra*) and formation of tris-bipyridine complexes during layer growth. The resulting multilayers are rougher than the BiPE/PSS layers, as indicated by ellipsometry, AFM and the absence of Kiessig fringes in XRR. The BiPE layers are readily disassembled by immersion in a solution containing EDTA.

Probing metal ion coordination with atomic force microscopy. The influence of metal ion coordination on the adhesion of adjacent BiPE layers is demonstrated by force-distance curves measured with AFM using a borosilicate particle in order to have a well-defined contact area. The surface of the particle is coated with a PEI/PSS/PAH/PSS/BiPE multilayer. Likewise, a silicon wafer is modified with a PEI/PSS/PAH/PSS/BiPE multilayer. Both interfaces are terminated with BiPE to facilitate metal ion coordination upon contact of the particle modified AFM tip and the surface. Force-distance curves are measured in aqueous solution containing 10 mM NaCl.

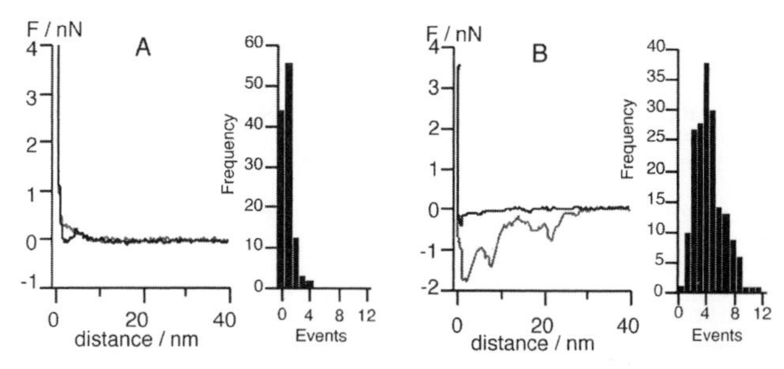

Figure 6. Exemplary force distance curve before (A) and after (B) incubation of the sample in a 0.1 M $NiCl_2$ solution measured in 10 mM NaCl. The histograms show the distribution of the adhesion events.

A representative force-distance curve in the absence of transition metal ions is shown in Figure 6 (A). At large separation between the two surfaces there is no physical interaction and as a result no change in the cantilever position. During the approach of the tip we observe a long-range repulsive force between the interfaces, which is attributed to electrostatic interactions causing the cantilever to bend. Upon withdrawal we observe no or very little attractive interactions between the two surfaces, which are attributed to unspecific steric and electrostatic interactions of possibly entangled polyelectrolytes under the influence of the applied force during the approach. To examine the influence of metal ion coordination, the silicon wafer is immersed in a solution containing 0.1 M $NiCl_2$. After rinsing the wafer with water we record another force-distance curve. A typical example is shown in Figure 6 (B). We notice a short-range attractive interaction during the approach of the tip to the surface, which causes the cantilever to bend. Upon withdrawal of the tip we observe several rupture events, which are attributed to metal ion ligand as well as electrostatic interactions between the two interfaces. We interpret this observation as follows. The metal ions yield a binding force between the two interfaces through metal ion coordination. Most likely, binding results predominantly through ligand exchange reactions because the interface was rinsed with water after metal ion deposition. If we assume that metal ion coordination is much stronger than electrostatic interactions, withdrawal of the tip will first pull apart and stretch the crosslinked polyelectrolyte network between the two interfaces until eventually the metal ion ligand bonds begin to rupture. As a result, the rupture events in the force-distance curves are generally broad typical for elastically deformable connected networks. The overall procedure is repeated for a defined number of experiments and added together, yielding a histogram. We show the number of rupture events for repeated approach-withdrawal cycles at one spot on the surface. Clearly, the presence of $NiCl_2$ has a profound influence on the number and distribution of rupture events. Therefore, we conclude that the attractive interactions revealed in the force-distance curves result from multiple interactions including electrostatic interactions of entangled polyelectrolytes as well as metal ion coordination. With this procedure we, therefore, measure the overall adhesion between the two interfaces. Clearly, the attractive interactions observed in the presence of transition metal ions confirm the importance of metal ion coordination to the formation of the multilayers.

Conclusions

We present a versatile and modular approach to incorporate metallo-units into thin films. A polyelectrolyte (BiPE) is employed that has permanent positive charges and metal ion receptors. Transition metal ions can be incorporated into the BiPE/PSS multilayers either by adsorption from solution or μ-contact stamping. The latter approach results in spatially confined regions of metallo-units. Through a competing complexing agent, such as EDTA, the metal ions are removed from the film. Multilayer formation is also possible by metal ion coordination between layers. Film growth is linear, however, the resulting interface is not as smooth as that of the

ELSA multilayers. The quality of these metal ion coordinated LbL films can be improved in several ways, including the length of the polymer, the reversibility of metal ion coordination, and the assembly protocol. Strong binding between layers through metal ion coordination is confirmed by force-distance curves measured by AFM. The presence of metal ions has a profound influence on the number of rupture events. Although it is not possible with this approach to investigate single interactions it is possible to study the overall effect of strong interactions, such as metal ion coordination, because the receptors are firmly attached through multiple bonds to the underlying substrate. These metal ion coordinated layers readily disassemble upon exposure to EDTA, which opens interesting opportunities to controlled release applications. While our initial purpose was meant as a proof of principle it become clear at the end of this study that metal ion coordination adds a rich flavor to ELSA multilayers in terms of fabrication, characterization, and function.

Acknowledgement. Financial support through the Deutsche Forschungsgemeinschaft is greatly appreciated. Helmuth Möhwald is acknowledged for valuable discussions. The authors especially thank Christa Stolle for the synthesis of the compounds. Hauke Schollmeyer is acknowledged for performing the X-ray experiments.

[1] Decher, G. *Science* **1997**, 277, 1232–1237.
[2] Holliday B. J.; Mirkin C. *Angew. Chem. Intern. Ed.* **2001**, 40, 2022-2043.
[3] Kurth, D. G. *Annals of the New York Academy of Sciences* **960** (2002) 29-39.
[4] Schütte, M.; Kurth, D. G.; Linford, M. R.; Cölfen, H.; Möhwald H. *Angew. Chem. Int. Ed.* **1998**, *37*, 2891-2893.
[5] Gutlich, P.; Hauser, A.; Spiering, H. *Angew. Chem. Intern. Ed.* **1994**, 20, 2024-2054.
[6] Liu, S., Kurth, D. G., Möhwald, H., Volkmer D. *Adv. Mater.* **2002**, 14, 225-228.
[7] Krass, H.; Papastavrou, G.; Kurth, D. G. *Chem. Mater.* **2003**, 15, 196-203.
[8] Kiessig, H. *Ann. d. Physik* **1931**, 10, 769-788.
[9] Constable, E. C. *Advances in Inorganic Chemistry*, **1989**, 34, 1-63.
[10] Bryant, G. M.; Fergusson, J. E.; Powell, H. K. *J. Aust. J. Chem.* **1971**, 24, 257-273.
[11] Jander Blasius, *Einführung in das anorganisch-chemische Praktikum*, 12. Auflage, S. Hirzel Verlag: Stuttgart **1987**,p 323.

Investigation of Surface and Interfacial Properties of Modified Epoxy Films Formed under Different Conditions

Ayzada Magsumova, Liliya Amirova, Makhmoud Ganiev*

Material Science Department, Tupolev Kazan State Technical University, Karl Marx 10, 420111 Kazan, Russia
E-mail: ajzamags@mail.ru

Summary: The surface free energies and their components of the modified amine-cured epoxy polymers were estimated using the geometric mean and acid-base approaches. The surface and interfacial properties of the modified epoxy films depend on the modifying agent percentage and on the conditions of the polymer formation.

Keywords: adhesion; interfaces; modification; resins; surfaces

Introduction

The epoxy resins are widely used as adhesives, surface coatings, polymer composite matrices, and encapsulators of electronic devices. In general the final adhesive properties of composite materials are determined by interfacial interaction on the epoxy/solid interface. Understanding, description and ultimate prediction of such interactions require, among other things, the knowledge of the energetic characteristics of the materials involved.

The analysis of numerous works shows, that experimental and theoretical methods of qualitative and quantitative estimation of the surface and interfacial properties are various enough. In papers[1-3] different semiempirical group contribution methods for an estimation of surface characteristics of epoxy-amine[1,2] and polyimide[3] polymers have been applied. Among the experimental methods inverse gas chromatography[4,5], flow microcalorimetry[6], atomic force microscopy[7,8], ellipsometry[9], NMR[9] and IR-spectroscopy[9,10] may be noted.

Among such variety of methods an indirect estimation of γ_S values, based on measurement of a contact angle of test liquids[11] is most applicable. The technique of a contact angle goniometry is one of the best according to the following reasons: ease of data collection and treatment, and it is an excellent indicator of relative changes occurring on a surface, including the results of modification. One of these methods, a geometric mean method (Fowkes approach) is based on

 DOI: 10.1002/masy.200450636

the idea of the additive contribution of dispersion (γ_S^d) and polar (γ_S^p) components.[11,12] C.J. van Oss, R.J. Good and M.K. Chaudhury have offered other method to estimate γ_S[13] according to which polar components of a surface energy are brought about by short-range interactions which are donor-acceptor or acid-base by the nature (the vOGC approach or acid-base approach). However some features on interpretation of donor-acceptor components remain debatable.[14]

In order to improve some of important operating properties of epoxy resins (e.g. viscosity, wetting ability, shrinkage) the modifying agents active on the solid surface are added to the oligomer.

The purpose of this work was to study the surface and interfacial properties of oligomer and polymer systems of epoxy resin modified by triglycidyl phosphate.

Experimental Section

Aromatic epoxy oligomer consisting mainly of diglycidyl ether of bisphenol A was chosen as an epoxy compound. Glycidyl esters of phosphorus acids were used for modification of epoxy polymers with the purpose of imparting them some valuable operating properties (high strength and adhesion, low inflammability, adjustable optical properties).[15] In the present study the triglycidyl phosphate (TGP), which is a transparent colorless low viscous liquid was taken as an example from available series of glycidyl esters of phosphorus acids (GEPA). Presence of three epoxy groups in TGP molecule allows it to graft into a polymer network.

We also investigated the surface and interfacial properties of cured modified epoxy polymers. The amine compound, 4,4'-diaminodiphenyl methane, was selected as a hardener. After heating up to 60°C the two-step curing carried out as follows: first stage at 20°C for 6 h and postcuring at 150°C for 1 h. For obtained polymers the surface free energy and its components were estimated using contact angles of test liquids. The surface tension components of the test liquids are listed in table1.

Contact angles were measured using the sessile drop method using the microscope accompanied with eyepiece goniometer, in an atmosphere of saturated vapor of test liquids in the thermostatted chamber at 20°C. The test liquids were chosen according to criteria suggested in the paper.[16] All these liquids are inactive towards the surface of epoxy polymer.

The surface of used metal substrates was cleaned from contaminants using solvents and further staying under vacuum. The oxide films on the metal surface remained undestroyed.

Table 1. Surface tension and its components of test liquids

Test liquids	Surface tension and its components, mJ/m^2							
	According to [11,12]			According to [13]				
	γ	γ^d	γ^p	γ	γ^{LW}	γ^{SR}	γ^+	γ^-
Water	72.2	22.0	50.2	72.8	21.8	51.0	25.5	25.5
Glycerol	64.0	34.0	30.0	64.0	34.0	30.0	3.92	57.4
Formamide	58.3	32.3	26.0	58.0	39.0	19.0	2.28	39.6
Diiodmethane	50.8	48.5	2.3	50.8	50.8	0	0	0
Ethylene glycol	48.3	29.3	19.0	48.0	29.0	19.0	1.92	47.0

Results and Discussion

At the first stage influence of TGP on surface tension γ_{LV} of an epoxy oligomer has been investigated. Surface tension of the modified epoxy oligomers on the low energy surface (polytetrafluoroethylene, PTFE) was determined by the du Nouy's ring method and the sessile drop method. The most frequently used surfactants are known to adsorb on the liquid/vapor interface, thus reducing surface tension up to 1.5 times. It is evident from Figure 1 (rhombus) that used organophosphorus modifying agent does not reduce γ_{LV} magnitude, but even slightly rises it. Hence, triglycidyl phosphate is surface inactive on the liquid/vapor interface.

At the same time, the addition of TGP to the epoxy oligomer led to essential decrease of contact angles on solid surfaces (Figure 1, triangles, squares, circles). The greatest effect of TGP addition is observed at its content up to 10 mass-% that can be explained by adsorption of TGP on the solid substrate. It is well known that the interactions at the interface of solid (S), liquid (L) and vapor (V) may be expressed using Young's equation:

$$\gamma_{SV} = \gamma_{LV} \cdot \cos\theta + \gamma_{SL} \tag{1}$$

When the same substrate is used (γ_{SV} = const) addition of the modifying agent results in an increase of $\gamma_{LV} \cdot \cos\theta$. Accordingly this lowers the interfacial tension γ_{SL}. Such influence of the modifying agent on the values of γ_{LV} and γ_{SL} gives an evidence of modifying agent (TGP) adsorption on the solid/liquid interface.

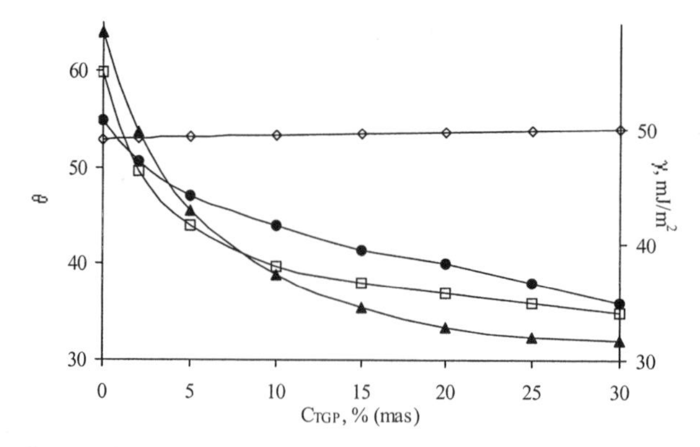

Figure 1. Dependence of contact angle θ on the interface metal/epoxy oligomer and surface tension γ_{SV} of the epoxy oligomer (◊) as a function of TGP concentration; ● = steel (USt 37-2); □ = titanium (TiAl6V4); ▲ = aluminum (AlCuMg2).

According to[11,12] the relationship between components of a surface free energy and equilibrium contact angle for solid and liquid phases is determined by Owens-Wendt equation:

$$(\cos\theta + 1)\,\gamma_L = 2(\gamma_S^d\,\gamma_L^d)^{1/2} + 2(\gamma_S^p\,\gamma_L^p)^{1/2}\,, \qquad (2)$$

where γ_L is the surface tension on the liquid/vapor interface; γ_S^d and γ_L^d are dispersion components of the surface free energy of a solid and liquid; γ_S^p and γ_L^p are the polar components of the surface free energy of a solid and liquid.

The total surface free energy γ_S can be expressed by the sum:

$$\gamma_S = \gamma_S^d + \gamma_S^p \qquad (3)$$

The estimation of acid γ_S^+, base γ_S^- and specific, polar or acid-base γ_S^{AB} components of the surface free energy of cured polymers were carried out with the consideration of values of non-polar Lifshitz-van der Waals component γ^{LW} which is practically identical to dispersion component γ^d as is offered in paper.[13] According to this approach the surface energy polar component γ^{SR} is determined by donor-acceptor (acid-base) interaction and can be estimated by the equation

$$\gamma^{AB} = 2\,(\gamma^-\gamma^+)^{\,S}, \qquad (4)$$

where γ^- and γ^+ are donor (base) and acceptor (acid) component of a surface free energy, respectively.

The basic equation connecting acid-base and non-polar interactions with value of a contact angle θ, looks as follows:

$$(\cos\theta + 1)\,\gamma_L = 2\,[(\gamma_S^{LW}\,\gamma_L^{LW})^{1/2} + (\gamma_S^{+}\,\gamma_L^{-})^{1/2} + (\gamma_S^{-}\,\gamma_L^{+})^{1/2}] \quad (5)$$

In Figure 2(a) (rhombus, circles, triangles) represent the dependence of the surface free energy values of epoxy polymers, formed on the polymer/vapor interface, obtained according to eqs (2) and (3) with the different content of the modifying agent.

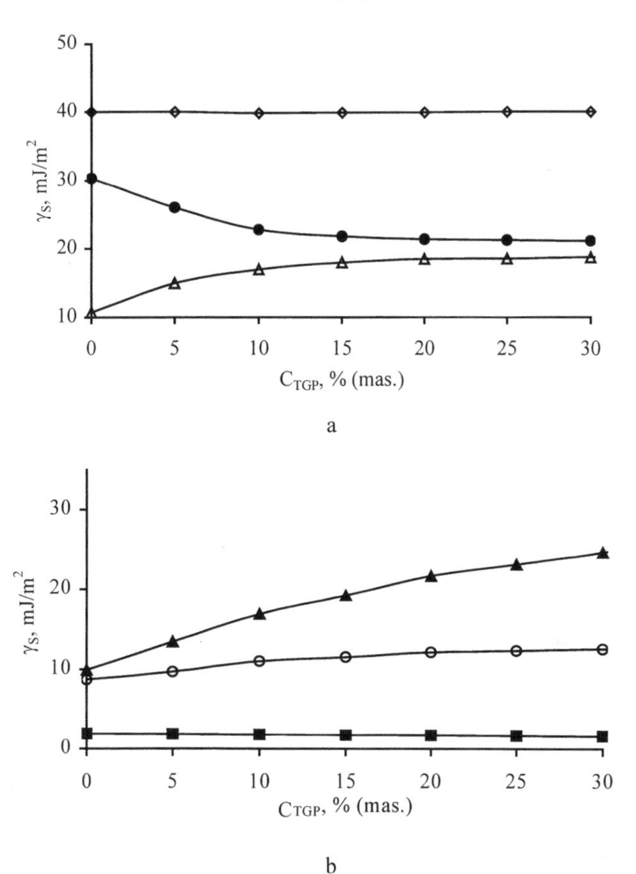

Figure 2. Dependence of a surface energy and its components as a function of TGP contents in amine-cured epoxy polymer: a - ◊ = surface energy, γ_S; ● = dispersion component, γ_S^D; Δ = polar component, γ_S^P; b - ○ = polar (acid-base) component, γ_S^{AB}; ■ = acid component, γ_s^+; ▲ = base component, γ_s^-.

It is seen that γ_S does not practically depend on TGP concentration up to 30 % (mas.). At the same time in the 0-10 % (mas.) range of the modifying agent the considerable reduction of the dispersion component γ_S^d and increase of the polar component γ_S^p are observed.

It is known, that the surface of the amine-cured epoxy polymers is practically monopolar and has basic character.[17,18] According to our data calculated using the vOGC approach (Figure 2b) addition of TGP to epoxy polymer composition does not change the nature of a surface, but strengthens its basicity. The last can be apparently explained by the increase of quantity of oxygen atoms in the base polymer network due to the rise of number of ester bonds and appearance of phosphoryl groups from TGP.

To study the influence of the substrate energetic properties on surface characteristics of the polymers formed on this substrate interface, two solid surfaces were used - polytetrafluoro ethylene (PTFE) and aluminum (AlCuMg2). It is seen from Figure 3 that the surface free energy of polymer formed on the PTFE and aluminum interfaces have different dependence on TGP concentration.

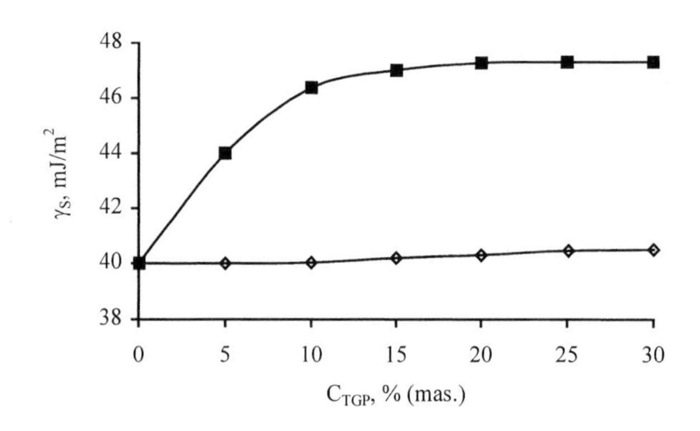

Figure 3. Change of surface energy of the modified epoxy polymer, formed on the PTFE (◊) and aluminum – AlCuMg2 (■) interface as a function of TGP concentration.

When the polymer is formed on the PTFE interface (Figure 3, rhombus), dependence analogous to behavior of polymer on the air interface (Figure 2a, rhombus) is observed. This means that the surface of polymer cured on the PTFE interface as well as on the air interface is formed predominantly by the non-polar groups. Samples formed on the high-energy surface aluminum interface (Figure 3, squares) have higher γ_{SV} values. The Figure 3 shows that addition of the modifying agent up to 10 % (mas.) is already sufficient for formation of polymer with high γ_{SV} value. Further, at the greater concentrations of the modifying agent the curve (Figure 3, squares) reveals saturation (plateau).

Thus, the results obtained reveal that increase of the substrate surface energy results in adsorption of triglycidyl phosphate on the solid substrate.

The study of adhesion properties of glues prepared from epoxy oligomer modified by TGP, shows, that addition of the modifying agent up to 10 % (mas.) considerably increases the adhesion strength of bonds (Figure 4). Apparently, it is also related with preferential adsorption of triglycidyl phosphate on a high-energetic solid surface.

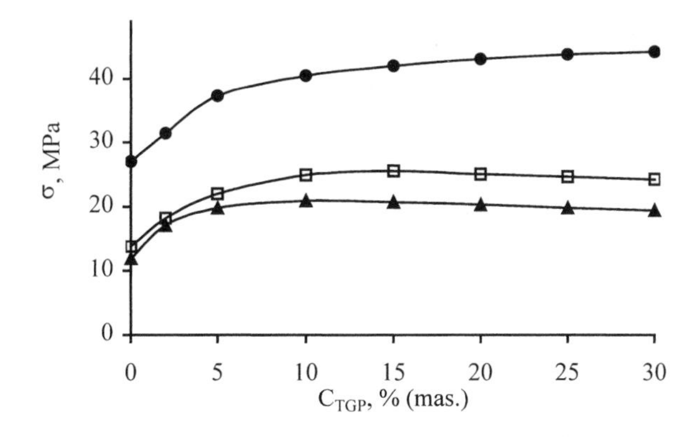

Figure 4. Strength at steady separation σ of the metal/(modified adhesive)/metal adhesion bonds as a function of the TGP content: ● = steel (USt 37-2); □ = titanium (TiAl6V4); ▲ = aluminum (AlCuMg2).

Conclusions

Triglycidyl phosphate shows activity on the high-energy solid/oligomer interface. Such distribution of the modifying agent allows increasing the wetting ability of oligomer to the solid substrate and adhesion of forming polymer to the substrate.

[1] S. A. Page, R. Mezzenga, L. Boogh, J. Berg, J-A. E. Manson, J. Coll. Interf. Sci. **2000**, 222, 55.
[2] R. Mezzenga, S. A. Page, J-A. E. Manson, J. Coll. Interf. Sci. **2002**, 250, 121.
[3] R. R. Thomas, L. E. Stephans, J. Coll. Interf. Sci. **2002**, 251, 339.
[4] S. J. Park, M. Brendle, J. Coll. Interf. Sci. **1997**, 188, 336.
[5] A. Saada, E. Papirer, H. Balard, B. Siffert, Coll. Interf. Sci. **1995**, 175, 212.
[6] M. F. Finlayson, B. A. Shah, J. Adhes. Sci. Technol. **1990**, 4, 431.
[7] M. Mantel, Y. I. Rabinovich, J. P. Wightman, R.-H. Yoon, J. Coll. Interf. Sci. **1995**, 170, 203.
[8] C. S. Hodges, Adv. Coll. Interf. Sci. **2002**, 99, 13.
[9] F. M. Fowkes, J. Adhes. Sci. Technol. **1990**, 4, 669.
[10] F.M. Fowkes, J. Adhes. Sci. Technol. **1987**, 1, 7.
[11] D. K. Owens, R. C. Wendt, J. Appl. Polymer Sci. **1969**, 13, 1741.
[12] F. M. Fowkes, Ind. Eng. Chem. **1964**, 56, 40.
[13] C. J. van Oss, M. K. Chaudhury, R. J. Good, Chem. Rev. **1988**, 88, 927.
[14] M. Morra, J. Coll. Interf. Sci. **1996**, 182, 312.
[15] L. M. Amirova, V. F. Stroganov, E. V. Sakhabieva, Intern. J. Polym. Mater. **2000**, 47, 43.
[16] D. Y. Kwok, H. Ng, A. W. Neumann, J. Coll. Interf. Sci. **2000**, 225, 323.
[17] M. Dogan, M. S. Eroglu, H. J. Erbil, J. Appl. Polym. Sci. **1999**, 74, 2848.
[18] T. S. Chung, W. Y. Chen, Y. H. Lin, K. P. Pramoda, J. Polym. Sci. **2000**, 38, 1449.

Solid-Supported Biomimetic Membranes with Tailored Lipopolymer Tethers

Anton Förtig,[1] *Rainer Jordan,* *[1,2] *Karlheinz Graf,*[3] *Giovanni Schiavon,*[4]
Oliver Purrucker,[5] *Motomu Tanaka**[5]
[1] Lehrstuhl für Makromolekulare Stoffe, TU München, Lichtenbergstr. 4, 85747 Garching, Germany,
E-mail: rainer.jordan@ch.tum.de
[2] Department of Chemistry, Chemical Engineering and Materials Science, Polytechnic University, Six Metrotech Center, Brooklyn NY 11201, USA.
[3] Max Planck Institut für Polymerforschung, Postfach 3148, 55021 Mainz, Germany
[4] Lehrstuhl für Anorganische und Analytische Chemie, TU München, Lichtenbergstr. 4, 85747 Garching, Germany
[5] Lehrstuhl für Biophysik E22, TU München, James-Franck-Str. 1, 85747 Garching, Germany
E-mail: mtanaka@ph.tum.de

Summary: Stable lipid membranes with controlled substrate-membrane spacing can be prepared using well-defined lipopolymers as a tether. Based on the living cationic ring-opening polymerization of 2-methyl- or 2-ethyl-2-oxazoline, lipopolymers can be synthesized bearing a lipid head group as well as a silanol reactive coupling end group. Using a “grafting onto” procedure these polymers can form dense, brush like monolayers, whose layered structures can be obtained by x-ray reflectivity measurements. By transfer of a pre-organized monolayer that is followed by vesicle fusion, stable polymer supported lipid membranes can be prepared. The substrate-membrane spacing can be controlled via the degree of polymerization, while the lateral diffusion of lipids within the membrane depends on the density of polymer tethers. Preliminary experiments implied that the membrane with long (N = 40) polymer tethers could reside trans-membrane receptors homogeneously, suggesting a large potential of this strategy.

Keywords: lipopolymer; poly(2-alkyl-2-oxazoline); tethered lipid bilayer membrane

Introduction

The construction of supported lipid membranes on solid substrates has become a popular research topic within the last several years. Especially a planar configuration of the bilayer allows the application of numerous different surface and interface sensitive experiments to obtain a detailed insight of the plasma membrane structure and membrane associated transport processes. Such membrane models can not only be used to study the membrane behavior in detail but ultimatively to incorporate transmembrane proteins, attach lipid associated recognition and signaling sites and

 DOI: 10.1002/masy.200450637

thus, investigate isolated functions and properties of biological cell membranes. Although several supported lipid membrane models were presented in the literature, a stable biomimetic membrane model, which enables meaningful investigation of various membrane associated diffusion and transport processes, is still at large. For membrane bilayers directly deposited onto a substrate the close proximity of a membrane and the solid support in the equilibrium state has a typical distance of 5 - 20 Å.[1] This thickness does not provide a sufficient internal water reservoir which results in nonspecific adsorption of incorporated membrane proteins to the substrate and their denaturing as well as a significant slower lateral diffusion in the lower lipid leaflet.[2] These limitations can be overcome by increasing the thickness of the lubricating water layer by introducing a hydrophilic polymer layer between the lipid membrane and the substrate as introduced by *Ringsdorf* [3] and *Sackmann*.[4] This polymer interlayer should also increase the mechanical stability of the supported membrane bilayer. E.g., black lipid membranes preserve their structure for only short time and are very sensitive to mechanical stress. Presumably, a good stabilization of the membrane can be obtained if the polymer interlayer is not only functioning as a cushion but also tethers the membrane to the supporting solid substrate. Such a polymer should feature a lipid head group for incorporation into the lipid layer by hydrophobic interactions and, on the other side, a chemical function for a permanent covalent attachment. This would form a defined brush-like lipopolymer monolayer, covalently bound to the surface and on the other side incorporated into the membrane (see Figure 1).

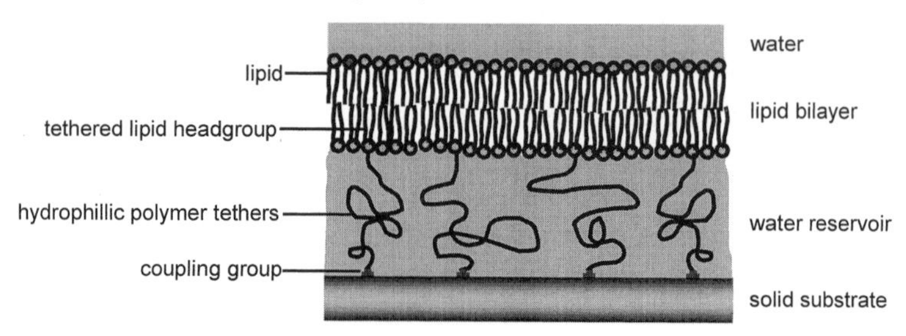

Figure 1. Solid-supported lipid bilayer tethered by lipopolymers.

However, if the grafting density of such polymer tethers is too high, the free mobility of the lipids in the under layer may be obstructed. So, for every application, a compromise between stability and fluidity needs to be found. Additionally, the type of polymer is crucial for the structure and function of the supported membrane and incorporated proteins. The polymer must fulfill certain requirements: First, it should be a hydrophilic polymer which interacts neither with membrane proteins nor with the lipid bilayer. Regarding the stability aspects, the polymer should be hydrolytically stable. Since the introduction of the concept of polymer cushions/tethers between the lipid membrane and the solid support, different approaches for the construction of such membrane models were studied. *Tamm* et al.[5] used linear polyethylene glycol (PEG) with a lipid head group (lipopolymer) and a triethoxysilane end group. Although they varied the grafting density of the lipopolymer tethers, they limited their study to a single polymer length. *Naumann et al.* [6] reported on the use of poly(2-ethyl-2-oxazoline)s with a lipid head group. The lipopolymers were photochemically coupled to a self-assembled monolayer of benzophenone. Since the grafting reaction is unspecific, the morphology of the polymer cushion may vary strongly. *Schiller et al.*[7] used a tetrameric oligomer to avoid a helical lipopolymer backbone and a disulfide coupling group to ultra flat gold surfaces. Within this amphiphilic multilayer, the lipids displayed a good short-range mobility but a very low lipid diffusion coefficient over lager areas. The short oligomeric spacers were grafted at extremely high densities to maintain the layer morphology. Incorporation of larger proteins into such constructs might be difficult, if not impossible.

In order to obtain full control upon the grafting density of the polymer tether, the interaction between the lipopolymer and the membrane lipids as well as the fixation onto the solid substrate, we developed the direct synthesis of silane end-functionalized lipopolymers.[8] We used a poly(2-methyl- or 2-ethyl-2-oxazoline) backbone as the hydrophilic tether. The living character and the variability of the cationic ring-opening polymerization of 2-alkyl-2-oxazolines result in linear polymer chains,[9] with an adjustable degree of polymerization, low polydispersity, and quantitative end-functionalization at both ends of the polymer (Figure 2).[8] The hydrophilic/lipophilic balance (HLB) can be fine-tuned by the choice of the monomer, the degree of polymerization and the lipid head group. Using different degrees of polymerization for the hydrophilic tether, the intermediate polymer/water layer in the future model membranes might be adjustable.

To be able to vary the lipid head group, different lipids have been synthesized and used as initiators for the polymerization. This allows us to fine-tune the polymer architecture to construct stable and functional biomimetic membranes and to overcome the present limitations.

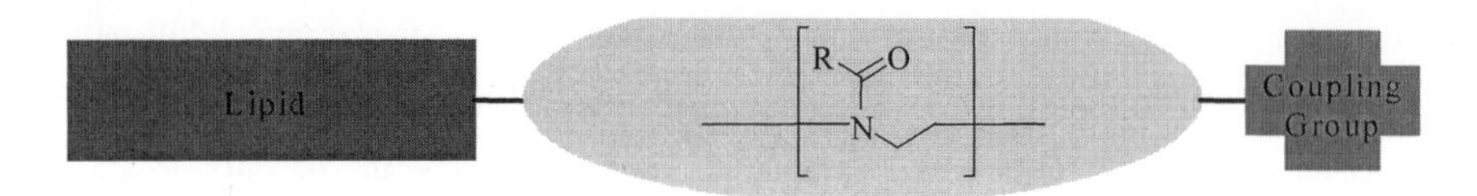

Figure 2. Cartoon of the 2-alkyl-2-oxazoline lipopolymer with a lipid head group, a hydrophilic polymer spacer and a silane coupling end function

Lipopolymer Synthesis

The potential variability of the lipopolymers structures and preservation of high control of the polymer architecture was the key criterion for the developed lipopolymer synthesis. The various lipid moieties were introduced into the polymer by the initiator method.[8] Alkyl moieties with a primary hydroxyl group were quantitatively converted into the corresponding triflates and then used as the initiator for the living cationic polymerization. This ensures a quantitative functionalization of the poly(2-oxazoline) with the lipid function. Most lipids are commercially available or can be readily synthesized. E.g. phytanol was obtained from phythol by reduction with Raney-Nickel.[10] The corresponding triflate was then directly used as the initiator or as a tosylate reacted with benzyl glycerol to yield, after cleavage of the benzyl protection group and triflatization the double chain analog lipid-initiator diphytanyl glycerol (Scheme 1).[11]

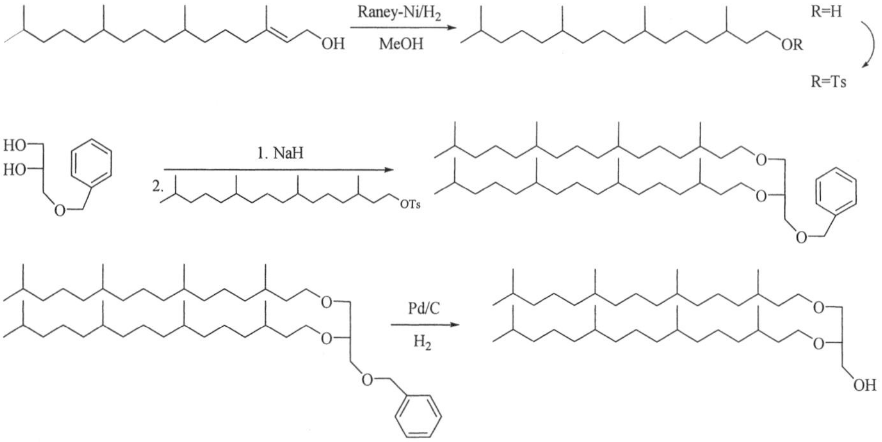

Scheme 1. Hydration of the phythol and synthesis of the diphytanyl glycerol lipid.

Other single and double chain lipids, such as *n*-hexadecyl triflate and 1,2-di-*n*-octadecyl-*s,n*-glyceroyl triflate were obtained directly from the alcohols. The choice of the lipids were motivated by (1) variation of the sterical need relative to the attached hydrophilic polymer to vary the critical packing parameter of the lipopolymer amphiphile, (2) matching of the alkyl chain length with the co-lipids in the future membrane construct, (3) good long-term stability (no ester bonds) and compatibility with the living polymerization.

The living cationic polymerization was then performed in chloroform to ensure good solubility of the initiator, the monomer and the lipopolymer during the entire polymerization. Each lipid was equipped with different polymer chain lengths (n=10, 20, 30, 40) simply by changing the initial initiator/monomer feed ratio for the living cationic polymerization. By this, a library of lipopolymers with different HLBs (HLB=hydrophilic/lipophilic balance) were obtained. The surface grafting function was finally introduced by a quantitative termination reaction using α-ω-amino alkylalkoxysilanes (Scheme 2).

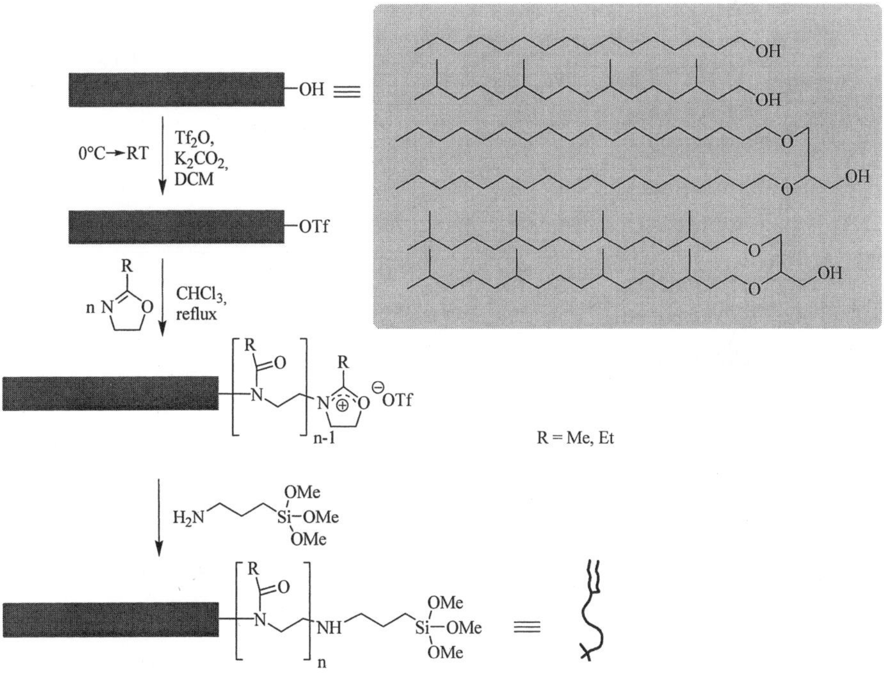

OH ≡ OH OH O O OH O O OH

0°C→RT Tf_2O, K_2CO_2, DCM

OTf

R n N O $CHCl_3$, reflux

R O N R N ⊕ O ⊖ OTf n-1

R = Me, Et

H_2N OMe Si–OMe OMe

R O N NH OMe Si–OMe OMe n ≡

Scheme 2. One-pot polymerization procedure resulting in end-functionalized lipopolymers.

The detailed procedures for the lipopolymer synthesis are published elsewhere.[8,12,13] The pure end-functionalized polymers can be obtained in good yields and narrow molecular weight distributions (see Table 1).

Table 1. Synthesized lipopolymers.

Polymer	**Lipid**	**Monomer**	$[M]_0/[I]_0$	**PDI** [1)]
$C_{16}PEOx_{10}Si$	n-hexadecyl-	2-ethyl-2-oxazoline	10	1.17
$C_{16}PEOx_{20}Si$			20	1.33
$PhyPMOx_{10}Si$	phytanyl-	2-methyl-2-oxazoline	10	1.12
$PhyPMOx_{20}Si$			20	1.16
$PhyPMOx_{30}Si$			30	1.14
$2C_{18}PMOx_{10}Si$	1,2-di(*n*-octadecyl)-*sn*-glyceroyl-		10	1.04
$2C_{18}PMOx_{20}Si$			20	1.33
$2C_{18}PMOx_{40}Si$			40	1.42
$2C_{18}PEOx_{10}Si$		2-ethyl-2-oxazoline	10	1.16
$2C_{18}PEOx_{20}Si$			20	1.31
$2PhyPMOx_{10}Si$	1,2-di(phytanyl)-*sn*-glyceroyl-	2-methyl-2-oxazoline	10	1.04 [2)]
$2PhyPMOx_{20}Si$			20	1.07 [2)]
$2PhyPMOx_{40}Si$			40	1.16 [2)]

1) determined by GPC (solvent: chloroform, polystyrene standards)
2) determined by GPC (solvent: dimethylacetamide, polymethylmethacrylate standards)

Lipopolymer monolayer

The first layer in a polymer tethered lipid membrane can either be prepared by a 'grafting from' [14] or a 'grafting onto' [15] process. Both procedures resulted in brush-like polymer monolayers with high grafting densities. Immediately after the grafting reaction the polymer layer was found to be homogenous and unstructured. A single submersion in water and consecutive exposure towards air at room temperature causes a self-organization of the lipopolymer layer into a hydrophilic water swollen polymer interlayer and an upper alkyl layer (Figure 3a).

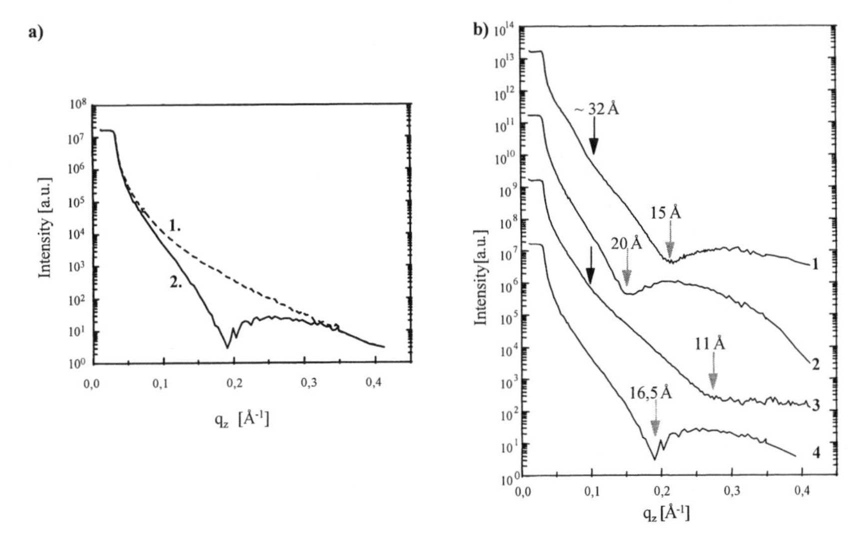

Figure 3. Specular x-ray reflection measurements of grafted lipopolymer monolayers on Si/SiO_2 wafers: **a)** **1.** Unstructured lipopolymer monolayer of $C_{16}PEOx_{10}Si$, **2.** same sample after exposure to water. **b)** X-ray reflectivity curves for lipopolymer monolayers **1.** $2C_{18}PEOx_{20}Si$, **2.** $2C_{18}PEOx_{10}Si$, **3.** $C_{16}PEOx_{20}Si$, **4.** $C_{16}PEOx_{10}Si$. Bright arrows indicate minima corresponding to the alkyl layer, the dark arrow the position of the minima for the total layer thickness.

Depending on the critical packing parameters of the lipopolymer amphiphile, the thickness of the alkyl layer varies.[16] This can be explained by the difference in the cross-sectional area occupied by the polymer and the therefore given area for the alkyl part to tilt. In all cases, no crystallization of alkyl chains were observed in the polymer supported alkyl monolayer. Parallel x-ray reflectivity measurements of the bulk lipopolymers revealed for one candidate ($2C_{18}PEOx_{10}Si$) the spontaneous formation of double layers (layer spacing d = 6.4 nm). This multilayer formation could also be observed at substrate surfaces (Figure 4). Here, $2C_{18}PEOx_{10}Si$ was grafted onto a silicon wafer. However, the cleaning step to remove ungrafted polymer was not complete. Investigation of the surface by scanning probe microscopy (SPM; digital instrument, multimode, Nanoscope IIa controller, tapping mode) revealed a distinct terrace profile reflecting the height difference of single lipopolymer layers. The step height of ~3 nm for the single layer, as found by SPM, fits nicely to the layer thickness of a grafted lipopolymer layer of 3.2 nm as determined by x-ray reflectivity (Figure 3) and is about half of the found double layer spacing in bulk (6.4 nm).

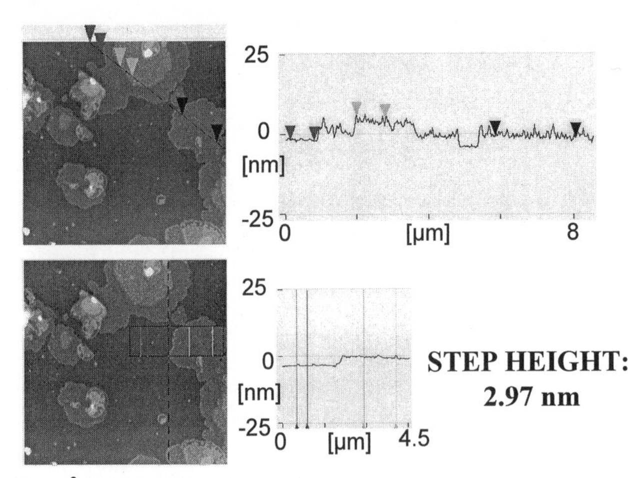

Figure 4. 10x10 μm^2 SPM scan (tapping mode) of a grafted $2C_{18}PEOx_{10}Si$ with topography.

Polymer Tethered Lipid Bilayers

For construction of the supported bilayer, air dried polymer monolayers prepared by the 'grafting from' or 'grafting onto' method were immersed to a solution of lipid vesicles in water. However, a formation of lipid bilayers could not be observed. Also numerous modifications of the procedure using different lipids and lipopolymers failed. Presumably, the exposure to water reorganizes the amphiphilic lipopolymer monolayer and a vesicle fusion did not occur. To build up stable polymer tethered lipid membranes, the procedure had to be modified in such a way that surface reconstruction of the first layer is suppressed. Regarding the mismatch in the cross-sectional area occupied by the polymer and the alkyl moiety, the alkyl layer has to be completed by additional lipids. This can be achieved by a step-wise preparation in which the first layer consisting of lipids and lipopolymers is pre-organized at the air/water interface of a Langmuir-Blodgett trough, compressed and then transferred onto the substrate. After an annealing step to complete the silanization reaction, a stable lipopolymer/lipid layer was obtained. Optimization of the LB-transfer procedure yielded homogenous and uniform layers for different lipopolymer/lipid compositions. This procedure allows us the deposition of pre-organized composite layers with defined grafting densities of the lipopolymer tether. The consecutive deposition of the upper lipid leaflet by fusion of lipid vesicles was successful.

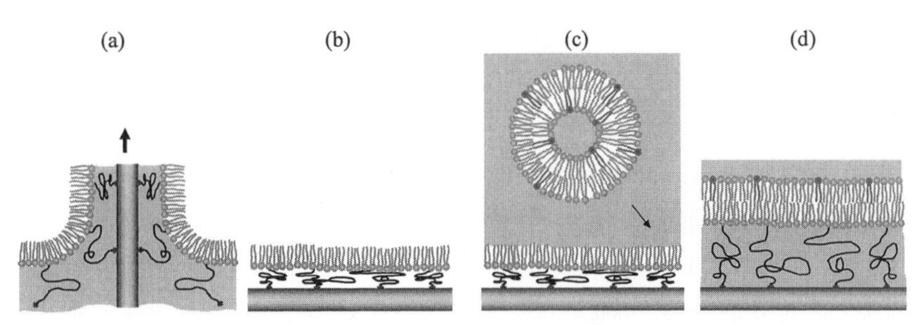

Figure 5. Schematic illustration of the stepwise preparation of a polymer-tethered membrane: (a) Langmuir-Blodgett (LB) transfer of a lipid / lipopolymer monolayer, (b) grafting of lipopolymers by annealing, and (c, d) spreading of the upper monolayer by vesicle fusion.

The additionally incorporated lipids in the first layer complete the outer hydrophobic layer and enable the vesicle fusion without the previously observed surface reconstitution in water.

The resulting polymer tethered lipid bilayers were investigated by fluorescence microscopy and their diffusion constants and mobile fractions by fluorescence recovery after photobleaching (FRAP). For these investigations, the stepwise construction procedure allowed the selective incorporation of fluorescent labeled lipids in the upper or lower layer. Inspection of the complete polymer supported bilayer showed a homogenous fluorescence for the labeled upper as well as the lower layer (Figure 6).

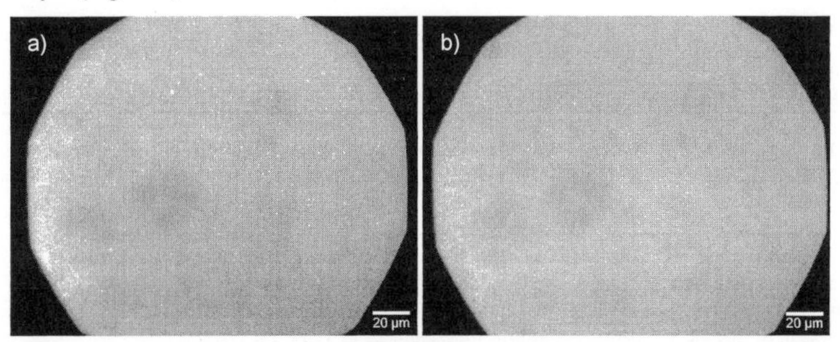

Figure 6. Homogenous polymer supported membrane with fluorescent labeled lipids in the a) lower and b) upper layer. (a) lower layer: 5 mol% $2C_{18}PMOx20$ / 94.8 mol% SOPC / 0.2 mol% Texas Red-PE; top layer: SOPC. b) lower layer: 5 mol% $2C_{18}PMOx20$ / 95 mol% SOPC; top layer 99.8 mol% SOPC / 0.2 mol% Texas Red-PE).

In continuous bleaching experiments, a good macroscopic diffusivity within both lipid layers was observed. Finally, FRAP measurements were performed to investigate the mobility and diffusion constant of the lipids within both leaflets. In all cases, a high percentage (96-98%) of mobile lipids was observed. The diffusion constant, within the upper leaflet were found to be constant with D = 1-2 $\mu m^2 s^{-1}$. The diffusivity in the lower leaflet varied between 1.6 $\mu m^2 s^{-1}$ for low grafting densities of the lipopolymer (5 mol% $2C_{18}PMOx_{10}Si$) and 0.4 $\mu m^2 s^{-1}$ for high grafting densities (50 mol%). Long term stability tests by repeated FRAP measurements and inspection with fluorescence microscopy, performed with some samples, showed no significant change in appearance or lipid mobility. Such membrane constructs seem to keep their structural integrity over weeks.

The above mentioned results show the potential of the here presented method to construct polymer tethered lipid membranes with high structural integrity. Currently, detailed studies of the influence of different lipid head groups of the lipopolymers as well as various matrix lipids are carried out. Preliminary fluorescence interference contrast microscopy experiments showed a clear dependence of the substrate-membrane spacing from the degree of polymerization of the used lipopolymer tether. In fact, our preliminary experiments revealed that the membranes with long (n=40) polymer spacers can reside transmembrane proteins (platelet integrins) homogeneously, demonstrating the significant influence of the spacer length.[13]

Acknowledgement: This work is financially supported by the Deutsche Forschungsgemeinschaft through the SFB 563 *'Bioorganic Functional Systems on Solids'*.

[1] S.J. Johnson, T.M. Bayerl, D.C. McDermott, G.W. Adam, A.R. Rennie, R.K. Thomas, E. Sackmann, *Biophys. J.* **1991**, *59*, 289.
[2] R. Merkel, E. Sackmann, E. Evans, *J. Phys. (France)* **1989**, *50*, 1535.
[3] L. Häußling, W. Knoll, H. Ringsdorf, F.-J. Schmitt, J. Yang, *Macromol. Chem. Macromol. Symp.* **1991**, *46*, 145.
[4] E. Sackmann, *Science* **1996**, *271*, 43.
[5] M.L. Wagner, L.K. Tamm, *Biophys. J.* **2000**, *79*, 1400.
[6] C.A. Naumann, O. Prucker, T. Lehmann, J. Rühe, W. Knoll, C.W. Frank, *Biomacromolecules* **2002**, *3*, 27.
[7] S.M. Schiller, R. Naumann, K. Lovejoy, H. Kunz, W. Knoll, *Angew. Chem. Int. Ed.* **2003**, *42*, 208.
[8] R. Jordan, K. Martin, H. J. Räder, K. K. Unger, *Macromolecules* **2001**, *34*, 8858.
[9] S. Kobayashi, S. Iijiama, T. Iijiama, T. Saegusa, *Macromolecules* **1987**, *20*, 1729.
[10] A. Bendavid, C.J. Burns, L.D. Field, K. Hashimoto, D.D. Ridley, K.R. Sandanayake, L. Wieczorek, *J. Org. Chem* **2001**, *66*, 3709.
[11] T. Eguchi, K. Arakawa, T. Terachi, K. Kakinuma, *J. Org. Chem* **1997**, *62*, 1924.
[12] A. Förtig, R. Jordan, O. Purrucker, M. Tanaka, *Polymer Preprints* **2003**, 44, 850.
[13] O. Purrucker, A. Förtig, R. Jordan, M. Tanaka, *ChemPhysChem* **2004** *in print.*
[14] a) R. Jordan, A. Ulman, *J. Am. Chem. Soc.* **1998**, *120*, 243. b) R. Jordan, N. West, A. Ulman, Y.- M. Chou, O. Nuyken, *Macromolecules* **2001**, *34*, 1606.
[15] F. Rehfeldt, M. Tanaka, L. Pagnoni, R. Jordan, *Langmuir* **2002**, *18*, 4908.
[16] R. Jordan, K. Graf, H. Riegler, K.K. Unger, *Chem. Commun.* **1996**, *9*, 1025.

Polypropylene Surface Peroxidation with Heterofunctional Polyperoxides

Natalya Nosova,[1] *Yuri Roiter,*[1] *Volodymyr Samaryk,*[1] *Sergiy Varvarenko,*[1] *Yuri Stetsyshyn,*[1] *Sergiy Minko,*[2] *Manfred Stamm,*[2] *Stanislav Voronov**[1]

[1] National University "Lvivska Polytechnica", 12 S. Bandera Str., Lviv 79013, Ukraine
E-mail: stanislav.voronov@polynet.lviv.ua
[2] Institute of Polymer Research, Hohe Strasse 6, D-01069 Dresden, Germany

Summary: Method of polyolefin surface activation via covalent grafting of polyperoxide nanolayer by free radical mechanism has been presented. The features of such the nanolayer formation under the thermoprocessing conditions, i.e.: formation of 3D crosslinked network in polyperoxide bulk; and its grafting with complete coating of polyolefin surface, -- is considered. The method provides an availability of uniformly placed peroxide groups of one type over the polyolefin surface activated, which may further be utilized for the tailored modification of polymer surfaces using the "grafting to" and "grafting from" techniques in that time when it is necessary.

Keywords: covalent grafting; crosslinking; nanolayers; peroxide copolymer; peroxidation of polymer surfaces

Introduction

The cases of contradiction in demands for certain materials, i.e.: between their bulk and surface properties, -- arise in the industry often enough. Such a situation occurs when material bulk properties, its availability and economic benefits allow an implementation of goods and surface properties of these materials do not meet the necessary demands. Especially often, this problem becomes apparent in the case of polymer materials. Their low free surface energy, hydrophobicity, scanty ability to be glued and the number of other factors restrict essentially their utilization despite of all useful physical-mechanical characteristics intrinsic to these materials due to the features of bulk structure and composition.

One of the general methods of this problem solution is a modification of surface both of virgin material and of goods on its basis. At this, the necessary change of material surface properties may be achieved under the condition of the retention of its bulk properties.

 DOI: 10.1002/masy.200450638

Theoretical grounds

The majority of polymers relate to the materials possessing low free surface energy. That is why, a modification of their surface at the expense of physical forces of sorption is hindered. In this connection, a lot of techniques elaborated for surface modification are oriented to the covalent grafting of modifiers to the surface. In the case of polyolefins, this problem is sharpened by that their surface incorporates no functional groups capable of facilitating such the grafting. A number of techniques have been elaborated that are based, for instance, on the preliminary generation of free radicals at the surface. It provides the possibility of the creation of modifying layers via the grafting of modifier preliminary created to the surface or via the initiation of monomer polymerization from such the surface.

Among the methods, which are applied for the radical generation, one can list an irradiation with electron beam radiation, treatment with plasma and corona discharges [1-3]. Aside of the utilization complexity, restrictions imposed by surface profiles, and ecological hazard, one can add a non-uniformity of surface modification connected with the zones of different reactivity present at the surfaces to the drawbacks of these methods [4].

A method proposed here, which provides the possibility of radical generation at the polymer surface, is freed significantly from those drawbacks. This method consists in the grafting of polyperoxide layer to the surface using a free radical mechanism, and its application with further surface modification can be represented as follows:

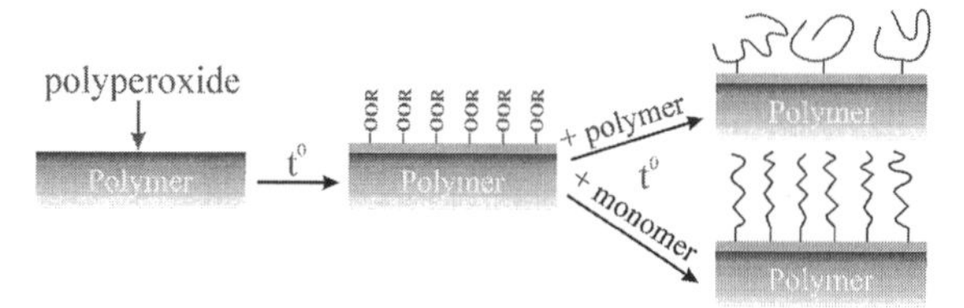

In accordance with the method proposed, grafting to the surface can be shown in general by the Scheme 1. Grafting of surface polyperoxide nanolayer proceeds at partial (not more than 50 %) decomposition of peroxide groups. The latter fact provides further possibility of free radical generation just in that time, when it is necessary, and to graft special polymers (dextran sulfate, dextran, heparin) or to initiate a polymerization of functional monomers (acrylic acid, vinyl acetate, acrylonitrile etc.) from the surface.

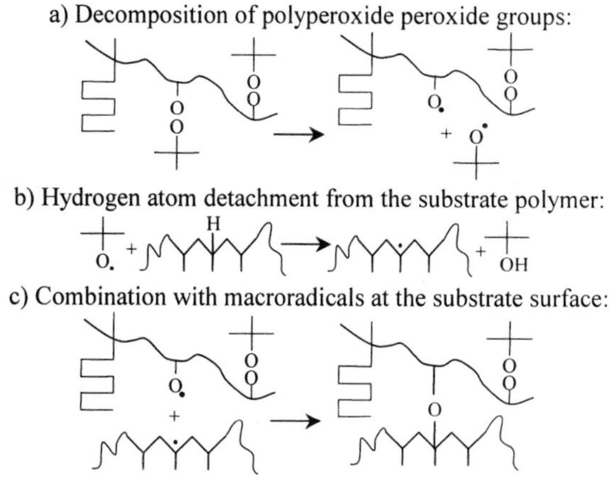

a) Decomposition of polyperoxide peroxide groups:

b) Hydrogen atom detachment from the substrate polymer:

c) Combination with macroradicals at the substrate surface:

Scheme 1. Process of polyperoxide macromolecule grafting to polypropylene surface.

The demands to surface profile, necessity of the application of complicated protective and special equipment, are lowered at the utilization of the proposed method of surface peroxidation: and the sources of free radicals are placed uniformly over the surfaces that relieve the problem of their different reactivity at further modification.

Materials and methods

Heterofunctional copolymer VEP-OMA based on 2-tert-butylperoxy-2-methyl-5-hexene-3-yne and octyl methacrylate has been utilized for the surface peroxidation. Its structure can be presented with the following scheme:

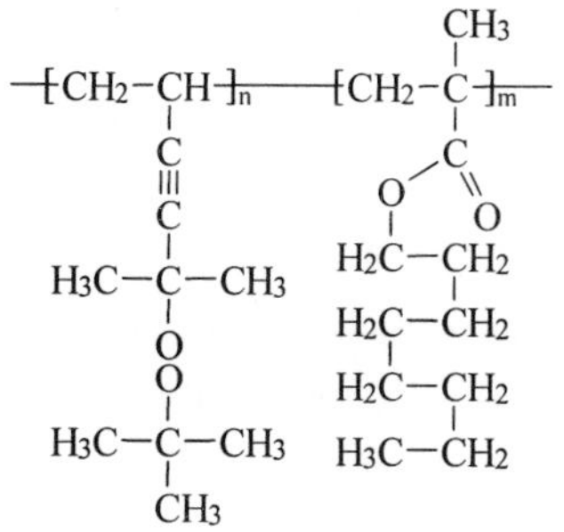

Scheme 2. Structure of VEP-OMA polyperoxide.

This copolymer synthesis has been reported earlier [5]. Similar synthesis technique has been applied for preparation of other copolymers of VEP with higher (meth)acrylates (lauryl methacrylate (LAMA); hexyl acrylate (HAT); butyl acrylate (BA)) utilized in work for comparison purposes.

Viscosity averaged molecular weight (M_v) of copolymers was measured after determination of the constants for Mark–Houwink equation [6] via the fractionating of copolymer samples and determination of fraction molecular weights by gel permeating chromatography using a HP 5890 CasChromat equipped with HP 7694 detector (Hewlett Packard, Japan). Characteristic viscosities of the copolymer samples were determined using an Ubbelohde viscometer.

Conversion of peroxide groups from copolymer composition at the heating in bulk was measured via chromatographic analysis of decomposition products using a gas-liquid chromatograph Selmichrom-1 (Selmi, Ukraine).

Polypropylene (PP, Montel Profax) and thermoplastic polyolefin (TPO, Solvay) as substrates of $20\times20\times3$ mm^3 size were chosen for the studies. For the ellipsometric studies, the model substrates of polypropylene (Aldrich) and polystyrene (Aldrich) attached covalently to silicon wafers (Wacker-Chemitronics) of $10\times10\times1$ mm^3 size were utilized in accordance with known technique [6].

Peroxidation of polymer substrates was performed as follows. Substrates were washed in Soxhlet apparatus with propanone for 4 hours in order to remove the impurities and technological additives from their surface. Then, substrates were dried under vacuum till constant weight. Application of peroxide copolymer solution of certain concentration was performed using a spincoating technique [6]. Peroxide-containing copolymer solution in heptane (0.08 ml) or toluene (model substrates) was applied to the substrate surface and substrate was kept at 2000 rpm for 1 min. Control substrates were treated with respective solvent. Then, substrates were thermostatted in hermetic box under argon atmosphere or in vacuum oven (model substrates) at defined temperature for certain time (temperature range at the performance of experiments was of $80\div130\ ^0C$). After the thermoprocessing, substrates were placed into Soxhlet apparatus and extracted with propanone for 4 hours and dried under vacuum till constant weight.

The changes of free surface energy components were determined using a known technique [7] measuring the contact angles for two liquids.

Null-ellipsometry studies and ellipsometric mapping were performed on model substrates of hydroxyl terminated polystyrene, PS-COOH (M_n ca. 16 000), grafted to silicon wafers (Wacker-Chemitronics) using a High-Speed In-Situ-44-Wave Lengths Ellipsometer (Woollam Co.) under ambient conditions. Surface area of single measurement covered by laser spot was of $0.1 \times 0.2\ cm^2$. PP refractive index (1.490) was taken from ellipsometer software data. Cross-linked VO refractive index (1.482) was determined for separately obtained silicon wafers with correspondingly heated VO layers of approximately 60÷120 nm performing the measurements using the same ellipsometer and solution of equation system using an ellipsometer software. After every experiment stage samples were placed onto ellipsometer working table with accuracy of ±0.01 cm. Measurements were performed in the middle part of the samples (in order to avoid the edge effects of PP and VO application) in 100 positions for total area of $0.9 \times 0.9\ cm^2$ with step of 0.1 cm in the direction of both abscissa and ordinate axes. The results obtained for ellipsometric mapping were transferred to matrix form and processed using a Microcal Origin 5.0 software.

Experiment and discussion

Structure formation and properties of heterofunctional polyperoxide nanolayer at polymer surface as well as a completeness of surface coating have to be dependent upon a type of peroxide group used in polyperoxide, their conversion and a height of applied nanolayer. Curve 1 in Figure 1 shows a dependency of ellipsometric height of polyperoxide layer applied to model substrates upon its concentration in solution utilized for spincoating. Application of this technique allows the formation of polyperoxide layer of 7 to 200 nm height. In accordance with literature data [8, 9] obtained using a low angle X-ray scattering, homopolymers and copolymers of higher esters of (meth)acrylic series display anisotropy in their bulk caused by the orientation of macromolecules skeleton chains and interaction of side branches. A determination of constants for Mark–Houwink equation [10] was performed in this work (Table 1) in comparison with mentioned literature data for higher (meth)acrylate homopolymers. Good accordance of α constant between these data allows to conclude that incorporation of peroxide links into higher (meth)acrylate polymers does not affect essentially the conformation state of their macromolecules. This, in turn, allows to suppose that the copolymers synthesized retain their anisotropy in bulk.

Schematically a polyperoxide layer at the substrate surface is represented in Scheme 3.

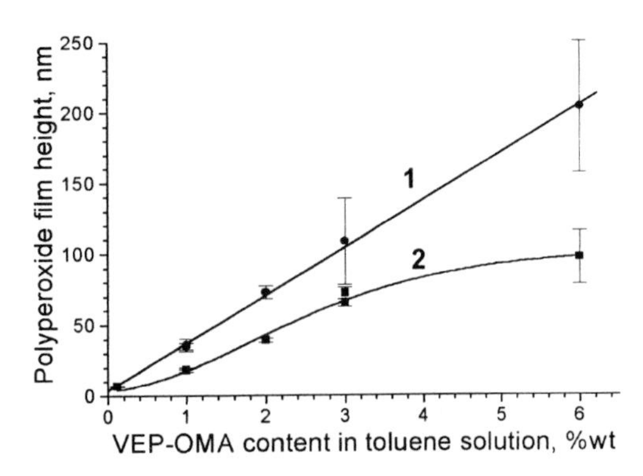

Figure 1. Dependence of the height of polyperoxide layer applied (**1**) and grafted (**2**) upon its concentration in toluene solution for spincoating.

Table 1. Characteristics of peroxide-containing copolymers and literature data for homopolymers [7,11] of higher (meth)acrylates. Measurements were performed at 20 ^{0}C.

(Co)polymer	VEP links %mol	Solvent	$K \cdot 10^4$	α	M_v kg/mol	$[\eta]$ dl/g
VEP-OMA	15.3	butanone	1.76±0.05	0.71±0.03	6.2	0.0818±0.001
VEP-LAMA	15.2	butyl acetate	3.31±0.07	0.65±0.04	12	0.0127±0.0005
VEP-HAT	15.0	butanone	1.05±0.02	0.83±0.03	9	0.182±0.002
VEP-BA	15.1	butanone	0.84±0.07	0.85±0.05	11	0.291±0.001
PolyOMA	0	butanone	0.447	0.69	–	–
PolyLAMA	0	butyl acetate	0.846	0.64	–	–
PolyHAT	0	butanone	0.212	0.78	–	–
PolyBA	0	butanone	0.156	0.81	–	–

A layer structure as indicated in Scheme 3 is of course not rigorously conclusive from our investigations. It is, however, quite probable also in view of literature data [8, 9] that at least a short range order is present which resembles the arrangement from Scheme 3 over some distance. The free surface and interface with the substrate can induce order over larger scale and the layer structure therefore is easier to achieve in thinner films. Coming from the opposite direction, using the parameters of molecule placement shown above and the layer height measured, and accounting for the most probable distance between chain end of 4.5 nm (estimated in accordance with [11]), a weight applied to the surface should be of $(2.7\pm0.1)\times10^{-5}$ g/cm^2. Satisfactory coinciding of these results confirms the layered structure of polyperoxide over the surface.

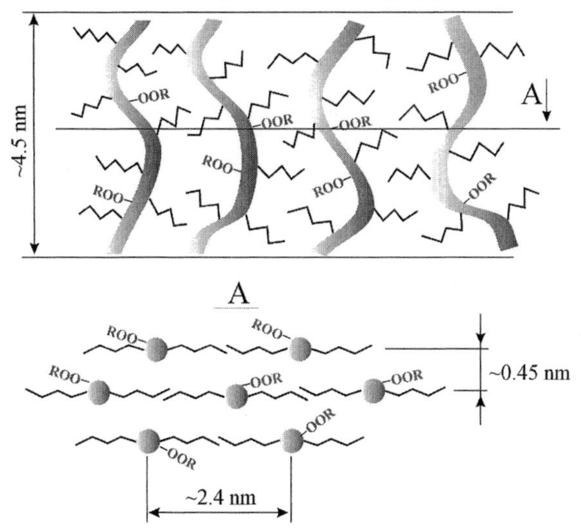

Scheme 3. Bulk copolymer layer of higher esters of (meth)acrylic series in accordance with literature data [8, 9].

Comb-like polymers with side aliphatic chains display adsorption activity to the polyolefin surfaces [12]. Thermal treatment of substrate with applied polyperoxide leads to the thermolysis of peroxide groups. The curves of peroxide group thermolysis obtained for polyperoxide decomposition in its bulk, are shown in Figure 2a. Peroxide group thermolysis leads to the crosslinking of polyperoxide layer that is witnessed by the change of polyperoxide molecular weight upon a conversion of peroxide groups shown in Figure 2b. Complete crosslinking to 3D network takes place at peroxide group conversion exceeding 30 % and it is weakly dependent upon a thermolysis temperature.

Layer grafting to the surface proceeds along with it crosslinking. The latter results in that the layer grafted cannot be soxhlet extracted from the substrate surface. A series of ellipsometric mappings obtained after extraction of the specimens heated for different times has been shown in Figure 3. Besides, curve 2 in Figure 1 demonstrates a height of grafted crosslinked layer in comparison with the height of applied layer (curve 1, Figure 1). Height of grafted layer at 50 % conversion of peroxide groups is of 50÷80 % in respect to the applied layer.

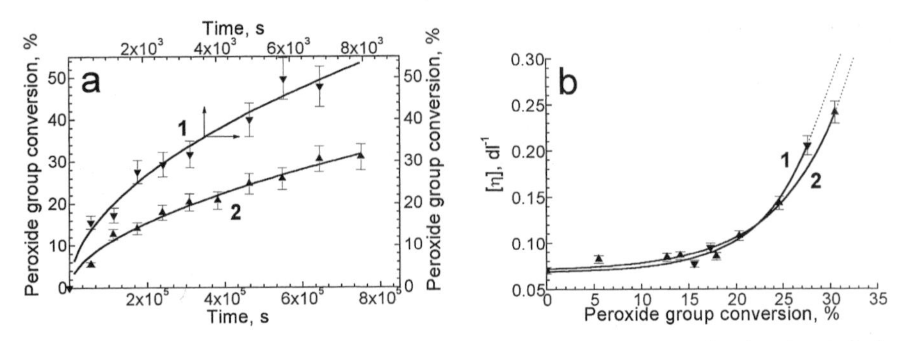

Figure 2. (**a**) Dependence of VEP-OMA peroxide group conversion upon a heating time in bulk. (**b**) Dependence of intrinsic viscosity of VEP-OMA solution in hexane upon a conversion of peroxide groups. Dotted lines point to the formation of crosslinked polyperoxide insoluble in organic solvents after ~30 % peroxide group conversion. VEP-OMA heating was performed in bulk at (**1**) 144 ^{0}C and (**2**) 109 ^{0}C.

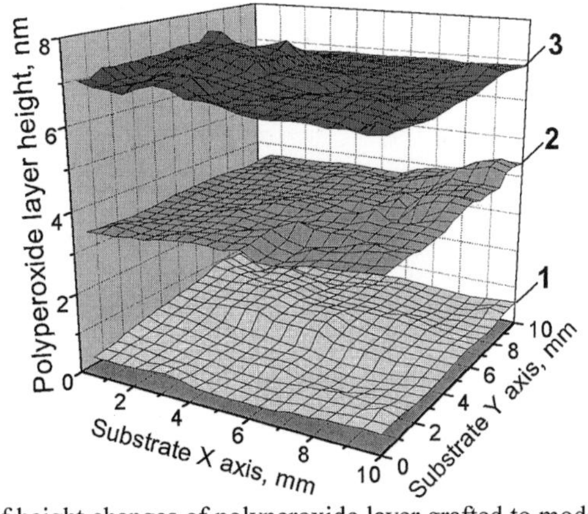

Figure 3. Dynamics of height changes of polyperoxide layer grafted to model layer of polystyrene upon a time of its thermoprocessing. Layer was spincoated from 0.12 %wt solution in toluene; grafting was performed at 130 ^{0}C for: (**1**) 5 hours; (**2**) 10 hours; and (**3**) 20 hours.

Experimental data presented above allow to suppose that polyperoxide grafting to polymer surface proceeds with simultaneous cross-linking in bulk. This bulk cross-linking has a higher rate than the grafting to the surface. At the initial stages, when grafting is not significant, the extraction of polyperoxide layer with solvent (acetone) cleans completely almost the substrate (polypropylene) surface and distribution of measured water contact angles corresponds to one for virgin PP (Curve 1, Figure 4). With further accumulation of grafted sites, distribution of contact angles shifts (Curves 2, 3) to the one of peroxide-containing copolymer (Curve 5). Average height of the grafted layer at this stage is of 8÷45 nm -- much higher than polyperoxide macromolecule sizes. This witnesses that not single VEP-OMA molecules but their cross-linked ensembles are grafted to the surface. The process is finished by the complete practically coating of the surface by cross-linked layer (Curve 4, Figure 4) grafted covalently to the polymer surface. Joint analysis of ellipsometric mapping and contact angle changes has shown that at 50 % of peroxide group conversion, complete coating takes place for the case of applied layer height of 130 nm (under conditions of the spincoating described, it relates to the utilization of 4 % VEP-OMA solution). It may be of interest that it is provided by about 240 layers of polyperoxide macromolecules.

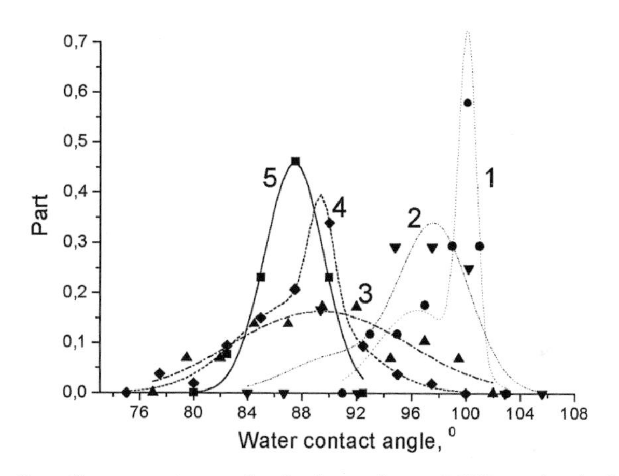

Figure 4. Distribution of water contact angles for the surfaces of different level of modification. **1** -- virgin PP surface, VEP-OMA layer grafted to PP surface (ellipsometric height): **2** -- 8÷10 nm; **3** -- 40÷45 nm; **4** -- 82÷90 nm; **5** -- model surface of cross-linked VEP-OMA.

As a result of further investigations performed, the modification of peroxidized polymer surfaces has been conducted successfully applying both “grafting to” (heparin, dextran, dextran sulfate) and “grafting from” (acrylic acid, acrylonitrile, vinyl acetate) techniques.

Conclusions

Surface grafted 3D network is formed in the result of heterofunctional polyperoxide grafting to polyolefin surface at the expense of free radical process proceeding. Since grafting is conducted with incomplete thermolysis of peroxide groups from polyperoxide structure, the network incorporates peroxide groups that may be used for the further surface modification. Formation of grafted 3D network is dependent upon a conversion of peroxide groups and a height of applied polyperoxide layer.

[1] US Patent 6,210,516 (2001), NeoMecs Inc., invs.: R. S. Nohr, J. G. MacDonald.
[2] US Patent 5,972,176 (1999), 3M Innovative Properties Company, invs.: S. M. Kirk, C. S. Lyons, R. L. Walter.
[3] US Patent 6,203,850 (2001), NeoMecs Inc., inv.: H. Nomura.
[4] US Patent 6,023,025 (2000), Nissin Electric Co., Ltd., invs.: T. Nakahigashi, A. Doi.
[5] S. A. Voronov, V. Ya. Samaryk, S. M. Varvarenko, N. H. Nosova, Yu. V. Roiter, *Dopovidi NAN Ukrayiny (Reports of National Academy of Sciences of Ukraine)* **2002**, *6*, 147.
[6] S. Minko, S. Patil, V. Datsyuk, F. Simon, K.-J. Echhorn, M. Motornov, D. Usov, I. Tokarev, M. Stamm, *Langmuir* **2002**, *18*, 289.
[7] D. W. Van Krevelen, *"Properties of polymers correlations with chemical structure"*, Khimiya, Moscow, 1976, p. 413, in Russian; transl. from: D. W. Van Krevelen, *"Properties of polymers correlations with chemical structure"*, Elsevier, Amsterdam, London, New York, 1972.
[8] N. A. Plate, V. P. Shybaev, *Vysokomol. soedinen.* **1971**, *13A(3)*, 410.
[9] N. A. Plate, V. P. Shybaev, V. P. Talroze, *Vysokomol. soedinen.* **1977**, *18B(5)*, 110.
[10] "*Compendium of Chemical Terminology. The Gold Book*", 2nd ed., A. D. McNaught, A. Wilkinson, Eds., Blackwell Science, Oxford 1997.
[11] Yu. S. Lipatov, A. E. Nesterov, T. M. Gritsenko, *"Polymer chemistry hand-book"*, Naukova dumka, Kiev, 1971, in Russian.
[12] L. Leger, E. Raphael, H. Hervet, *Adv. Polym. Sci* **1999**, *138*, 185.

Macromol. Symp. **2004**, *210*, 349-358

Improvement in Strength of the Aluminium/Epoxy Bonding Joint by Modification of the Interphase

Lan H. Phung,[*1] *Horst Kleinert,*[1] *Irene Jansen,*[2] *Rüdiger Häßler,*[3] *Evelin Jähne*[4]

[1] Dresden University of Technology, Institute of Production Engineering, George Bähr-Straße 3c, D-01062 Dresden, Germany
[2] Fraunhofer Institute of Material and Beam Technology, Winterbergstr. 28, D-01277 Dresden, Germany
[3] Institute of Polymer Research Dresden, Hohe Str. 6, D-01069 Dresden, Germany
[4] Dresden University of Technology, Institute of Macromolecular Chemistry and Textile Chemistry, Mommsenstr. 4, D-01062 Dresden, Germany

Summary: This contribution describes the influence of different surface pre-treatments including self-assembly of phosphoric acid mono alkyl ester as adhesion promoter (AP) for adhesive bonding of aluminium alloy AlMg3. The investigations were performed using a cold hardening two components epoxy-adhesive. The pre-treated surfaces, the interphase structures and the joints were characterized by: SEM/EDX, surface tension, XPS, DMA and the determination of mechanical parameters. The results interestingly show that the test sample with three step pre-treatment (degreasing in acetone, then anodic oxidation in phosphoric acid and adsorption of AP) has the highest adhesive strength and durability.

Keywords: adhesion promoter; aluminium-bonding; epoxy-adhesive; interphase; pre-treatment

Introduction

Adhesive bonding of aluminium components is widely used in the aerospace industries and is becoming more common in the automotive and architectural sectors. When epoxy adhesives are applied onto aluminium substrates, an interphase between the coating part and the aluminium surface is created. Chemical, physical and mechanical properties of the formed interphase depend on the substrate nature and its surface treatment, the nature of the hardener and the curing cycle.

The impact mechanisms of surface pre-treatments like degreasing, etching, anodic oxidation are the elimination of organic contamination, the improving of the wetting behaviour, the decreasing of magnesium : aluminium ratios, the increasing of the surface roughness and the increasing of the oxide/hydroxide layer thickness. Some current pre-treatments offer an excellent durability but introduce toxic chemicals such as hexavalent chromium.

DOI: 10.1002/masy.200450639

Therefore, recent studies have used self-assembled molecules (SAMs) as AP to study how to replace the present chromated procedure on aluminium and how to improve the lacquer adhesion and corrosion inhibition [1]. The combination of different pre-treatments and SAMs application can be considered as an alternative for environmentally friendly surface treatments that bring not only a higher durability but also an improvement of the adhesive strength. Besides, the fracture behaviour has some changes due to the different surface treatments.

The aim of this work is to investigate the effect of different pre-treatments and AP application on the surface state of an aluminium substrate, and how these treated surfaces influence the formation of the polymer network structure in the interphase and the mechanical properties of the aluminium/epoxy joints.

Experimental

Materials and sample preparation

The substrates used in this study are sheets of the aluminium alloy AlMg3. The AlMg3 samples were pre-treated, using different methods (table 1). Phosphoric acid mono-(12-hydroxy-dodecyl) ester was used as AP. The pre-treated substrates were exposed to a 10^{-3}M solution of the AP in water. The exposure time was 5min at room temperature (RT). The substrates were rinsed with distilled water and dried in vacuum or in a stream of dry nitrogen. The adhesive was a two components cold curing epoxy resin (Delo-Duopox 1891). This adhesive does not contain fillers and primers.

Table 1: Surface pre-treatment procedures

Pre-treatment	Description
degreasing	degreased in aceton applying ultrasonic waves , 10min
etching 1	degreasing, alkaline degreased in NaOH 10%, 60°C, 2min
etching 2	degreasing, alkaline degreased in NaOH 10%, 60°C, 2min pickled in HNO_3 15%, RT, 2min
etching 3	degreasing, etched in 10g/l Bonder[a] V338M, 50°C, 20sec pickled in 5 g/l H_2SO_4, 11.7 g/l $Al_3(SO_3)_3{*}H_3O$+ 3.4g/l HF 40%, 50°C, 30sec
etching 4	degreasing, alkaline degreased in P3 almeco 20[b]: 30g/l, 60°C, 10min etched in P3 almeco 40[b]/NaOH: 15g/l P3 almeco 40, 50g/l NaOH, RT, 1 min pickled in HNO_3 15%, RT, 2min
PAA	degreasing, anodic oxidation in phosphoric acid 15%, 15V, 28°C, 23min
SAA	degreasing, anodic oxidation in sulphuric acid 15%, 15V, 28°C, 23min

[a)] Bonder V338M is from Chemetal
[b)] P3 almeco 20 und P3 almeco 40 are from Henkel

Test procedure:

The interphase aluminium surface/polymer, where the aluminium surface was differently pre-treated was investigated by scanning electron microscopy (SEM), surface roughness, contact angle, X-ray photoelectron spectroscopy (XPS), dynamic mechanical analysis (DMA) and some strength tests.

- Scanning electron microscopy (SEM/EDX): Electron micrographs were obtained with a Zeiss DSM 982 Gemini SEM. The energy dispersive X-ray spectroscopy (EDX) was operated with the primary electron energy of 3keV. The element composition of the near–surface layer of the aluminium (information depth approx. 1µm) was investigated by the EDX which was coupled to the SEM. In addition, other SEM and EDX cross sections of samples, that were embedded into epoxy resin, polished and steamed with gold, were performed to measure the thickness of oxide/hydroxide layer.
- Contact angle measurements: Contact angles were measured by the sessile drop techniquewith DSA 10 from Krüss GmbH, Hamburg. Water was used as a test liquid.
- Roughness measurements: With a HOMMEL TESTER T8000 the roughness of different pre-treated surfaces was measured.
- X -ray Photoelectron Spectroscopy (XPS): The very top surface region with information depth of 3–5 nm of the samples was studied by core level XPS. It was performed using a PHI 5702 spectrometer with a lateral resolution 30 µm, viewed samples diameter 800 µm area. The take-off angle was 45°.
- Dynamic Mechanical Analysis (DMA): The dynamic mechanical analyzer DMA 983 of the company TA instrument / USA was used under nitrogen atmosphere. The measurements were carried out in the resonance mode with an oscillation amplitude of 0.2 mm. For the dynamic measurements the heating rate was 3 K/min in the temperature range between –50°C and + 250°C. After mixing the adhesive it was applied between the aluminium substrates and clamped vertically into the oscillate system of the dynamic mechanical analyzer. The storage time represents the period between adhesive application (assembling) and time of the measurement. The glass transition temperature is definable by the maximum of the dynamic viscosity (the main

relaxation region of the mechanical damping) [2].

- Strength tests: The shear tension test according to the European standard test procedure (DIN EN 1465), the shear stress-shear strain-test (DIN 54451) and the floating roller peel test (DIN EN 1464) were performed with a shear tension machine AGS-G (Shimadzu). The strength of adhesive joints was measured after the curing and after the aging-test according to the standard procedures VDA 621-415. The VDA 621-415 consists of 10 cycles, in which one cycle is composed of the following segments:
    - 24h salt-spray-test DIN 50021
    - 96h condensation water alternating atmosphere DIN 50017
    - 48h RT DIN 50014

Results and discussion

Figures 1-3 show the results of the shear tension test, the floating roller peel test and the shear stress-shear strain-test after the VDA 621-415 relative to the initial strength depending on surface treatments.

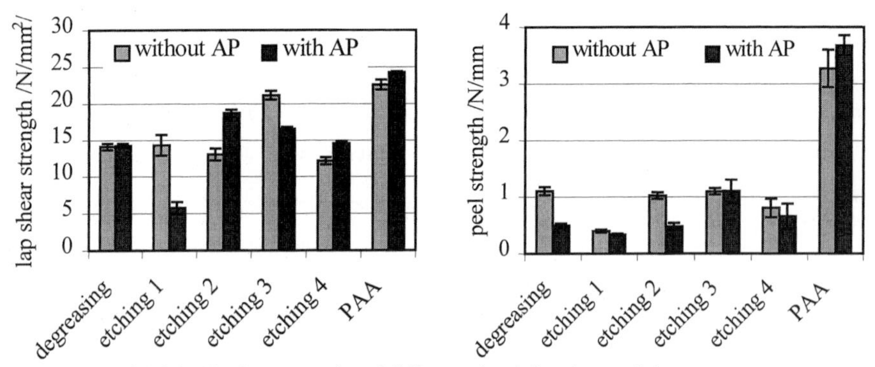

Figure 1. The initial adhesive strengths of different aluminium/epoxy-joints

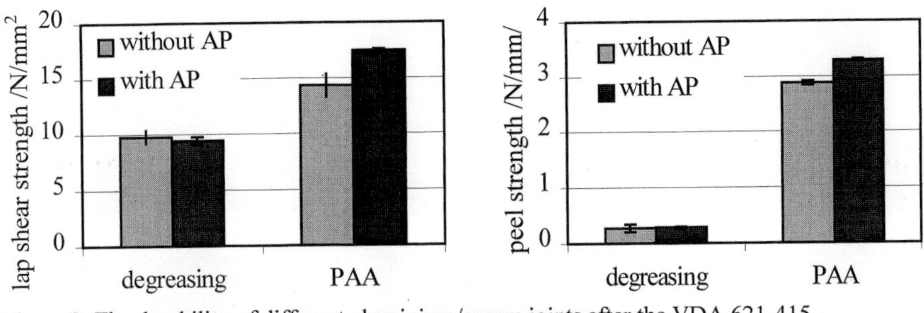

Figure 2. The durability of different aluminium/epoxy-joints after the VDA 621-415

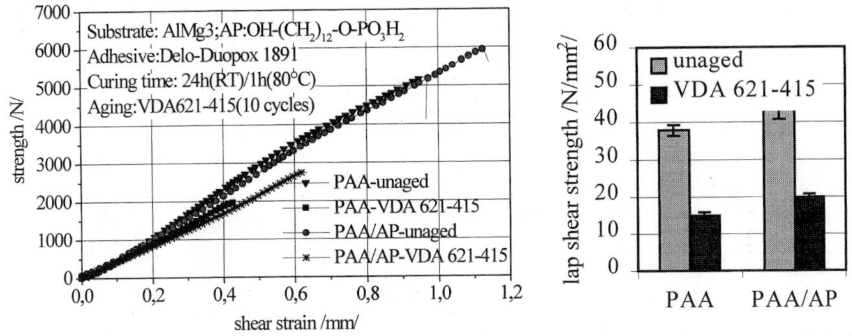

Figure 3. The shear stress-shear strain-diagrams of anodized aluminium/epoxy joints (PAA) and anodized with applied AP aluminium/epoxy-joint (PAA/AP) (DIN 54451)

It attracts attention that bonded specimens, which are degreased in acetone or alkaline- or acid-etched show clearly deteriorations of mechanical properties in comparison with PAA. The combination of those pre-treatments with AP have different effects on the adhesive strength of the joints. The highest improvement of the initial adhesive strength and the durability of the joints due to the impact of AP is achieved when the substrates were anodized. Therefore, PAA is considered as the preferable pre-treatment for applying AP to enhance the durability of the joints. This appeared to be interesting for further investigations concerning the influence of different surface pre-treatments and the role of the AP on the durability of aluminium/epoxy-joints.

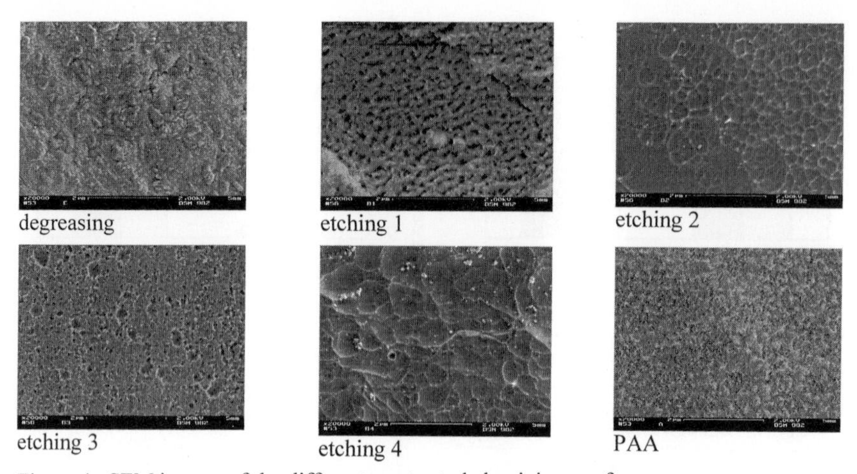

Figure 4. SEM images of the different pre-treated aluminium surfaces

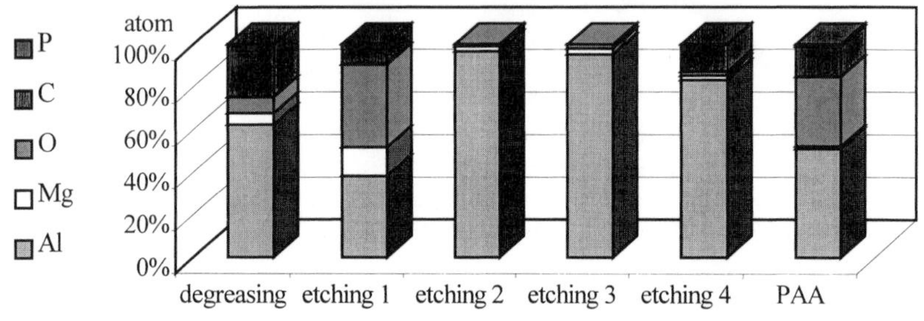

Figure 5. EDX linescans recorded from differently pre-treated aluminium surfaces

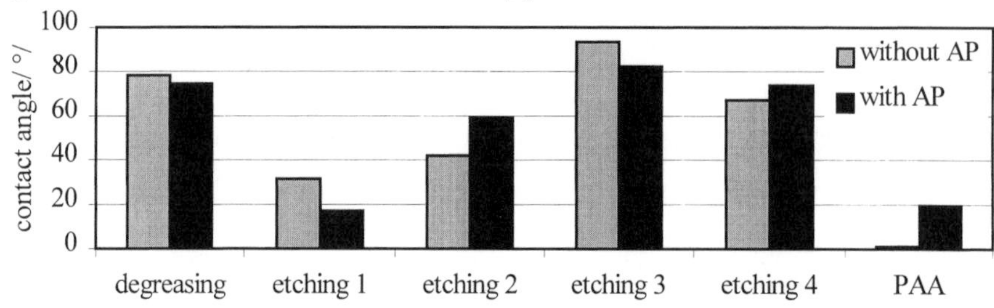

Figure 6. Water contact angles measured on pre-treated Al immersed in 10^{-3} M solution of AP

The effect of the various pre-treatments on the topography structures and the chemical state of aluminium surfaces was shown in Figs. 4 and 5, respectively. Degreasing in acetone can not completely eliminate the organic contamination and the roll texture that comes from the production process was found on degreased aluminum surfaces. Etch pits formations were seen on etched samples because etching dissolves some layers of the metal [3]. Alkaline etching (etching 1) removes the organic contamination and the original oxide layer but is unable to remove the magnesium. Other alkaline etchings and acid picklings eliminate the organic contamination, decrease the Mg : Al ratios, and form a very thin oxide/hydroxide structure. This oxide/hydroxide layer is at least three times smaller than the anodized oxide/hydroxide-layer (about 1 μm), shows poor corrosion resistance and a high degree of organic contaminations. PAA leads to columnar cell structure of the aluminium oxide/hydroxide layer [4].

The adsorption of AP on the aluminium surface was firstly investigated by the contact angle measurement (Fig. 6). If adsorption happens, the contact angle must be increased. It is evident that the degreasing, etching 1 and etching 3 are not suitable pre-treatments of aluminium surface for adsorption of AP. These pre-treatments induce the thin and unhomogeneous oxide layers, that contain high magnesium and organic contamination. It is supposed, that those surface properties prohibit the adsorption of AP on aluminium surfaces. The 2nd etched, 4th etched and anodized surfaces result in the increased contact angles after adsorption of AP. It proves that AP may adsorb on those pre-treated surfaces. Nevertheless, the contact angle measurements on those surfaces are not reproducible, because of the high surface roughness of the samples (Table 2), that significantly influences the water contact angle. This is the reason why the additional surface measurements were carried out to evaluate the ability of adsorption of AP on the different pre-treated aluminium surfaces.

Table 3 shows XPS results that represent the percentage of aluminium, oxygen, carbon and phosphorus atoms at the different pre-treated aluminium surfaces after an adsorption in AP solution. Highly curve fittings of the Al 2p core level yield an increase in hydroxyl group concentration from 64% to 70% on the anodized (PAA) and anodized/adsorbed AP (PAA/AP), respectively (Fig. 7). The curve fittings of the O 1s core level give the same information. In addition, the carbon concentration on anodized and applied AP surface was significantly increased and the atomic ratios, like [C]:[O], [P]:[O] or [Al]:[O], were changed after using AP. They confirmed the presence of AP molecules on the anodized sample. It suggests that the phosphoric acid was fixed on the AlMg3 surface after the PAA process. Hence, a small additional amount of phosphorus was only found after the adsorption of AP on those surfaces. It is supported by the appearance of phosphorus after applying AP of the SAA sample. That is a proof that AP is adsorbed on the anodized aluminium surface.

Table 2. Roughness measurements of the different pre-treated AlMg3 surfaces

	degreasing	etching 1	etching 2	etching 3	etching 4	PAA
arithmetical average roughness value /μm/	0.24	0.25	0.22	0.25	0.19	0.29
maximum roughness profile height /μm/	2.33	2.89	2.12	2.37	2.06	3.56

Table 3. XPS analyses of the AlMg3 surfaces depending on the different pre-treatments

Sample	C1s	O1s	P2p	Al2p	Mg2p	Rest	[C]:[O]	[P]:[O]	[Al]:[O]	[Mg]:[O]
degreasing	44.7	35.57	-	7.82	6.69	5.22	1.26	-	0.17	0.19
PAA	9.28	59.44	5.19	24.97	1.12	-	0.16	0.087	0.42	0.019
PAA/AP	24.4	50.17	5.02	20.42	-	-	0.49	0.1	0.40	-
SAA	23.41	51.98	-	20.64	-	3.44	0.45	-	0.39	-
SAA/AP	42.77	39.01	3.1	13.67	-	0.49	1.10	0.079	0.35	-

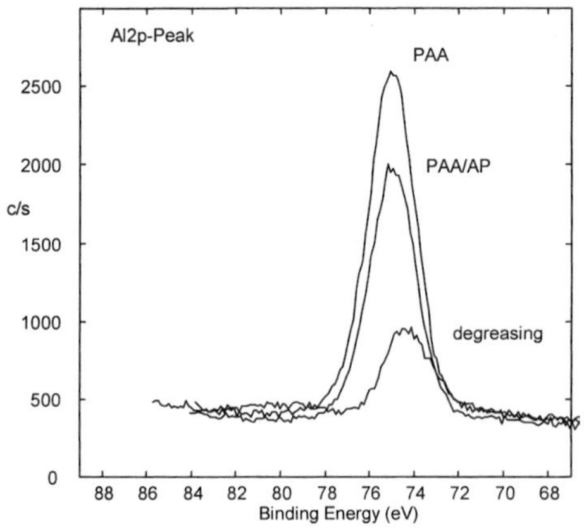

Figure 7. XPS Al 2p core level scan from degreased, anodized and anodized/adsorbed AP aluminium surfaces

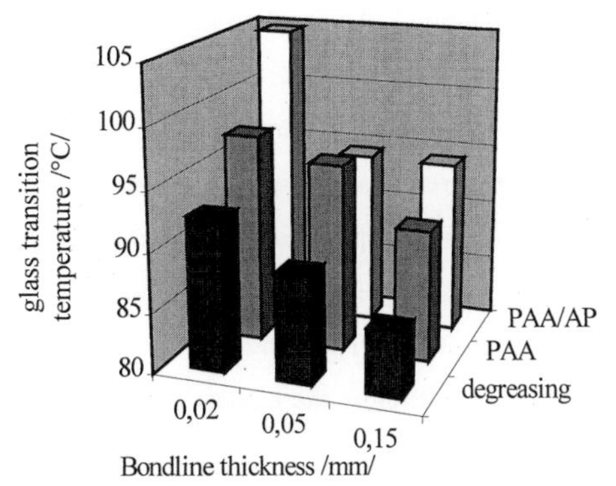

Figure 8. DMA results to determine the glass transition temperatures of the joints depending on the pre-treatment of the aluminium and bonded thickness

By the Dynamical Mechanical Analysis (DMA) several different glass transition temperatures of the joints depending on the surface treatment and the bondline thickness were found (Fig.8). The measured glass temperatures increase more especially in the substrate – near phase compared with the bulk phase of the polymer. The different temperature between those two phases is about 10 to 15 K. Thus, it may be concluded that the treated surface has a significant impact on the cross-linking of the epoxy resin. After adsorption of AP the anodized samples lead to the highest glass transition temperatures. It can be assumed that reactions between AP and oxirane of the adhesive can be considered as an initial reaction step in the curing process that causes to a formation of a epoxy network of higher density in the interphase of the aluminium/epoxy joints.
As a consequence, in the course of the curing process a continuous molecule growth takes place, oligomers and macromolecules are formed. This decreases the molecular mobility in the epoxy system which in turn results in an increasing glass transition temperature [5].
The differences of resulting surface properties in the scale of micrometers and nanometers and of resulting polymer network structures in the interphase due to different pre-treatments and the application of AP show a good correlation with the above presented strength and durability performances of aluminium/epoxy joints.

Conclusion

The adsorption of AP on differently pre-treated surfaces has been developed to improve adhesive strength and durability of the aluminium/epoxy-bonding joints. The results have proved that the AP has adsorbed better on the anodized surfaces than on other pre-treated surfaces. This treatment method leads changes both, the topography and more important, the chemical state of the aluminium surface. The PAA pre-treated aluminium oxide layers have more porous, thickness and hydroxyl group concentration than other pre-treated oxide layers. Hence, the phosphoric acid head-group of the AP was able to adsorb spontaneously onto the anodized surface, that has been proved by X-ray photoelectron spectroscopy. These induced chemical surface states influence the curing reaction of the epoxy system. Therefore the highest glass transition temperature is found for the near-interphase region on the anodized aluminium surfaces after adsorption of the AP by the DMA measurement. It is convincing that the different network structures in turn should influence the mechanical properties and the ageing behaviour of the epoxy/aluminium bonds. This has been confirmed by the industrial adhesion– and corrosion tests.

In conclusion, the idea using SAMs as adhesion promoter and corrosion protection layers for anodized aluminium surfaces appears to be promising. It is shown for the investigated systems that the effect of the anodic oxidation of aluminium in phosphoric acid and then adsorption of phosphoric acid mono-(12-hydroxy-dodecyl) ester as adhesion promoter on the interphase properties of the aluminium/epoxy joints has to be considered a very complex system. This interphase is characterised by a gradient of properties which vary from the pre-treatment ones to the bulk properties of the adhesive. The thickness, composition and structure of this surface treatment/adhesive system play a major role in the control of the adhesion performance and the joint durability.

Acknowledgment

The results presented here were obtained during research supported by the Gottlieb-Daimler- and Karl Benz-Foundation. The author is grateful to Mrs. Dr. Adolphi, Mrs Dziewiencki, Mrs Kern, Mrs Barzan for performing the XPS, contact angle, REM, DMA measurements. Furthermore, thanks are given to Prof. Füssel and the colleagues in the IPE for their support and cooperation.

[1] Maege I., Jaehne E., Henke A., Adler H.J.P., Bram C., Jung C., Stratmann M.: Progress in Organic Coatings 1998; 34: 1-12
[2] Häßler R., Kleinert H., Bemmann U.: Adhäsion kleben & dichten 1997; 9: 41-43
[3] Lewington T.A., Alexander M.R., Thompson G. E., and McAlpine E.: Surface Engineering 2002; Vol.18. No.3: 228-232
[4] Jansen I., Simon F., Häßler R., Kleinert H. : Macromol. Symp. 2001: 465-478
[5] Bockenheimer C., Valeske B., Possart W.: Int J Adhes Adhes 2002; 22: 349 – 356

Conformational Transitions and Aggregations of Regioregular Polyalkylthiophenes

Nataliya Kiriy,[*1] Evelin Jähne,[1] Anton Kiriy,[2] Hans-Juergen Adler[1]*

[1]Institute of Macromolecular Chemistry und Textile Chemistry, Mommsenstr.13, 01062 Dresden, Germany
[2]Institute of Polymer Research Dresden, Hohe Str. 6, 01069 Dresden, Germany
E-mail: nataliya.kiriy@chemie.tu-dresden.de

Summary: Diverse conformational transitions and aggregations of regioregular poly (3-alkylthiophene)s (PATs) in different environment have been studied by means of AFM and UV-vis-spectroscopy. In methanol, which is a non-solvent for both alkyl side groups and aromatic backbone at low polymer concentration, PATs chains fold into compact poorly ordered flat structures. At higher polymer concentration PATs molecules undergo 3D aggregation into near spherical particles. In hexane, which is a selective solvent for alkyl side chains, PATs molecules undergo ordered main-chain collapse followed by 1D aggregation. Concentration-independent red shift of λmax and good resolved fine vibronic structure in the electronic absorption spectra indicate that planarization occurs on the single-molecule level.

Keywords: atomic force microscopy (AFM); conducting polymers; helical conformation; one-dimensional aggregation; solvatochromism

Introduction

Poly(3-alkylthiophehe)s belong to one of the most studied families of conductive polymers[1] having potential applications such as light-emitting diodes,[2] thin film transistors[3] and chemical sensors.[4] It is well-known that optical properties,[2,5] conductivity,[6] and field-effect mobility[3] strongly depend on single-chain conformation and solid-state packing mode. The presence of flexible side groups does not only enhance the solubility of PATs, but also leads to novel optical effects. Indeed, PATs with regioregular head-to-tail (RRHT) structure show reversible color changes response to the temperature[5,7] or to altering of the solvent quality.[8] The observed red shift in electronic absorption spectra is due to a reversible transition between a nonplanar (less conjugated) and more planar (more conjugated) conformation of the main chain. Despite of extensive experimental and theoretical studies, conformations of PATs in different solvents as

 DOI: 10.1002/masy.200450640

well as exact structures of their aggregates remain unknown and still extensively discussed. It is generally accepted in the literature to connect the dramatic color change of PATs with the transition from twisted conformation of polymer chains in good solvent into the *planar* rigid-rod one with all-*anti* configuration of aromatic rings (Figure 1a) no matter either such a transformation occurs during evaporation of *good solvents* or upon addition of *non-solvents.*[8]

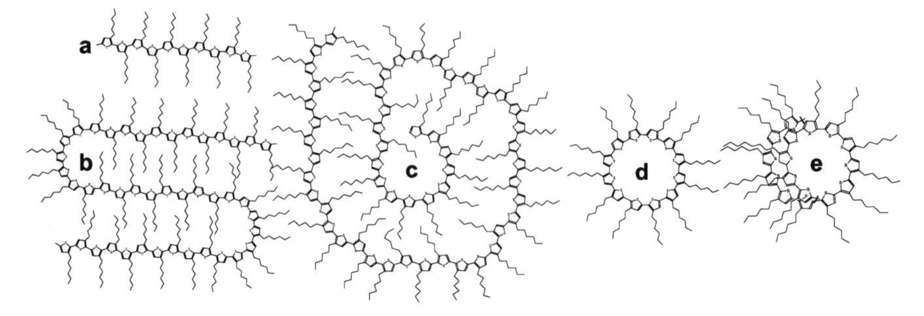

Figure 1. Possible planar conformations of regioregular head-to-tail poly(3-alkyllthiophene): all-anti (rigid-rod) (a); hair-pin (b); spool-like (c); cyclo[12]thiophene (d); all-syn helix (e).

Recently it was shown by scanning tunneling microscopy that polythiophenes backbone capable to adopt some *folded* but still *planar* conformations.[9] For example, an intramolecular hair-pin fold of PATs is composed of seven thiophenes rings in all-*syn* conformation[10] (Figure 1b), whereas larger curvatures include additional *anti*-conformations in the fold (spool-like conformation, Figure 1c). On the other hand, molecular mechanics optimizations[11] of a simple PAT model with 24 thiophene rings show the existence of a helical form with twelve *syn*-connected thiophene units per turn. All sulfur atoms of the helix are directed inside the cavity, whereas hydrocarbon groups oriented outside the helix (Figure 1e). For all structures listed in Figure 1, one can expect considerably red shifted UV-vis spectra (comparably with the spectra taken in good solvents), because they have either completely planar conformation or slightly deviated from planarity.

Here we report on a combined AFM[12] and spectroscopic investigation of the diverse conformation transitions and aggregations of RRHT poly(3-octylthiophene) (POT) and poly(3-hexylthiophene) (PHT) in polar and unpolar solvents. We will show that planar conformations of PATs with high content of syn-configuration of thiophene rings (Figures 1b, c, e) play an important role in the chromic behavior of PATs.

Experimental Section

Materials. POT was purchased from Aldrich. PHT was synthesized via McCullougth method.[13]

Sample preparation. Samples were prepared according procedure reported before.[11]

UV-vis Measurements. UV-vis measurements were carried out using Perkin Elmer UV/vis Spectrometer Lambda 19).

AFM measurements. Multimode AFM instrument (Digital Instruments, Santa Barbara) operating in the tapping mode was used. Silicon tips with radius of 10-20 nm, spring constant of 30 N/m and resonance frequency of 250-300 KHz were used after calibration with the gold nanoparticles (diameter 5 nm) to evaluate the tip radius. The dimensions of structures obtained from AFM images were corrected (decreased) by the tip radius.

Results and discussion

In this study we used RRHT polymers: relatively long POT (GPC-data: M_w = 142 kg/mol; PDI = 2.6) and shorter PHT (GPC-data: M_w = 24 kg/mol; PDI = 1.6). On the base of MALDI-TOF data contour length of polymers used in this study equal approximately to L_w = 124nm and L_N = 48nm for POT and L_w = 28 nm and L_N = 18 nm for PHT. In this study we used RRHT polymers: relatively long POT (GPC-data: Mw = 142 kg/mol; PDI = 2.6), and shorter PHT (GPC-data: Mw = 24.5 kg/mol; PDI = 1.4).

Aggregation in a good solvent (chloroform). Spin coating of a relatively concentrated PHT solution (0.1g/l) in chloroform results in a smooth film (root mean square (RMS) = 0.6 nm, Figure 2a) with a lamellae morphology that evidently from AFM phase image. Slow evaporation of the solvent from PHT solution at lower polymer concentration (0.02 g/l) leads to a lamellar network about 2 nm in the height (H) (Figures 2b, d). Elongated domains (H = 2-4 nm and L = 30-50 nm) form at slow evaporation of the solvent from extremely diluted solution of PHT (0.001g/l) (Figures 2e, c). Even longer structures (length up to 120nm) are formed at the same conditions from the solution of higher molecular weight POT (not shown). Such an observation

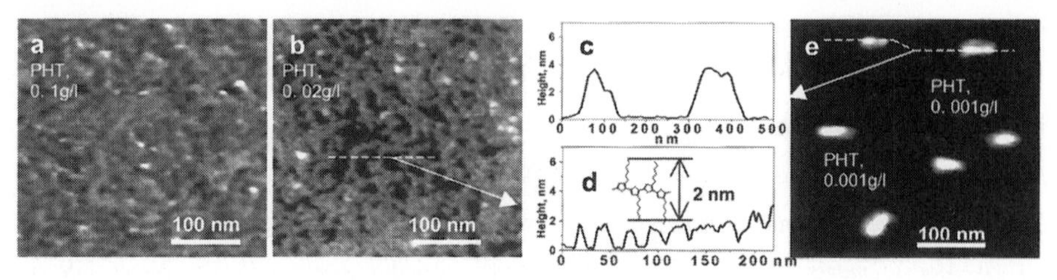

Figure 2. AFM phase (a), topography (b, e), images and cross-sections of PHT (c, d) deposited from CHCl3 solutions at different polymer concentration: 0.1g/l (a); 0.02g/l (b, d); 0.001g/l (c, e).

consistent with the face-to-face aggregation of PATs molecules adopted the rod-like (all-*anti-*) conformation and oriented perpendicularly to the surface (Figure 1a) and is in agreement with previously reported data.[3,14] Red shifted absorption maximum in UV-vis spectra of the PHT film spin-coated onto the quartz slide (λ max = 530 nm and weak shoulder at 600 nm) reflects distinct planarization of the backbone and moderate order of the molecular packing.

Solvatochromism in polar solvent (methanol). It is well-documented that addition of methanol to PATs solution in chloroform affords a colloidal solution.[8] Such a transformation accompanies with a concentration independent bathochromic shift of the π-π^* adsorption band from λmax = 450 nm (in chloroform) to λmax = 520 nm with two weak shoulders at 560 and 610 nm (in chloroform-methanol (CM) mixture - 1/1(v/v)) (Figure 3a).

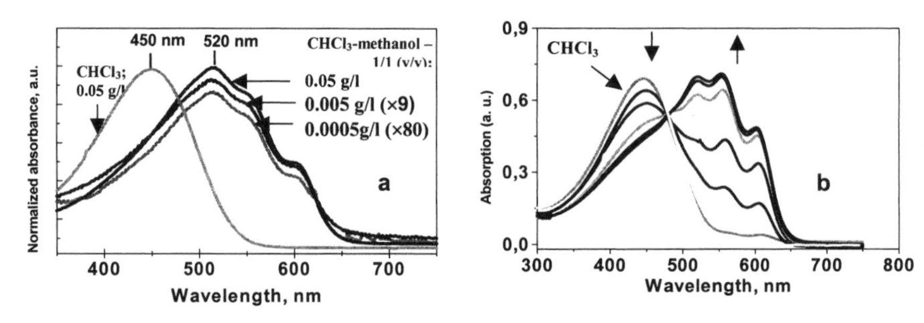

Figure 3. UV-vis spectra of PHT taken in chloroform at concentration 0.05g/l (gray line); taken in chloroform – methanol mixture (1/1(v/v)) at different concentration: 0.05g/l; 0.005g/l (intensity of adsorption have been increased in 9 time); 0.0005g/l (intensity of adsorption have been increased in 80 time) (a); UV-vis spectra of PHT (0.05g/l) at different chloroform-hexane ratio (vertical arrows show evolution of the spectra upon increasing of hexane content)(b).

Figures 4a, b show representative AFM images and cross-sections of adsorbed particles about H=20nm and 100-200nm in diameter (D) formed in CM solution (1/1 v/v) o POT at relatively high polymer concentration (0.015 g/l).

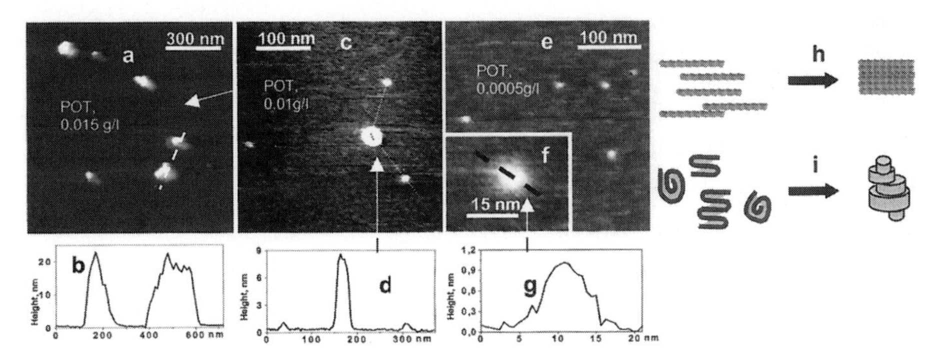

Figure 4. AFM topography images and cross-sections of POT deposited from chloroform-methanol solution (1/1(v/v)) at different polymer concentration: 0.015g/l (a, b); 0.01g/l (c, d); 0.0005g/l (e, f, g); Schematic representation of different aggregation modes of PATs: the face-to-face packing of chains in all-anti conformation (h); aggregation of collapsed hair-pin and spool-like conformations (i).

Diameter of the particles is larger than calculated contour length of POT molecules, therefore, any conformation listed in the Figure 1 from extended to collapsed, in principle, can fit the aggregates in Figure 4a. Deposition from extremely diluted PATs solution (0.0005 g/l, CM mixture - 1/1v/v) immediately after addition of methanol leads to small uniform disk-shaped particles with H = 0.4 - 0.8 nm and D = 15 - 20 nm that correspond to D = 5 - 10 nm after deconvolution (AFM image not shown). Slightly larger particles with the similar morphology were adsorbed from POT solution of the same concentration and solvents composition (apparent values: H = 0.5 - 0.9 nm, D = 25 – 40 nm; diameter after deconvolution D= 10 – 25 nm, Figures 4e, f, g). In both cases length of the particles is considerably less than the calculated contour length of corresponding PAT molecules that displays their *collapse*. The volume of these particles are close to the volume of corresponding PAT molecules that clearly reflects lack of the aggregation at this concentration regime. At a concentration higher than 0.005 g/l PATs molecules start to aggregate. Figures 4c, d display the coexistence of small particles (similar to the one in Figure 4c) and larger aggregates (apparent values: H = 8 – 10 nm; D = 45-65 nm; after deconvolution D = 20 – 40 nm) adsorbed from POT solution (0.01 g/l). On the other hand, *concentration independence* of the UV-vis absorption spectra (Figure 3a) indicates that efficient planarization of the backbone and increase of the conjugation length occurred upon addition of methanol to solution of PATs in chloroform are *single-molecule events*. From obtained experimental data we can conclude that, at least, at low polymer concentration, upon addition of the solvent, which is poor for alkyl side groups, PATs molecules *firstly undergo collapse transition into the compact and planar state and then aggregate.*

Our observation disagrees with the most accepted model for the aggregation of PHT and other substituted polythiophenes occurring in poor solvents. Accordingly with this model, PATs aggregates consist of face-to-face stacked molecules having a *planar rod-like* conformation similarly to the organization of PATs films (Figure 4h).[8] In this case, at least, one dimension of such a particle should be equal (or more) to the contour length of PAT molecules (120-50 nm for POT) that contradict AFM data. As seen from AFM images not only single-molecule particles in Figure 4e but also primary aggregates in Figure 4c have dimensions less than the contour length of POT molecules. Therefore, we proposed that in chloroform-methanol mixture PATs molecules undergo solvophobically driven collapse into the structures with high content of cis-configuration of thiophene units (hair-pin or spool-like conformation, Figures 1b, c). Such conformations reduce a solvent-accessible surface area and decrease an unfavourable interaction between alkyl groups and polar environment. Further aggregation of flat spool-like structures leads to poorly ordered particles (Figure 4i).

It was previously reported[3] that, diffractogram of RR PHT cast films deposited from chloroform has sharp reflections at 2θ = 5.3°; 10.7° and 16.2° which can be assigned to the (100), (200), and (300) planes of a lamellar phase with a spacing 16.7 Å. In that case alkyl chains act as spacers between stacks of closely packed planar main chains. The peak at 23.1° (3.8 Å) corresponds to thiophene face-to-face stacking distance. In contrast, a powder X-ray diffraction pattern of RR PHT precipitated from methanol-chloroform mixture,[8] has considerably less intense peak at 2θ = 5.3°, weak peaks at 10.7°, and 16.2°, and broad galo at 15-25° that reflects *similar but significantly less ordered* molecular packing. Such an observation fits the proposed model of the PHT collapse followed by the aggregation occurred in methanol-chloroform mixture. The spool-like structure shown in Figure 1c presents the PHT molecule with the degree of polymerization (DP) equal to 52. More than half of monomer units of this structure (32 units) have *anti*-configuration and form slightly distorted lamellae with the spacing 16.7 Å. Remaining 20 monomer units in the structure (Figure 1c) adopt syn-conformation and can be responsible for the broad galo centered at 20°. Even higher fraction of anti-configuration and more intense reflection at 5.3° one can expect for PHT molecules of higher DP.

Solvatochromism in unpolar solvents (hexane). Although solvatochromism of PATs induced by polar solvents have been widely studied, their behavior in unpolar solvents is considerably less investigated.[11] High molecular weight PATs are insoluble in hydrocarbons, however, we found that a stable colloidal dispersion could be obtained upon addition of hexane to the solution of PATs in chloroform. This accompanies with considerable red shift of UV-vis absorption maximum from λmax = 448 nm in pure chloroform to λmax = 559 nm (for POT) and λmax = 553 nm (for PHT) in chloroform-hexane - 1/5 mixture. Appearance of the fine vibronic structure (resolved transitions at 522 nm; 559 nm; 605 nm for POT and 520 nm; 553 nm; 602 nm for PHT) indicates the rigidification of the conjugation system. The clear isobestic point at 480 nm (for PHT) reflects a discontinuous character of the transition and the coexistence of two distinct conformational structures for these polymers (Figure 3 b).

Figure 5 shows the AFM images of particles adsorbed onto mica from solution at different PHT concentrations and chloroform-hexane (CH) ratios.

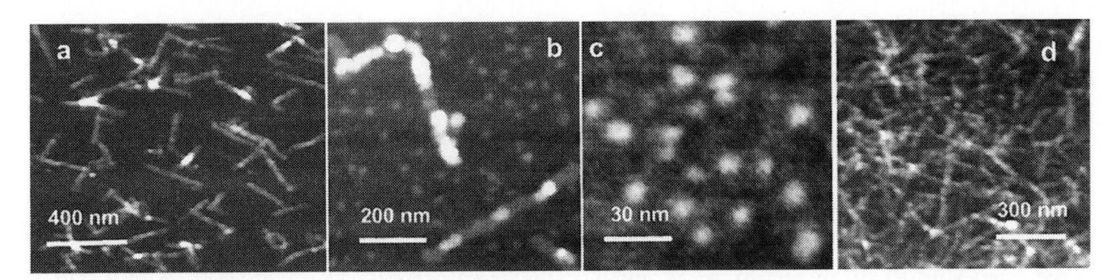

Figure 5. AFM topography images of PHT deposited from chloroform-hexane solution (1/5(v/v)) in 15 minutes after addition of hexane at different polymer concentration: 0.01g/l (a); 0.4g/l (d); immediately after addition of hexane: PHT concentration - 0.01g/l (b, c).

We performed the statistical analysis of geometrical parameters extracted from the AFM images and profiles including number and weight average lengths for one-dimensional (1D) structures, as well as diameter and height for spherical particles. Narrow distributed round-shaped particles with D = 3-5 nm and H = 1-2.5 nm (all data given after deconvolution) were adsorbed from the PHT solution at concentration 0.001 g/l and CH ratio - 1/5(v/v), which had been stirred during 15 minutes before deposition (not shown). Adsorption onto mica from the solution of POT at the same conditions resulted in particles with H = 2-10 nm and D = 3-5 nm (not shown). At higher polymer concentration (0.01 g/l) PHT molecules exhibited selective 1D growth into rod-like structures with the length up to 700 nm (Figure 5a). In contrast, adsorption from PHT solution of the same concentration (0.02 g/l) immediately after addition of hexane resulted in coexistence of small particles (H = 1.0-2.0 nm; D = 3-5 nm) and 1D aggregates (L > 200 nm, Figures 5b, c). Such a picture is an obvious evidence for a non-equilibrium state of the aggregation process. Spin-coating of a relatively concentrated solution of PHT (0.4 g/l) in chloroform-hexane mixture (1/5) leaded to well-distinguished network of rod-like structures (RMS = 4.6nm, Figure 5d). Well resolved UV-vis spectra of the PHT film deposited onto the quartz slide from chloroform-hexane mixture (1/5) (λmax = 523, 556, and 601 nm) differed considerably from the spectra of PHT film obtained by spin-coating from chloroform solution and closes to those taken in chloroform-hexane mixture (1/5). Obtained results show that there is no significant altering of solution conformation of PATs molecules during adsorption and solvent evaporation. Thus, in the chloroform-hexane mixture PATs undergo conformational transition from random twisted into a more planar and ordered state followed by the 1D-aggregation. Two following possible mechanisms for such transformations including a formation of rod-like micelles and the helical conformation have been recently proposed and discussed in our previous paper.[11]

Conclusion

Diverse conformational transitions and aggregations of RRHT PATs in different environments have been studied by means of AFM and UV-vis-spectroscopy. Aggregation from chloroform solution which is a good solvent for PATs, occurred upon slow evaporation of the solvent leads either to the lamellar network or to the elongated domain structures depending on polymer concentration. Red shifted and poorly resolved electron absorption spectra of PHT thin films reflect distinct planarization of the backbone and moderate order of molecular packing. Such observations consistent with the face-to-face packing mode of PATs molecules adopted the rod-like (all-*anti-*) conformation. In presence of methanol, which is *non-solvent* for both alkyl side groups and aromatic backbone at *low polymer concentration*, PATs chains fold into compact poorly ordered flat structures without aggregation. Concentration-independent UV-vis spectra confirm single molecule origin of solvatochromism of PATs. At higher polymer concentrations PATs molecules undergo 3D aggregation into near spherical particles. In hexane, which is the *selective solvent* for alkyl side chains, PATs molecules undergo *ordered main-chain collapse* followed by 1D aggregation. The *concentration-independent* red shift of λmax and good resolved fine vibronic structure in the electronic absorption spectra indicate that planarization occurs on the *single-molecule level*. The *helical conformation* of the main chain of PATs with twelve thiophenes rings per each helical turn has been proposed.[11]

Acknowledgment. We are grateful for the financial support provided by DFG foundation (Project DFG-Sachb. AD 119/6-1).

[1] Skotheim, T. A.; Elsenbaumer, R. L.; Reynolds, J. R. *Handbook of Conducting Polymers*; Marcel Dekker: New York, 1998.
[2] Sirringhaus, H.; Kawase, T.; Friend, R. H.; Shimoda, T.; Inbasekaran, M.; Wu, W.; Woo, E. P. *Science* **2000**, *290*, 2123.
[3] Sirringhaus, H.; Brown, P. J.; Friend, R. H.; Nielsen, M. M.; Bechgaard, K.; Langeveld-Voss, B. M. W.; Spiering, A. J. H.; Janssen, R. A. J.; Meijer, E. W.; Herwig, P.; de Leeuw, D. M. *Nature* **1999**, *401*, 685.
[4] McQuade, D. T.; Pullen, A. E.; Swager, T. M. *Chem. Rev.* **2000,** *100,* 2537.
[5] Rughooputh, S. D. D. V.; Hotta, S.; Heeger, A. J.; Wudl, F. J. *Polym. Sci., Polym. Phys. Ed.* **1987**, *25*, 1071.
[6] McCullough, R. D.; Tristram-Nagle, S.; Williams, S. P.; Lowe, R. D.; Jayaraman, M. *J. Am. Chem. Soc.* **1993**, *115*, 4910.
[7] Langeveld-Voss, B. M. W.; Janssen, R. A. J.; Christiaans, M. P. T.; Meskers, S. C. J.; Dekkers, H. P. J. M.; Meijer, E. W. *J. Am. Chem. Soc.* **1996**, *118*, 4908.
[8] Yamamoto, T.; Komarudin, D.; Arai, M.; Lee, B.-L.; Suganuma, H.; Asakawa, N.; Inoue, Y.; Kubota, K.; Sasaki, S.; Fukuda, T.; Matsuda, H. *J. Am. Chem. Soc.* **1998**, *120,* 2047.
[9] Mena-Osteritz, E.; Meyer, A.; Langeveld-Voss, B. M. W.; Janssen, R. A. J.; Meijer, E. W.; Bäuerle P. *Angew. Chem.* **2000**, *112*, 2791.
[10] Krömer, J.; Rios-Carreras, I.; Fuhrmann, G.; Musch, C.; Wunderlin, M.; Debaerdemaeker, T.; Mena-Osteritz, E.; Bäuerle P. *Angew. Chem.* **2000**, *112*, 3623.
[11] Kiriy, N.; Jähne, E.; Adler, H.– J.; Schneider, M.; Kiriy, A.; Gorodyska, G.; Minko, S., Jehnichen, D.; Simon, P.; Fokin, A. A.; Stamm, M. *Nano Lett.* **2003**, *3*, 707.
[12] Kiriy, A.; Minko, S.; Gorodyska, A.; Stamm, M. *J. Am. Chem. Soc.* **2002**, *124,* 10192. Kiriy, A.; Gorodyska, A.; Minko, S.; Jaeger, W., Štěpánek, P.; Stamm, M. *J. Am. Chem. Soc.* **2002**, *124*, 13454.
[13] Loewe, R. S.; Khersonsky, S. M.; McCullough, R. D. *Adv. Mater.* **1999**, *11*, 250.
[14] Sandberg, H. G. O.; Frey, G. L.; Shkunov, M. N.; Sirringhaus, H.; Friend, R. H.; Nielsen, M. M.; Kumpf, C. *Langmuir* **2002**, *18*, 10176.

Macromol. Symp. **2004**, *210*, 369-376

Preparation of Filled Temperature-Sensitive Poly(*N*-isopropylacrylamide) Gel Beads

Dirk Kuckling,[*1] *Thomas Schmidt,*[2] *Genovéva Filipcsei,*[3]
Hans-Jürgen P. Adler,[1] *Karl-Friedrich Arndt*[2]

[1] Institut für Makromolekulare Chemie und Textilchemie, Technische Universität Dresden, D-01062 Dresden, Germany
E-mail: dirk.kuckling@chemie.tu-dresden.de

[2] Institut für Physikalische Chemie und Elektrochemie, Technische Universität Dresden, D-01062 Dresden, Germany

[3] Department of Physical Chemistry, Budapest University of Technology and Economics, H-1521 Budapest, Hungary

Summary: Temperature-sensitive filled poly(*N*-isopropylacrylamide) (PNIPAAm) gel beads with diameters in the range of millimeters were prepared using the alginate technique. The polymerization and cross-linking reaction of NIPAAm in the presence of inorganic filling particles was performed in spherical networks of Ca-alginate forming interpenetrating networks (IPN). Thermo-sensitive gel beads could be obtained by washing these IPN with EDTA solution. The PNIPAAm gel beads were analyzed by optical methods to observe there swollen diameter in dependence on the temperature. The diameters of the swollen gel beads were in the range of 0.1 – 2 mm. The influence of the monomer to cross-linker ratio (MCR) and the filling materials (ferrofluid, $BaTiO_3$, TiO_2, and Ni,) were studied. The phase transition temperature (T_{pt}) was only weakly influenced by the MCR and the filling material remaining at around 34 °C.

Keywords: alginate technique; ferrofluid; filler; gel beads; PNIPAAm hydrogels; temperature-sensitive polymers

Introduction

Novel polymer gels that are responsive to external stimuli have been developed in the past decade [1-3]. The stimuli that induces change in polymer gels include temperature, pH, solvent and ionic composition, electric field, light intensity and the recognition of specific molecules. The discovery of a discontinuous volume phase transition in gels, which is often called the collapse transition, has made such soft materials of technological interest [4,5]. The behavior of these gels can be utilized in mechanical devices, controlled release delivery and separation systems [6-8].

 DOI: 10.1002/masy.200450641

Poly(*N*-isopropylacrylamide) (PNIPAAm) hydrogel is one of the most frequently studied temperature-responsive gel [4]. Above the transition temperature the network chains are in the collapsed state. This transition is called the lower critical solution temperature (LCST) behavior. For PNIPAAm gels swollen in water the LCST was found to be 34 °C. Several other gels show reversible swelling and shrinking transitions with different LCST as well. These gels were often used to immobilize enzymes and as carriers of certain functional groups important for biochemical or biomedical applications [5]. For those applications gel beads were usually used in the size range from approximately 0.1 μm to several mm. These gels were heated by a surrounding heat source to control the degree of swelling. A more efficient heating method was proposed recently. Takahashi and his coworkers immobilized needle like iron oxide (γ-Fe_2O_3) powders with 0.5-0.8 μm length into polymer gels for magnetic heating [9,10]. Due to the hysteresis loss of the hard magnetic material in the presence of an alternating magnetic field, the magnetic energy was converted into heat inside the gel, which increased the temperature. This method is energetically efficient as only the gel beads are heated and not the entire environment.

Magnetic field sensitive gels were also developed to increase the rate of shape change [11-13]. Magnetic nanoparticles of Fe_2O_3, called magnetite, were incorporated into chemically cross-linked poly(vinylalcohol) hydrogels. It was established that the peculiar magnetic and magneto-elastic properties of these gels could be used to target magnetic gel beads to a certain place or to create a wide range of motions to control the shape of the gels.

Fe_2O_3 nanoparticles with a typical size of 10 nm are magnetically soft materials. They exhibit superparamagnetic behavior [14]. The mono-domain ferromagnetic particles of colloidal size are the elementary carriers of a magnetic moment in the ferrogel. In the absence of an applied field they are randomly oriented due to thermal agitation and thus the ferrogel has no net magnetization. As soon as an external field is applied, the magnetic moments tend to align with the field to produce a bulk magnetic moment (M). With ordinary field strengths the tendency of the dipole moments to align with the applied field is partially overcome by thermal agitation. As the strength of field increases, all the particles eventually align their moments along the direction of the field leading to the magnetisation saturates. If the applied field is turned off, the particles quickly randomize, and M is again reduced to zero. This means that the magnetization curve shows no hysteresis at all and can be fitted by a Langevin function corrected with the distribution of the magnetic dipole moments.

The properties of magnetic beads containing hard magnetic particles in their gel matrix differ significantly from those containing soft magnetic particles. If the gel is loaded with a magnetically hard filler material, it behaves like a permanent magnet. As a consequence - due to the magnetic interactions - the gel beads form aggregates, even without the application of an external magnetic field. Intensive stirring is required to counterbalance the magnetic interactions and prevent aggregation. If magnetic soft particles are introduced into the gel, then the beads have no permanent magnetization and as a result they form aggregates only in the presence of an external magnetic field.

The aim of the present work is the synthesis of temperature sensitive hydrogel beads by the alginate technique. They were characterized to their temperature sensitivity in dependence on the cross-linking ratio. Filled PNIPAAm gel beads have been synthesized by incorporating inorganic filling materials, too. The influence of the fillers on the swelling behavior was analyzed.

Preparation of Filled PNIPAAm Beads

The gel beads were prepared according to the method developed by Park and Choi [15]. An interpenetrated network (IPN) was prepared by the simultaneous gelation of Ca-alginate to form spherical bead shapes and the concomitant free radical polymerization of the NIPAAm and cross-linker (N,N'-methylene bisacrylamide) within the beads. The beads were then incubated in ethylenediamine tetraacetate (EDTA) solution to chelate Ca-ions and extract the alginate from the IPN beads. The molar cross-linking ratio (molar ratio of monomer to cross-linker (MCR)) was varied between 24 and 75 at a constant monomer concentration of 7.68 wt-%. At a constant cross-linking ratio (MCR 50) the monomer concentration was varied between 7.68 and 4.07 %. A schematic procedure is shown in Figure 1.

Ca-ions formed an ionotropic complex with the alginates backbones leading to the formation of a water swollen gel network. The monomer, cross-linker, filling material and the redox initiator were kept in these Ca-alginate gel beads, the co-initiator (ammonium persulfate) in the medium rapidly diffused into the gels and initiated the polymerization. The cross-linked PNIPAAm in the Ca-alginate gel could be described as an interpenetrating network (IPN) structure, in which the two polymer networks were entangled in each other. Homogeneous PNIPAAm gel beads could easily be formed from the IPN gel in large spherical shapes, when the alginate polymer was dissolved by addition of Ca-chelating agents.

A magnetic field sensitive gel is a special type of filler loaded gel, where the finely divided filler particles have strong magnetic properties. Preparation of such gels does not require a special polymer or a special type of magnetic particle. As a polymer network one may use every flexible chain molecule, which can be cross-linked. The filler particles can be obtained from ferro- as well as ferrimagnetic materials.

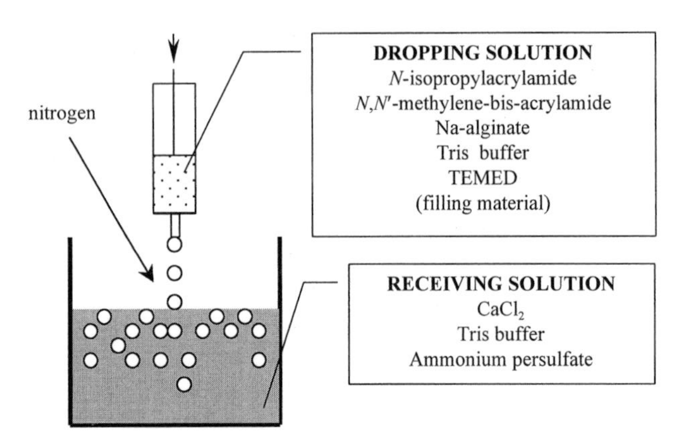

Figure 1: A schematic procedure for the preparation of PNIPAAm gel beads

The preparation of a magnetic PNIPAAm gel (MPNIPAAm) was similar to that of other filler-loaded elastomeric networks. Well-dispersed particles were precipitated in the polymeric material. This "in situ" precipitation can be made before, during and after the cross-linking reaction [16,17]. Firstly, a ferrofluid, which contained magnetite sol particles, was prepared from iron (II) chloride and iron (III) chloride in aqueous solution. In order to counter-balance the van-der-Waals attraction and the attractive part of the magnetic dipole interactions, colloidal stability was maintained by a small amount of perchloric acid, which induced peptization. The purified and stabilized magnetite sol with a concentration of 15.7 wt-% was used for further preparative work. During the preparation the dropping solution contained 0.5 ml of the ferrofluid (corresponding to 0.94 wt-% magnetite content) [18]. For the preparation of the with inorganic particles filled PNIPAAm gels the particles were suspended in the solution with a total concentration of 10 wt-%. The structure of the inorganic particles is shown in Figure 2. The MCR was varied between 24 and 50 at a constant monomer concentration of 7.68 wt-%.

The bead size can to some extent be controlled by using different sized needles on the string. In this manner beads ranging from 0.5 to 4 mm may be conveniently prepared. Here, gel beads with an average diameter of d = 2.0 mm and narrow size distribution were prepared to enable for accurate measurements. It should, however, be mentioned that none of these properties are size dependent. The preparation of smaller beads requires another technique.

Figure 2: SEM of the a) Ni particles (bar = 10 μm), b) $BaTiO_3$ particles (bar = 1 μm), and c) TiO_2 particles (bar = 0.5 μm)

In order to determine the temperature dependence of the degree of swelling the experiments were monitored using a digital video system. This method enables the measurement of very small changes in the diameter (one pixel on the screen) on the real time video image.

Collapse Transition of Non-Magnetic and Magnetic PNIPAAm Gel Beads

The temperature dependence of the PNIPAAm gel and MPNIPAAm gel beads was studied. The behavior of MPNIPAAm gel was similar to that of the non-magnetic PNIPAAm gels. An abrupt volume change in response to temperature change was observed in both cases (Figure 3). d_0 is the diameter of the gel beads at approximately 10 °C.

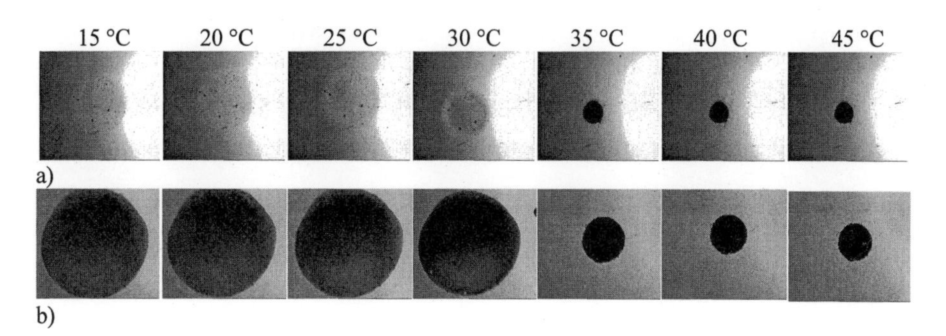

Figure 3: Thermo shrinking phenomenon of a) non-magnetic PNIPAAm gel bead, and b) magnetic PNIPAAm gel bead

An increase of the temperature led to a significant decrease in the volume of the gel beads (Figure 4). The swelling ratio changed from 1 to 0.2 in every case. It can be seen that the cross-linking ratio does not affect the phase transition temperature, which was around 34 °C.

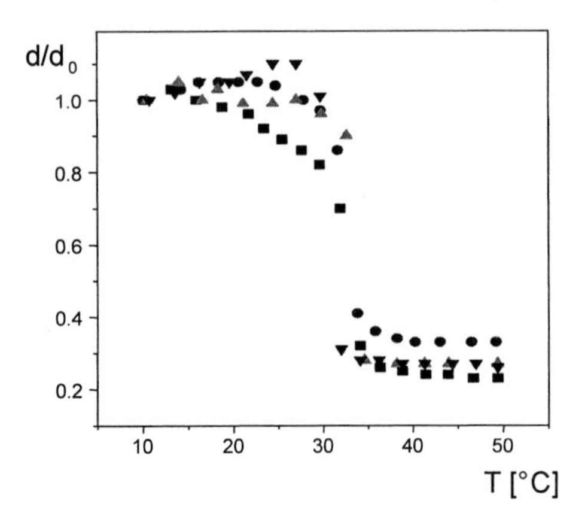

Figure 4: The effect of different cross-linking ratio on the LCST of PNIPAAm gel beads (■ - MCR 24, ● - MCR 32, ▲ - MCR 50, ▼ - MCR 75)

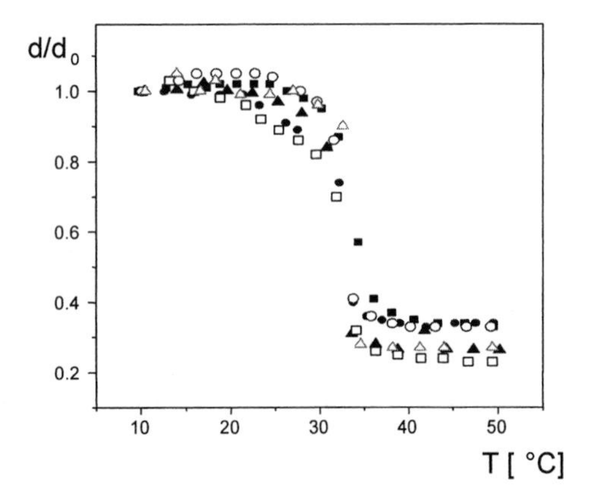

Figure 5: The effect of different cross-linking ratio on the LCST of PNIPAAm and MPNIPAAm gel beads (open symbols - PNIPAAm gel, solid symbols - MPNIPAAm gel) (■- MCR 24, ● - MCR 32, ▲ - MCR 50)

The influence of larger filling particles (in the range of some μm) was analyzed by incorporating other inorganic particles ($BaTiO_3$, TiO_2 and Ni). Figure 6 shows the temperature sensitive behavior of the filled gel beads. A shift in the phase transition temperature was not observed at the analyzed gels (T_{cr} = 34 °C) due to low interactions between the filling material and the cross-linked PNIPAAm. However, the differences in the degree of swelling were effected by the nature of the filler particles.

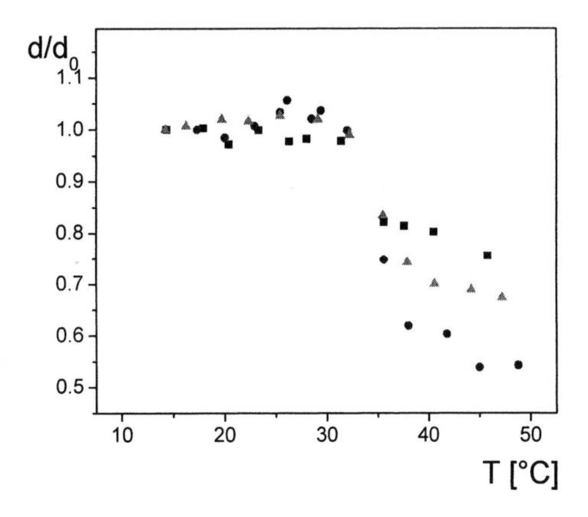

Figure 6: The effect of incorporated inorganic particles on the temperature sensitivity of the PNIPAAm gels (■- $BaTiO_3$, ● - TiO_2, ▲ - Ni)

Conclusion

This work was carried out in order to describe the preparation of filled temperature sensitive PNIPAAm gel beads. The temperature sensitivity of filled PNIPAAm gels is quite similar to that of the same gel containing no nanoparticles. Consequently the collapse transition is not affected by the presence of the nanoparticles. The special magnetic properties of these gels may be used to target and orient the temperature responsive gel beads and enables their easy separation from the environment through the use of non-uniform magnetic field.

Acknowledgements

The authors are grateful to Mr. D. Theiss and Mrs. E. Kern for the SEM measurements and to Mr. B. Wahl and Mr. D. Thiemig for the analysis of non-magnetic filled gels. The DFG (Deutsche Forschungsgemeinschaft) is gratefully acknowledged for their financial support of this work within the "Sonderforschungsbereich 287".

[1] N.A. Peppas, R.W. Korsmeyer, "*Hydrogels in Medicine and Pharmacology*", CRC Press, Boca Raton, Florida 1987.
[2] D. De Rossi, K. Kawana, Y. Osada, A. Yamauchi, "*Polymer Gels: Fundamentals and Biomedical Applications*", Plenum Press, New York London 1991.
[3] R.S. Harland, R.K. Prud'homme, "*Polyelectrolyte Gels*", ACS Symposium Series 480, 1992.
[4] T. Tanaka, *Phys. Rev. Lett.* **1978**, *40*, 820.
[5] T. Okano, "*Biorelated Polymers and Gels*", Academic Press, Boston, San Diego, New York, London, Sydney, Tokyo and Toronto 1998.
[6] M. Suzuki, O. Hirasa, *Adv. Polym. Sci.* **1993**, *110*, 241.
[7] A.S. Hoffman, *Adv. Drug Delivery Rev.* **2002**, *54*, 3.
[8] D. Kuckling, A. Richter, K.-F. Arndt, *Macromol. Mat. Eng.* **2003**, *288*, 144.
[9] F. Takahashi, Y. Sakai, Y. Mizutani, *J. Fermentation Bioeng.* **1997**, *83*, 152.
[10] N. Kato, Y. Takizawa, F. Takahashi, *J. Intell. Mat. Syst. Struct.* **1997**, *8*, 588.
[11] M. Zrínyi, L. Barsi, D. Szabó, H.-G. Kilian, *J. Chem. Phys.* **1997**, *106*, 5685.
[12] M. Zrínyi, *Trends Polym. Sci.* **1997**, *5*, 277.
[13] D. Szabó, G. Szeghy, M. Zrínyi, *Macromolecules* **1998**, *31*, 6541.
[14] R.E. Rosenweig, "*Ferrohydrodynamics*", Cambridge University Press 1985.
[15] T.G. Park, H.K. Choi, *Macromol Rapid Commun.* **1998**, *19*, 167.
[16] J.E. Mark, *Brit. Polym. J.* **1985**, *17*, 144.
[17] W. Haas, M. Zrínyi, H.-G. Kilian, B. Heise, *Coll. Polym. Sci.* **1993**, *271*, 1024.
[18] P.M. Xulu, G. Filipcsei, M. Zrinyi, *Macromolecules* **2000**, *33*, 1716-1979.

Macromol. Symp. **2004**, *210,* 377-384

Adjustable Low Dynamic Pumps Based on Hydrogels

Andreas Richter, Christian Klenke, Karl-Friedrich Arndt*

Institute for Physical Chemistry and Electrochemistry, Dresden University of Technology, 01062 Dresden, Germany
E-mail: andreas.richter@chemie.tu-dresden.de

Summary: This paper discusses the suitability of hydrogel actuators as drives of automatic pumps for long-term drug release. We demonstrate that such actuators can execute a defined task if several functional units are connected serially. Investigated functions of this pump are the adjustability of the time-delay from the initial operation up to the beginning of the drug release, opening the sterile drug ampoule, and the drug release at a specified timeframe. The described parameters to influence the pump behaviour are satisfying only for realization of continuous working devices. An outlook of the development of programmable pulsate pumps is given.

Keywords: drug release; fluidic component; hydrogel actuator; medical pump

1. Introduction

Smart hydrogels are able to change their volume by more than one magnitude in response to a lot of various sensitivities such as temperature, pH value, light, ion, and substance concentrations. Therefore, an enormous importance for many technological and scientific applications was expected [1]. Currently, some technological breakthroughs particularly in fluidics and sensorics could be indicated. Automatically [2-6] and electronically controllable [7] valves and microvalves have been described. Furthermore, a micro machined pH sensor based on smart hydrogels with a short insight of the sensor behaviour has been presented [8].

On the other hand, hydrogels without a smart behaviour also are important for many high-technology applications. Particularly the principle of osmotic pumps [9], introduced in 1975, allowed the development of a new generation of drug delivery systems. For example, modern therapies of gastrointestinal tract diseases are hardly conceivably without such systems.

Automatic long-term working drives could be an interesting option to substitute some electromechanical drive systems, which needs auxiliary energy and includes a complicate mechanical design.

In present paper we investigated the suitability of long-time working osmotic hydrogel actuators

 DOI: 10.1002/masy.200450642

as drives of drug release pumps. Such pumps could substitute the complex electromechanical medical pumps, which are commonly used.

Experimental

Materials

Sodium polyacrylate (PAAc-Na, HySorb C 7015 and M 3100) was obtained from BASF. Poly(*N*-isopropylacrylamide) (PNIPAAm) samples were prepared by following procedure. The crosslinking agent was *N,N′*-methylenebisacrylamide (BIS). The initiator and accelerator for the polymerization reaction were potassium peroxidisulfate (KPS) and *N,N,N′,N′*-tetramethyl-ethylenediamine (TEMED) respectively (both from Aldrich Chemical Co.). *N*-isopropylacrylamide (NIPAAm) and various amounts of BIS (1 mol-% to 10 mol-%, for example BIS4 prepared with 4 mol-%) were dissolved in deionized water. The total monomer concentration was 0.53 mol/l. To initiate the polymerization reaction 0.3 mol-% of KPS and TEMED were added to the oxygen free (bubbled with Nitrogen) solution. After polymerization (ca. 12 h at room temperature) the PNIPAAm gel was immersed in deionized water for about one week to wash out non-reacted reagents.

After drying the hydrogels particles with irregular shape were obtained by milling and subsequent fractionating into different particle sizes using test sieves.

Pump Design

Functional Requirements

Functional requirements of a pump for drug release are:

- a position independent operation;
- an adjustable time-delay from the initial operation up to the beginning the drug release;
- opening of the sterile drug ampoule immediately before starting the drug release;
- a drug release at a specified timeframe and at a defined volume.

To realize these features an automatic execution of a task sequence is necessary (see Fig. 1).

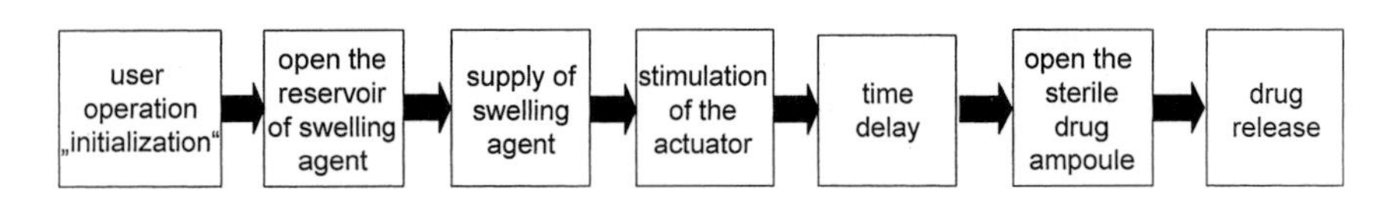

Figure 1. Automatic executable task sequence of a pump for drug release.

Pump Design and Functionality

The pump includes a trigger which allows their initialization (see Fig. 2). This switch can open the supply unit or swelling agent reservoir, respectively. A spring generates a pressure on the swelling agent reservoir. The resulting permanent hydrostatic excess pressure inside the swelling agent chamber enables a position independent loading of the actuator chamber or hydrogel actuator, respectively, with the swelling agent. The actuator chamber is encased by a high-flexible foil.

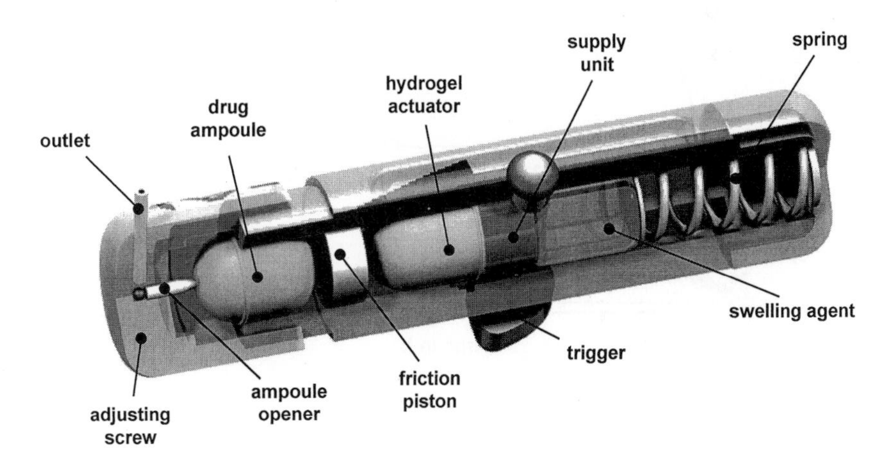

Figure 2. Design of a pump for drug release.

If the pump is initiated, the hydrogel actuator swells and stretches this foil. At the beginning of the swelling process the actuator must overcome a non-load distance or displaces a friction piston. The range of this distance is limited by the drug ampoule. The length of this way depends on the time behaviour of hydrogel actuator at the swelling process. Because the non-load distance can be adjusted using the adjusting screw, the time between pump initialization and starting the drug release is delayed. After passing the time-delay unit the actuator presses the drug ampoule at the opener to open the sterile ampoule. In the last step the hydrogel actuator crushes the ampoule content and the pump releases the drug.

Results and Discussion

Basic Characteristic of Hydrogel Actuator

The time behaviour of a hydrogel actuator depends on their smallest effective dimension. If the actuator material is based on particles, this dimension is the particle diameter. Furthermore, it will be influenced by the composition of hydrogel particles, their homogeneity, and the crosslinking density of hydrogel (see Fig. 3).

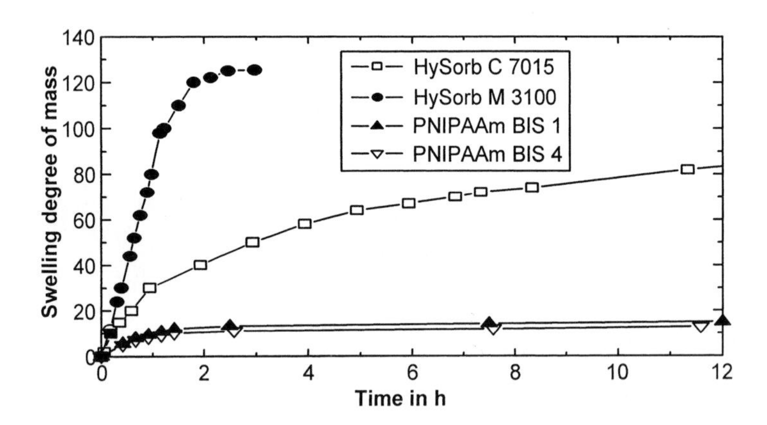

Figure 3. Swelling behaviour of various actuators based on hydrogel particles (diameter ≈ 500 μm).

The basic time behaviour of such actuators follows an exponential function. The volume changes at the beginning of the swelling process are very fast. Approaching the swelling equilibrium the swelling rate decreases significantly. This feature is causing some problems for the pump design because a linear characteristic would be better usable.

Time-Delay

Firstly, the swelling characteristic must be adjusted for realization the time-delay function. A fast swelling process at the beginning results in a very long non-load distance, which is unacceptable because the length of the pump is increasing strongly. Using a hydrogel with a slow swelling behaviour such as C 7015 (see Fig. 3) a 18 mm long non-load distance is required to realize a time-delay of 2 h (see Fig. 4b). However, an applied opposing force to the swelling force could provide a decrease of the time – distance function. Their influence of the swelling behaviour of a hydrogel actuator is shown in Figure 4a.

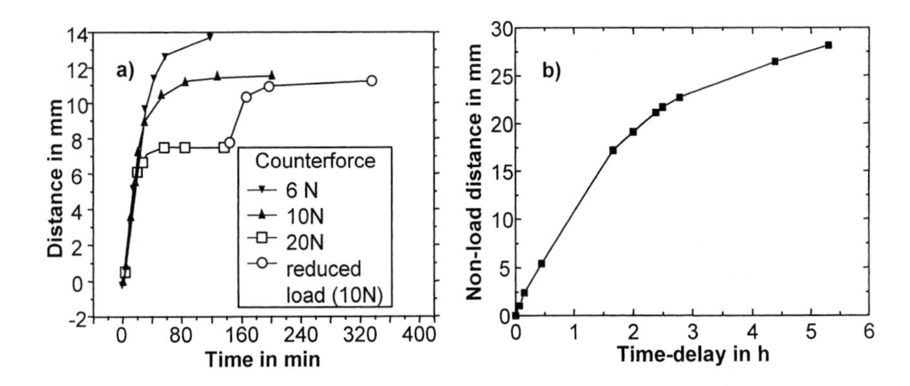

Figure 4. Dependency of time-delay from the distance: a) Time-delay distance depends on a applied opposing force; b) Time-delay as a function of a non-load distance.

The time-delay of 2 h can be realized with a short time-delay distance of about 8 mm using an opposing force of 20N. Furthermore, it can be seen that the hydrogel actuator obtains a swelling equilibrium which depends on the value of counterforce. The swelling process will restart if the opposing force is reduced. Hence, we utilize a friction piston which generates an opposing force to the swelling force. A nullifying of the counterforce is possible by increasing the inner diameter of pump body (strong decrease of the frictional force).

Release Performance

As it has been shown in Fig. 3, that the time and swelling behaviour of a hydrogel actuator is influenced by the composition of hydrogel particles, their homogeneity, the crosslinking density, and the particle size. Furthermore, the maximal volume change of the actuator depends strongly on their volume.

For a long-term release a characteristic nearly linear and a long-term swelling process of the hydrogel actuator is important. Both features can be realized through restriction of the supplying quantity of swelling agent (see Fig. 5). If the provided quantity of swelling agent is less than the hydrogel actuator needs for swelling to equilibrium, a linearization of the swelling characteristic as well as the release characteristic of the pump (see Fig. 5a) is resulted.

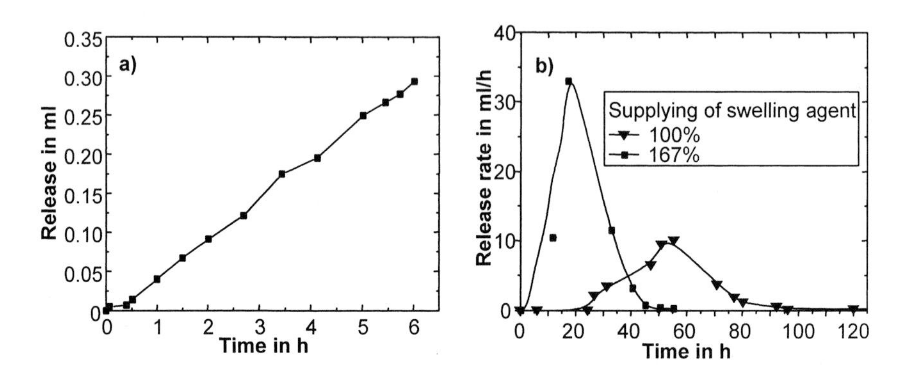

Figure 5. Release of a hydrogel driven pump depending on the supplied quantity of swelling agent: a) Linearization of the pump release; b) Release rate of pump depending on the supplying quantity of swelling agent.

On the other hand, an excess of the swelling agent decreases the maximal release of the pump and increases its release period (see Fig. 5 b).

Operational Behaviour

The behaviour of the drug release pump is shown in Figure 6. The time-delay function could successfully be realized. However, the use of a frictional force as a counterforce is problematic because the quality of the transition fit and the fabrication tolerance, respectively, must be very accurate. The ampoule opener process was also practicable. The membrane of the ampoule, which must be opened, should be relatively thin, otherwise the required opening force is all-too large. The pump can release a drug volume of about 500 µl in 2 h.

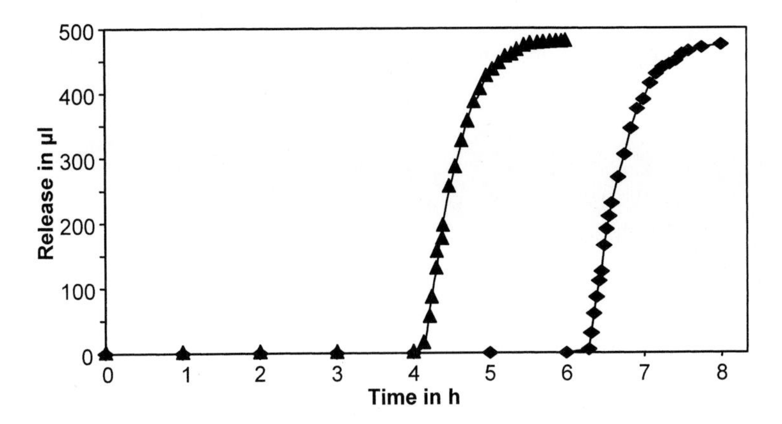

Figure 6. Operational behaviour of a drug release pump with adjustable time-delay.

Conclusion and Outlook

It was shown that hydrogel actuators can execute a defined task sequence if several functional units connected serially. Proved functions of a pump are the adjustable time-delay from the initial operation up to the beginning the drug release, opening of the sterile drug ampoule, and the drug release in a specified timeframe. The described parameters to influence the pump behaviour are satisfying only for realization of continuously working devices. Currently, such pumps will be developed ready to go into production.

To obtain a pulsating operational behaviour further requirements must be complied. Pulsations of drug release could be realized through serial connection of actuator materials with different swelling properties or by affecting the swelling agent supply. Interrupts are obtainable using water soluble swelling agent barriers. Further works will investigate the technical feasibility of programmable pulsatile working pump (see Fig. 7).

Such pump should include a programming unit which consists of some dials. These dials are showing several chambers which could be filled with various actuator materials, swelling agent barriers, etc. Through rotating the dials against each other task sequences are programmable.

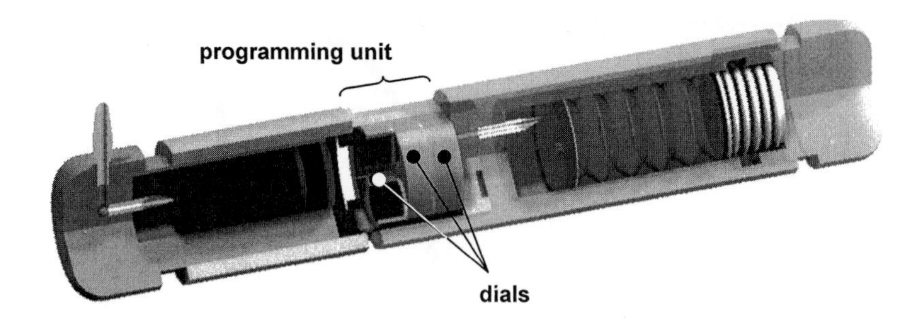

Figure 7. Programmable hydrogel driven pumps including a programming unit based on three dials.

Acknowledgment

Financial support has been provided by the Deutsche Forschungsgemeinschaft (SFB 287 "Reactive Polymers"). The authors would like to thank D. Kuckling and K. Kretschmer for the synthesis of PNIPAAm.

[1] A.E. English, E.R. Edelman, T. Tanaka, in: T. Tanaka (ed.): Experimental methods in polymer science: modern methods in polymer research & technology. Academic Press, New York (2000), 547.
[2] K.-F. Arndt, D. Kuckling, A. Richter, *Polym. Adv. Technol.* **11** (2000) 8-12, 496.
[3] D. Kuckling, A. Richter, K.-F. Arndt, *Macromol. Mater. Eng.* **288** (2003) 2, 144.
[4] D.J. Beebe, J.S. Moore, J.M. Bauer, Q. Yu, R. H. Liu, C. Devadoss, B.-H. Jo, *Nature* **404** (2000), 588.
[5] A. Baldi, Y. Gu, P. E. Loftness, R. A. Siegel, and B. Ziaie, Proc. 15th. Int. IEEE Conference on Microelectromechanical Systems 2002, Las Vegas, NV, 105.
[6] S. Mutlu, Cong Yu, F. Svec, C.H. Mastrangelo, J.M.J. Frechet, Y.B. Gianchandani, *Transducers* **1** (2003), 802.
[7] A. Richter, D. Kuckling, K.-F. Arndt, T. Gehring, S. Howitz, *J. Microelectromech. Syst.* **12** (2003) 5, 748.
[8] R. Bashir, J.Z. Hilt, O. Elibol, A. Gupta, N.A.Peppas, *Appl. Phys. Lett.* **81** (2002), 3091.
[9] F. Theeuwes: Elementary Osmotic Pump. *J. Pharm. Sci.* **64** (1975), 1987.

Macromol. Symp. **2004**, *210*, 385-391

Mixed Conducting Polymers on Ion-Conducting Sensor Glasses – Charge Transfer in the System Polymer/Glass

Heimo Jahn, Heiner Kaden, Monika Berthold*

Kurt-Schwabe Institute for Measuring and Sensor Technology Meinsberg, Fabrikstraße 69, 04720 Ziegra-Knobelsdorf, Germany
E-mail: jahn@ksi-meinsberg.de

Summary: Mixed ion/electron conducting polymer layers based on polypyrrole have been used as internal reference electrodes in all-solid-state pH glass electrodes. The effect of the nature and composition of the polymer used and of the deposition technique applied on the performance of the resulting sensor has been studied. For this purpose, crucial sensor properties, e.g. parameters of the calibration function, response behaviour and complex impedance, have been determined experimentally at room temperature. The results show that several properties studied remained nearly uninfluenced by changes of the polymer composition. The zero potential point of the calibration line was found to be the most sensitive parameter. Principally, almost all mixed conducting polymers used seems to result in a stable charge transfer in the system polymer/glass.

Keywords: charge transfer; composites; conducting polymers; pH sensor; polymer/glass interface; polypyrroles

Introduction

The interface between a polymer film and a substrate controls the adhesion and the electrical behaviour of a variety of polymeric systems including coatings of polymers on glass fibre composites, in photovoltaic devices and coatings of conducting polymers on glasses. Intrinsically conducting polymers, amongst them polypyrrole (PPy), are interesting sensor materials, especially due to their mixed ionic and electronic conductivity. In previous works [1-6] we could show that mixed conducting polymer composites consisting of polypyrrole doped with $FeCl_3$ ($PPy[FeCl_3]$) and Nafion® are promising candidates for a new kind of internal reference systems in all-solid-state pH-sensitive glass membrane sensors. In this work the influence of the nature and composition of further mixed conducting polymers used in such reference systems on the performance of the resulting sensor has been studied.

 DOI: 10.1002/masy.200450643

Experimental

The set-up of the all-solid-state pH-sensitive glass sensors studied is given in Fig. 1 schematically. The solid-state internal reference system of these sensors is based on a mixed conducting polymeric interlayer between the ion-sensitive glass and a metal wire replacing the conventionally used liquid-based reference system: Ag/AgCl/buffer solution.

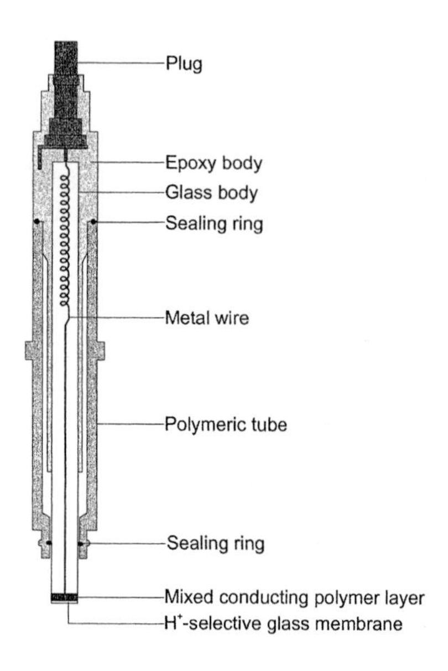

Figure 1. Schematic diagram of an all-solid-state pH glass sensor using a mixed conducting polymer layer as internal reference electrode.

Preparation

The preparation of the polymeric layer was performed by deposition of PPy on the inner side of the ion-sensitive glass membrane applying two different techniques:

(1) deposition from a suspension of PPy dispersed in a Nafion® solution [1, 6],

(2) direct precipitation from non-aqueous solutions of pyrrole employing an oxidizing agent.

In the case of the first way of preparation, chemically or electrochemically doped PPy was used. The doping agents applied were $FeCl_3$, $Fe(NO_3)_3$, $Na_2S_2O_8$ (chemical doping) or $NaClO_4$, polystyrenesulphonic acid sodium salt (NaPSS), Nafion® (electrochemical doping). Nafion in its protonated form (Nafion® 117) was utilized. In the case of the second way, $FeCl_3 \cdot 6H_2O$ or $H_3PMo_{12}O_{40} \cdot xH_2O$ were the oxidizing agents employed. The resulting polymer layers consist either of a PPy-Nafion® composite (technique 1) or of a composite containing PPy and the by-products of the reaction between pyrrole and the oxidizing agent (technique 2).

Measurements

Crucial characteristics of the resulting sensors, e.g. parameters of the calibration function (and their long-term stability), response behaviour and complex impedance, have been determined experimentally at room temperature (23 ± 3 °C). The measurements were carried out in a Faraday cage using an external Ag/AgCl reference electrode with saturated aqueous KCl filling (SSE). All potentiometric investigations were performed in the range $1.7 \leq pH \leq 9.2$. The response behaviour was determined by the method of quick change of buffer solutions[7]. Further, more detailed descriptions of the procedures of preparation and measurement have been given elsewhere [1, 5, 6].

Results and Discussion

The long-time behaviour of a critical parameter of the calibration function, the zero potential point (pH_0), is presented in dependence of quantitative (Fig. 2) and qualitative (Fig. 3) changes in the composition of PPy-Nafion® composites used and in the technique of polymer deposition applied (Fig. 4). As can be seen in the given figures both the change of the PPy/Nafion® ratio as well as the use of different dopants of the PPy applied in the composite are reflected by the values of this parameter. The minima observed during the first two weeks after preparation of the solid contacts indicates strong initial interactions between the components of the system PPy-Nafion®/glass for all PPy[$FeCl_3$]/Nafion ratios studied (Fig. 2). However, the long-term stability of pH_0 after this "conditioning" period was high. The most stable values were obtained for polymers containing 50 wt.% PPy. The long-term stability of pH_0 for PPy-Nafion® layers with generally 50 wt.% PPy but with varying dopants (Fig. 3) has been found to be high over very long

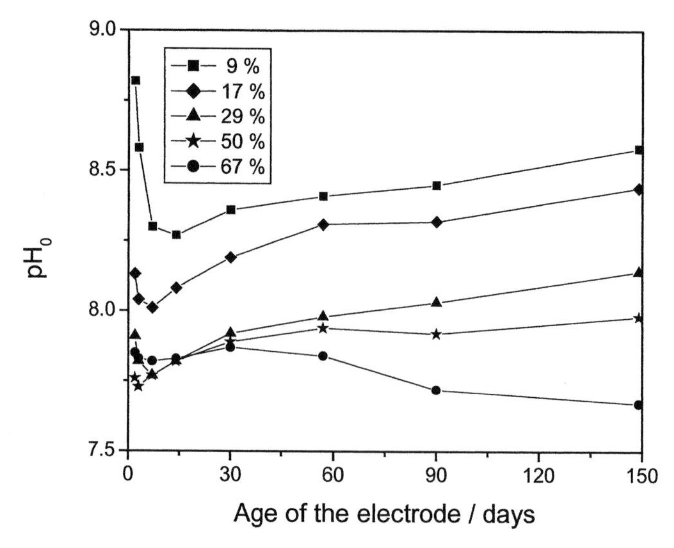

Figure 2. The influence of the PPy content (wt.%) of the mixed conducting PPy[$FeCl_3$]-Nafion® layer on the long-time behaviour of the zero potential point (pH_0) of the pH sensor at 25 °C.

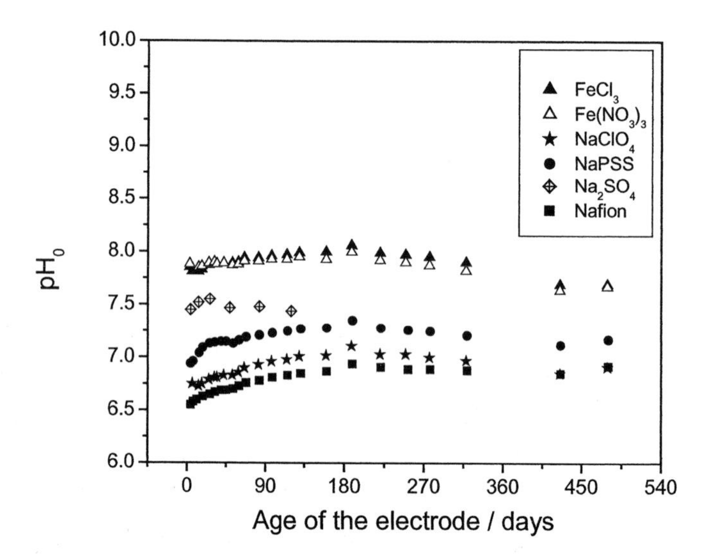

Figure 3. The effect of the dopant of PPy on the long-term stability of pH_0 for electrodes using PPy-Nafion® composites (50/50 w/w) as polymer layer.

periods (up to 16 month) in all cases studied. The direct precipitation of PPy from non-aqueous solutions of pyrrole (technique 2) employing the oxidizing agents $FeCl_3{\cdot}6H_2O$ and $H_3PMo_{12}O_{40}{\cdot}xH_2O$, respectively, resulted in solid contacts showing a relatively poor long-time behaviour compared to this based on PPy-Nafion® composites (Fig. 4). The differences in the absolute values of pH_0 for the various polymer layers are the consequence of the different doping agents used. In contrast to pH_0, the parameters slope and linearity of the calibration line have found to be nearly uninfluenced by the investigated changes in composition and technique of preparation. In the case of sensors based on directly precipitated PPy layers, however, an enlarged scatter of the values of these parameters was observed.

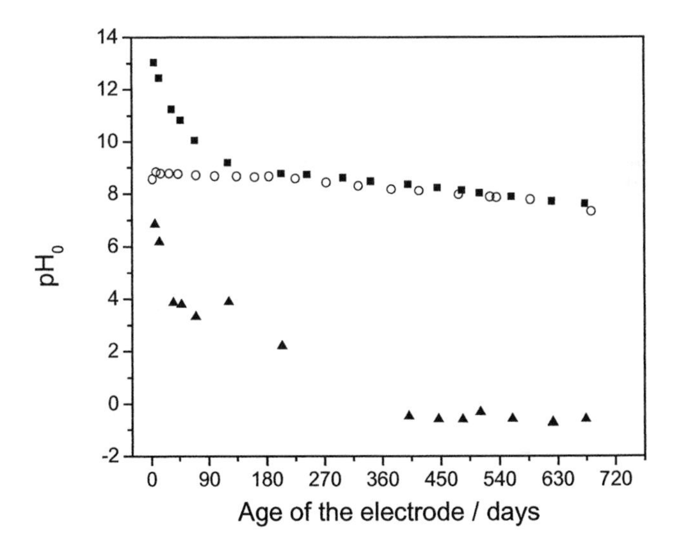

Figure 4. Comparison of the long-time behaviour of pH_0 for electrodes containing differently prepared polymer layers: PPy directly precipitated from non-aqueous solutions of $FeCl_3{\cdot}6H_2O$ (triangles) or $H_3PMo_{12}O_{40}{\cdot}xH_2O$ (squares), PPy[$FeCl_3$] applied as PPy-Nafion® composite containing 50 wt.% PPy (circles).

A typical example for the response *vs.* time dependence during pH changes of electrodes employing a PPy-Nafion® composite (50/50 w/w) based on PPy which was chemically doped by ferric nitrate (PPy[$Fe(NO_3)_3$] is given in Fig. 5. The response times for 95% of equilibration (t_{95})

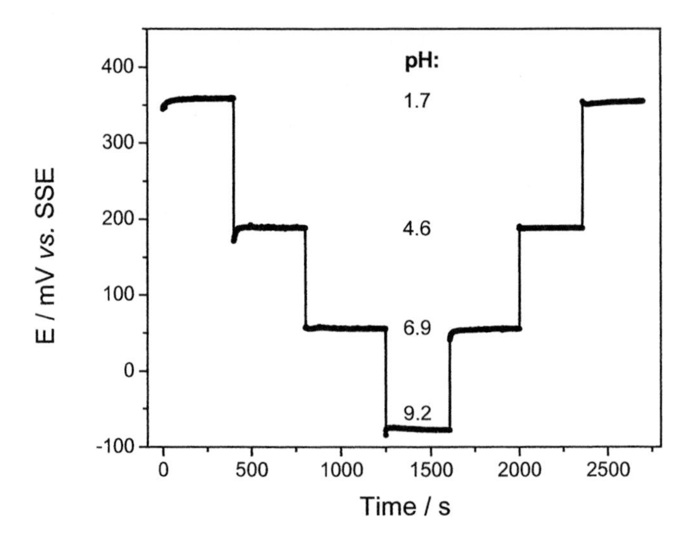

Figure 5. Response behaviour of a sensor with PPy[Fe(NO_3)$_3$]-Nafion®/ Pt as internal reference system during the change of pH of the test solution at 25 °C.

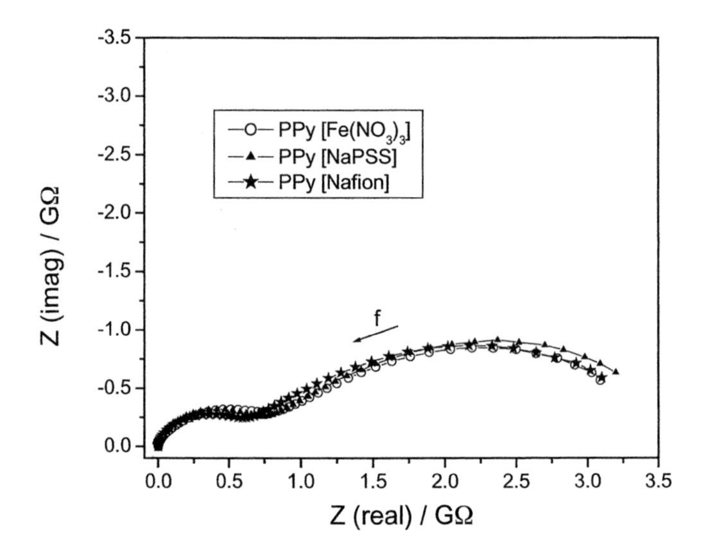

Figure 6. Nyquist plot of complex impedance of the system SSE, 1 M KCl/ **pH glass/ PPy-Nafion®**/ Pt containing differently doped PPy at 25 °C. Z (imag), Z (real): imaginary and real part of complex impedance. In direction of the arrow: higher frequencies f.

were estimated to be smaller than 10 s for all pH steps studied. This result is in good agreement with that found earlier ($t_{95} < 9$ s)[6] for a PPy[$FeCl_3$]-based sensor of the same type.

The complex impedance curves of three electrodes under investigation (Fig. 6) show two semicircles in the frequency range 0.001 Hz to 100 kHz. The impedance of the glass/polymer interface (included in right semicircle according to SANDIFER et al.[8]) is nearly uninfluenced by the dopant used.

Conclusions

It has been shown that the zero potential point was the parameter which was most sensitive to changes in the composition of the mixed conducting polymer layer. Further sensor characteristics like the slope and linearity of the calibration line, the response time as well as the impedance spectra were found to be nearly uninfluenced by the dopant applied. The long-term stability of the sensor response of the electrodes containing PPy-Nafion® composites was much higher than that of the electrodes containing directly precipitated PPy without admixture of Nafion®. Principally, almost all mixed conducting polymers used in this work exhibited a good performance as internal reference electrodes of pH selective glass sensors. This result indicates a stable charge transfer in the studied systems: Pt/mixed conducting polymer/ion-conducting glass.

[1] Ger. 10018750 (2000), Kurt-Schwabe-Institut für Mess- und Sensortechnik e.V. Meinsberg, invs.: H. Kaden, H. Jahn, M. Berthold, K.-H. Lubert; Chem. Abstr. **2001**, *134*, 109879e.
[2] H. Jahn, M. Berthold, H. Kaden, *Macromol. Symp.* **2001**, *164*, 181.
[3] H. Kaden, H. Jahn, M. Berthold, K. Jüttner, K.-M. Mangold, S. Schäfer, *Chem. Eng. Technol.* **2001**, *24*, 1120.
[4] H. Jahn, M. Berthold, H. Kaden, in: „Dresdner Beiträge zur Sensorik“, Vol. 16., J.P. Baselt, G. Gerlach, Eds., w.e.b. Universitätsverlag, Dresden 2002, pp.174-177.
[5] H. Kaden, H. Jahn, M. Berthold, *Solid State Ionics* (in press).
[6] H. Jahn, H. Kaden, *Microchim. Acta* (submitted for publication).
[7] H. Galster, „pH-Messung“, VCH Verlagsgesellschaft, Weinheim 1990, p.152.
[8] J.R. Sandifer, R.B. Buck, *Electroanal. Chem. Interfac. Electrochem.* **1974**, *56*, 385.

Macromol. Symp. **2004**, *210*, 393-401

Electrostimulated Shift of the Precipitation Temperature of Aqueous Polyzwitterionic Solutions

George Georgiev,[1] Anna Tzoneva,[1] Lyudmil Lyutov,[1] Stefko Iliev,[1] Irena Kamenova,[1] Ventseslava Georgieva,[1] Elena Kamenska,[1] Andreas Bund[2]*

[1]Faculty of Chemistry, University of Sofia, 1, J. Bourchier Avenue, 1164 Sofia, Bulgaria
[2]Institute of Physical Chemistry and Electrochemistry, Dresden University of Technology, D-01062 Dresden, Germany
E-mail: georgs@chem.uni-sofia.bg

Summary: The precipitation temperature (T_{pr}) value of aqueous poly(dimethylamino-ethoxyacryloyl-propylsulphonate) (PDMAPS) solutions decreases with the rise of electric field intensity both in the absence and in the presence of a low molecular salt. This electrostimulated T_{pr} shift is explained qualitatively by means of the model taking into account both the dominating intermacromolecular dipole-dipole interaction and the dipole cluster formation.

Keywords: phase separation; “sol-gel” transition; polyzwitterion

Introduction

The existence of both positive and negative electric charges in some or all the monomer units is a characteristic feature of polyzwitterions (PZ). The specific conformational transitions of PZ macromolecules and their aqueous solution properties (such as antipolyelectrolyte effect [1-3], electroviscosity [4, 5], phase separation [6-8]) are related to the strong dipole-dipole interactions [6-9]. The results obtained [4, 5] stimulated the investigation of the effect of the electric field on the reversible and thermally stimulated “sol-gel” transition in PZ aqueous solutions. The first results of this study are reported here. They are discussed in the light of the recently proposed qualitative model accounting the influence of the vector characteristic of the dipole moment on the dipole-dipole interaction, as a result of which the specific clusters from the antiparallel oriented dipoles (named “zip”-clusters) are formed.

Experimental

Poly(dimethylamino-ethoxyacryloyl-propylsulphonate) (PDMAPS, molecular weight $\overline{M_n} = 7.6 \times 10^4$ *g/mol*, molecular weight distribution $\overline{M_w}/\overline{M_n} = 1.3$) was obtained according to a method reported in [2]. The optical transmittance (*Tr*) in electric field

 DOI: 10.1002/masy.200450644

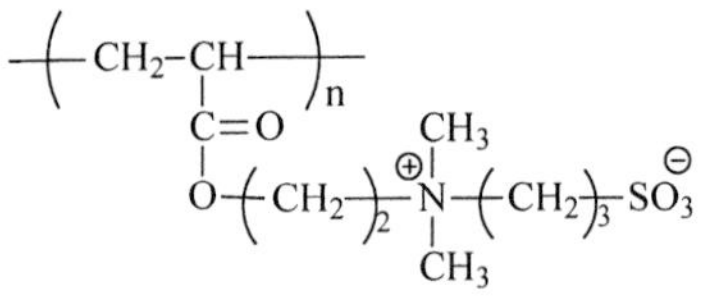

$$-\!\left(CH_2-\underset{\underset{\underset{O-\left(CH_2\right)_2-\overset{\oplus}{N}(CH_3)_2-\left(CH_2\right)_3-SO_3^{\ominus}}{|}}{C=O}}{CH}\right)_n-$$

(PDMAPS)

of the aqueous PDMAPS solutions was determined by means of a Specol apparatus (Carl Zeiss-Jena) in a glass thermostatic cell with effective sample thickness of 0.5 *cm*. The optical path through the medium was 1.8 *cm*. Tantalum electrodes were placed on the two opposite walls of the cell, thus allowing the control of the changes in the electric field acting on the PZ solution. The distance between the electrodes was 1.1 *cm*. In order to provide a homogeneous electric field in the sample, the electrodes were longer than the cell by 0.3 *cm* at both sides. The electric field applied was perpendicular to the optical path. The precipitation temperature (T_{pr}) of the PDMAPS aqueous solutions used was determined from the maximum of the derivative $d(Tr)/d\,T = 0$, where T is the temperature. The temperature dependence of the shear elasticity of the PDMAPS solutions was monitored with a piezoelectric thickness-shear mode (TSM) resonator. The corresponding measuring technique has been fully described earlier [10-12]. In short, the electrical admittance of a piezoelectric quartz crystal was monitored with a network analyzer (R3753BH, Advantest, Tokyo, Japan). Due to the piezoelectric effect the quartz transfers mechanical impedances into electrical ones.

Results and discussion

Electric and temperature effect on the "sol-gel" transition in PDMAPS aqueous solutions

The temperature dependencies of the optical transmittance of PDMAPS aqueous solutions at different PZ concentrations are shown in Figure 1. In the same temperature interval (24-45°C) the sharp transmittance increase is accompanied by a distinct increase of the storage shear modulus as indicated by the TSM resonator (Figure 2). Figure 2 shows only relative values in arbitrary units as no calibration of the TSM resonator was performed. The curves in Figure 1 are used as an origin to account the electric field effect on the "sol-gel" transition in the solutions. The obtained T_{pr} values are given in Table 1. As expected, they increase with the rise in PZ concentration. The shift to the right of the curves depicted in Figure 1 and the solution transmittance decrease with the rise in PDMAPS concentration are corollaries of the increased intermacromolecular interaction. According to the above mentioned model [6, 8, 9], the dipole-dipole interaction between monomer units is predominant, resulting in a dipole-dipole cluster formation according to the "zip"-mechanism ("zip"-clusters) [9].

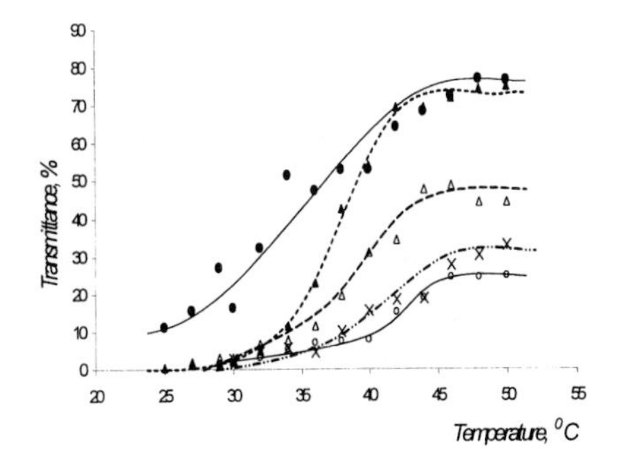

Figure 1. Temperature dependencies of the optical transmittance of PDMAPS aqueous solutions at different PZ concentrations: (●) 0.1, (▲) 0.2, (Δ) 0.3, (x) 0.4, and (○) 0.5 *wt. %*

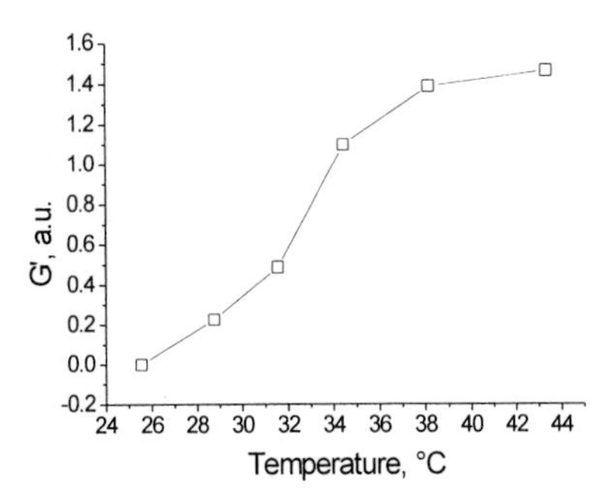

Figure 2. Temperature dependence of the storage shear modulus of 1.0 wt.% PDMAPS aqueous solution.

At higher temperatures, the number, length and structural perfection of these clusters decrease. The intermacromolecular "zip"-clusters form a PZ physical network and its destruction with the rise in temperature is the reason for the observed "sol-gel" transition, characterized by the T_{pr} value (Figure 1).

Table 1. Dependence of T_{pr} and transmittance reached at 50 °C of PDMAPS aqueous solutions on the PZ concentration.

PZ concentration (*wt. %*)	**T_{pr} (*^{0}C*)**	**Transmittance at 50 °C (%)**
0.1	33.8	76.6
0.2	38.2	73.2
0.3	40.8	48.8
0.4	42.0	32.8
0.5	43.8	24.8

With the rise in PZ concentration, the number of junction points in the physical network also increases. This increase of the network density is a reason for both the T_{pr} increase and the decrease of the transmittance reached at 50 °C of PDMAPS aqueous solutions at higher PZ concentrations (Table 1).

In the presence of electric field, the above mentioned dependencies are shifted to the left (Figure 3 compared to Figure 1). This response to the electric field is in an opposite direction to the increasing PZ concentration effect (Figure 1). The curves depicted in Figure 3 suggest that the higher the *E* value, the more significant is the displacement against the references obtained at $E = 0$. This is a general trend, observed also with the other investigated PDMAPS concentrations (0.1, 0.2, 0.3, and 0.4 *wt. %*). In Table 2, the T_{pr} values determined from these dependencies are included. They are decreasing with the rise in *E*, suggesting that the applied electric field, similarly to the increasing temperature, destroys the junction points ("zip"-clusters) of the physical network formed at lower temperatures. This is a consequence of dipole orientation along the field, which is impossible if the "zip"-clusters with random orientation are not destroyed.

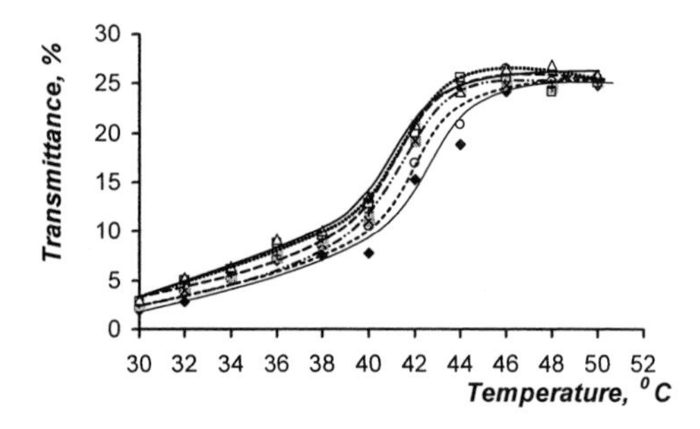

Figure 3. Temperature dependencies of the optical transmittance of 0.5 *wt. %* PDMAPS aqueous solutions at various intensities of the electric field: (♦) 0.0, (○) 1.0, (x) 2.0, (+) 2.4, (□) 3.0, and (Δ) 4.0 *kV/cm*.

Table 2. T_{pr} values of PDMAPS aqueous solutions as dependent of the electric field intensity (*E*) at different PDMAPS concentrations.

PDMAPS (*wt. %*)	**T_{pr} (°C)**					
	0.0 (*kV/cm*)	**1.0 (*kV/cm*)**	**2.0 (*kV/cm*)**	**2.4 (*kV/cm*)**	**3.0 (*kV/cm*)**	**4.0 (*kV/cm*)**
0.1	33.8	32.4	31.8	30.4	30.2	29.5
0.2	38.2	37.6	37.2	36.6	36.3	35.8
0.3	40.8	38.9	38.6	38.4	38.3	38.0
0.4	42.0	40.0	39.8	39.6	39.2	38.9
0.5	43.8	42.8	42.1	41.8	41.6	41.4

Comparison of the dependencies depicted in Figures 1 and 3 suggests that the concentration effect on the "sol – gel" transition is stronger than that of *E* in the investigated PDMAPS concentration (0.1 – 0.5 *wt. %*) and *E* ranges (0 – 4 *kV/cm*). This is clearly seen from the data included in Table 2; the variation along the rows is smaller in absolute value than that in the columns.

It is seen that the curves in Figures 4 and 5 are shifted to the left at higher NaCl concentrations, *i.e.*, the "sol – gel" transition is realized at lower temperatures. Comparison of the dependencies in Figure 4 with those in Figure 1 suggests that the ionic strength influence

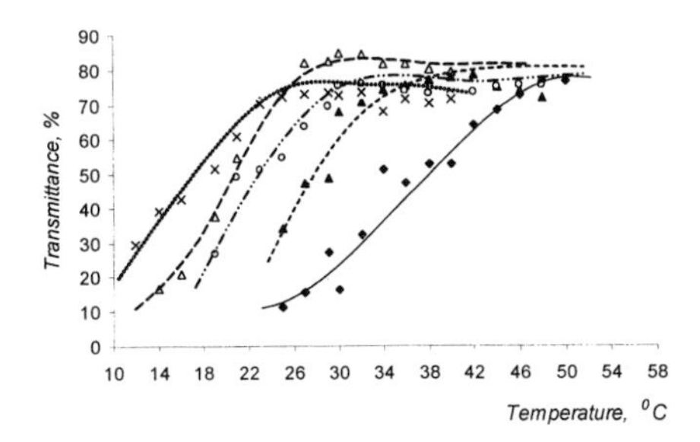

Figure 4. Temperature dependencies of optical transmittance of the 0.1 *wt. %* PDMAPS aqueous solution in the absence of electric field at different NaCl concentrations: (♦) 0.000, (▲) 0.010, (○) 0.050, (Δ) 0.075, and (x) 0.100 *wt %*

is in a direction opposite to that of the PZ concentration, both effects being however similar in absolute value. The temperature dependencies registered in the presence of electric field (E = 1.0, 2.0, 2.4, 3.0, and 4.0 *kV/cm*) are similar to those of Figure 4. Figure 5 shows the temperature dependencies obtained at the highest value of E = 4 *kV/cm*. They all confirm the conclusions drawn from the comparison of Figures 1 and 4.

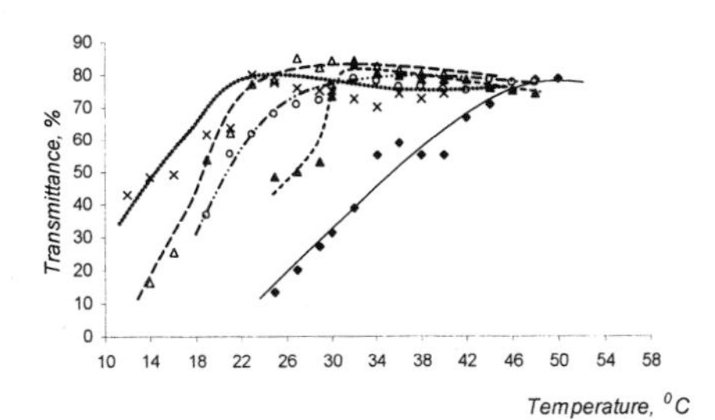

Figure 5. Temperature dependencies of the optical transmittance of the 0.1 *wt. %* PDMAPS aqueous solution in the presence of electric field (E = 4 *kV/cm*) at various NaCl concentrations: (♦) 0.000, (▲) 0.010, (○) 0.050, (Δ) 0.075, and (x) 0.100 *wt %*

Comparison of the results presented in Figures 4 and 5 with those of Figure 3 also leads to interesting conclusions. It is clear that the effect of ionic strength and E are unidirectional, but ionic strength plays a far more significant role, at least at the studied variation ranges. This conclusion is confirmed also by the observation that the change in E does not affect the maximal value of the PDMAPS aqueous solution transmittance (Figure 3) whereas changes in PDMAPS concentration influence considerably this transmittance (Figure 1).

Combined effect of temperature, electric field, and low molecular salt on the "sol – gel" transition in PDMAPS aqueous solutions

The characteristic antipolyelectrolyte effect [1-3] of PZ aqueous solutions is related to the influence of low molecular salts on the dipole-dipole interactions between the monomer units in the PZ macromolecule. For this reason, the analysis of the influence of ionic strength on the electric and thermal effect discussed in the previous section is important.

Figure 6 shows the curves for 0.1 *wt.%* NaCl and 0.1 *wt. %* PDMAPS at different E values. It is clearly seen, that with the rise in *E* values, the temperature dependencies are shifted to the left, *i.e.,* the "sol-gel" transition takes place at a lower temperature. A similar weak displacement of the curves is established also at the other NaCl concentrations. Table 3

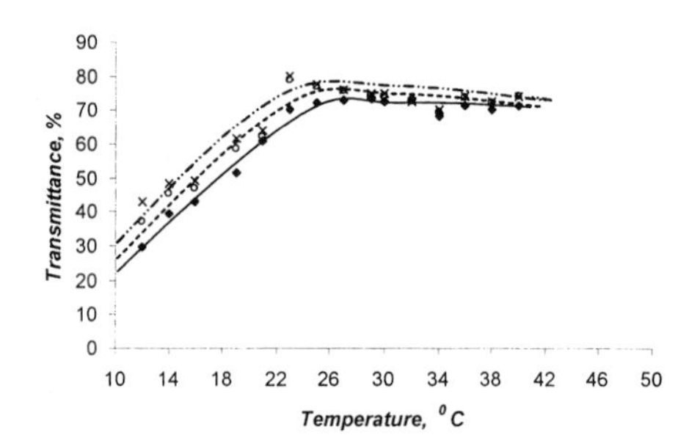

Figure 6. Temperature dependencies of the optical transmittance of the 0.1 *wt. %* PDMAPS aqueous solution in the presence of 0.10 *wt. %* NaCl and electric field with intensity: (♦) 0.0, (○) 2.4, and (x) 4.0 *kV/cm*

includes T_{pr} values determined from the curves of Figures 4 – 6 for the 0.1 *wt. %* PDMAPS aqueous solution. The changes in these numerical characteristics of the investigated "sol – gel" transition are in full agreement with the conclusions drawn from the graphical dependencies: unidirectional effects of ionic strength and *E* on the transitions, that of ionic strength being more significant.

The observed salt effect on the "sol-gel" transition should be considered as a new demonstration of the antipolyelectrolyte effect. Actually, the shielding action of the low molecular ions on the dipole-dipole interaction results in the destruction of intermacromolecular "zip"-clusters to a different extent and the transition is realized at lower temperatures; the curves in Figures 4 and 5 are shifted to the left, and the T_{pr} values in all columns of Table 3 decrease.

Table 3. T_{pr} values of the 0.1 *wt. %* PDMAPS aqueous solution as dependent on *E* and NaCl concentration.

NaCl concentration, (*wt. %*)	**T_{pr} *(°C)***					
	0.0 (*kV/cm*)	**1.0 (*kV/cm*)**	**2.0 (*kV/cm*)**	**2.4 (*kV/cm*)**	**3.0 (*kV/cm*)**	**4.0 (*kV/cm*)**
0.000	33.8	32.4	31.8	30.4	30.2	29.5
0.010	29.5	29.7	29.8	29.6	29.5	29.5
0.050	27.5	25.2	24.8	23.8	24.6	24.4
0.075	20.8	20.4	19.8	19.1	18.8	18.6
0.100	19.8	20.2	19.4	18.9	18.0	16.4

The application of an external electric field in the presence of a low molecular salt has a bilateral effect on the "sol-gel" transition in PDMAPS aqueous solutions. On the one hand, the above mentioned orientational effect of the field should cause additional destruction of intermacromolecular "zip"-clusters and displacement of the temperature dependencies to the left (decrease in T_{pr}), as shown in Figure 6 and Table 3. On the other hand, the higher mobility of low molecular ions compared to that of PZ monomer units results in a decrease of their concentration in the PZ globules, their shielding effect decreases; hence the curve displacement is to the right. The influence of this mobility difference on the electroviscous properties of PZ aqueous solutions has been discussed in more details in previous our works [4, 5]. The observed opposition of the consequences of dipole orientation and different mobilities of low molecular ions and macromolecular zwitterions, hence their compensation, is one of the likely explanations of the weak influence of *E* on the "sol-gel" transition. However, the fact that in the presence of NaCl (when there are no reasons for curve displacement to the right) the influence of *E* is again quite weak (first row in Table 3) shows that this explanation is at least fragmentary. One should assume that the effect of the dipole-dipole orientation on the dipole-dipole "zip"-cluster destruction is weak, at least within the limits of the *E* variations studied.

Unlike the dependencies shown in Figure 1, those in Figures 4 – 6 do not reveal substantial changes in the reached value of the optical transmittance. This means that neither the ionic strength (Figures 4 and 5) nor *E* (Figure 6) affect this value as the PZ concentration does. The observed weak (by no more than 10 %) decrease in optical transmittance at higher temperatures, ionic strengths (Figures 4 and 5), and *E* values (Figures 6), *i.e.*, when the dipole–dipole clusters (including the intermacromolecular ones) should be destroyed to a great extent, deserves attention. The explanation of this interesting result resides in the transformation of intramacromolecular dipole-dipole contacts into intermacromolecular ones under the conditions indicated. During the process of destruction of dipole–dipole "zip"-clusters, the size and shape of the PZ macromolecules change; the size increases because some intramacromolecular clusters are also destroyed. This means that the probability of intermacromolecular dipole-dipole contacts (not necessarily clusters) is increased. This is actually an increase in junction points and network density under these extremely conditions for "zip"-cluster formation.

Conclusions

It is established that the electric field intensity is a factor influencing the "sol-gel" transition of aqueous PDMAPS solutions in the absence (Figure 3) and in the presence (Figure 6) of low molecular salt. The electric field effect is lower than those of ionic strength (Figures 4 and 5) and PZ concentration (Figure 1), at least within the limits of variation ranges of these quantities used in the present study. It is essential that the effects of *E* and ionic strength are in the same direction and are opposite to that of PZ concentration. The proposed explanation of this observation is based on the recently developed qualitative model of the structural organization and transformations of PZ macromolecules [6, 8, 9]. According to this model, the higher the PZ concentration, the higher the intermacromolecular "zip"-cluster concentration and the T_{pr} value. The clusters act as junction points in the physical network formed [9]. With the ionic strength increase, the shielding effect of low molecular ions facilitates the destruction of these intermacromolecular dipole-dipole clusters. The density of the physical network decreases leading to a decrease in the T_{pr} value of the PDMAPS aqueous solution.. Nevertheless, the results of the present work add a new, easily realized and controlled factor to the classical approaches of affecting the "sol-gel" transition of PZ aqueous solutions (*e.g.*, ionic strength, PZ concentration), namely the electric field intensity.

Acknowledgement

The financial support of this work was provided by a grant from the Bulgarian National Scientific Foundation (Grant № X-1307). The authors would like to thank of Dr. Velin Spassov for his helpful discussions.

[1] J. C. Galin, in: "*Polyzwitterions, the Polymeric Materials Encyclopedia*", J. C. Salomone, Ed., CRS Press, Boca Raton, **1996**, *9*, 7189.
[2] V. M. Monroy Soto, J. C. Galin, *Polymer* **1984**, *25*, 254.
[3] T. A. Wielema, J. B. F. N. Engberts, *Eur. Polym. J.* **1990**, *26 (6)*, 639.
[4] G. S. Georgiev, A. A. Tzoneva, V. A. Spassov, *e-Polymers* **2002**, 037 (http://www.e-polymers.org).
[5] G. S. Georgiev, A. A. Tzoneva, V. A. Spassov, *J. Univ. Chem. Technology and Metallurgy* **2002**, *XXXVII (1)*, 109.
[6] J. L. Bredas, R. R. Chance, R. Siley, *Macromolecules* **1988**, *21*, 1633.
[7] D. N.Schulz, D. G. Peiffer, P. K. Agarwal, J. Larabee, J. J. Kaladas, L. Soni, B. Handwerker, R. T. Garner, *Polymer* **1986**, *27*, 1734.
[8] G. S. Georgiev 1st Int. Symp. Reactive Polymers, Dresden, Abstracts, 2000, P7.
[9] G. S. Georgiev, Z. P. Mincheva, V. T. Georgieva, *Macromol. Symp.* **2001**, *164*, 301.
[10] G. Schwitzgebel, Ber. Bunsenges, *Phys. Chem.* **1997**, *101*, 1960.
[11] A. Bund, G. Schwitzgebel, *Anal. Chem.* **1998**, *70*, 2584.
[12] A. Bund, H. Hmiel, G. Schwitzgebel, *Phys. Chem. Chem. Phys.* **1999**, *1*, 3938.

Application of Sensitive Hydrogels in Chemical and pH Sensors

Gerald Gerlach,[*1] *Margarita Guenther,*[1] *Gunnar Suchaneck,*[1] *Joerg Sorber,*[1] *Karl-Friedrich Arndt,*[2] *Andreas Richter*[2]

[1] Dresden University of Technology, Institute for Solid State Electronics, Mommsenstr. 13, 01062 Dresden, Germany
[2] Dresden University of Technology, Institute of Physical Chemistry and Electrochemistry, Mommsenstr. 13, D-01062 Dresden, Germany

Summary: In the present work, pH-sensitive poly(vinyl alcohol)/poly(acrylic acid) (PVA/PAA) blends as well as hydrogels based on poly(N-isopropylacrylamide) (PNIPAAm), which are sensitive to organic solvent concentration in aqueous solutions, were used in silicon micromachined sensors. A sensitivity of approximately 15 mV/pH was obtained for a pH sensor with a 50 μm thick PVA/PAA hydrogel layer in a pH range above the acid exponent of acrylic acid (pK_a=4.7). The output voltage versus pH-value characteristics and the long-term signal stability of hydrogel-based sensors were investigated and the measurement conditions necessary for high signal reproducibility were determined. The influence of the preparation conditions of the hydrogel films on the sensitivity and response time of the chemical and pH sensors is discussed.

Keywords: hydrogels; pH-sensitive; sensors; solvent composition; swelling;

Introduction

Hydrogels are used for a variety of applications where environmental sensitivity is needed. Hydrogels consist of a crosslinked polymer network that can contain large amounts of a solvent. By controlling the functional groups along hydrogel's backbone chains, the amount of solvent uptake, and thus the dimensions of the gel, can be affected by environmental stimuli, such as solvent composition, pH value, temperature, ionic strength, and electric field.

The swelling ability of pH-sensitive hydrogels depends on the functional acidic or basic groups on the polymer backbone. Due to the dissociation of these groups and the influx of counterions, the concentration of ions in the hydrogel is higher than in the surrounding solution. This causes a difference in osmotic pressure and results in a solution flux into the hydrogel and, consequently, a swelling. The interaction and repulsion of charges along the polymer chain also lead to an

 DOI: 10.1002/masy.200450645

increase in swelling. Hydrogels with acidic groups will be ionised at higher pH, thus an increase in pH leads to an increase in swelling. When basic groups are used, the hydrogel will swell by lowering the pH. The degree of swelling depends on the amount of groups, crosslinking density, and the hydrophilicity of the monomers.[1-4] Hydrogels are capable to reversibly convert chemical into mechanical energy and are used as active sensing components in micro-electro-mechanical systems (MEMS).[5-7]

In the present work we propose an approach for pH and chemical sensors based on such swellable hydrogels, using a mechano-electrical transducer with piezo-resistive elements. The characteristics and the long-term signal stability of such sensors are investigated.

Experimental

The following hydrogel systems were used in this work:

1. *Poly(vinyl alcohol)/poly(acrylic acid) (PVA/PAA) blends*: The swelling degree of these hydrogels changes steeply with the change of the pH value of the measuring species: it is at minimum in acids and takes its maximum in bases.[1, 2] PVA and PAA polymers obtained from Aldrich Chemical Co. were dissolved separately in distilled water under stirring at 80°C (PVA 15 wt% and PAA 7.5 wt%). For hydrogel formation, the solutions are then mixed in such a manner that 80 wt% were PVA and 20 wt% PAA. This mixture is stirred for 1 h at 60°C to manufacture a homogeneous solution. Thin films of PVA/PAA blends were deposited by spin-coating onto the silicon wafers covered with a 550 nm thick PECVD silicon oxide film which served as convenient substrates. Finally, the dried hydrogel films were isothermally annealed in an oven at 130°C for 20 min.
2. *Hydrogels on basis of the N-Isopropylacrylamides (NIPAAm)*: Hydrogels on PolyNIPAAm (PNIPAAm) basis react sensitively on changes of the concentration of organic components in aqueous solutions. [3] The crosslinked PNIPAAm-hydrogels were prepared by free radical polymerization of NIPAAm in water with *N,N'*-methylene-bisacrylamide (BIS; BIS content: 4 mol%) as the crosslinking agent. A solution of NIPAAm, BIS (overall monomer concentration: 0.53 mol/l) and potassium peroxodisulfate as the initiator ($3x10^{-3}$ mol/mol monomer) were cooled to 0°C and purged with nitrogen for 15 min. The *N,N,N',N'*-tetramethylethylenediamine as the accelerator

($3x10^{-3}$ mol/mol monomer) was added and the solution was immediately transferred into a Petri dish to get hydrogel foils. After 17 h reaction time at 20°C, the gels were separated and washed with water for one week.

For the design of chemical and pH sensors a sensor chip with a distortable thin silicon membrane is used (Fig. 1). The chip surface contains the electronic components: piezoresistors metallization and bonding islands. The hydrogel itself is brought into a cavity at the backside of the silicon chip and closed with a cover. This cavity on the backside of the chip is wet etched, a silicon nitride mask was used as an etch resist. Therefore, only the backside of the chip comes in contact with the measuring species, whereas the frontside with the electronic components is protected from it. In this case, the long-term stability of the sensor is solely determined by the stability of the hydrogel characteristics.

Figure 1 illustrates the operational principle of hydrogel-based sensors. The swelling or shrinking processes of the hydrogel are monitored by a corresponding change in piezoresistance of an integrated Wheatstone bridge inside a rectangular silicon membrane. The membrane deflection causes a stress state change inside the membrane and therefore a change of the resistivity of the resistors. This changes proportionally the sensor's output voltage.[8]

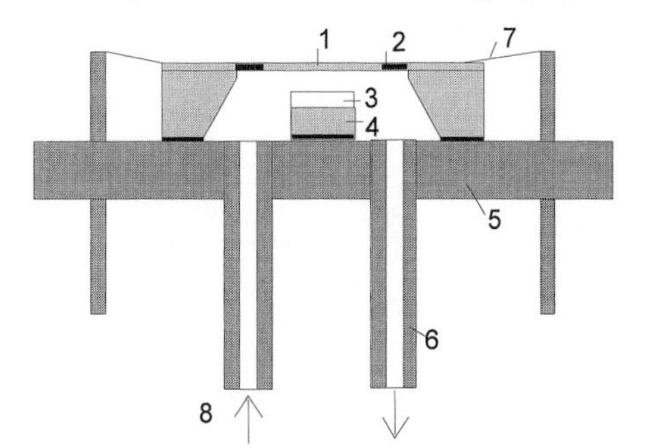

Figure 1. Operational principle of hydrogel-based sensors
1 bending plate; 2 mechano-electrical transducer (piezoresistive bridge);
3 swellable hydrogel layer; 4 Si platform;
5 socket ; 6 tube ; 7 interconnect; 8 solution.

The sensor chip is bonded to a socket with inlet and outlet flow channels. The aqueous solution to be measured is pumped through the inlet tubes into the silicon chip cavity. In order to obtain a sufficient measuring signal, the 50 μm thick PVA/PAA-hydrogel layer is located on a silicon platform to achieve a small enough gap between dry (absolutely unswollen) hydrogel and the silicon membrane. The hydrogel layers were spin coated on the Si wafer, dried and then crosslinked. For PNIPAAm chemical sensors, 250 μm thick hydrogel foil pieces were used. The dried PNIPAAm-foils were prepared by evaporation of water at room temperature and then cut into pieces of 1 mm x 1 mm.

Results and discussion

The sensor's output voltage was measured during the swelling of the PVA/PAA hydrogel layer under influence of solutions with different pH values, as shown in Fig. 2. A hysteresis was obtained for cycling from acidic to basic conditions and vice versa. The hysteresis loop widened with increasing number of measuring cycles. During the first pH increase the hydrogel layer abruptly swelled at pH values above the acid exponent of acrylic acid (pK_a=4.7). Then the threshold value pH_1 shifted to higher pH values with increasing number of measuring cycles. The downward threshold pH_2 for pH value decrease processes shifted to lower values. The hysteresis and the widening the hysteresis loop with increasing number of cycles could be caused by a screening effect.[9] With increasing the pH values by adding base, the counterions H^+ in the gel can be replaced by cations. At higher pH values, the hydrogel includes excess cations, which screen the ionized groups. This effect is disappearing only at low pH values.

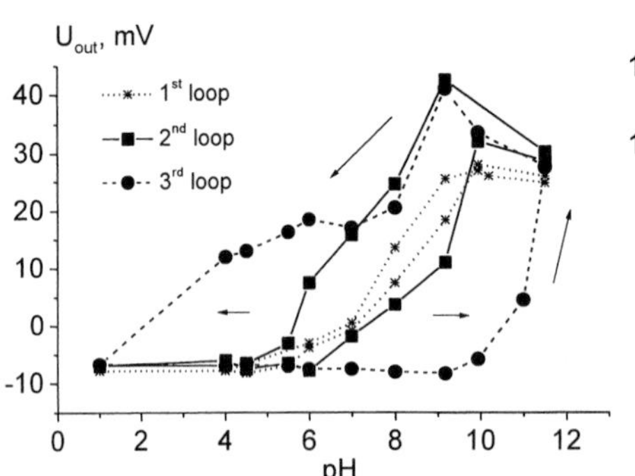

Figure 2. Sensor output voltage as a function of pH value.

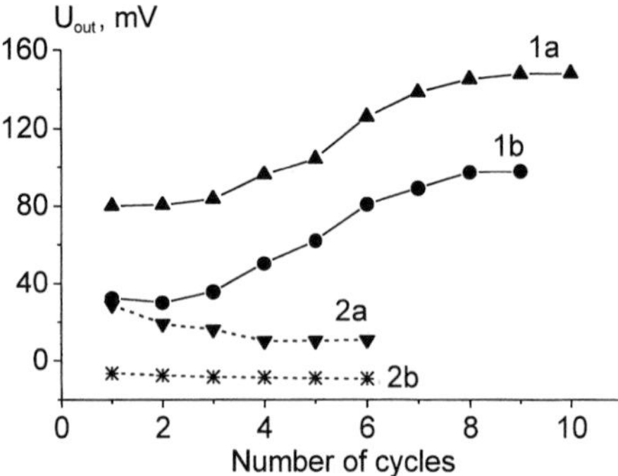

Figure 3. Sensor output voltage:
1 - at pH=11 (a) and at pH=7 (b) for alternate pH 11…7; 2 - at pH=11 (a) and at pH=1 (b) for alternate pH 11…1.

In order to investigate the reproducibility of the sensor response, pH1 and pH11 or pH7 and pH11 solutions were pumped alternatively. The resulting sensor response is shown in Fig. 3. On the one hand, the amplitude of the signal at pH11 (curve 2a) became smaller after the first cycles in acid solution until it became almost constant. On the other hand, the signal amplitude at both pH11 and pH7 (curves 1a und 1b) increased steadily from cycle to cycle at cycles between values pH=7 and pH=11. After several cycles of alternating immersions in the pH7 and pH11 solutions, the signal at pH11 reached its saturation. This behaviour shows clearly that hydrogel based sensors can favorably be used only in certain limited pH value region besides the region of the hysteresis loop. Preferably, sensors for low and for high pH values are useful. As expected by this behaviour, the sensor response is reversible at low pH values after initial conditioning in deionized water (Fig. 4). For a high signal reproducibility, the dry hydrogel layer should be placed in deionized water for more than 1 day. Before starting the measurements, the water uptake of hydrogel must reach its full saturation.

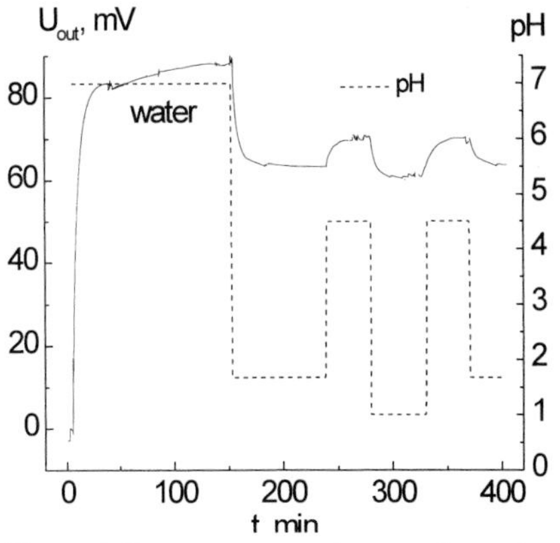

Figure 4. Sensor output voltage at alternating pH changes.

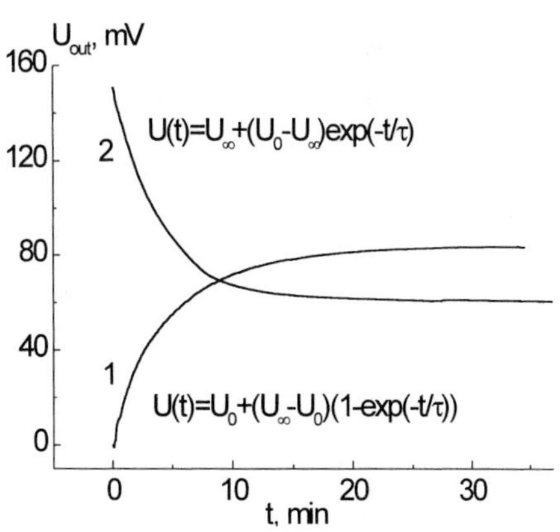

Figure 5. Sensor output voltage during: 1) swelling process of a dry hydrogel in deionized water, 2) deswelling process (from pH11.5 to pH1).

Beside the sensitivity and signal reproducibility, the response time is one of the most crucial aspects for sensor implementation. The sensor response time is determined by the polymer layer thickness as well as by the hydrogel swelling kinetics. The latter depends on the rates of diffusion of three species in the gel structure. First, solutes which are not excluded from the gel will diffuse into the gel. Second, base or acid must diffuse into the gel to spark its swelling or collapse. Third, water must enter or leave the gel. The kinetics is complicated because three these different mass transfer processes occur, often simultaneously. The mechanisms for swelling and shrinkage are different, since shrinkage occurs more quickly than swelling.[1, 10, 11] It was found that the time dependence of the deswelling process is identical with that one of the swelling process of a dry gel in deionized water.[1] Figure 5 shows the sensor's output voltage during the rapid swelling of dry hydrogel layer in deionized water (curve 1). The behaviour can be described by means of an exponential increase, which gives a time constant τ_1=5 min. The time dependence of the rapid deswelling process from pH11.5 to pH1 yields also τ_1=5 min (Fig. 5, curve 2), indicating that both processes are determined by the diffusion of water into the gel and out of the gel, respectively. It can be shown, that hydrochloric acid diffuses into the gel much faster than the rate at which the gel shrinks, indicating that the diffusion of water out of the gel, rather than ion exchange, limits the shrinkage.[10]

In contrast, during the slow swelling of the dry hydrogel layer in solution with pH=11.5, water and sodium hydroxide uptake occur at the same rate suggesting that ion exchange is the rate limiting step during swelling. In this case the swelling behaviour is described by a time constant of τ_2=80 min. Time constants in the same order of magnitude $\tau \geq 80$ min were found also for slow swelling processes both from pH1 to pH11.5 and from pH7 to pH11.5 as well as for slow deswelling process from pH11.5 to pH7. The latter is determined by the decay of the electrostatic repulsion between the polycarboxylate groups within the network. The sensor response time can be shortened by using a thinner hydrogel layer which needs also thinner silicon membrane of the chip for a sufficient sensor output signal.

The same sensor concept was also applied to chemical sensors for the measurement of solvent concentrations in aqueous solutions. For this purpose, the pH sensitive PVA/PAA was replaced by PNIPAAm. Fig. 6 shows the

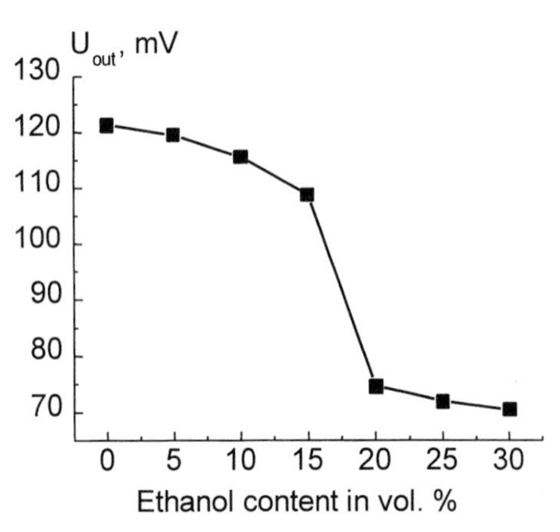

Figure 6. Output voltage of PNIPAAm based sensor in aqueous mixtures of ethanol.

characteristic of such a organic solvent concentration sensor. The sensor's output voltage was measured during the swelling of the hydrogel under influence of water solutions with different ethanol contents. The volume phase transition occurs at about 20 vol% of ethanol. This value corresponds to that one of PNIPAAm which was found for the swelling degree defined as the ratio between the masses of swollen and dry hydrogel.[3] For this sensor, the time constant describing swelling behaviour was determined to be $\tau \geq 60$ min.

Conclusions

We have demonstrated the operational principle of hydrogel-based chemical and pH sensors. To realize pH sensors, PVA/PAA blends which show a pH value dependant swelling behaviour were used as chemo-mechanical transducers. The hydrogel swelling leads to a bending of a thin silicon membrane and, by this, to an electrical output voltage of the sensor chip. For a pH sensor with a 50 µm thick PVA/PAA hydrogel layer, a sensitivity of approximately 15 mV/pH was obtained in a pH range above the acid exponent of acrylic acid (pK_a=4.7). Sufficient signal reproducibility was found both in the low and the high pH value regions besides the hysteresis loop. A similar sensor design was also used for chemical sensors on PNIPAAm basis for measuring the content of organic solvents in aqueous solutions. It was found that an initial conditioning in deionized water is necessary for high signal reproducibility. In order to shorten the sensor response time, the hydrogel film preparation and the film thickness have to be optimized.

Acknowledgments

The authors gratefully acknowledge support for this work from the Deutsche Forschungsgemeinschaft (Collaborative Research Center (SFB) 287, Project C11).

[1] K.-F. Arndt, A. Richter, S. Ludwig, J. Zimmermann, J. Kressler, D. Kuckling, H.-J. Adler, *Acta Polym.* **1999**, *50*, 383-390.
[2] A. Richter, *„Quellfähige Polymernetzwerke als Aktor-Sensor-Systeme für die Fluidtechnik"*, Fortschrittberichte Reihe 8 Nr. 944, VDI – Verlag, Düsseldorf 2002.
[3] K.-F. Arndt, D. Kuckling, A. Richter, *Polym. Adv. Technol.* **2000**, *11*, 496-505.
[4] F. Jiangi, G. Lixia, Europ. Polym. J. **2002**, *38*, 1653-1658.
[5] R. Bashir, J.Z. Hilt, O. Elibol, A. Gupta, N. A. Peppas, *Appl. Phys. Lett.* **2002**, *81*, 3091-3093.
[6] S. Herber, W. Olthuis, P. Bergveld, *Sensors and Actuators B*, **2003**, *91*, 378-382.

[7] J. Cong, X. Zhang, K. Chen, J. Xu, *Sensors and Actuators B,* **2002**, *87*, 487-490.
[8] German Patents DE 101 29 985C2, DE 101 29 986C2, DE 101 29 987C2, June 12, **2001**.
[9] A. Suzuki, H. Suzuki, *J. Chem. Phys.* **1995**, *103*, 4706 – 4710.
[10] S. H. Gehrke, E. L. Cussler, *Chem. Eng. Sci.* **1989**, *44*, 559-566.
[11] K. L. Wang, J. H. Burban, E. L. Cussler, *Adv. Polym. Sci.* **1993**, *110*, 67-79.

Synthesis and Characterization of Polypyrrole Dispersions Prepared with Different Dopants

Yan Lu, Andrij Pich, Hans-Juergen P. Adler*

Institute of Macromolecular Chemistry and Textile Chemistry, Dresden University of Technology, 01062 Dresden, Germany
E-mail: yan.lu@chemie.tu-dresden.de

Summary: Stable polypyrrole dispersions were prepared by chemical oxidative polymerization of pyrrole in an aqueous medium containing different anionic salts – sodium benzoate, potassium hydrogen phthalate and sodium hydrogen succinate. Results of the elemental analysis and FT-IR spectroscopy confirmed that the anionic salts are incorporated in the conducting polymers and functioned as the dopants. The retardation of pyrrole polymerization was observed when a certain amount of the salt was used as dopant. SEM images of polypyrrole dispersions indicate large spherical particles (150-180nm). The conductivity of polypyrrole composites has also been investigated.

Keywords: conducting polymer; dopant; morphology; polypyrrole

Introduction

In the last two decades a lot of research work has been done in the field of conducting polymers, such as polypyrrole (PPy) and polyaniline (PANI). As we know the general intractability of conducting polymers presents a serious problem, the preparation of polypyrrole in its colloidal forms has been sought to improve its processibility[1]. And in this case the selection of an appropriate stabilizer to provide the effective stabilizaton of colloidal system and to control the morphology and size of polymer particles is very important. Recently, a lot of polymers have been reported on being used as stabilizer for successful preparation of polypyrrole colloids by chemically oxidizing pyrrole in either water or other media, such as poly(vinyl pyrrolidone) (PVP)[2], poly(vinyl alcohol-co-acetate) (PVA)[3], poly(ethyleneoxide) (PEO)[4], poly(styrenesulfonate) (PSS)[5], poly(vinyl methyl ether) (PVME)[6], ethylhydroxycellulose[7], etc. In our pervious paper, we investigated the morphology of obtained polypyrrole particles, which were prepared by oxidative polymerization in the presence of PVME or crosslinked PVME microgels using water or aqueous ethanol as a dispersion medium. Spherical and needle-like polypyrrole particles were found when PVME and crosslinked PVME microgels were used as stabilizer,

 DOI: 10.1002/masy.200450646

respectively[8]. In this paper, three different anionic salts – sodium benzoate, potassium hydrogen phthalate and sodium hydrogen succinate were used as dopants in the polymerization of pyrrole in the presence of PVME as stabilizer. The properties of the polypyrrole with respect to morphology, conductivity, stability, and processabiility were affected by the addition of these salts.

Experimental

Material

Pyrrole(Py) was purchased from Aldrich Chemical Crop., distilled under vacuum and stored in a refrigerator before use. Sodium persulfate ($Na_2S_2O_8$), sodium benzoate (Benzoate), potassium hydrogen phthalate (Phthalate) and sodium hydrogen succinate (Succinate) were obtained from Aldrich Chemical Corp. and were used as received. PVME was obtained as a 50 wt% aqueous solution (BASF, Lutonal® M40) and used as supplied. The weight-average molecular weight, M_w, was determined by static light scattering in butanone to be 57,000 g/mol.

Polymerization of pyrrole

PVME was dissolved in appropriate amount of water at 25°C in a reactor equipped with stirrer. Then pyrrole was added to the stirred solution and the reactor was purged with nitrogen for 15 min. Oxidant, which was first dissolved in water, was then added dropwise and the polymerization was allowed to proceed for 24 hours. The resulting stable dispersions were cleaned by dialysis (Millipore membrane 100 000 MWCO) to remove oxidant and by-products.

Measurement

The nitrogen content of PPy was determined by elementary analysis. IR spectra were recorded with Mattson Instruments Research Series 1 FTIR spectrometer. Dried polymer samples were mixed with KBr and pressed to form a tablet. The morphology of the samples was examined by scanning electron microscopy (SEM) using a Zeiss DSM 982 Gemini microscope. The conductivity measurements were made at room temperature using the standard two-point probe method.

Results and discussion

Table 1 gives detailed information of Py dispersion polymerization at room temperature. And in all runs stable PPy colloids were formed in the presence of different salts. By following the polymerization process, the results indicate that the effect of the presence of benzoate, phthalate and succinate on the rate of PPy formation is marginal (Fig. 1). Gill et at.[9] reported that the oxidation of aniline proceeds slowly when sodium dodecylbenzenesulfonate (DBSNa) was added to the reaction mixture. But, Kudoh[10] noted that Py polymerization was faster in the presence of an anionic surfactant. Our results show that retardation of the polymerization has been observed in all runs with salts, possibly as a consequence of the limited miscibility of monomer and salt solutions. Furthermore, it is also easy to see that the retardation of pyrrole polymerization is henced with the increase of the dopant´s amount, which indicates a more active involvement of the anionic salt in the reaction course.

Table 1. The recipe of the polymerization pyrrole in water at 25°C.

Sample	Dopant	Dopant (g)	Pyrrole (g)	PVME (g)	$Na_2S_2O_8$ (g)	Water (g)	Stability
WP1-1	-	-	0,1	0,1	0,5	75	Stable
WP1-2	Na-benzoate	0,21	0,1	0,1	0,5	75	Stable
WP1-2a	Na-benzoate	0,043	0,1	0,1	0,5	75	Stable
WP1-2b	Na-benzoate	1,08	0,1	0,1	0,5	75	Stable
WP1-3	KH- phthalate	0,31	0,1	0,1	0,5	75	Stable
WP1-4	NaH-succinate	0,21	0,1	0,1	0,5	75	Stable

Table 2 shows the elemental composition of final products. And the results of elemental analysis show the reduction of the nitrogen content in the final products, which demonstrates the incorporation of dopants into PPy after Py polymerization.

Table 2. Elemental composition(wt%) of PPy prepared in the presence of dopants.

Sample	Dopant	C	H	N	S
WP1-1	None	56,46	4,98	12,90	0,59
WP1-2	Benzoate	49,80	5,64	7,69	3,11
WP1-3	Phthalate	56,09	5,24	11,12	1,48
WP1-4	Succinate	51,20	6,32	7,94	2,91

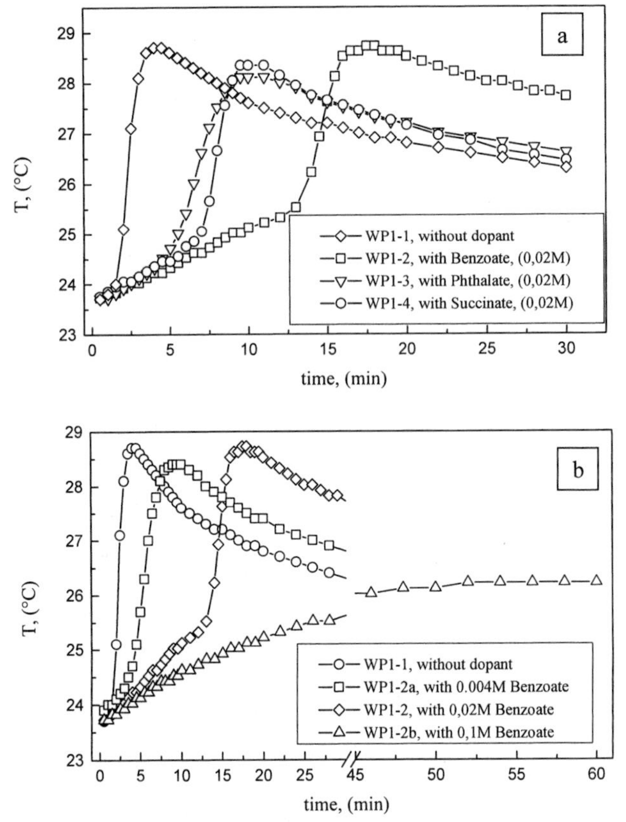

Figure 1. Temperature profile of the exothermic oxidation of pyrrole in the presence of various salts (a) and various amount of salts (b).

Moreover, the presence of anionic salts in PPy as dopants is also confirmed by a comparison of the infrared spectra of PPy prepared in the presence and absence of the salt (Fig.2). The peak at about 1452 cm^{-1} belonging to the carboxylate stretching of neat Na-benzoate is observed also in the spectra of polypyrrole polymer (WP1-2) prepared in presence of this salt. And the peaks at about 926 and 791 cm^{-1} are responsible for the δ(CH) and γ(CH) vibration of pyrrole, respectively. Similar results have been also investigated in the infrared spectra of polypyrrole prepared in the presence of other anionic salts.

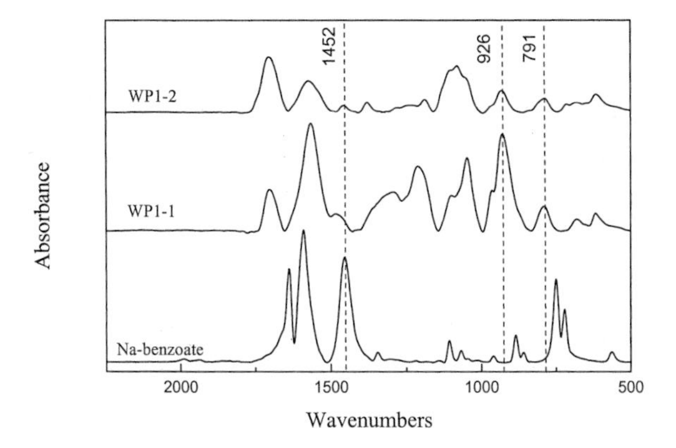

Figure 2. Infrared spectra of polypyrrole prepared with and without Na-benzoate and the spectrum of Na-benzoate.

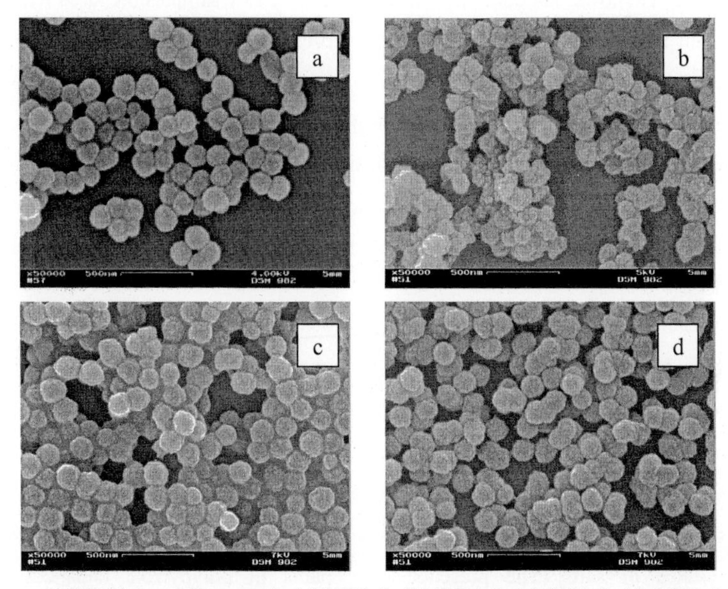

Figure 3. SEM images for sample (a) WP1-1, (b) WP1-2, (c) WP1-3, (d) WP1-4.

Electron microscopy images of polypyrrole colloids are shown in Figure 3. SEM investigations indicate that the morphology of obtained polypyrrole particles prepared in the presence of anionic salts are quite similar to the particles prepared in the absence of salt. In all cases, the spherical particles exhibit a “cauliflower” morphology, which was reported in other reviews[5,8]. Additionally, comparing the particle size of

PPy samples prepared with and without salts, it can be observed that larger particle size (about 180nm) is obtained by PPy samples prepared with salts and this observation can be explained by the slower rate of oxidation of Py polymerization in the presence of salt, which has been shown in Figure 1.

Tab. 3. Influence of presence of the dopants on conductivity of the PPy dispersions.

Sample	Py, (g)	PPy, (%)	σ, (Scm-1)
WP1-1	0,1	61,73	$8,5*10^{-9}$
WP1-2	0,1	36,80	$1,17*10^{-5}$
WP1-3	0,1	53,21	$4,55*10^{-7}$
WP1-4	0,1	40,00	$1,5*10^{-7}$

Table 3 shows the conductivity of PPy samples, which was measured at room temperature. The lower conductivity value obtained by PPy samples prepared without any dopant may due to the reason that PPy particles were covered by insulator out layer of stabilizer PVME, which resulted in the difficulties to transport the charge carriers. In the case of PPy prepared in the presence of dopants, an improvement of the conductivity has been observed even at much lower PPy loads. While phthalate and succinate were used as dopant, the conductivity of the final products increased by 1-2 orders of magnitude, and while benzoate was used, the conductivity increased by 3-4 orders.

Conclusions

Results of the elementary analysis, IR spectroscopy confirmed that the anionic salts were incorporated in the conducting polymers and functioned as the dopants. The retardation of pyrrole polymerization was observed when a certain amount of the salt was used as the dopant. SEM images show larger spherical PPy particles prepared in the presence of salts. The presence of dopants in polypyrrole enhances the conductivity, especially in the case of sodium benzoate, the conductivity of the final product has been improved by 3 orders.

Acknowledgements

The authors are thankful to Mr. E. Kern for SEM measurements and Deutsche Forschungsgemeinschaft (DFG, Sonderforschungbereich 287 'Reactive Polymers') for financial support.

[1] J. Stejskal, *J. Polym. Mater.*, 2001, 18, 225.
[2] S.P. Armes, B. Vincent, *J. Chem. Soc., Chem. Commun.*, 1987, 288
[3] N. Cawdery, T.M. Obey, B. Vincent, *J. Chem. Soc., Chem. Commun.*, 1988, 1189
[4] R. Odegard, T.A. Skotheim, H.S. Lee, *J. Electrochem. Soc.*, 1991, 138(10), 2930
[5] Z. Qi, P. G. Pickup, *Chem. Mater.*, 1997, 9, 2934
[6] M.L. Digar, S. N. Bhattacharyya, B. M. Mandal, *Polymer*, 1994, 35(2), 377
[7] T. K. Mandal, B. M. Mandal, *Polymer*, 1995, 36(9), 1911
[8] A. Pich, Y. Lu, H.-J. P. Adler, T. Schmidt, K.-F. Arndt, *Polymer*, 2002, 43, 5723
[9] M.T. Gill , S. E. Chapman, C.L. DeArmitt, F. L. Baines, C. M. Dadswell, J.G. Stamper, N. C. Billingham, S. P. Armes, Synth. Met., 1998, 93, 227
[10] Y. Kudoh, Synth. Met., 1996, 79,17

Novel Reactive Thermosensitive Polyethers – Control of Transition Point

Andrzej Dworak,[1] *Barbara Trzebicka,*[1] *Wojciech Wałach,*[1] *Alicja Utrata,*[1,2] *Christo Tsvetanov*[3]

[1] Polish Academy of Science, Institute of Coal Chemistry, Sowinskiego 5, Gliwice, Poland
[2] Silesian University of Technology, Faculty of Chemistry, Strzody 9, Gliwice, Poland
[3] Bulgarian Academy of Sciences, Institute of Polymers, Sofia, Bulgaria

Summary: A new class of thermosensitive polymers based on polyethers is discussed. Using living anionic polymerisation techniques a series of homo- and block copolymers of 2,3-epoxypropanol-1 (the glycidol), ethoxy ethyl glycidol ether, its hydrophobic derivative, and ethylene oxide of different molar masses and topology (linear and comb-like) was obtained. By simple chemical modification of hydroxyl groups in polyglycidol segments hydrophobic elements were introduced into polymer chains, which allowed to control the transition point related to the lower critical solution temperature between 0 to 100°C. The relation between the transition temperature and the structure of obtained polymers is discussed.

Keywords: anionic polymerisation; cloud point; polyethers; polyglycidol; temperature sensitive polymers

Introduction

Stimuli sensitive polymers are structures, which exhibit distinct and reversible change of properties in response to the action of external stimuli. Different stimuli may be applied to invoke the change of properties, temperature being one of the most frequently studied[1-3]. Many polymers exhibit the so called lower critical solution temperature: a temperature above which the polymer precipitates from the solution, in most studied cases from water solution. The control of the transition temperature is important for all actual or envisaged application of such polymers[4-6]. This temperature is the function of the balance of hydrophobic and hydrophilic groups of the macromolecules undergoing transition and of polymer architecture.

 DOI: 10.1002/masy.200450647

The aim of this work is to present a new group of temperature sensitive polymers based upon reactive polyethers, the transition point of which may be easily controlled within wide limits.

Experimental

Synthesis of linear polyglycidol

Polyglycidol of moderate molar mass (below 20 000), further referred to as PGl_L was obtained by the anionic polymerisation of ethoxy ethyl glycidyl ether using potassium tert-butoxide as initiator. The subsequent removal of the acetal group, as described previously[7] is necessary to obtain the desired polyglycidol. In order to obtain high molar mass polyglycidol the initiator of Vandenberg[8], the partially hydrolysed diethyl zinc ($ZnEt_2/H_2O$ 1:0.8) was used[9]. This high molar mass linear polyglycidol is further referred to as PGl_H.

Synthesis of polyglycidol-graft-polyglycidol

Linear polyglycidol was ionised by potassium tert-butoxide and used to initiate the anionic polymerisation of glycidol acetal. The acetal groups were removed using formic acid. Details are given in[10]. This comb-like polyglycidol is further referred to as PGl_G.

Synthesis of block copolymers of PEO and glycidol

The polyoxyethylene glycol cesium dialcoholates (M_n=10000, M_n=6000) and monoalcoholate (M_n=5000) were used as the initiators of the polymerisation of glycidol acetal. Details are given in[11]. The block copolymers are denoted as PGl_x-b-PEO_y-b-PGl_x or PEO_y-b-PGl_x; x, y denotes the number average degree of polymerisation of the polyglycidol and PEO blocks, respectively.

Esterification of hydroxyl group of polyglycidol

Obtained linear, graft and block polymers of glycidol were reacted with acetic anhydride in pyridine/DMF. The degree of esterification was controlled by varying the amount of acetic anhydride. The details of esterification are given in[12]. Ester groups are denoted by Ate.

Results and Discussion

The synthesis of the linear homopolymer of glycidol and its block copolymers with PEO

It is known that the cationic and anionic polymerisation of 2,3-epoxypropanol-1 (glycidol) are difficult to control and lead to branched products with not well defined structure. That is why the hydroxyl group of this monomer has to be protected before the polymerisation if linear chains of controlled DP are aimed at. In all our experiments with glycidol we applied the acetalization of the hydroxyl group, a protection method proposed by Spassky[13].

Low molar mass polyglycidol (PGl_L) was obtained by the anionic polymerisation of ethoxy ethyl glycidyl ether and subsequent removal of the protecting group under acidic conditions[7] (scheme 1). When degree of polymerisation of ca. 300 is not exceeded, the system is close to living, allowing to control the molar masses and yielding polymers with narrow molar mass distribution.

1 step n CH_2-CH-CH_2-O-CH(CH_3)-O-C_2H_5 (t-BuOK; THF or $ZnEt_2/H_2O$(1:0,8); bulk) → -(CH_2-CH(-CH_2-O-CH(CH_3)-O-C_2H_5)-O)$_n$- **A**

2 step -(CH_2-CH(-CH_2-O-CH(CH_3)-O-C_2H_5)-O)$_n$- (HCOOH; KOH or HCl) → -(CH_2-CH(CH_2OH)-O)$_n$-

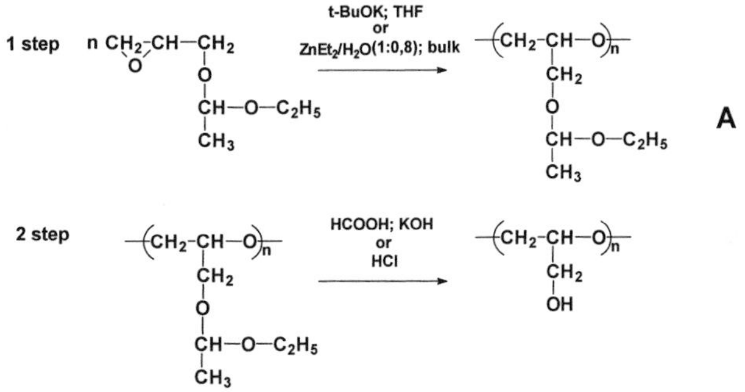

Scheme 1. Synthesis of linear polyglycidol.

Higher molar masses are difficult to be achieved using the anionic polymerisation, probably because of the influence of adventitious electrophilic impurities. However, the use of the Vandenberg catalyst[8], the partially hydrolysed diethylzink ($ZnEt_2/H_2O$=1:0.8) leads to polymers of high molar masses (PGl_H) (up to 800 000), however, at the expense of lost control and broader molar mass distribution.

t-BuO-(CH_2-CH(CH_2OH)-O)$_n$-H (1. < 0.1 n t-BuOK; 2. m CH_2-CH-CH_2-O-CH(CH_3)(OC_2H_5); 3. hydrolysis) → t-BuO-(CH_2-CH(CH_2OH)-O)$_{n-x}$-(CH_2-CH(CH_2-O-(CH_2-CH(CH_2OH)-O)$_{m/x}$H)-O)$_x$-H

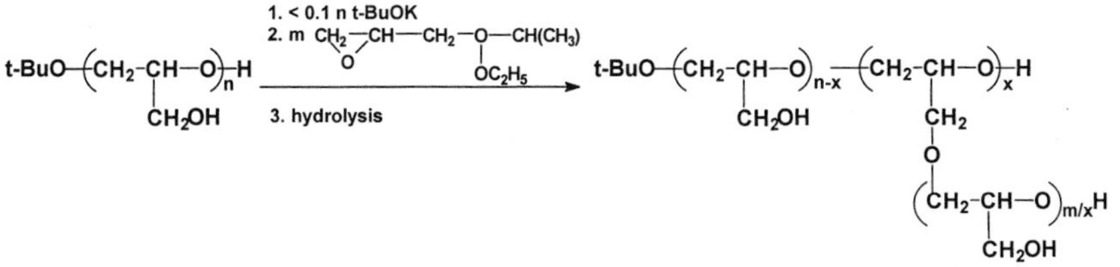

Scheme 2. Synthesis of comb-like polyglycidol.

The comb-like polyglycidol (PGl_G) was obtained[10] using an anionic „grafting from" polymerisation. Linear polyglycidol, obtained as described below, was partially transformed into its cesium alcoholate. This macroinitiator was used to initiate the polymerisation of protected glycidol. After deprotection, polyglycidol–graft–polyglycidol was obtained (scheme 2).

The living character of the polymerisation of protected glycidol makes the synthesis of block co-polymers possible. Mono- or dialcoholates of polyethylene glycols were used as the initiators of the polymerisation of ethoxy ethyl glycidyl ether. A number of di- and triblock copolymers with poly(ethylene oxide) block, flanked on both or one side with polyglycidol chains were obtained (PGl_x-b-PEO_y, PGl_x-b-PEO_y-b-PGl_x) (scheme 3).

1 step $CH_3O(CH_2\text{-}CH_2\text{-}O)_nH$, $HO(CH_2\text{-}CH_2\text{-}O)_nH$ $\xrightarrow{CsOH}$ $CH_3O(CH_2\text{-}CH_2\text{-}O)_nO^-Cs^+$, $^+Cs^-O(CH_2\text{-}CH_2\text{-}O)_nO^-Cs^+$

2 step $CH_3O(CH_2\text{-}CH_2)_nO^-Cs^+$ or $^+Cs^-O(CH_2\text{-}CH_2)_nO^-Cs^+$ + $CH_2\text{-}CH\text{-}CH_2\text{-}O\text{-}CH(CH_3)\text{-}O\text{-}C_2H_5$ (epoxide) $\longrightarrow$

$CH_3O(CH_2\text{-}CH_2\text{-}O)_n(CH_2\text{-}CH(CH_2\text{-}O\text{-}CH(CH_3)\text{-}O\text{-}C_2H_5)\text{-}O)_m$

diblock copolymer

or **B**

$(CH_2\text{-}CH(CH_2\text{-}O\text{-}CH(CH_3)\text{-}O\text{-}C_2H_5)\text{-}O)_m(CH_2\text{-}CH_2\text{-}O)_n(CH_2\text{-}CH(CH_2\text{-}O\text{-}CH(CH_3)\text{-}O\text{-}C_2H_5)\text{-}O)_m$

triblock copolymer

3 step **B** $\xrightarrow{\text{1. HCOOH; 2. KOH, dioxane, methanol}}$ $CH_3O(CH_2\text{-}CH_2\text{-}O)_n(CH_2\text{-}CH(CH_2OH)\text{-}O)_m$

or

$(CH_2\text{-}CH(CH_2OH)\text{-}O)_m(CH_2\text{-}CH_2\text{-}O)_n(CH_2\text{-}CH(CH_2OH)\text{-}O)_m$

Scheme 3. Synthesis of di- and triblock copolymers of glycidol and ethylene oxide.

Due to the living character of the polymerisation the length of blocks was well defined and the molar mass distribution (therefore the block length distribution) remained narrow (table 1).

Table 1. Studied polymers of glycidol.

Sample	Molar mass[1)]	M_w/M_n
Homopolymers of glycidol		
PGl_L (linear)	20 000	1.21
PGl_H (linear)	186 000	1.67
PGl_G (graft)	82 000	1.25
Triblock copolymers PGl_x-PEO_y-PGl_x[2)]		
PGl_{68}-b-PEO_{227}-b-PGl_{68}	21 000	1.06
PGl_{44}-b-PEO_{227}-b-PGl_{44}	15 500	1.04
PGl_{28}-b-PEO_{227}-b-PGl_{28}	14 000	1.04
PGl_{80}-b-PEO_{136}-b-PGl_{80}	25 600	1.07
PGl_{34}-b-PEO_{136}-b-PGl_{34}	11 400	1.03
PGl_{17}-b-PEO_{136}-b-PGl_{17}	8 300	1.03
Diblock copolymers PEO_y-PGl_x[2)]		
PEO_{113}-b-PGl_{40}	8 000	1.02
PEO_{113}-b-PGl_{72}	11 100	1.02
PEO_{113}-b-PGl_{110}	12 600	1.10
PEO_{113}-b-PGl_{220}	20 800	1.16

(1) From GPC with refractive index and multiangle laser light scattering detectors
(2)Indices denote the number average degree of polymerisation of the polyglycidol blocks (x) and poly(ethylene oxide) blocks (y)

Temperature sensitivity of the studied polymers

Hydrophobic modification of the homopolymers

The polyglycidol is a water soluble, highly hydrophilic polymer. It is a close analogue of poly(ethylene oxide). However, unlike the PEO, it has reactive hydroxyl groups in the chain units, which in case of their hydrophobic modification may open the route to induce the lower critical solution temperature behaviour and allowing to control the transition point.

Already the intermediate product of the synthesis of linear glycidol polymers, the poly(ethoxy ethyl glycidyl ether) (polymer A in scheme 1) is water soluble below +10°C and precipitates when this temperature is exceeded[14, 15]. Following the route of Laschewsky[16] we have chosen the simplest hydrophobic modification of the hydroxyl group, the esterification of the polymers with acetic acid.

In the case of the homopolymers of glycidol, the degree of substitution between 30 and almost 100% was obtained (table 2) The aim was to study the relationship between the degree of substitution, the molar mass and the chain topology (linear and comb-like) and the lower critical solution temperature, for which the cloud temperature was taken as a measure.

Table 2. Thermosensivity of hydrophobically substituted homopolymers of glycidol.

Sample	Degree of esterification [%]	Cloud point [°C]
Esterified linear polyglycidol of molar mass 20 000 (PGl_L, table 1)	90	4
	63	33
	53	49
	51	57
Esterified linear polyglycidol of molar mass 186 000 (PGl_H, table 1)	66	41
	55	55
	45	>100
Esterified comb-like polyglycidol (PGl_G, table 1)	84	10
	70	23
	50	56
	46	57

For all studied samples, the cloud point depends upon the degree of esterification. The higher the degree of substitution with the hydrophobic ester groups, the lower the transition temperature.

The degree of substitution influences the transition temperature most significantly, but the molar mass and the topology of the macromolecules is also important. For the linear samples of higher molar mass PGl_H, the cloud point at the same degree of esterification is higher then for the sample of lower molar mass PGl_L.

Although the molar mass of the comb-like polyglycidol PGl_G is four times higher than of the linear polyglycidol PGl_L, its cloud point is almost identical with the cloud point of the latter for every degree of esterification, but much lower then for the esterified high molar mass polyglycidol PGl_H. This indicates that at the same degree of esterification the comb-like macromolecules appears more hydrophobic, then the linear chains of comparable molar mass. Most likely reason is that during the esterification of the comb-like polyglycidol the outer hydroxyl groups of the branched macromolecules are esterified preferably, causing an increased concentration of the hydrophobic groups in the outer sphere. The outer sphere decides about the interaction with water.

Hydrophobic modification of the block copolymers of ethylene oxide and glycidol

The hydrophobic ester groups were introduced into the di- and triblock copolymers by esterification (tables 3 and 4).

Table 3. Composition of block copolymers of glycidol acetate (PAte) and ethylene oxide (PEO). Numbers denote number average degree of polymerisation of the blocks.

Sample	Degree of esterification [%]
$PAte_{68}$-b-PEO_{227}-b-$PAte_{68}$	100
$PAte_{44}$-b-PEO_{227}-b-$PAte_{44}$	100
$PAte_{28}$-b-PEO_{227}-b-$PAte_{28}$	85
$PAte_{80}$-b-PEO_{136}-b-$PAte_{80}$	96
$PAte_{34}$-b-PEO_{136}-b-$PAte_{34}$	95
$PAte_{17}$-b-PEO_{136}-b-$PAte_{17}$	100

The cloud point of the hydrophobically modified triblock copolymers poly(glycidol acetate)–block–poly(ethylene oxide)–block–poly(glycidol acetate) may be varied within wide limits (Figure 1) and depends upon both the ratio of the hydrophilic (poly(ethylene oxide)) and hydrophobic (poly(glycidol acetate)) units and upon the length of the central hydrophilic PEO block (Figure 1). The shorter hydrophilic central block makes the cloud point of the copolymer less sensitive to the hydrophobic modifications: the cloud point vs degree of esterification curve for the copolymer with the longer central block lies significantly below the curve for the copolymer with shorter PEO sequences.

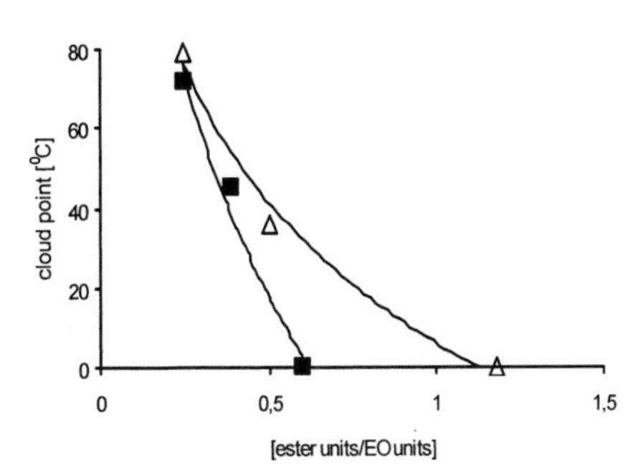

Figure 1. Cloud point vs content of acetate groups to EO units in PAte-b-PEO-b-PAte with DP_{PEO}=227 (■) and DP_{PEO}=136 (Δ).

Table 4. Polymer composition and cloud point of water solutions (5 wt%) of the diblock copolymers poly(glycidol acetate)–block–poly(ethylene oxide).

Sample	Degree of esterification [%]	Cloud point [°C]
PEO_{113}-b-$PAte_{40}$	100	95
PEO_{113}-b-$PAte_{72}$	100	95
PEO_{113}-b-$PAte_{110}$	95	87
PEO_{113}-b-$PAte_{220}$	98	Insoluble

Also the diblock copolymers, consisting of a hydrophilic poly(ethylene oxide) block and a hydrophobic poly(glycidol acetate) block are thermosensitive, their solution exhibit a cloud point due to the lower critical solution temperature (table 4). If the hydrophilic and the hydrophobic blocks are of similar length or if the hydrophilic block is longer then the hydrophobic one, the cloud point does not depend upon the copolymer composition and the cloud point is rather high, between 87 and 95°C. The probable reason is the micelle formation in water solution, which is conformed by the preliminary results of the fluorescence spectroscopy study with pyrene probe. The organization of the diblock copolymers in water solutions will be the subject of a separate study.

Conclusions

Polyglycidol, an itself highly hydrophilic polymer, may easily be transformed into a polymer exhibiting lower critical transition temperature by simple hydrophobic modification of the hydroxyl groups. After the modification temperature sensitive homo- and copolymers of glycidol are obtained. The transition point may easily be controlled within wide limits. The presence of reactive hydroxyl groups in the temperature sensitive macromolecules obtained makes further modifications possible and prospective applications possible.

Acknowledgment

This work was supported by the Polish Ministry of Scientific Research and Information Technology, grant no. 4 T09A 052 25.

[1] A. Hoffman, in *Polym. Mater. Encyclopedia*, CRC Press, **1996**, 3282-3291
[2] X. Zhang, J. Zang, R. Zhou, Ch. Chu, *Polymer*, **2002**, 43, 4823
[3] T. Nonaka, K. Makinose, S. Kurihara, *Polym. Prepr.*, **1999**, 40(2), 287
[4] B. Jeong, S. W. Kim, Y. H. Bae, *Adv. Drug Delivery*, **2002**, 52, 37
[5] F. Yam, X. Y. Wu, *Polym. Prepr.*, **1999**, 40, 312
[6] R. Langer, *Accounts of Chemical Research,* **2000**, 33, 94
[7] A. Dworak, I. Panchev, B. Trzebicka, W, Wałach, *Macromol. Symp.* **2000**, 153, 233
[8] E. J. Vandenberg, *J. Polym. Sci., Part A: Polym. Chem. Ed.,* **1985**, 23, 915
[9] A. Haouet, M. Sepulchre, N. Spassky, *Eur. Polym. J.*, **1983**, 19(12), 1089
[10] W. Wałach, A. Kowalczuk, B. Trzebicka, A. Dworak, *Macromol. Rapid Comm.*, **2001**, 22, 1272
[11] A. Dworak, K. Baran, B. Trzebicka, W. Wałach, *Reactive & Functional Polymers*, **1999**, 42, 31
[12] A. Dworak, B. Trzebicka, A. Utrata, W. Wałach, *Polym. Bull.*, **2003**, 50, 47
[13] D. Taton, A. Le Borgne, M. Spelchure, N. Spassky, *Macromol. Chem. Phys.,* **1994,** 195, 139,
[14] Ph. Dimitrov, S. Rangelov, A. Dworak, Ch. Tsvetanov, *Macromolecules*, in press
[15] A. Dworak, B. Trzebicka, W. Wałach, A. Utrata, *Polimery,* **2003**, 48, 484
[16] A. Laschewsky, E. D. Rekai, E. Wischerchoff, *Macromol. Chem. Phys.*, **2000**, 17, 201

Macromol. Symp. **2004**, *210*, 427-436

Synthesis and Aqueous Solution Properties of Functionalized and Thermoresponsive Poly(D,L-lactide)/Polyether Block Copolymers

Aleksandra Porjazoska,[2] *Philip Dimitrov,*[1] *Ivaylo Dimitrov,*[1] *Maja Cvetkovska,*[2] *Christo B. Tsvetanov**[1]

[1]Institute of Polymers, Bulgarian Academy of Sciences, 1113 Sofia, Bulgaria
[2]St. St. Cyril and Methodius University, Faculty of Technology and Metallurgy, 1000 Skopje, Macedonia

Summary: Tri- and pentablock amphiphilic copolymers containing hydrophobic poly(D,L-lactide) block(s) and hydrophilic polyethers were synthesized in order to obtain new precursor architectures suitable for drug delivery systems. Polyglycidol-*b*-poly(ethylene oxide)-*b*-poly(D,L-lactide) possess high hydroxyl functionality provided by the linear polyglycidol block. Thus very stable hydroxyl functionalized micelles in aqueous media were obtained. On the other hand poly(D,L-lactide)-*b*-poly(ethylene oxide)-*b*-poly(propylene oxide)-*b*-poly(ethylene oxide)-*b*-poly(D,L-lactide) form temperature sensitive aggregates. The copolymers obtained were analyzed by SEC and NMR, and their aqueous solution properties were followed by cloud point measurements and determination of critical micellization temperature. TEM was used for particles visualization.
Keywords: biopolymers; block copolymers; polyethers

Introduction

Amphiphilic block copolymers containing poly(D,L-lactide) (PL) hydrophobic blocks are promising materials for advanced drug delivery systems.[1-5] Recently block copolymers of PL and poly(ethylene oxide) (PEO) have attracted great attention due to the unique biocompatibility of PEO combined with an absolute biocompatibility and biodegradability of PL.[6-15] Accordingly, Kataoka et al. have developed a new approach to aldehyde-functionalized PL-PEO micelles in order to engineer their surface properties.[16-18] Thus drug targeting agents can be easily attached to micelles consisting of hydrophilic PEO shell and a drug-loaded hydrophobic PL core.

Polyglycidol (PG) is a very hydrophilic and highly hydroxyl-functional polymer.[19-21] Linear PG is obtained by anionic polymerization of ethoxyethyl glycidyl ether (EEGE) followed by cleavage of the ethoxyethyl groups.[22-24] The PG block could be relatively easily

 DOI: 10.1002/masy.200450648

incorporated into a common PEO-PL copolymer thus obtaining a precursor material for highly functional micelles.

It is well-known that PL-PEO copolymers form very stable micelles in aqueous media and in most cases temperature has no significant effect on the equilibrium of the system.[8,12,15] Kohori et al. have reported on thermo-sensitive micelles comprised of diblock copolymers of PL and poly(N-isopropylacrylamide) (PNIPAM).[1] This strategy can be extended by use of other than the PNIPAM polymers, which exhibit lower critical solution temperature (LCST) such as the oligomeric poly(propylene oxide).[25-27] The micellization of PEO-PPO-PEO (Pluronic, BASF; Synperonic, ICI) copolymers is a temperature induced process, which begins at a given critical micellization temperature (*cmt*) or critical micellization concentration (*cmc*).[28] The incorporation of PL outer blocks into a PEO-PPO-PEO macromolecule represents a convenient way of obtaining new PL-based thermosensitive materials. In this system the PEO blocks will provide kinetic stability for the aggregates above the LCST of PPO thus extending the micellar region.

The main goal of the present work is to expand the concept of PL-based amphiphilic block copolymers by introducing new multiblock copolymers containing highly functional or thermo-responsive blocks. PL-PEO-PPO-PEO-PL and PG-PEO-PL copolymers were synthesized with the aid of anionic polymerization. Their self-association properties in aqueous media were studied by cloud point measurements, determination of *cmc*, and transmission electron microscopy (TEM).

Experimental

A. Materials. All solvents were purified by standard methods. D,L-lactide (L) was purchased from Polysciences and purified by recrystallization from toluene. Pluronic P123 with the composition $EO_{20}PO_{70}EO_{20}$ (Aldrich) was freeze-dried before use. $Sn(Oct)_2$ (Aldrich), Ca (Aldrich), CaH_2 (Aldrich) and $CsOH.H_2O$ 99.5 % (Acros Organics) were used as received. EEGE was synthesized according to procedure, described elsewhere[29] and purified by vacuum distillation. Fractions of purity exceeding 99.0% (GC) were used for polymerizations. 1-methoxy-2-ethanol was purified by vacuum distillation. Ethylene oxide (EO) (Clariant) was used as received.

B. Synthesis of Block Copolymers. 1. PEEGE-PEO-PL. To $CsOH.H_2O$ (9 mmol), magnetically stirred in a reaction vessel equipped with argon and vacuum line, an equimolar amount of 1-methoxy-2-ethanol was added at 90°C. After stirring for 1 hour, the system was switched to the vacuum line for 2 hours. After flushing the vessel with argon another 9 mmol of additional alcohol were added. The polymerization of the PEEGE block was started by adding an appropriate amount of the EEGE monomer to the fresh initiator mixture at 60°C. The formation of the first block was completed in 24 hours. After removing a sample for analysis, the reaction temperature was raised to 90°C and EO was bubbled through the system for 2, 3, or 5 hours, depending on the desired length of the PEO block. After EO polymerization the PEEGE-PEO diblock copolymers were carefully purified and analyzed. The polymerization of L was performed in THF at 40 °C for 4 hours. The PEEGE-PEO-$CaNH_2$ macroinitiator was obtained by the procedure, described by Piao et al.[30]

2. PL-PEO-PPO-PEO-PL. The building of the PL outer blocks was performed by polymerization of L initiated by P123 precursor with one of the following reagents: $Sn(Oct)_2$, CaH_2, $CsOH.H_2O$, and $Ca(NH_3)_6$. To 1 mmol of P123, magnetically stirred in a reaction ampoule equipped with argon and vacuum line, $Sn(Oct)_2$, CaH_2, or $CsOH.H_2O$ were added at 90°C. The molar ratio $[OH]/Sn(Oct)_2$ was $1x10^3$,[31] whereas it was 1 for $[OH]/CaH_2$[6] and $[OH]/CsOH.H_2O$. After stirring for 1 hour, the system was switched to the vacuum line for 2 hours. An appropriate amount of L was added and the polymerization continued for 24 hours at 120°C. The polymerization of L via calcium hexaammoniate was performed in THF at 40 °C for 4 hours.

Copolymers Purification. The samples were dissolved in methylene chloride and filtered through Hylfo Super Cel® (diatomaceous earth). After precipitation from freshly distilled dry diethyl ether the copolymers were extensively dried under vacuum.

C. Cleavage of Ethoxyethyl Protective Groups. The described procedure below is similar to a recent method for deprotection of tetrahydropyranyl ethers.[32] A given amount of PEEGE-PEO-PL was dissolved in MeOH. Then $AlCl_3.6H_2O$ was added and the reaction was kept for 0.5 hours at room temperature. The EEGE:$AlCl_3$:MeOH molar ratio was 100:1:800. The reaction product was filtered through diatomaceous earth and the solvents were evaporated under reduced pressure.

D. Analyses. SEC analyses were performed on Waters system equipped with four Styragel columns with nominal pore sizes of 100, 500, 500, and 1000 Å and with a differential refractometric detector. THF was used as the solvent at 40 °C at an elution rate of 1ml/min.

Toluene was used as an internal standard for indication of elution volumes. Polystyrene standards were used for the molecular weight calibration.

The ^{1}H and ^{13}C NMR spectra were recorded at 250 MHz and 62.5 MHz respectively, on a Bruker WM 250, using $CDCl_3$ as solvent.

Cloud point (CP) transitions of 2% aqueous solutions of the samples were followed on a Specord UV-VIS spectrometer (Carl Zeiss, Jena) switched to transmittance regime at $\lambda = 500$ nm using a thermostated cuvette holder. The solutions were initially equilibrated at 0 ° C before heating them gradually (0.1 °C/min).

The determination of *cmc* by 1,6-diphenyl-1,3,5-hexatriene (DPH) dye solubilization was done in a way similar to Ref. 24 and 28.

Samples for TEM were prepared by direct dissolution of the samples in bidistilled water (C=10 g/l) followed by slow evaporation of the solvent. Measurements were performed on JEM 200 CX apparatus.

Results and Discussion

Synthesis of Multiblock Copolymers. The cesium initiating system was preferred for the sequential polymerization of EEGE and EO leading to a PEEGE-PEO precursor. It is well-known that PL is not stable in strong basic media.[33] By the formation of the third block of PL a dark brown tint was observed which was most likely due to unwanted side reactions. Hence the formation of the PL block was initiated by $Ca(NH_3)_6$, which was found to be very efficient for polymerization of cyclic esters.[30,34-36] In general Ca initiators are favored for the ring-opening polymerization of lactides, because they show: i) very high efficiency at temperature as low as 40°C; ii) low toxicity, which is very important in drug delivery.

For the synthesis of PL-PEO-PPO-PEO-PL, CsOH, $Sn(Oct)_2$, CaH_2 and $Ca(NH_3)_6$ initiators were used for the purpose of comparison. As seen from Table 2, L polymerization initiated by $Ca(NH_3)_6$ excels the other systems, proceeding at the lowest temperature and giving 100% yield. The copolymers molecular weight characteristics, determined by SEC and ^{1}H NMR, and their compositions determined by ^{1}H NMR are summarized in Table 1 and Table 2.

The ethoxyethyl groups of the PEEGE block were successfully cleaved to hydroxyl ones (Figure 1), thus converting it to a linear PG block, which number of functionalities is equal to its degree of polymerization. Apparently the PG-PEO-PL copolymers represent very suitable precursors to hydroxyl-functional micelles composed of a biodegradable PL core and a highly hydrophilic PG-PEO shell (Scheme 1).

Table 1. Composition and molecular weight characteristics of the PEEGE block copolymers.

Composition (NMR)	Mn (NMR)	MWD (SEC)
$EEGE_7$	1000	1.25
$EEGE_7EO_{29}$	2300	1.10
$EEGE_7EO_{29}L_{13}$	3200	1.18
$G_7EO_{29}L_{13}$	2700	-
$EEGE_{16}$	2300	1.18
$EEGE_{16}EO_{137}$	8300	1.13
$EEGE_{16}EO_{137}L_{17}$	9500	1.32
$EEGE_{16}EO_{137}L_{40}$	11000	1.24
$G_{16}EO_{137}L_{40}$	9500	
$EEGE_{16}EO_{246}$	13000	1.35
$EEGE_{16}EO_{246}L_{44}$	16000	1.33
$EEGE_4$	600	1.48
$EEGE_4EO_{159}$	7500	1.45
$EEGE_4EO_{159}L_{25}$	9300	1.22

Table 2. Composition and molecular weight characteristics of pentablock copolymers of PL and Pluronic P123.

Composition (NMR)	Ininitiator	L Conversion (%, NMR)	Mn (NMR)	MWD (SEC)
L_4P123L_4		66	6400	1.4
L_5P123L_5	CsOH	83	6500	1.3
$L_{10}P123L_{10}$		100	7200	1.5
$L_{22}P123L_{22}$		73	8900	1.7
L_3P123L_3		50	6200	1.4
L_4P123L_4	$Sn(Oct)_2$	66	6400	1.3
L_8P123L_8		80	7000	1.2
L_5P123L_5	CaH_2	83	6500	1.5
L_6P123L_6		60	6700	1.3
$L_{10}P123L_{10}$	$Ca(NH_3)_6$	100	7200	1.5
$L_5EO_{30}PO_{34}EO_{30}L_5$		100	5400	1.4

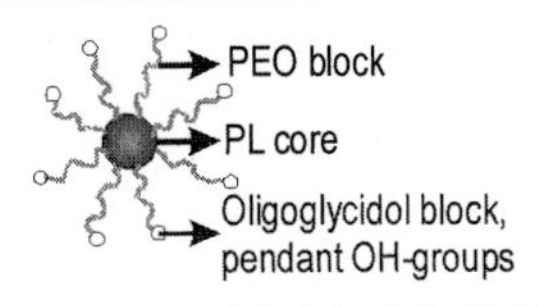

Scheme 1. Model of the PG-PEO-PL micelle

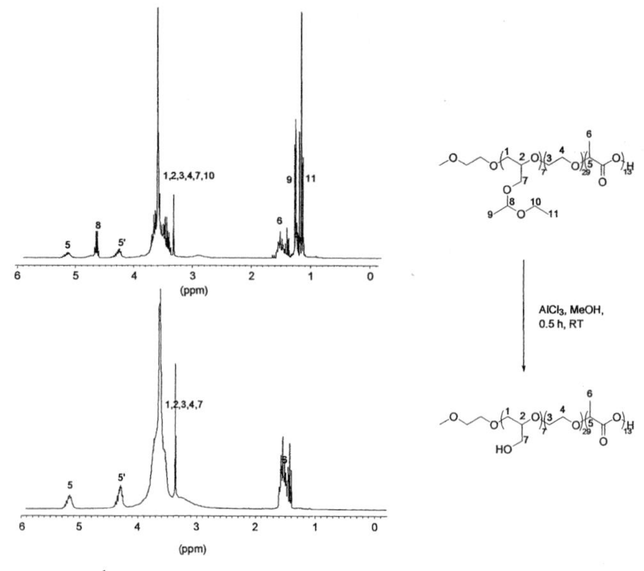

Figure 1. ^{1}H NMR Spectra of $EEGE_7EO_{29}L_{13}$ (1) and $G_7EO_{29}L_{13}$ (2)

Aqueous Solution Properties

Cloud Points. Clouding of aqueous dispersions of PG-PEO-PL series was not observed in the temperature range accessible for measurements. The same was observed for PEO-PL copolymers by other authors.[8] The stabilizing role of the protective PEO shell, which prevents dispersion particles from collapsing, is emphasized here by the presence of highly hydrophilic PG blocks positioned at the shell surface.

On the contrary, all the copolymers of the PL-PEO-PPO-PEO-PL series clouded at some temperature, which depended on the length of the PL blocks (Table 3). As seen from Figure 2, the copolymers with higher L content clouded at lower temperatures. The relatively long PPO block plays an important role for the phase separation since it is water soluble at low temperature only. The presence of a PPO block provides the system with thermo-sensitive properties, similar to PNIPAM block in PNIPAM-PL copolymers, described recently.[1]

Table 3. *Cmc* and CP data for tri- and pentablock copolymers

Composition	CP	cmc	
		20°C	37°C
$EEGE_7EO_{29}L_{13}$		3.46	3.46
$G_7EO_{29}L_{13}$		5.00	5.00
$EEGE_{16}EO_{137}L_{17}$		6.10	3.97
$EEGE_{16}EO_{137}L_{40}$		1.10	1.10
$G_{16}EO_{137}L_{40}$		4.50	4.50
$EEGE_{16}EO_{246}L_{44}$		2.67	2.67
$EEGE_4EO_{159}L_{25}$		4.69	3.86
$L_5EO_{30}PO_{34}EO_{30}L_5$	45	4.60	1.14
L_3P123L_3	36	0.40	0.09
L_4P123L_4	23	0.24	0.11
L_5P123L_5	16	0.33	0.13
L_6P123L_6	13	0.10	0.05
L_8P123L_8	12	0.18	0.11

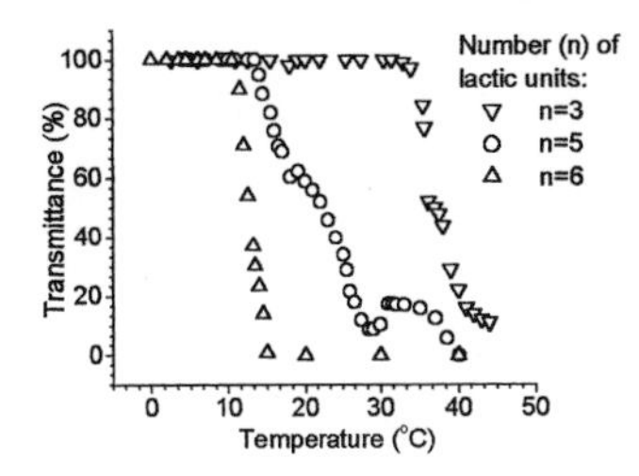

Figure 2. Clouding Curves of $L_nEO_{20}PO_{70}EO_{20}L_n$, C= 20 g/l

Cmc Measurements. Hydrophobic dye solubilization method was used extensively for the determination of *cmc*[8] and *cmt*[15] of amphiphilic copolymers containing hydrophobic PL block(s). The experimental *cmc* values of our multiblock copolymers are summarized in Table 3. *Cmc* of the copolymers depended predominantly from the length of the PL blocks – the higher the DP of the PL block(s), the lower the *cmc*.

PG-PEO-PL. The *cmc* values of the PEEGE-PEO-PL and the PG-PEO-PL copolymers were not affected by temperature, thus indicating for the appearance kinetically "frozen" systems due to the glassy state of the PL core.[4] The *cmc* of PG-PEO-PL are consistently higher than *cmc* of PEEGE-PEO-PL as a result of the increased hydrophilicity of the PG block (Figure 3a). By varying the length of the PEO blocks, one can adjust the hydrophilic-hydrophobic balance of the system. Copolymers with longer hydrophilic PEO blocks have weaker self-association tendency, hence their *cmc* appear at higher concentrations.

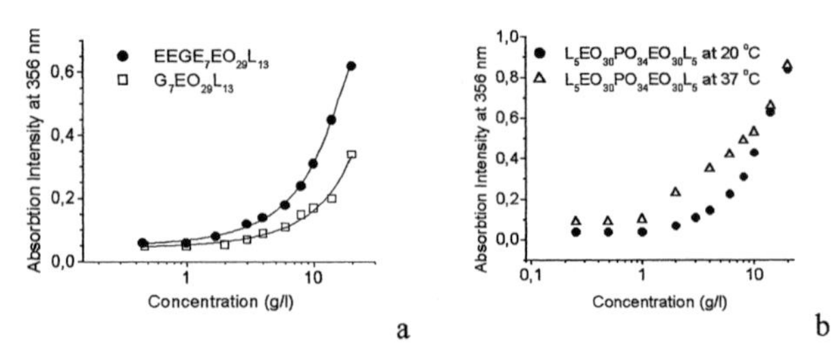

Figure 3. a) *cmc* curves of $EEGE_7EO_{29}L_{13}$ and $G_7EO_{29}L_{13}$ at 37°C ; b) *cmc* curves of $L_5EO_{30}PO_{34}EO_{30}L_5$ at 20 and 37°C

PL-PEO-PPO-PEO-PL. As a consequence of the PPO thermosensitive block, the micellization of PL-PEO-PPO-PEO-PL was observed to be temperature dependent in all cases (Figure 3b). Obviously the fluctuations of the *cmc* values listed in Table 3 are most probably due to the relatively high polydispersities of the samples. As can be seen from the proposed model in Scheme 2 the temperature-driven aggregation of PL-PEO-PPO-PEO-PL proceeds through two stages at least. Above *cmc* and at temperatures below the LCST of PPO flower-like micelles having hydrophobic PL and PEO-PPO-PEO shell are formed. Above the LCST, PPO becomes dehydrated, and as a result the PPO chains form a second hydrophobic domain.

Scheme 2.

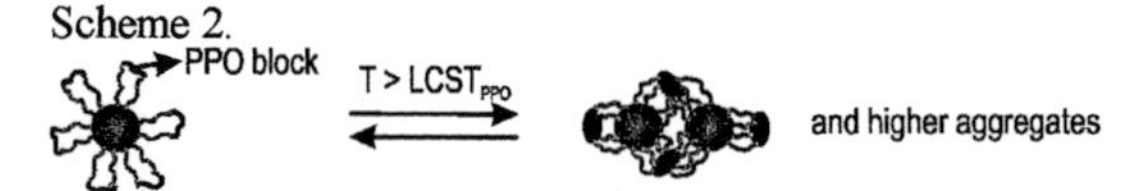

TEM Measurements

Figure 4a shows a micrograph of aggregates prepared from the copolymer $G_7EO_{29}L_{13}$ by the direct dissolution method. Granular PL cores of 40 to 50 nm are built of spherical particles of 15-30 nm of diameter. Most probably a secondary aggregation process occurred during the solvent evaporation. Particles obtained by $L_{10}EO_{20}PO_{70}EO_{20}L_{10}$ (Figure 4b) are considerably larger (from 100 to 170 nm) and are also of complex architecture.

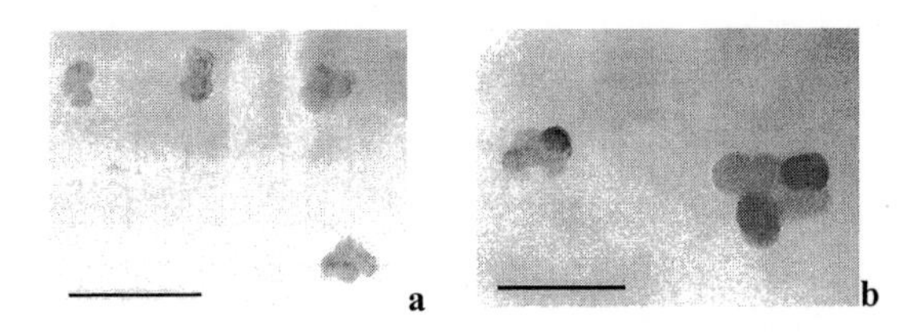

Figure 4. TEM micrographs of a) $G_7EO_{29}L_{13}$, (bar is 100 nm) and b) $L_{10}EO_{20}PO_{70}EO_{20}L_{10}$, bar is 200 nm.

Conclusions

Novel tri- and pentablock copolymers comprised of hydrophobic PL and hydrophilic polyoxiranes were synthesized mainly with the aid of anionic polymerization.

PG-PEO-PL copolymers formed stable functionalized micelles in aqueous media, while PL-PEO-PPO-PEO-PL solutions clouded, and cmc values were affected by the temperature. The hydrophilic-hydrophobic balance may be effectively varied by altering the number of L units.

The copolymers described in the present work are candidates for carriers in drug delivery systems. The hydroxyl - functionalized PG block opens an attractive approach for attachment of drug-targeting agents on PG-PEO-PL micelle surface. Thermosensitive PL-PEO-PPO-PEO-PL micelles can participate in immobilizing of two drugs – one in the PPO domains and the other – in PL core.

Acknowledgement

A.P. would like to acknowledge Macedonian Academy of Sciences as well as Bulgarian Academy of Sciences for the opportunity to collaborate and specialize in the Institute of Polymers, BAS.

[1] F. Kohori, K. Sakai, T. Aoyagi, M. Yokoyama, Y. Sakurai, T. Okano, *J. Contr. Rel.*, **1998**, *55*, 87.
[2] P. Jarret, C. B. Lalor, L. Chan, M. P. Redmon, A. J. Hickey, *Coll. Surf. B: Biointerfaces,* **2000**, *17*, 11.
[3] H. S. Yoo, T. G. Park, *J. Contr. Rel.*, **2001**, *70*, 63.
[4] T. Riley, S. Stolnik, C. R. Heald, C. D. Xiong, M.C. Garnett, L. Illum, S. S. Davis, S. C. Purkiss, R. J. Barlow, P. R. Gellert, *Langmuir,* **2001**, *17*, 3168.
[5] T. Riley, C. R. Heald, S. Stolnik, M.C. Garnett, L. Illum, S. S. Davis, S. M. King, R. K. Heenan, S. C. Purkiss, R. J. Barlow, P. R. Gellert, C. Washington, *Langmuir,* **2003**, *19*, 8428.
[6] I. Rashkov, N. Manolova, S. M. Li, J. L. Espartero, M. Vert, *Macromolecules,* **1996**, *29*, 50.
[7] S. M. Li, I. Rashkov, J. L. Espartero, N. Manolova, M. Vert, *Macromolecules,* **1996**, *29*, 57.
[8] S. Tanodekaew, R. Pannu, F. Heatley, D. Attwood, C. Booth, *Macromol. Chem. Phys.*, **1997**, *198*, 927.
[9] S. Tanodekaew, J. Godward, F. Heatley, C. Booth, *Macromol. Chem. Phys.*, **1997**, *198*, 3385.
[10] X. Chen, S. P. McCarthy, R. A. Gross, *Macromolecules*, **1997**, *30*, 4295.
[11] T. Fujiwara, M. Miyamoto, Y. Kimura, S. Sakurai, *Polymer*, **2001**, *42*, 1515.
[12] T. Fujiwara, M. Miyamoto, Y. Kimura, T. Iwata, Y. Doi, *Macromolecules,* **2001**, *34*, 4043.
[13] C. R. Heald, S. Stolnik, K. S. Kujawinski, C. De Matteis, M. C. Garnett, L. Illum, S. S. Davis, S. C. Purkiss, R. J. Barlow, P. R. Gellert, *Langmuir,* **2002**, *18*, 3669.
[14] S. H. Lee, S. H. Kim, Y. K. Han, Y. H. Kim, *J. Polym. Sci. A: Polym. Chem.*, **2002**, *40*, 2545.
[15] S. Y. Park, B. R. Han, K. M. Na, D. K. Han, S. C. Kim, *Macromolecules*, **2003**, *36*, 4115.
[16] Y. Yamamoto, Y. Nagasaki, M. Kato, K. Kataoka, *Coll. Surf. B: Biointerfaces*, **1999**, *16*, 135.
[17] K. Emoto, Y. Nagasaki, M. Iijima, M. Kato, K. Kataoka, *Coll. Surf. B: Biointerfaces*, **2000**, *18*, 337.
[18] E. Jule, Y. Nagasaki, K. Kataoka, *Langmuir*, **2002**, *18*, 10334.
[19] R. Tokar, P. Kubisa, S. Penczek, A. Dworak, *Macromolecules,* **1994**, 27, 320.
[20] A. Dworak, W. Walach, B. Trzebicka, *Macromol. Chem. Phys.* **1995**, *196*, 1963.
[21] A. Sunder, R. Hanselmann, H. Frey, R. Mülhaupt, *Macromolecules*, **1999**, *32*, 4240.
[22] D. Taton, A. Le Borgne, M. Sepulchre, N. Spassky, *Macromol. Chem. Phys.* **1994**, *195*, 139.
[23] A. Dworak, G. Baran, B. Trzebicka, W. Walach, *React. Funct. Polym.*, **1999**, *42*, 31.
[24] Ph. Dimitrov, E. Hasan, S. Rangelov, B. Trzebicka, A. Dworak, Ch. B. Tsvetanov, *Polymer,* **2002**, 43, 7171.
[25] M. Malmsten, P. Linse, K.-W. Zhang, *Macromolecules,* **1993**,*26*, 2905.
[26] K. Mortensen, D. Schwahn, S. Janssen, *Phys. Rev. Lett.*, **1993**, *71(11)*, 1728.
[27] H.G. Schild, D.A. Tirell, *J. Phys. Chem.*, **1990**, *94*, 4352.
[28] P. Alexandridis, J.F. Holzwarth, T.A. Hatton, *Macromolecules,* **1994**, *27*, 2414.
[29] A. Fitton, J. Hill, D. Jane, R. Miller, *Synthesis,* **1987**, 1140.
[30] L. Piao, Z. Dai, M. Deng, X. Chen, X. Jing, *Polymer*, **2003**, *44*, 2025.
[31] G. Schwach, J. Coudane, R. Engel, M. Vert, *J. Polym. Sci. A: Polym. Chem.*, **1997**, *35*, 3431
[32] V. V. Namboodri., R.S. Varma, *Tetrahed. Lett.*, **2002**, *43*,1143.
[33] H. Tsuji, Y. Ikada, *J. Polym. Sci. A: Polym. Chem.*, **1998**, *6*, 59.
[34] L. Piao, M. Deng, X. Chen, L. Jiang, X. Jing, *Polymer*, **2003**, *44*, 2331.
[35] Z. Zhong, M. J. K. Ankone, P. J. Dijkstra, C. Birg, M. Westerhausen, J. Feijen, *Polym. Bull.*, **2001**, *46*, 51.
[36] Z. Zhong, M. J. K. Ankone, P. J. Dijkstra, C. Birg, M. Westerhausen, J. Feijen, *Macromolecules*, **2001**, *34*, 3863.

Macromol. Symp. **2004**, *210*, 437-446

Complexes of Cellulose and Trypsin

Nina E. Kotelnikova, Sofia A. Mikhailova, Elena N. Vlasova*

Institute of Macromolecular Compounds, Russian Academy of Sciences, Bolshoy pr. 31, St. Petersburg, 199004 Russia
E-mail: kotel@mail.rcom.ru

Summary: Adsorption of trypsin to microcrystalline cellulose has been determined as functions of protein concentration and pH of the aqueous medium. The study of adsorption at several pH values indicates that interaction of trypsin to the microcrystalline cellulose interface is controlled by the electrostatic effect. The FTIR, desorption, and SEM data reveal that a part of trypsin is strongly bound to the microcrystalline cellulose matrix. Resulting complexes consist of microcrystalline cellulose, trypsin, and water.

Keywords: biopolymers; proteins

Introduction

The adsorption of proteins to solid interfaces and interactions between surfactants and proteins at liquid/solid interfaces are processes of significance to our daily life and in such fields as medicine, food processing, biotechnology, etc. Protein adsorption is also scientifically intriguing. The use of cellulose, a natural polymer, as a carrier for biologically active compounds is widely known [1]. However, pure cellulose has limited application as a matrix for proteins in spite of the fact that the combination of this natural polymer and some proteins can be promising for various medical applications. Powder microcrystalline cellulose (MCC) is used for the manufacture of drug forms of many preparations due to its valuable sorption properties [2].

It is known that protein adsorption from solutions to solid interfaces is a complicated process and adsorption value depends on many factors. The most important of them are surface properties of adsorbent, such as, its chemical composition, hydrophilicity or hydrophobicity, its charge, specific area and porosity.

The goal of this paper is to obtain multicomponent complexes on the basis of diffusion-adsorption interaction between MCC (matrix) and proteolytic ferment trypsin. Our results on adsorption of trypsin from aqueous solutions to the MCC interface are presented at various values of pH and concentration of initial ferment solutions. The possibility of trypsin to release from the interface appears to be also an important result of this study.

DOI: 10.1002/masy.200450649

Experimental

Microcrystalline cotton cellulose had particle size < 0.05 mm and was bone-dried up to moisture content not exceeding 1.0 wt %. The DP_v of MCC was 190. Pore volume, pore radius, specific area determined by method of water vapour sorption were 2.16 cm^3/g, 20 mcm, and 230 m^2/g, respectively. Content of the end carbonyl groups in the MCC chains is $3.2 \cdot 10^{-5}$ mol/g, of the carboxyl groups - $5.8 \cdot 10^{-6}$ mol/g. MCC is negatively charged at pH range 3.0-10.0 due to its carboxyl groups. Its isoelectric point (IP) is lower than 4.0 [3].

Crystallized bovine trypsin (SPOFA) was analytical grade and its MM is 23.8 kDa. Its IP is 10.6. Trypsin molecules are positively charged in solutions at pH<IP, and negatively charged at pH>IP (see Scheme 1). The relative activity of the initial trypsin sample determined by Erlanger method [4] is 76 %.

Scheme 1. Charge dependence on pH

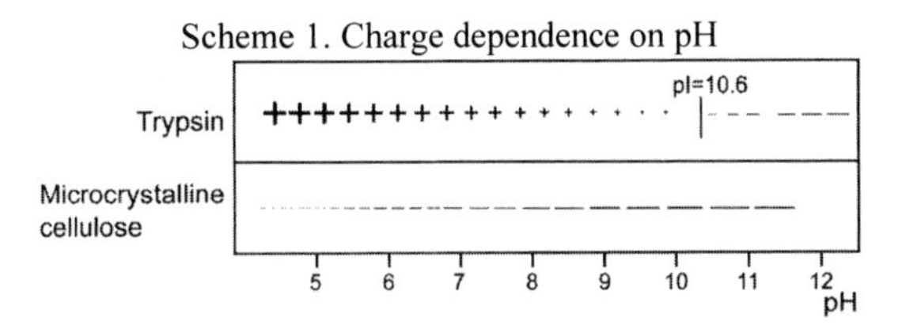

Adsorption interaction of MCC with trypsin in aqueous solutions and its desorption were carried out at pH range 4.2-12.0 as described elsewhere [5]. The initial ferment concentration (C_0) before adsorption as well as equilibrium trypsin concentration (C_{eq}) were estimated by Lowry method [6]. Adsorption values were calculated from difference between C_0 and C_{eq}. The quantity of desorbed trypsin was calculated in mass. % to the content of adsorbed ferment. The properties of the MCC-trypsin adsorbates in solid state were studied by FTIR and WAXS [1]. SEM study was performed to characterise changes on the MCC surface after adsorption of trypsin and its desorption. Samples were observed using raster electron microscope (Jeols JSM-35 CF).

Results and Discussion

Effect of pH of trypsin solutions on the adsorption values. Kinetics study

Dependence of adsorption values (AV, mg/g) on pH exhibits a hyperbola shape over all concentration range with a maximum at pH 8.0 - 8.5 (AV reaches 21.0 mg/g and 33.0 mg/g at C_0 2.3 mg/ml and 4.0 mg/ml, respectively). AV is markedly lower at pH 4.0 (AV=4.0 mg/g and 16.0 mg/g at C_0 2.3 mg/ml and 4.0 mg/ml, respectively) and at pH range 11.0-12.0 (AV= 3-4 mg/g and 13-16 mg/g at C_0 2.3 mg/ml and 4.0 mg/ml, respectively) (Fig. 1).

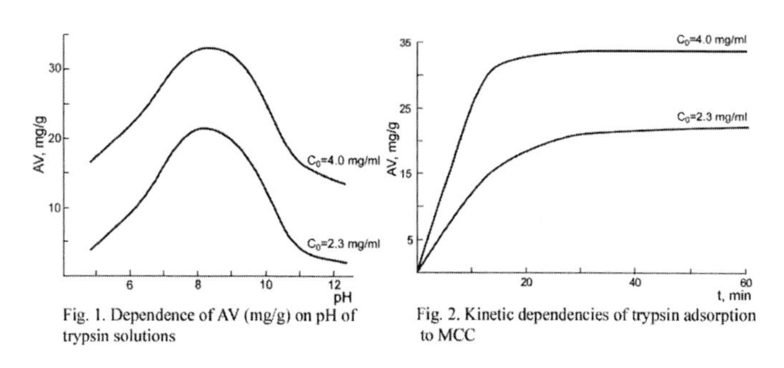

Fig. 1. Dependence of AV (mg/g) on pH of trypsin solutions

Fig. 2. Kinetic dependencies of trypsin adsorption to MCC

Kinetic dependencies of adsorption values of trypsin on MCC at different concentrations of trypsin initial solutions at pH 8.25 have a typical shape for adsorption processes taking place from solutions on porous matrices and consist of two main stages: the first one proceeds at a high rate and the second one proceeds at a slow rate (Fig. 2). The adsorption values depend on trypsin concentration. The maximum AV is 21 mg/g (at C_0=2.3 mg/ml) and 33 mg/g (at C_0 4.0 mg/ml), i.e. when C_0 increases 1.7 times AV increases 1.6 times.

Semilogarithmic adsorption anamorphoses confirm that this is two-stage process. Each stage is described by an equation of the pseudo-first-order reaction. The apparent rate constants of the first stage (K_1) are listed in Table 1. It can be seen that K_1 increases 1.1 times when C_0 of trypsin solutions increases 1.7 times. The K_1 value is lower than in the case of human serum albumin (HSA) adsorption to MCC [5].

Table 1. Maximum adsorption values, apparent rate constants, and Freundlich's constants of trypsin adsorption onto MCC

C_0, mg/ml	AV, mg/g	$K_1 \cdot 10_3$, min $^{-1}$	Freundlich's constants	
			1/n	$K_f \cdot 10_2$
2.3	21	7.5	1.24	6.53
4.0	33	8.5		

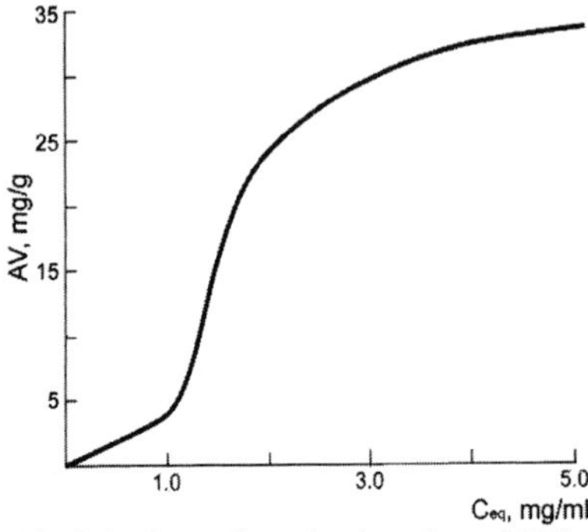

Fig. 3. Isotherm of trypsin adsorption to MCC

Isotherm of trypsin adsorption at pH 8.25 is given in Fig. 3. The AV is plotted against protein equilibrium concentration C_{eq}. Adsorption time was 24 hours. The adsorption isotherm does not exhibit a Langmuir's shape. In agreement with adsorption of other polymers and their complexes onto MCC [1, 5], it exhibits a pronounced S-shape, which is known to indicate the stepwise character of adsorption. The isotherm can be assigned to type IV, which characterises the combination of physical adsorption and chemisorption on the matrix and can not be described by Langmuir's equation but can be satisfactorily described by Freundlich's equation [1]. Freundlich's constants are listed in Table 1. Compare these constants with those obtained under adsorption of high molecular mass PVP [1] and HSA [5] from their solutions to the MCC, it can be seen that trypsin affinity (constant 1/n) to MCC is similar to that for PVP and higher than that for HSA. The relative ability of MCC to adsorb trypsin (constant K_f) is much higher than those for PVP and HSA.

Thus, complexes of MCC and trypsin under their adsorption interaction have been prepared.

Desorption of trypsin from the complexes

The ability of the complexes to emit trypsin into the solutions determines their properties as biologically active samples. It is important for their practical application.

The release values (RV) of trypsin under desorption at pH 2.0 and 8.25 depend on its concentration in the adsorbate: the higher the concentration, the higher the RV (Table 2). The maximum

RV is 69 mass. %. It is of interest that the retention of trypsin in cellulose matrix does not depend on pH solutions under desorption. Moreover, the content of ferment retained in cellulose matrix after desorption is a constant value (~10 mg/g, i.e. ~1 mass. %) and does not also depend on AV. It means that adsorption is partly reversible, and a part of adsorbed trypsin is strongly bound to the MCC matrix.

Table 2. Desorption of trypsin from the adsorbates and ferment activity in solutions after desorption

AV, mg/g	pH of solutions under desorption	RV, mass. %	Content of trypsin in matrix after desorption, mg/g	Solutions of trypsin after desorption	
				Concentration of trypsin, mg/ml	Relative activity of trypsin, %
21	2.0	51	9.8	0.34	16
33	2.0	69	10.2	0.76	-
21	8.25	52	9.6	0.30	18
33	8.25	69	10.2	0.70	18

The relative activity of trypsin in solutions does not almost depend on its concentration after desorption. This unexpected result should be explained in future. The relative activity is lower than that in the original trypsin sample. This can be due to the capacity of trypsin to autolysis in solutions. However, taking into account the high maximum RV of trypsin under desorption (69 mass. %) and its independence on pH, one should propose that trypsin can principally exhibit its ferment properties in both parts of the digestive tract of animals and people.

The FTIR study of the MCC-trypsin complexes

FTIR study of complexes MCC-trypsin reveals no changes in the chemical structure of the initial MCC. The most specific absorption bands in the protein spectra are located in the absorption range of 1640-1660 cm^{-1} (the amide I) and 1530 cm^{-1} (the amide II). FTIR spectra of the MCC (1), of the complex MCC-trypsin (2), of the initial trypsin in solid state (3), and of the subtraction spectrum of trypsin in the complex (4) in the fingerprint 1500-1775 cm^{-1} are given in Fig. 4. Compare spectra 3 and 4, it can be seen that there is no shift in the position of the absorption band in the range of 1532 cm^{-1}. The shape of this band is also not changed. This means that conformational changes of ferment molecules are not occurred. It is known that adsorption of proteins also strongly depends on the shape of molecules. Trypsin molecules have a globular shape. The existence of globular structures decreases contact capacities of macromolecules and surfaces and leads to preferable solvent adsorption. Therefore, strong competition between the MCC, trypsin, and a solvent (water) appears to be in the adsorption process. It was established that the resulting complexes are the intercalates of MCC, trypsin,and water due to high hydrophilicity of cellulose matrix. The presence of retained wa-ter (which has an absorption band in the range of 1640-1660 cm^{-1}) as in the MCC matrix as in ferment after adsorption does not allow to make any definite conclusion on the changes of ferment absorption band at 1640-1660 cm^{-1}.

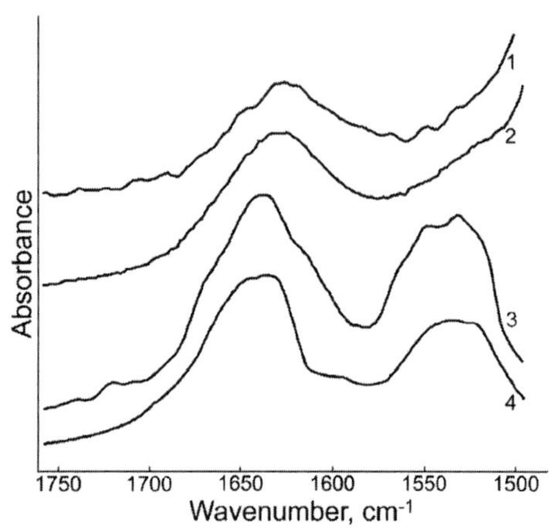

Fig. 4. FTIR spectra of the initial MCC (1), the complex MCC-trypsin (2), the initial trypsin in solid state (3), and a subtraction spectrum of trypsin in the complex (4) in the fingerprint 1500-1750 cm

Effect of the charge on the interaction of the MCC and trypsin macromolecules

The MCC The trypsin molecules are positively charged at pH range lower than its IP (pH 10.6) (see Scheme 1). The maximum AV under adsorption of trypsin to MCC is achieved at pH range 8.0-8.5. AV becomes much lower at pH range higher than IP of trypsin (pH 10.6) where its molecules are negatively charged. Thus, the adsorption of trypsin is determined by the charge of interacted molecules (at pH range lower than 8.5). This interaction can be performed via electrostatic bonding between carboxylic groups of MCC and amine groups of ferment.

Effect of trypsin adsorption on the MCC supramolecular and morphological structure (WAXS and SEM data)

WAXS study performed on the samples of adsorbates and desorbates in solid state show that supramolecular structure of cellulose under adsorption and subsequent desorption is not changed. The structure of cellulose modification I and its crystallinity in the MCC-trypsincomplexes are similar to those of the initial MCC sample. However, one can conclude that the length of cellulose crystallites diminishes and their shape changes.

According to SEM data (Fig. 5.A-D), all complexes MCC-trypsin after trypsin adsorption to MCC reveal remarkable surface disordering compare to the initial MCC fibre. It can be concluded that trypsin deeply penetrates into the fibres. Thus, fibres of the initial MCC sample (5.A) treated with trypsin solution exhibit a loose shape with destroyed fibril structure on the surface (5.B and 5.C). This is especially noticeable in the complexes after partial desorption of adsorbed trypsin (5.D). This corresponds to the high release of trypsin under desorption. Consequently, the mutual effect of adsorption interaction between cellulose and trypsin is noticeable. Thus, some trypsin properties (for instance, its solubility) depend on adsorption process. This ferment becomes partly insoluble after adsorption onto the cellulose matrix. On the other hand, the strong effect of trypsin adsorption on the MCC morphological structure is shown by SEM study of the complexes MCC-trypsin.

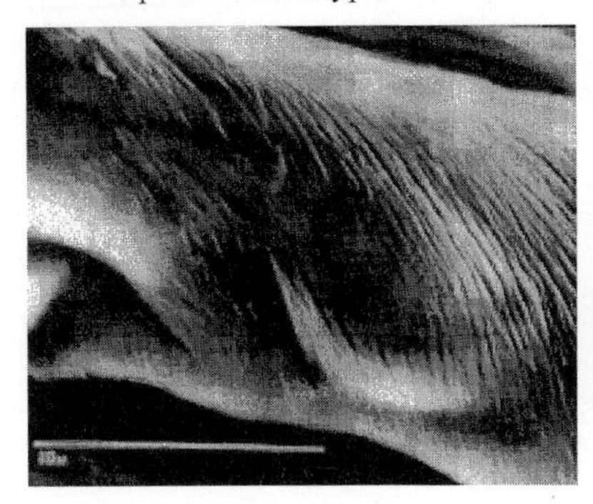

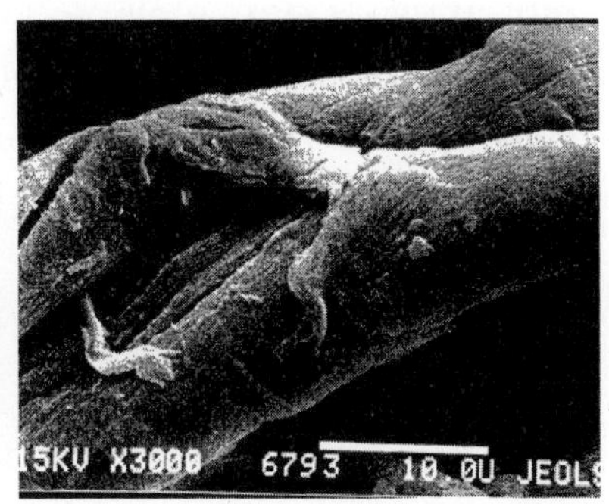

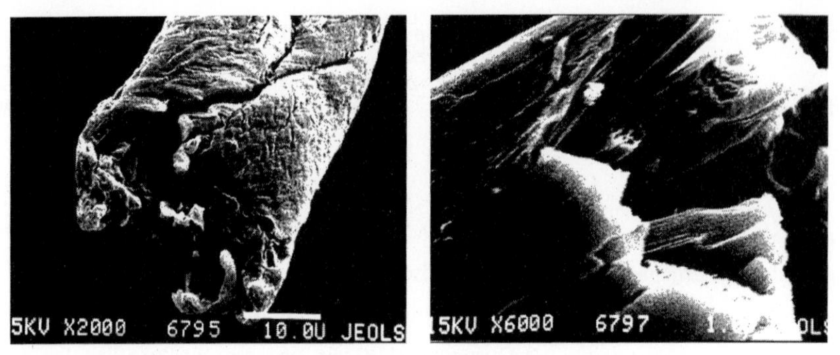

Fig. 5 (A-D). SEM micrographs of the initial MCC fibre (A) and the complexes MCC-trypsin (B, C - after trypsin adsorption to MCC; D - after trypsin desorption from the complexes)

Mechanism of the trypsin interaction with cellulose matrix under adsorption

The correlation between the amount of ferment and cellulose in the complexes was estimated, taking into account AV of trypsin after adsorption to MCC matrix and subsequent desorption from the complexes (Table 3). It is seen that the amount of cellulose chains corresponding to lmole of trypsin depends on the AV of trypsin. It has been already shown (Table 2) that the amount of trypsin retained in the MCC matrix after desorption is a constant value. This corresponds to 1 mole of ferment strongly bound to 77 moles of the MCC (Table 3, sample 5). The MCC surface is highly hydrophilic because it is enriched with OH groups. This favours multipoint ionic interaction between the amino-and OH-groups. As a result, stable bonds (=ÑO(H)···H^{+}NR and =ÑO^{-}···H^{+} NR) can be formed. Thus, three main ways for the MCC-trypsin complexes formation can be proposed. Stable bonds between MCC and trypsin are formed due to the interaction of ferment amine groups with the end aldehyde groups and surface carboxyl groups of MCC as well as with cellulose OH surface groups (Schemes 2-4).

Table 3. Correlation between the amount of trypsin molecules, cellulose chains, and active groups of cellulose in adsorbates MCC-trypsin at different adsorption values

Number of sample	AV, mg/g	Amount of cellulose chains, aldehyde- and carboxyl- groups to 1 mole of trypsin		
		MCC chains	-HC=O groups	-COOH groups
1	3	256	191	45
2	13	59	59	11
3	21	36	36	7
4	33	23	23	4
5 (after desorption)	9.9	77	77	14

Scheme 2. Formation of aldimine bonds via interaction of trypsin amino-groups and the end aldehyde groups of MCC.

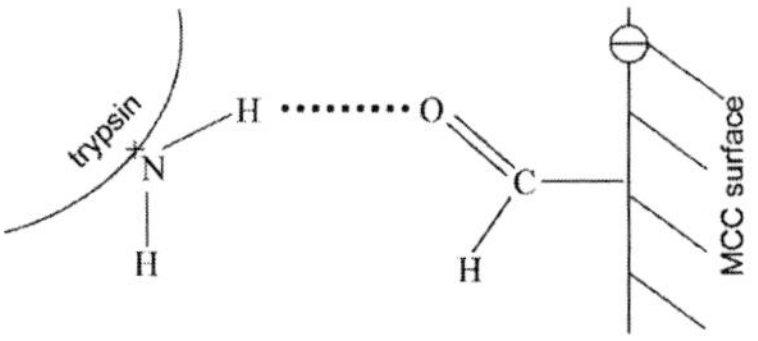

Scheme 3. Electrostatic interaction of trypsin amino- groups and surface carboxyl groups of MCC

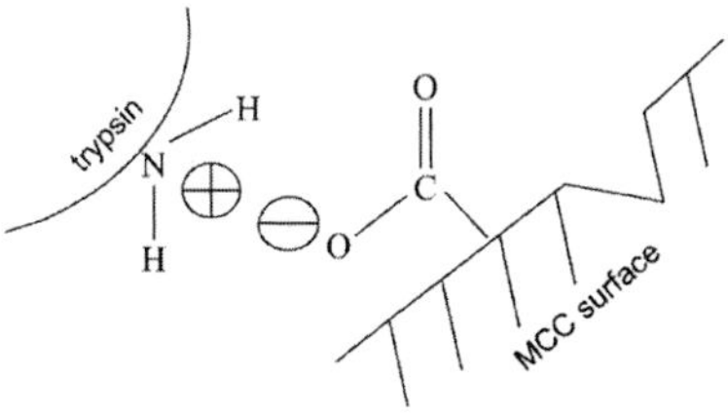

Scheme 4. Formation of H-bonds via interaction of trypsin amino- groups and OH groups of MCC

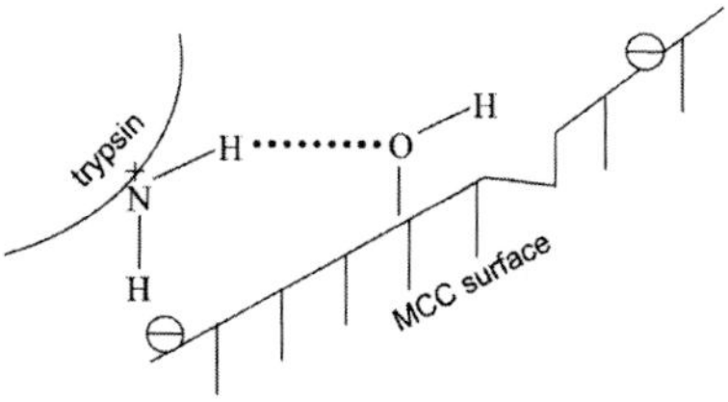

The important conclusion of this study is that the strong competition between MCC, trypsin, and a solvent (water) occurs in the adsorption process.

Conclusion

1. Adsorption of trypsin to microcrystalline cellulose matrix has been determined as functions of ferment concentration and pH of its solutions. The maximum adsorption value is reached at the pH range 8.0-8.5, i.e. lower than the isoelectric point of trypsin. This indicates that interaction of trypsin with the MCC interface is controlled by the electrostatic effect.
2. The isotherm with respect to the adsorption of trypsin is S-shaped and described by Freundlich's equation.
3. Desorption data reveal that a part of adsorbed trypsin is strongly bound to the MCC matrix. Mechanisms of bonding of trypsin to cellulose matrix are proposed.

Acknowledgement

The authors acknowledge Mr. V. Lavrentiev and Dr. N. Saprikina performing WAXS measurements and SEM experiments, Dr. B. Volchek for helpful discussions on FTIR results, and Prof. G. Vlasov for valuable comments.

(1) N. E. Kotelnikova, E. F. Panarin, A. V. Shchukarev, R. Serimaa, T. Paakkari, K. Jokela, S. V. Shilov, N. P. Kudina, G. Wegener, E. Windeisen, I.S. Kochetkova. Carbohydrate Polymers. 38(3), 239 (1999)
(2) S.G. Veinshtein, I.V. Julkevich, N.E. Kotelnikova, G.A. Petropalovsky. Bull. Experim. Biolog. Mediciny (russ). 2, 167 (1987)
(3) M.P. Sidorova, L.E. Ermakova, N.E. Kotelnikova, N.P. Kudina. Colloid J. (russ.). 63, 1, 106 (2001)
(4) B.F. Erlanger, N. Korowsky, W. Cohen. Arch. Biochem. Biophys. 95, 271 (1961)
(5) N.E. Kotelnikova, O.V. Lashkevich, E.F. Panarin. Macromol. Symp. 166, 147 (2001)
(6) O.H. Lowry, N.J. Rosenbrough, A.L. Farr, R.J. Randall. J. Biol. Chem. 193, 265 (1951)

Macromol. Symp. **2004**, *210,* 447-456

Automatically and Electronically Controllable Hydrogel Based Valves and Microvalves – Design and Operating Performance

Andreas Richter, Steffen Howitz, Dirk Kuckling, Katja Kretschmer, Karl-Friedrich Arndt*

Institute of Physical Chemistry and Electrochemistry, Dresden University of Technology, 01062 Dresden, Germany
E-mail: Andreas.Richter@chemie.tu-dresden.de

Summary: This paper describes automatically and electronically controlled valves and microvalves based on smart hydrogels. The operating performance of such devices will be discussed in dependence on various design parameters. Furthermore, it will be shown that hydrogel based valves are showing an outstanding possibility of miniaturization, a leakage free switching behavior up to a pressure drop of 8.4 bar, and a pronounced particle tolerance.

Keywords: automatic control; electronic control; hydrogel; microvalve; operating performance; valve

Introduction

In the last two decades a lot of hydrogels which are sensitive to temperature, ion and substance concentrations were developed. The distinction of the so-called stimuli-responsive or smart hydrogels is the property to change their volume reversible and reproducible by more than one order of magnitude even through very small alterations of certain environmental parameters. Therefore, an enormous importance for many technological and scientific applications was expected [1]. Particularly special mostly bio medical applications such as drug delivery systems suggest the rightness of this prediction. A first technological breakthrough at the development of hydrogel based valves and microvalves could be observed. An automatic valve for process engineering applications with sensitivities against the temperature, pH value, and contents of organic solvents was presented [2, 3]. In [4] an automatic microvalve was presented which possesses an automatic function to control a micro flow in dependence of pH value. Other hydrogel based microvalves can regulate micro flows as a function of pH value and

 DOI: 10.1002/masy.200450650

concentrations of glucose [5], and temperature [6]. However, a detailed description of the operating performance of hydrogel based valves, the influence of design parameters, and phenomena at volume phase transition of hydrogel actuators are mostly outstanding. However, their knowledge is absolutely essential for developing well functioning systems. The paper summarizes our research on the development of automatically and electronically controllable hydrogel based valves.

Experimental

Synthesis of Gel

The actuator material poly(*N*-isopropylacrylamide) (PNIPAAm) was prepared at following procedure. The crosslinking agent was *N,N′*-methylenebisacrylamide (BIS). The initiator and accelerator for the polymerization reaction were potassium peroxidisulfate (KPS) and *N,N,N′,N′*-tetramethyl-ethylenediamine (TEMED) (both from Aldrich Chemical Co.). NIPAAm and various amounts of BIS (1mol% to 10mol%, BIS4 – 4mol%) were dissolved in deionized water. The total monomer concentration was 0.53 mol/l. To initiate the polymerization reaction 0.3 mol-% of KPS and TEMED, respectively, were added to the oxygen free (bubbled with N_2) solution. After polymerization (ca. 12 h at room temperature) the PNIPAAm gel was immersed in deionized water for about one week to wash out non-reacted reagents. After drying the PNIPAAm BIS 4 gel the particles were obtained by milling and subsequent fractionating into different particle sizes using test sieves. The particles are irregular shaped.

The photo crosslinkable hydrogels were prepared from PNIPAAm copolymers bearing 4.5 mol-% light sensitive chromophores based on dimethyl maleimide. The copolymer solution (20 wt-% in butanone), containing 2 wt-% thioxanthone with respect to the polymer weight as photo sensibilisator was spin coated onto the SiO_2-support pretreated with 1,1,1,3,3,3-hexamethyldisilazane (HMDS) as an adhesion promoter. The film was subsequently dried and irradiated with a UV lamp (Hg lamp 400 W, wavelength 360 to 450 nm). Irradiation of the polymer resulted in an irreversible crosslinking by a [2+2]-cycloaddition. The non-crosslinked polymer was removed with a water alcohol mixture (20 wt-% ethanol, 80 wt-% water).

Microvalve Fabrication

The microvalves (see Fig. 1) are consisting of a channel structure support (5), a Pyrex glass cover (invisible in Fig. 1), and a circuit card (7) for electrical contacting. The channel geometry (3) and the actuator chamber (4 in Fig. 1a) are generated by a two-side etching process. Applied materials were Si wafers, or Pyrex glass wafers. Heating elements (110 nm in thickness, resistance 50 Ω, not shown in the Fig. 1a) and temperature sensors (8) were prepared by a platinum-thin-film system with lift-off patterning. A heating element (6) is located below the actuators on the channel structure support, while the temperature sensor is placed on the rear of the Pyrex glass cover. The microvalve can be controlled electronically by these elements. All layers were coupled by a combinated flip-chip and gluing technology.

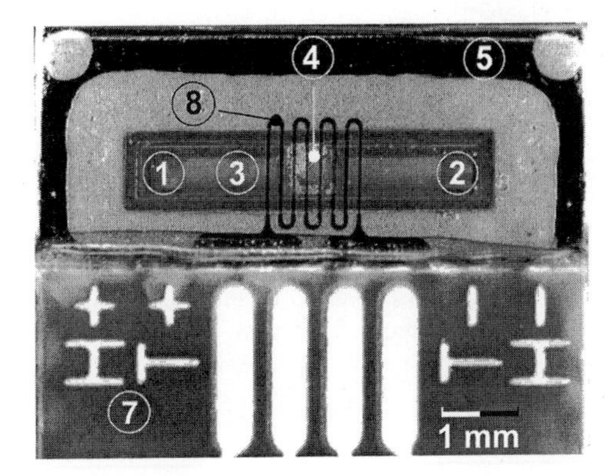

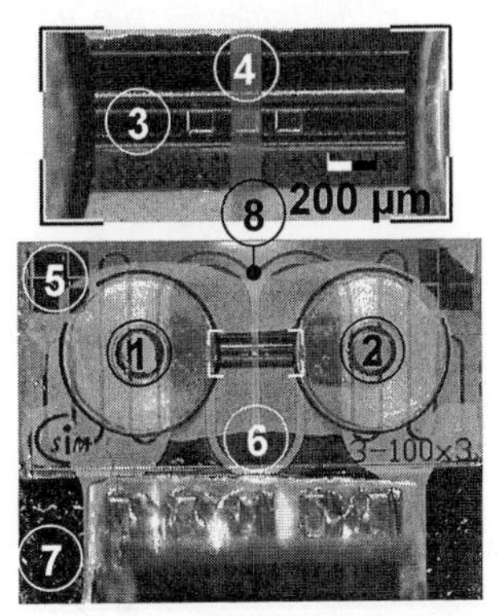

Figure 1. Photographs of hydrogel based microvalves; a) (left) – particle based microvalve, b) (right) - photopatterned microvalve; 1 – inlet; 2 – outlet; 3 – flow channel; 4 – actuators; Fig. 1a: actuator chamber filled with hydrogel particles, Fig. 1b: three actuator dots; 5 – structure layer; 6 – heating meander; 7 – circuit card; 8 – temperature sensor.

To assemble the microvalve body with the actuator hydrogel particles based on the homopolymer PNIPAAm (Fig. 1a) were manual incorporated into the actuator chamber. The microvalve shown in Fig. 1b contents three hydrogel dots. They were placed directly into the channel by photopatterning using the photo crosslinkable PNIPAAm copolymer. This valve set-up does not require an actuator chamber. A satisfying adhesion of the hydrogel dot to the underground structure was achieved by an adhesion promoter [7].

Macrovalve Design

This valve is made from stainless steel (see Fig. 2). The design is consisting of a big tube, which is the flow channel with inlet (1) an outlet (2). The actuator chamber, which is filled with hydrogel particles (3), is placed inside the flow channel. Two steel meshes (4) on both sides of the actuator chamber prevent the outflow of particles. A small tube is used as a bypass (5). All parts are connected via welding or screwing.

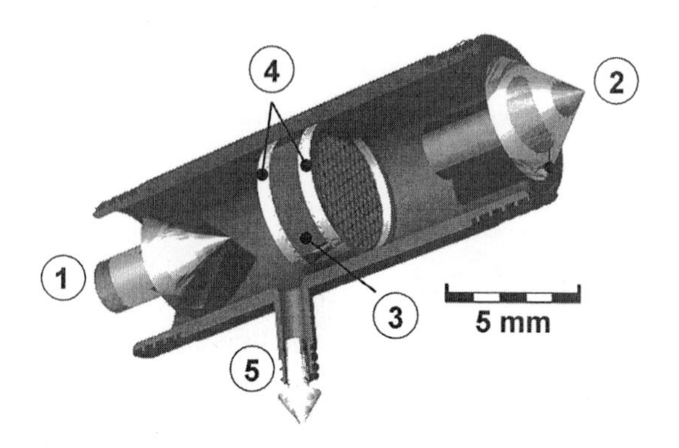

Figure 2. Schematic design of a automatic hydrogel based valve; 1– inlet, 2 – outlet, 3 – actuator chamber filled with hydrogel particles, 4 – steel mesh, 5 – bypass.

Results and Discussion

Operation Performance

A. Electronically Controllable Microvalves

By the direct placement of the hydrogel actuator in a flow channel the actuator has incessantly contact with the process medium, which acts as swelling agent as well. PNIPAAm and its photo crosslinkable copolymers exhibit lower critical solution temperature (LCST) behavior with a volume phase transition temperature (T_C) of approximately 33 °C (PNIPAAm, see Fig. 3a) or 21 °C to 29 °C (photo crosslinkable NIPAAm copolymers), respectively. Below T_C, e. g. at room temperature, the hydrogel is swollen, and above T_C the hydrogel is deswollen. In the normal case the medium temperature is below the T_C and the hydrogel seals the flow channel completely ("normally closed" function).

An electronic control of the valve shown in Fig. 1 was achieved by the heating element. To open the valves the gel actuators was warmed up above T_C with the heating element. The hydrogel actuator deswells and allows the fluid to flow through the channel. In order to obtain controllability of the valve, between completely opened and closed, a temperature sensor was integrated into the set-up to maintain a standard temperature.

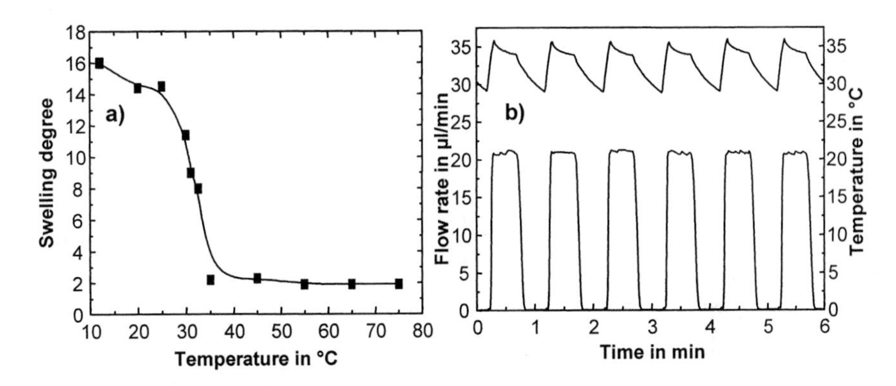

Figure 3. Principal behavior of PNIPAAm BIS 4; a) Swelling degree of PNIPAAm BIS 4 in dependence on temperature; b) Behavior of an electronically controllable microvalve with PNIPAAm BIS 4 particle actuator (size 500x500x50 μm^3) at power modulated work (power peak at 380 mW for 500 ms, retaining power 150 mW). Upper curve: temperature vs. time, lower curve: flow rate vs. time.

Fig. 3b shows the opening and shut-off behavior of an electronically controllable microvalve with an actuator chamber size of 500 µm x 500 µm x 200 µm filled with PNIPAAm BIS 4 particles (diameter of (82 ± 8) µm). In order to control the water flow rate the valve was temporarily warmed up with 380 mW to 35 °C. Subsequently the temperature was kept constant within 1 K with 150 mW. The fastest switching times, obtained for the particle based microvalves, were 300 ms at 350 mW for opening. Presently, the spontaneous shut-off time was approximately 2 s. The valve closes within 1 s when an external fan for cooling was used. The response times of the photo patterned valves are slower. This circumstance is probable caused by a smaller amount of groups which induce the phase transition behavior. A microvalve with dots of the size (250 µm x 250 µm x 50 µm is opening in 4 s and closing in 10 s.

B. Automatically Controlled Valve

The macrovalve which is illustrated in Fig. 2 is designed for an automatic function of hydrogel actuator. Hence, a large valve chamber and a bypass to the source of process medium is supposed. In the open state the process medium is passing the actuator chamber. If the valve is closed the solvent can flow back to the inlet using the bypass. This intake shunt guarantees a persistent presence of the actual stimulus at the hydrogel actuator.

The homopolymer PNIPAAm is likewise showing sensitivities to contents of alcohols in water (see Fig. 4a).

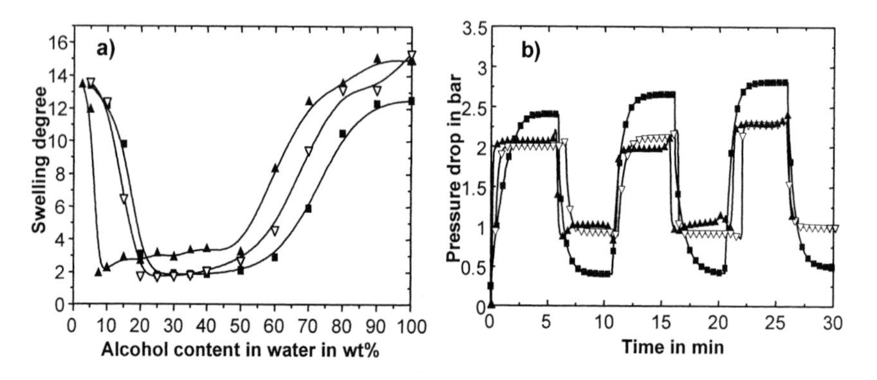

Figure 4. Behavior of PNIPAAm BIS 4 in dependence of alcohol concentration in water at room temperature (21 °C); a) Swelling degree of PNIPAAm depends on alcohol type and content; b) Behavior of a macrovalve loaded with PNIPAAm particles in dependence of alcohol content in water which was switched between 0 and 40 wt-% (■ - methanol, ▽ - ethanol, ▲ - 1-propanol).

It can be seen from Fig. 4a that two regions exist, which owe a volume phase behavior. The region at lower alcohol contents shows a decrease of the swelling degree with increasing alcohol concentration. The larger the carbon number of alcohol, the lower the concentration to obtain similar swelling degrees. The region at higher alcohol contents shows an inverse behavior.

This volume phase transition behavior of PNIPAAm can be used to obtain an automatic function of hydrogel based valves towards alcohol content in aqueous solutions. As shown in Fig. 4b in the range of the low concentrated phase transition the behavior of the valve is reproducible. The opening time is about 40 s while the closing time is ca. 25 s. The response time increases with lowering the carbon number of alcohol. At the higher concentrated volume phase transition the valve is closing at higher alcohol contents in water, and opening at lower alcohol contents.

Material, Design, and Operational Parameters

The swelling and deswelling process of hydrogels is diffusion controlled. Hence, the effective dimensions of hydrogel actuators influence strongly their switching behavior. To obtain small effective dimensions we use thin films of hydrogels or hydrogel particles as actuator material. In the last case the switching time is stronger dependent on the dimensions of particles than on the total actuator volume. However, only for macrovalves the particle size has a strong influence on the switching time (see Fig. 5a). For microvalves with a ratio of particle size to chamber size between 0.075 and 0.15 is the influence on the particle size of switching time negligible.

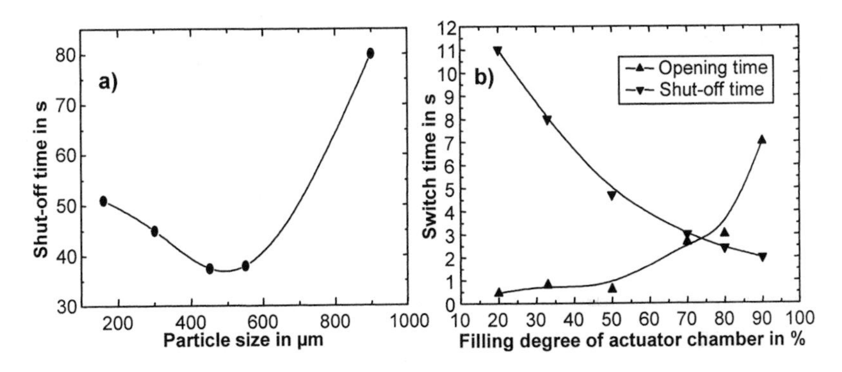

Figure 5. Switching time of hydrogel particle based valves in dependence of various design parameters. a) Shut-off time of an thermal stimulated valve with an actuator chamber of Ø 5mm x 4 mm, particle material PNIPAAm BIS 4; b): Switching time of a particle based microvalve in dependence on the filling degree of the actuator chamber, size of actuator chamber (800x800x200)μm^3, actuator material PNIPAAm BIS 4, dry particle size (82.5 ± 7.5) µm.

The most important parameter which influences the switching time of hydrogel actuator is the filling degree of the actuator chamber with dry hydrogel particles (see Fig. 5b). A small filling degree allows to obtain short opening times but the closing time is high. Up to a filling degree of 50% the opening time remains constant. A further increase of this parameter is resulting in small shut-off times while the opening time increases. In dependence on the priority of opening and shut-off time this parameter has to be optimized.

Another weighty design parameter is the size of the actuator chamber. The larger the actuator chamber the higher the switching time of a valve (see Fig. 6b). This figure shows also the effect of another design parameter, which is important for temperature stimulation. An increase of the heat capacity of valve body (valve with the actuator chamber size (800x800x200)μm³ induces a higher consumption of heating power to obtain fast opening times.

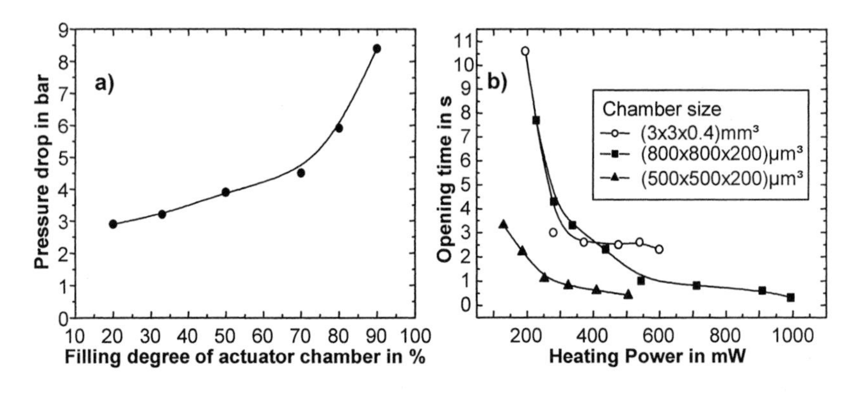

Figure 6. Leakage free pressure resistance in dependence on filling degree (a) and opening time in dependence on applied heating power for various hydrogel particle based microvalves (b).

The softness of hydrogels suggests that a leakage flow should be avoided because the actuator chamber can be fully filled with the swollen hydrogel. In fact, the maximal possible pressure drop, which is not showing a leakage flow, is a strong function of the filling degree of the actuator chamber with hydrogel particles (see Fig. 6a). At a filling degree of 90 % a leakage free pressure drop of 8.4 bar is obtainable. If the maximal pressure drop is exceeded the valve is mostly irreversible breaking through because the hydrogel particles were flushed out. Only at a very small filling degree the leakage flow is proportional to the increase of pressure drop.

For electronically controllable microvalves, which are using a thermal electronic interface, the temperature gradient is the most important operational parameter. Particularly the deswelling process and the opening time, respectively, are strongly dependent on the applied heating power (see Fig. 6b). Though, this dependence decreased after exceeding an angular point, which depends on the heat capacity of valve body. A further increase of heating power induces only a slight opening time decrease.

The softness of hydrogel actuator indicates a pronounced particle tolerance of hydrogel valves. A process medium, which includes splinters from polystyrene (irregular shape particle diameter less than 60 µm), did not impair the shut-off function of a microvalve with a chamber size of (800x800x200) μm^3. A leakage flow could not be observed. However, single splinters can remain in the actuator chamber. To remove such particles a flushing step must be executed.

At constant environmental and process conditions the behavior of hydrogel based valves is reproducible and shows a maximum error in reproducibility of less than 1 %. However, such stable conditions are not given. For neutral gels which include the most temperature sensitive hydrogels such as PNIPAAm a lot of cross-sensitivities are known. By contact with the process medium particularly salts and pH may shift the phase transition temperature of hydrogel actuators. Also cross-sensitivities to a number of organic solvents are known. This might cause serious malfunction if the composition of the process medium is not properly chosen. It is necessary to choose an appropriate hydrogel for a specific fluid.

Conclusion and Outlook

Presented results are showing that smart hydrogel based valves can offer four advantages:

- automatic sensor – actuator functions to various environmental parameters,
- outstanding possibility of miniaturization,
- leakage free switching behavior up to a pressure drop of 8.4 bar, and
- pronounced particle tolerance.

The actual switching times of hydrogel based microvalves (300 ms for opening and 2 s for shut-off) are sufficient for a lot of applications. However, switching frequencies higher than 10 Hz are hardly realizable. Hence, hydrogel based valves and microvalves cannot be used in highly dynamic applications.

The valve design which includes an actuator chamber could be very easy loaded with particles based on any hydrogel. Hence, it is a basic design usable to realize automatically controlled valves. However, integrated sensor – actuator functions could be only successfully realized if possible cross-sensitivities caused by composition of process media or environmental parameters will be respected. We believe that hydrogel based actuators can generate manifold developments in microfluidics, bio technology, chemical, and medical engineering.

[1] A.E. English, E.R. Edelman, T. Tanaka, "Polymer hydrogel phase transitions", In: T. Tanaka (ed.): *Experimental methods in polymer science: modern methods in polymer research & technology*. Academic Press, New York (2000), pp 547-589.
[2] K.-F. Arndt, D. Kuckling, A. Richter, *Polym. Adv. Technol.* **11,** 496 (2000).
[3] D. Kuckling, A. Richter, K.-F. Arndt, *Macromol. Mater. Eng.* **288,** 144 (2003).
[4] D.J. Beebe, J.S. Moore, J.M. Bauer, Q. Yu, R. H. Liu, C. Devadoss, B.-H. Jo, *Nature* **404**, 588 (2000).
[5] A. Baldi, Y. Gu, P. E. Loftness, R. A. Siegel, and B. Ziaie, Proc. 15th. Int. IEEE Conference on Microelectromechanical Systems 2002, Las Vegas, NV, 105 (2002).
[6] S. Mutlu, Cong Yu, F. Svec, C.H. Mastrangelo, J.M.J. Frechet, Y.B. Gianchandani, *Transducers* **1**, 802 (2003).
[7] J. Hoffmann, M. Plötner, D. Kuckling, W.-J. Fischer, *Sens. Actuators* 77, 139 (1999).

Macromol. Symp. **2004**, *210,* 457-464

Versatility of Potential Biomedical Use of Functional Polysuccinates

Jan Łukaszczyk, Piotr Benecki, Katarzyna Jaszcz, Monika Śmiga*

Silesian University of Technology, Department of Physical Chemistry and Technology of Polymers, 44-100 Gliwice, ul. M. Strzody 9, Poland

Summary: Polysuccinates with pendant allyl groups (PSAGE) were synthesized by melt copolymerization of succinic anhydride with allyl glycidyl ether and eventually other glycidyl ethers. It was found that PSAGE could be crosslinked by radical copolymerization with methyl methacrylate. Oxidized PSAGE considered as multifunctional epoxy resin was cured with use of glutaric anhydride to form solid material susceptible to hydrolytic degradation to water-soluble non-toxic products. Comb-like amphiphilic polysuccinates containing both pendant poly(oxyethylene) chains and epoxy groups have been synthesized as well and checked for their solubility in water. Properties of PSAGE-type polymers suggests their potential use as biomaterials and polymeric drug carries.

Keywords: biodegradable biomaterials; drug carriers; functional polysuccinates

Introduction

Increasing use of synthetic biodegradable polymers in medicine and pharmacy is a result of intensive studies on their synthesis and characteristics as well as on their new applications. Besides well known practical applications of the polymers in medical equipment, apparatus, single use devices etc. at the present they are used for more sophisticated purposes as biomaterials and polymer drug carriers. Polymer biomaterials are both termoplasts and thermosets, while for preparation of polymeric prodrugs non-crosslinkable polymers are used almost exclusively. Generally biodegradable polymers for medical application should respond to the same criteria of biocompatibility and biofunctionality as non-degradable polymers. Additionally the rate and the mechanism of biodegradation and the nature of degradation products must be taken into consideration.

At the present the most often used biodegradable polymers are thermoplastic polyesters derived from lactic or glycolic acids or caprolactone [1]. Since some time however growing interest in functional biodegradable polymers could be observed [2].

Recently we have described the synthesis of functional polysuccinates with pendant allyl groups [3,4], which could be utilized directly or after oxidation of double bonds. Polyesters composed of succinic acid, naturally present in living tissues, and of allyloxyglycerin could be

 DOI: 10.1002/masy.200450651

considered as potentially biodegradable and useful in medical applications.

The aim of this work was to recognize and to demonstrate some of possible modes of the use of functional polysuccinates in different areas of medicine.

Results and discussion

The functional polysuccinates with pendant allyl groups (PSAGE) were obtained by melt copolymerization of succinic anhydride (SA) and allyl glycidyl ether (AGE) at 120^0C in the presence of benzyltrimethylammonium chloride (BTMAC) and some water added [3,4]. Their alternating structure, i.e. lack of oligoether blocks was confirmed by NMR and MALDI-TOF spectra as well as by determination of double bonds content [3].

The content of allyl groups i.e. functionality of PSAGE could be easily adjusted by replacing a part of AGE with other glycidyl ethers, i.e. by changing AGE/GE feed ratio, where GE is the sum of glycidyl ethers. Biodegradable aliphatic polyester chain and pendant allyl groups enable consideration of potential biomedical application of PSAGE-type polymers or oligomers: in biodegradable bone cements, as biodegradable termosets for temporary implants and as amphiphilic polymeric drug carriers.

Biodegradable bone cements

In spite of low reactivity and even retarding effect of allyl groups in radical copolymerization [6], which is due to easy abstraction of allyl hydrogen and resonance stabilization of resulting radical, allyl ethers are used in some industrial systems, e.g. unsaturated polyester resins and coatings [7,8] cured by radical mechanism. This enable consideration of the solutions of PSAGE-type polyesters in low viscosity monomers as potential biodegradable thermoset resins, which could be cured in site of application like classical acrylate bone cements [9] or bone substitutes proposed so far [10]. Composition of PSAGE with some MMA is viscous liquid, which could be self-cured after addition of small amount of benzoyl peroxide (BPO) and N,N-dimethylamino-p-toluidine (DMPT) in reasonable time and with moderate exothermic effect, i.e. with peak temperature below 90^0C, which is acceptable for acrylic bone cement [11]. Selected results of initial study of curing of PSAGE-MMA compositions initiated by redox system: BPO-DMPT are shown in tab.1. Due to susceptibility of polysuccinate chain to hydrolytic degradation, the composition of PSAGE and MMA or other monomer could be considered eventually as biodegradable bone cement for temporary, resorbable support of damaged bone tissue, though improving of some properties of cured materials requires further studies.

Table1. Influence of the cement composition and initial curing temperature on hardening and on selected properties of the material

Components/Curing parameters/Properties		Cement composition/Properties		
PSAGE M_n=14800, MWD = 1.26 [g]		1	1	1
MMA [g]		0.75	0.75	1
BPO [wt.%]		4.37	2.23	1.96
DMPT [wt.%]		0.53	0.27	0.23
Initial temp.23°C	Setting time [min.]	3	4.25	8
	Peak temperature [°C]	48	45.6	-
Initial temp.37°C	Setting time [min.]	1.5	3.25	-
	Peak temperature [°C]	48.1	42.1	-
Compressive stress [MPa] at strain 20%		5.87	5.47	9.91
Extractable fraction [%]		25.1	40.0	16.9

Thermoset resins for pressureless casting

Polysuccinates with various degree of unsaturation were synthesized by replacing a part of AGE with butyl glycidyl ether. Polyesters containing 40-91 mol% of unsaturated repeating units derived from AGE were oxidized to respective poly(epoxypolyester)s. Pendant allyl groups in PSAGE were epoxidized quantitatively by *m*-chloroperbenzoic acid (MCPBA) in CH_2Cl_2 solution at room temperature as described elsewhere [5]. It was found that duration of the process required for full epoxidation (24-96h) increases, while decreasing degree of unsaturation, i.e. the content of AGE in the feed and thus in the polymer obtained. In spite of disappearance of double bonds observed in ^{1}H NMR spectra, the epoxide content (EC) determined for the resins obtained (EC = 0.14-0.41 eq/100g) was always lower than calculated theoretical one ($EC_{theor.}$ = 0.18-0.48).

Epoxyfunctional polysuccinates obtained (EPSAGE) have been considered as multifunctional epoxy resin, which could be cured with use of dicarboxylic acid anhydrides. One may expect that the resin cured with succinic or glutaric anhydrides is susceptible to hydrolytic degradation producing diglycerin ether or glycerin and succinic acid or succinic and glutaric one.

All end products of the hydrolysis, especially both acids naturally present in the body could be considered as well tolerated by living tissue [12].

Selected epoxyfunctional polysuccinates were cured with both anhydrides mentioned above, but due to technical problems with homogenization of the resin and SA ($Tm = 120^0$C), in systematic studies only GA was used due to its lower melting temperature ($Tm = 54$-55^0C). Appearant mechanical properties as well as other properties of various samples appeared to be diversified and dependent on functionality of initial resin and on the content of anhydride hardener. Selected properties of cured resins are gathered in tab.2.

Table 2. Influence of the composition of epoxyfunctional polysuccinates cured with glutaric anhydride (GA)* on their properties.

Composition of cured resin		Shore hardness	Tg	Water sorption	Swelling in CH_2Cl_2 vapors	Weight loss [%] during degradation**	
AGE/GE feed ratio	Amount of GA [mol/1mol of epoxy groups]	[^{0}Sh A or D]	[^{0}C]	[%]	[%]	28 days	70 days
1.0	0.0	82D	36.2	14.1	76	76.7	99.7
	0.2	-	31.2	12.0	-	74.4	97.1
	0.4	71D	-	7.4	79	62.1	96.4
	0.6	73D	34.9	2.6	79	11.9	98.5
	0.85	79D	46.7	2.1	69	82.4	99.8
0.6	0.0	-	8.6	10.5	203	48.2	99.9
	0.2	78A	18.4	7.7	169	-	-
	0.4	83A	-	5.6	149	-	-
	0.6	85A	14.9	3.2	129	81.6	98.4
	0.85	86A	17.6	3.0	106	-	-
0.4	0.0	-	1.0	9.8	214	29.6	99.3
	0.2	62A	6.0	12.6	194	27.5	100
	0.4	74A	-	9.9	192	11.5	95.0
	0.6	81A	3.7	6.8	155	30.6	97.9
	0.85	75A	4.7	7.1	151	92.0	99.9

*resins cured at elevated temperature at 150^{0}C according to procedure described elsewhere [4, 5],
**accelerated degradation at 70^{0}C in PBS, pH = 7.4

As could be expected, the resins with higher functionality and cured with higher amount of GA were harder and displayed higher Tg values and lower sorption capacity in swelling experiments. Cured resins of low functionality expressed with EC values after curing were soft and weak. This was observed also in fractographic analysis of the samples before and after partial hydrolytic degradation in phosphate buffer solution (pH 7.4). Only the sample of cured resin with highest functionality displayed fracture surface typical for glassy material, while the same material after some time of degradation as well as samples of cured resins with lower functionality displayed fracture surface typical for soft, weak material [5].

Unexpectedly thermal hardening of epoxyfunctional polysuccinates was observed even without addition of an anhydride, probably due to the polymerization and/or other reactions of pendant epoxy groups. There were observed also differences in the rate of hydrolytic degradation followed by determination of weight loss, but there was no simple relation between the composition and rate of degradation. The rate of degradation is probably connected with the network density (related closely to AGE/GE ratio in the initial resin), with the character of crosslinks (ether or ester ones) and additionally with hydrophilicity of the samples. As could be expected the highest rate of weight loss was observed for the samples with highest content of ester bonds, i.e. for those cured using 85% of GA in relation to EC. All the samples studied appeared to be susceptible for full degradation to water-soluble products. NMR studies of degradation products confirmed the presence of succinic and glutaric acids as well as glycerin and butanol [5], though the presence of diglycerin ether and soluble oligomers cannot be excluded.

The results obtained so far suggest that optimized compositions of EPSAGE and GA could be used for fabrication of resorbable, temporary implants, when thermoplastic biodegradable polymers cannot be used due to high costs of the mold.

Drug carriers for designing polymeric prodrugs

The comb-like epoxyfunctional polysuccinates with pendant poly(oxyethylene) chains were obtained in two ways:

1) in direct polyreaction with use of poly(ethylene glycol) macromonomers
2) by covalent grafting of poly(ethylene glycol) derivatives onto functional polymer

On the first way, the comb-like polyesters with poly(oxyethylene) and allyl pendant groups were synthesized from SA, AGE and ω-methoxy poly(ethylene glycol) glycidyl ether (MPEGGE). MPEGGEs were obtained in the reaction of epichlorohydrin with sodium alcoholate groups of ω-methoxy poly(ethylene glycol)s (MPEGs) with different molecular weight (nominal MW=750-5000). Amphiphilic comb-like polyesters were synthesized in melt as described above for PSAGE. The GEs were used in 20 % molar excess in relation to SA. MPEGGE/GE feed ratio was changed within the range of 0.1-0.7.

(n + m) CH_2—CH_2 (SA) + n CH_2—CH–CH_2OCH_2CH=CH_2 (AGE) + m CH_2—CH–$CH_2O(CH_2CH_2O)_x$–CH_3 (MPEGGE)

cat.
120°C

$\left[OCOCH_2CH_2COOCH_2CH(CH_2OCH_2CH{=}CH_2) \right]_n \left[OCOCH_2CH_2COOCH_2CH(CH_2O(CH_2CH_2O)_xCH_3) \right]_m$

MPEGGE/GE ratio in the resulting polymers (found by ^{1}H NMR and calculated from iodine number (IN)) was equal to MPEGGE/GE feed ratio for polyesters synthesized with use of macromonomers with lower molecular weight (MW of MPEG equal to 750 or 1100). In the polyesters synthesized with use of macromonomers with higher molecular weight, MPEGGE/GE ratio in polymers were higher than in feed (tab.3).

Table 3. Characteristics of comb-like polyesters with pendant allyl groups

MW of MPEG [Da]	MPEGGE/GE* feed ratio [mol/mol]	IN [g I_2/100g]	M_n (GPC) [Da]	M_w/M_n	MPEGGE/GE* in polyester calculated from		Solubility in water**
					^{1}H NMR	IN	
750	0.7	10.43	16700	1.30	0.73	0.71	+
1100	0.1	65.87	18700	1.43	0.13	0.12	+/-
	0.5	24.14	16100	1.22	0.44	0.40	+
2000	0.1	47.33	22300	1.65	0.14	0.13	+/-
	0.3	7.08	22700	1.27	0.66	0.61	+/-
	0.5	4.62	22400	1.21	0.74	0.71	+
	0.7	3.05	21250	1.22	---	0.79	+
5000	0.1	18.86	33300	1.38	0.21	0.18	+/-
	0.5	10.47	39800	1.21	---	0.30	+

*GE=MPEGGE+AGE, ** + water-soluble, +/- water-dispersible

The molecular weights (M_n values, GPC), of the comb-like polyesters were within the range of ca. 16700-39800 Da (tab.3) and were dependent on the length of the poly(oxyethylene) grafts, but practically did not depend on the degree of grafting.

The length and content of poly(oxyethylene) grafts in the polyesters was found to influence the solubility in water (tab.3).

The allyl groups in comb-like polyesters were oxidized to epoxy ones. Oxidation led to respective polyesters with oxirane groups in side chains.

$$\left[OC(O)CH_2CH_2C(O)OCH_2CH(CH_2OCH_2\overset{O}{CH{-}CH_2}) \right]_n \left[OC(O)CH_2CH_2C(O)OCH_2CH(CH_2O(CH_2CH_2O)_xCH_3) \right]_m$$

The length of the poly(oxyethylene) grafts and degree of grafting influenced the conversion of double bonds. Oxidation of the polyesters with the longest graft chains and with the highest degree of grafting was the most difficult. Elongation of the reaction time, increase of the polymer concentration in reaction solution or higher excess of MCPBA allowed however to achieve complete oxidation of allyl groups. Oxidized polyesters exhibited solubility characteristics similar to that before oxidation. In the second way the comb-like functional polyesters were obtained by covalent grafting of $MPEGNH_2$ onto epoxidized PSAGE. $MPEGNH_2$ was obtained (yield 33%) from ω-metoxy-poly(ethylene glycol), nominal MW=2000, by means of modified Gabriel synthesis in the reaction of tosylated MPEG with potassium phthalimide and next with hydrazine [13].

The grafting of $MPEGNH_2$ onto functional polymer was carried out in methylene chloride, at room temperature for 48 or 96 hours.

$$\left[OC(O)CH_2CH_2C(O)OCH_2CH(CH_2{-}O{-}CH_2\overset{O}{CH{-}CH_2}) \right]_n + CH_3O{-}(CH_2CH_2O)_x{-}CH_2CH_2{-}NH_2$$

$$\downarrow$$

$$\left[OC(O)CH_2CH_2C(O)OCH_2CH(CH_2OCH_2CH(OH)CH_2NHCH_2CH_2{-}(OCH_2CH_2)_x{-}OCH_3) \right]_y \left[OC(O)CH_2CH_2C(O)OCH_2CH(CH_2OCH_2\overset{O}{CH{-}CH_2}) \right]_z$$

The molecular weight of the poly(epoxyester) before grafting (calculated from end groups analysis (acid value: AV=37 mg KOH/1g, hydroxyl value: HV=12 mg KOH/1g)) was 2289 Da. The conversion of epoxy groups calculated from ^{1}H NMR was 6% when the reaction was run 48h and 20% after 96h. Calculated molecular weight of obtained comb-like polymers was ca. 3740 in the first case and 6740 in the second case. The products were water-dispersible.

Conclusions

Pendant allyl groups in the polysuccinates could be utilized directly (in radical copolymerization with low viscosity monomers) or could be utilized indirectly (after transformation into other functional groups, e.g. epoxy ones).

Polysuccinates with allyl pendant groups could be used probably in formulation of biodegradable, injectable bone cements, but problems with low reactivity of functional groups must be solved.

Epoxidized polysuccinates, i.e. respective poly(epoxyester)s could be useful as thermoset resins for presureless casting of temporary medical implants of desired individual shape.

Poly(epoxyester)s derived from functional polysuccinates could be modified in such a way, that they would become water-soluble, while preserving a part of their epoxy groups useful for coupling drugs to form polymeric prodrugs.

Biocompatibility of the end products of hydrolytic degradation may be predicted in case of medical implants made of poly(epoxyester)s cured with glutaric anhydride (succinic acid, glutaric acid, glycerin), while only expected for other biomaterials or drug carriers based on polysuccinates.

Acknowledgement

Financial support of European Graduate College "Advanced Polymer Materials" is gratefully acknowledged.

[1] D. Bendix, *Polym. Degrad. Stab.*, **1998**, *59*, 12; B.L. Seal, T.C. Otero, A. Panitch, *Mater. Sci. Eng.* **2001**, *R34*, 147.
[2] H. Chung, D. Xie, A.D. Puckett, J.W.Mays, *Eur. Polym. J.*, **2003**, *39*,1817; H. Uyama, M. Kuwabara, T. Tsujimoto, S. Kobayashi, *Biomacromolecules*, **2003**, *4*, 211.
[3] J. Łukaszczyk, K. Jaszcz, *React. Funct. Polym.*, **2000**, *43*,25.
[4] J. Łukaszczyk, K. Jaszcz, *Macromol. Chem. Phys.*, **2002**, *203*, 301.
[5] J. Łukaszczyk, K. Jaszcz, *Polym. Adv. Technol.*, **2002**, *13*, 871.
[6] G. Odian, *Principles of Polymerization*, 2nd ed., Wiley-Interscience, New York 1981, p.251; J.C. Bevington, T.N. Huckerby, B.J. Hunt, A.D. Jenkins, *J. Macromol. Sci.-Pure Appl. Chem.*, **2001**,*38*, 981.
[7] H.J. Traenckener, H.U. Pohl, *Angew. Makromol. Chem.*, **1982**, *108*, 619.
[8] G. Rokicki, E. Szymańska, *J. Appl. Polym. Sci.*, **1998**, *70*, 2031.
[9] D.H. Kohn, P. Ducheyne, "Materials for Bone and Joint Replacement", in: Cahn R.W., Haasen P., Kramer E.J. (Eds.), "*Materials Science and Technology*", vol.14, "*Medical and Dental Materials*", VCH, Weinheim 1992.
[10] M.D. Timmer., S.B. Jo, C.Y. Wang, C.G. Ambrose, A.G. Mikos, *Macromolecules*, **2002**, *35*, 4373; A.K. Burkoth, K.S. Anseth, *Biomaterials*, **2000**, *21*, 2395.
[11] ISO/FDIS 5833:2001 "*Implants for surgery – Acrylic resin cements*".
[12] S.Rose, S.Bullock, *The Chemistry of Life*, Penguin Books, London 1991.
[13] S. Furukawa, N. Katayama, T. Lizuka, I. Uwabe, H. Okada, *FEBS Lett.* **1980**, *2*, 239.

Macromol. Symp. **2004**, *210*, 465-474

Temperature-Sensitive Poly(vinyl methyl ether) Hydrogel Beads

*Daniel Theiss, Thomas Schmidt, Karl-Friedrich Arndt**

Institut für Physikalische Chemie, Technische Universität Dresden, Mommsenstr. 13, 01062 Dresden, Germany
Fax: +49-351-463-32013;
E-mail: karl-friedrich.arndt@chemie.tu-dresden.de

Summary: Temperature-sensitive hydrogel beads were prepared by radiation crosslinking of poly(vinyl methyl ether) PVME spheres wrapped in Ca-alginate. The obtained gel beads have diameters in the sub-millimeter or millimeter range (depending on the PVME concentration). They were characterized by sol-gel analysis, swelling measurements, and differential scanning calorimetry.
The gel content g increases with increasing radiation dose D. The swelling degree Q_V decreases with increasing PVME concentration c_p and increasing D. In comparison to PVME bulkgels the phase-transition temperature of the synthesized PVME gel beads is a little decreased.

Keywords: alginate technique; electron beam irradiation; poly(vinyl methyl ether); temperature-sensitive hydrogel beads

1. Introductions

'Smart' hydrogels are changing their volume and their mechanical properties as a result of small changes in the properties of the surrounding medium [1-3]. Hydrogels with a lower critical solution temperature LCST are in the swollen state below and in the shrunken state above this temperature. A well-known polymer with LCST behavior in aqueous medium is poly(vinyl methyl ether) PVME with $T_{cr} \approx 34°C$ [4]. High-energy radiation induces in aqueous PVME solutions a radical process and the polymer chains crosslink [5,6].
PVME hydrogels as bulk material were synthesized by electron beam or γ-ray irradiation of its aqueous solution to form temperature-sensitive hydrogels [7-15]. These gels were applied for

 DOI: 10.1002/masy.200450651

mechanical devices [16,17], for thrombogenicity studies [18], or as fiber material [19].

The synthesis of hydrogels with reduced dimensions by using radiation techniques was reported [20-23]. Pulsed electron beam irradiation of diluted polymer solution leads to intramolecularly crosslinked macromolecules, the so-called nanogels [20]. Microgels can be formed by a radiation induced polymerization and crosslinking of emulsified monomer solutions [21,22], or by the irradiation of phase-separated structures of a temperature-sensitive polymer [23].

Another well-investigated method of the synthesis of micro-sized hydrogels is the alginate technique [24-29]. Calcium-crosslinked alginate spheres are targets for the polymerization and crosslinking of monomers. The dimension of the so-called gel beads is influenced by the droplet size of the alginate beads.

The aim of the work was the synthesis of temperature-sensitive hydrogel beads in the sub-millimeter and millimeter range. We will demonstrate, that micro-hydrogels can be formed by irradiation of Ca-alginate stabilized PVME bead suspensions above the phase-transition temperature of PVME. After crosslinking the alginate layer was removed by washing with EDTA solutions. The possibilities to regulate the diameter of the beads by the polymer concentration was investigated. The gel beads were characterized by sol-gel analysis, swelling measurements, and DSC measurements.

2. Experimental

2.1 Materials

PVME was obtained as an aqueous solution (50 *wt.%*) from BASF (Lutonal M40). Its molecular weight was measured by static light scattering in 2-butanone to M_w = 57,000 *g/mol*. In the experiments PVME solutions were used without further purification. Sodium alginate (Aldrich), ethylene diamine tetraacetate EDTA (Grüssing), calcium chloride $CaCl_2$ (Grüssing), and acetone (Merck) were all used as received.

2.2 Synthesis

The gel beads were prepared according to the method developed by Park and Choi [24]. An interpenetrated network (IPN) was prepared by the gelation of Ca-alginate to form spherical bead shapes. Temperature-sensitive gel beads were obtained by radical polymerization and crosslinking of poly(*N*-isopropyl acrylamide) PNIPAAm.

The alginate technique should be used for the synthesis of PVME gel beads, too. PVME was incorporated into the alginate beads and subsequently irradiated with electron beam. Sodium alginate (1-2 *wt.%*) and PVME (1-8 *wt.%*) were dissolved in bidestilled water and degassed with nitrogen. The solutions were injected by using a syringe into 300 *ml* aqueous $CaCl_2$ solution (3 *wt.%*) heated to $T = 40°C$. The Ca-ions crosslink the alginate and globular PVME/alginate beads are formed. The scheme of the gel bead preparation is shown in fig. 1.

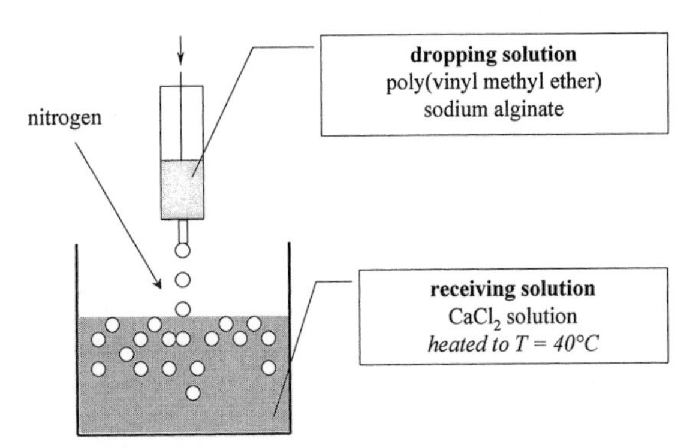

Fig. 1: Scheme of the synthesis principle of temperature-sensitive PVME gel beads. The aqueous PVME/alginate solutions were dropped into the $CaCl_2$ solution ($T = 40°C$) and subsequently irradiated.

The gel beads suspensions were kept at this temperature (at low temperatures parts of PVME diffuse out of the beads) and irradiated with accelerated electrons. The irradiation experiments with an electron beam were carried out with an electron accelerator ELV-2 (Budker Institute of Nuclear Physics Nowosibirsk, Russia). The energy of the electrons was 1.5 *MeV* at a beam power of 20 *kW*. At constant value of beam current the absorbed dose depends on the exposure time (typical < 1 *min*). The radiation dose *D* was varied from 60 *kGy* to 120 *kGy*. The stability of the alginate-shell during high-energy irradiation (possible degradation processes induced by the radiation) was proofed.

2.3 Characterization of gel beads

Sol-gel-analysis

After radiation crosslinking the samples were dried in vacuum for several days and weighted. The uncrosslinked polymer has to be removed. The beads were put into a Soxhlett thimble and the sol was extracted with acetone in a Soxhlett extractor for 5 *d*. The crosslinked alginate was not removed by this method. After the extraction the gel was weighted again. The gel content *g* of PVME is determined as ratio of the mass of the gel after extraction (m_{gel}) to the mass before removing the sol content ($m_{gel} + m_{sol}$).

$$g = \frac{m_{gel}}{m_{gel} + m_{sol}}; \qquad s = 1 - g \tag{1}$$

The alginate shell was removed by washing with EDTA solution, and temperature-sensitive PVME gel beads were obtained.

Degree of swelling and differential scanning calorimetry (DSC)

The degrees of swelling (Q_m) in water were measured by weighting the swollen, extracted gel ($m_{swollen}$) and the non-swollen (dry) extracted sample (m_{dry}).

$$Q_m = \frac{m_{swollen}}{m_{dry}} \tag{2}$$

$$Q_V = \frac{V_{swollen}}{V_{dry}} = \left(\frac{d_{swollen}}{d_{dry}}\right)^3 \tag{3}$$

In order to determine the temperature dependence of the degree of swelling in water (Q_V) the change of dimension of the beads was monitored using a digital video system. A JVC (TK C 1380) camera was connected to a PC through a real time video digitalizer card. The diameter *d* of the gel beads, both in the swollen ($d_{swollen}$) and in the dry state (d_{dry}), was followed on the magnified picture by an image analyser program (analysis Doku 2.11.007, Soft Imaging Systems GmbH 1986-97, VGA Driver, Version 1.1a). This method enables us to measure of very small changes in diameter (one pixel on the screen) on the real time video image. The ratio of the dimension of the swollen gel (heating rate appr. 15 *min* / 2 *K*) to the dimension in the dry state is proportional to the volume degree of swelling Q_V (equ. 3). For DSC measurements the 2920 Modulated DSC (TA Instruments) was used. The heating rate was 5 *K/min*.

3. Results and Discussions

3.1 Sol-gel analysis

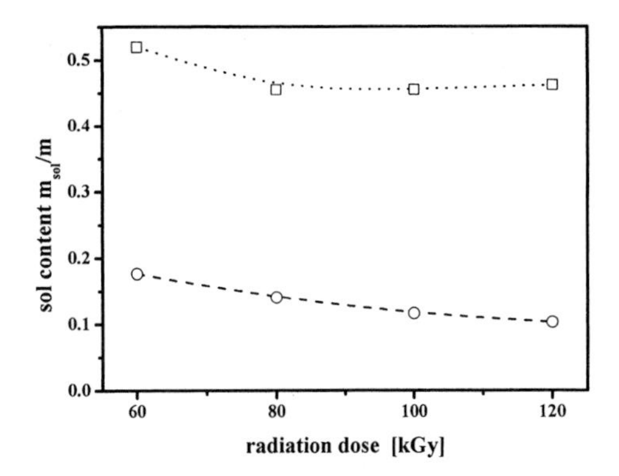

Fig. 2: Sol-gel analysis of the PVME gel beads after extraction with acetone (O) and EDTA (□) solution ($c_p = 6\ wt.\%$, $c_{alginate} = 2\ wt.\%$).

The sol contents were analyzed according to their different extraction steps. First, the uncrosslinked PVME was removed by Soxhlett extraction with acetone. Second, the alginate shell was removed by washing with EDTA solution.

The whole sol content (uncrosslinked PVME and alginate) is relatively high (about 50%). However, alginate network is only the target for the crosslinking process. In the literature it is described that alginate as bulk material or in aqueous solution mainly undergo degradation due to the high-energy radiation [30-32]. Parts of the alginate shell can be removed during the irradiation experiment or extraction. Analyzing the sol content of PVME with regard to the PVME gel without alginate the sol content decreases with increasing dose. The values of the sol content are in the same range like for bulkgels [13,14]. These sol contents were analyzed according to CHARLESBY-PINNER [33] (equ. 4).

$$s + \sqrt{s} = \frac{p_0}{q_0} + \frac{1}{q_0 \cdot u_1 \cdot D} \tag{4}$$

where p_0 is the fracture density per unit dose, q_0 the density of crosslinked units per unit dose, u_1 is the initial number average degree of polymerisation and D the irradiation dose. The gelation dose D_g is determined for $s = 1$. The plot according equ. 4 is shown in fig. 3.

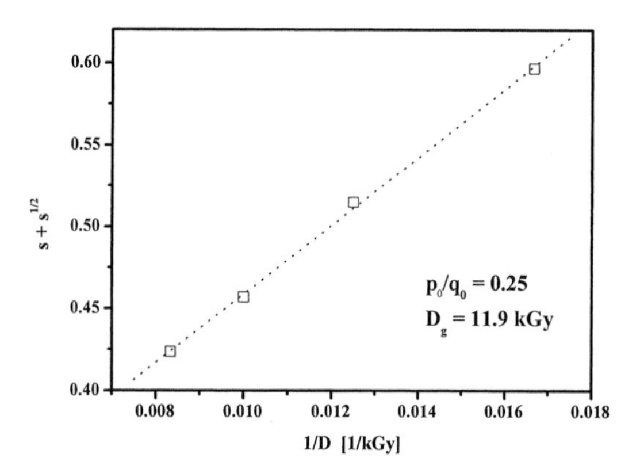

Fig. 3: Results (c_p = 6 *wt.%*, $c_{alginate}$ = 2 *wt.%*) of the analysis according to CHARLESBY-PINNER [33]. The gelation dose D_g is 11.9 *kGy* and the value of p_0/q_0 is 0.25.

Evaluating the data of the sol contents in dependence on the radiation dose leads to a linear behavior. The values of D_g = 11.9 *kGy* and p_0/q_0 = 0.25 were calculated and are typical for PVME (PVME solutions (20 *wt.%*) γ-ray irradiated D_g = 10.9 *kGy* and p_0/q_0 = 0.25 [14]).

3.2 Swelling measurements

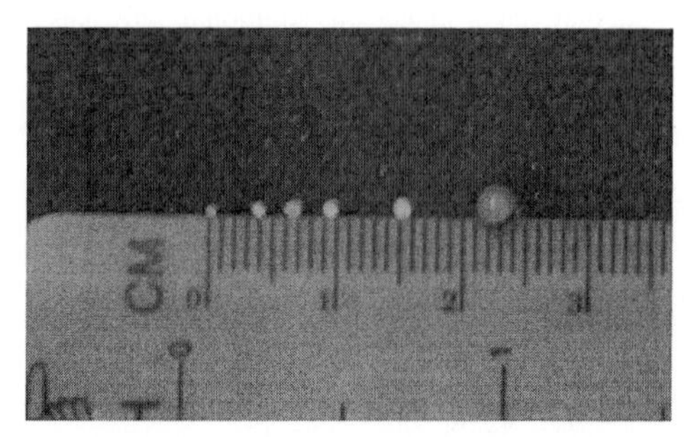

Fig. 4: Photograph of the PVME gel beads in dependence on PVME concentration c_p (from the left to the right side: 1 *wt.%*, 2 *wt.%*, 4 *wt.%*, 6 *wt.%*, 8 *wt.%* in the dry state, and 8 *wt.%* in the swollen state, constant alginate concentration $c_{alginate}$ = 2 *wt.%*).

The degrees of swelling were obtained, both by measuring the mass and the volume of the gel beads, in dependence on the temperature *T*.

Fig. 4 shows a photograph of the gel beads in dependence on the concentration of PVME (constant alginate concentration). The diameters (in the range of *mm*) of the dry gel beads increase with increasing PVME concentration. In the swollen state the dimensions of the beads strongly increase compared to the dry gel beads, but the same tendency was observed.

The degrees of swelling have been analyzed in dependence on the radiation dose, too. Fig. 5 shows the dose dependence of the degree of swelling (both Q_m and Q_V, independently determined). As expected, the degrees of swelling decrease (Q_V from 20 to 9) with increasing radiation dose. At higher doses more radicals are formed and the crosslinking density increases. The volume degree of swelling is almost higher than the mass degree of swelling. The same tendency was obtained as in the case of PVME bulkgels in the same range of dose, but at a higher polymer concentrations (20 *wt.%*).

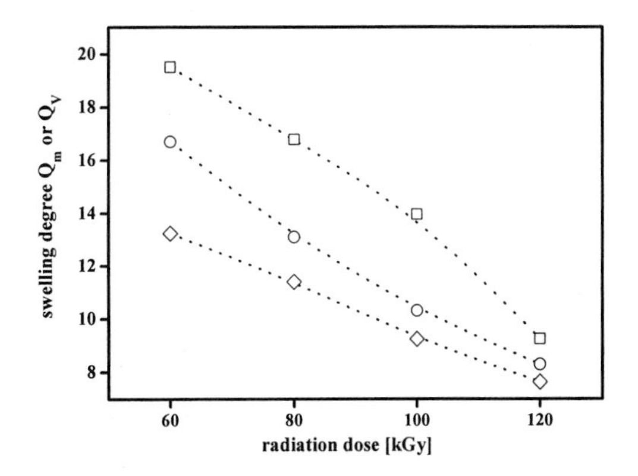

Fig. 5: Swelling degrees Q_V (□) and Q_m (○) of PVME gel beads (c_p = 6 *wt.%*, $c_{alginate}$ = 2 *wt.%*) in dependence on the radiation dose *D* at *T* = 20°*C* (Q_m of PVME bulkgels (◇) synthesized with 80 *kGy* electron beam were added for comparison).

3.3 Temperature-sensitivity

The temperature dependent properties of the PVME gel beads were determined by swelling measurements and by DSC measurements.

Swelling measurements (Q_V) in dependence on the temperature show a decreasing Q_V with increasing T (fig. 6). The same dose dependence is shown like in fig. 5. The phase-transitions are not sharp and can not clearly be determined by swelling measurements.

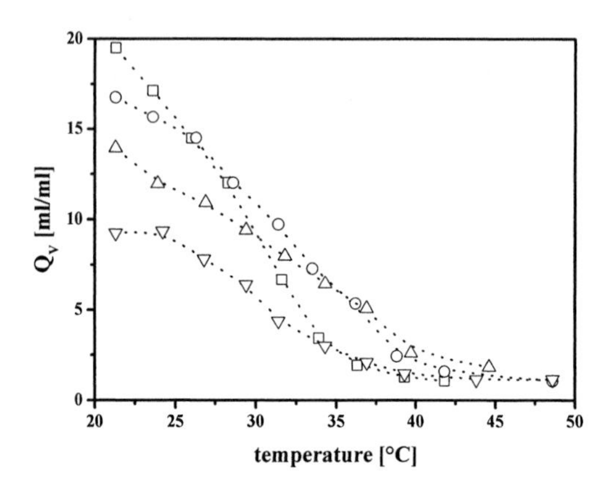

Fig. 6: Swelling degree Q_V of the PVME gel beads (c_p = 6 *wt.%*, $c_{alginate}$ = 2 *wt.%*) in dependence on the temperature T and the radiation dose D (□ - 60 *kGy*, O - 80 *kGy*, △ - 100 *kGy*, ▽ - 120 *kGy*).

For a correct analysis the phase-transition temperature DSC measurements were performed. Fig. 7 shows the results of the DSC measurements of the PVME beads in dependence on the ratio PVME to alginate. PVME bulkgels were used to compare these results.

The DSC graphs show a small decrease of phase-transition temperature of the gel beads in comparison to the bulkgels ($T_{max} \approx 37°C$). This effect can be caused by the crosslinking of small fragments of degradation products of the alginate. The ratio PVME to alginate does not influence the phase-transition temperature.

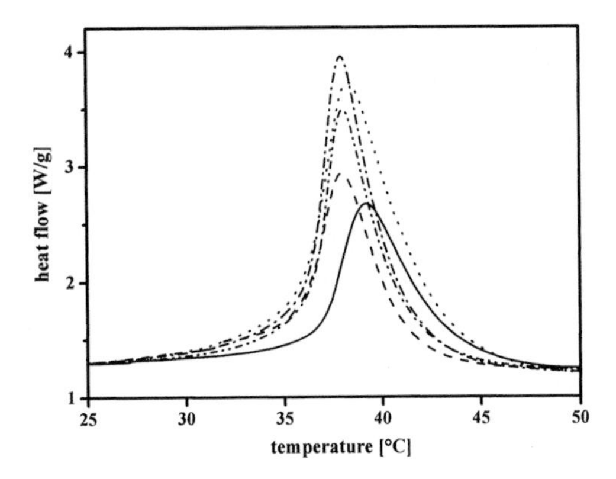

Fig. 7: DSC traces of PVME gel beads in dependence on the ratio PVME to alginate (solid – PVME bulkgel, dash – 1 *wt.%* : 1 *wt.%*, dot 2 *wt.%* : 2 *wt.%*, dash dot – 2 *wt.%* : 4 *wt.%*, dash dot dot – 2 *wt.%* : 6 *wt.%*) .

4. Conclusions

Temperature-sensitive PVME hydrogel beads were synthesized by electron beam irradiation of Ca-alginate PVME networks. The obtained diameters of the beads were varied by the PVME concentration in the range of *mm* (increasing d with increasing c_p). The crosslinking reaction in the beads was not influenced by the alginate shell. The parameters of the CHARLESBY-PINNER equation (by analyzing the PVME content) have nearly the same values like were obtained for PVME bulkgels. The gel beads show a temperature-sensitive behavior.

Acknowledgement

The authors are grateful to Mr. H. Dorschner and Mr. G. Neubert (Institut für Polymerforschung Dresden e.V.) for the electron beam, and to Mrs. I. Poitz (Institut für Makromolekulare Chemie und Textilchemie) for the DSC measurements. The financial support of this work by the Deutsche Forschungsgemeinschaft within the SFB 287 "Reaktive Polymere" is gratefully acknowledged.

(1) T. Tanaka, *Phys. Rev. Lett.* **1978**, *40*, 820-823.
(2) Y. Hirokawa, T. Tanaka, *J. Chem. Phys.* **1984**, *81*, 6379-6380.
(3) M. Shibayama, T. Tanaka, *Adv. Polym. Sci.* **1993**, *110*, 1-62.
(4) M. Schäfer-Soenen, R. Moerkerke, R. Koningsveld, H. Berghmans, K. Dušek, K. Šolc, *Macromolecules* **1997**, *30*, 410-416.
(5) I. Janik, P. Ulański, J.M. Rosiak, C. von Sonntag, *J. Chem. Soc., Perkin Trans. 2* **2000**, 2034-2040.
(6) I. Janik, P. Ulański, K. Hildenbrand, J.M. Rosiak, C. von Sonntag, *J. Chem. Soc., Perkin Trans. 2* **2000**, 2041-2048.
(7) B.K. Kabra, M.K. Akhetar, S.H. Gehrke, *Polymer* **1992**, *33*, 990-995.
(8) M. Suzuki, O. Hirasa, *Adv. Polym. Sci.* **1993**, *110*, 241-261.
(9) R. Kishi, H. Ichijo, O. Hirasa, *J. Int. Mat. Syst. Struct.* **1993**, *4*, 533-537.
(10) X. Liu, R.M. Briber, B.J. Bauer, *J. Polym. Sci. B – Polym. Phys.* **1994**, *32*, 811-815.
(11) R. Moerkerke, F. Meussen, R. Koningsveld, H. Berghmans, W. Mondelaers, E. Schacht, K. Dušek, K. Šolc, *Macromolecules* **1998**, *31*, 2223-2229.
(12) R. Kishi, O. Hirasa, H. Ichijo, *Polym. Gels Networks* **1997**, *5*, 145-151.
(13) K.-F. Arndt, T. Schmidt, H. Menge, *Macromol. Symp.* **2001**, *164*, 313-322.
(14) I. Janik, E. Kasprzak, A. Al-Zier, J.M. Rosiak, *Nucl. Instrum. Methods Phys. Res. B* **2003**, *208*, 374-379.
(15) T. Schmidt, C. Querner, K.-F. Arndt, *Nucl. Instrum. Methods Phys. Res. B* **2003**, *208*, 331-335.
(16) R. Kishi, H. Ichijo, O. Hirasa, *J. Int. Mat. Syst. Struct.* **1993**, *4*, 533-537.
(17) H. Ichijo, O. Hirasa, R. Kishi, M. Oowada, K. Sahara, E. Kokufata, S. Kohno, *Rad. Phys. Chem.* **1995**, *46*, 185-190.
(18) C.A. Aziz, M.V. Sefton, J.M. Anderson, N.P. Ziats, *J. Biomed. Mat. Res.* **1996**, *32*, 193-202.
(19) O. Hirasa, Y. Morsishita, R. Onomura, H. Ichijo, A. Yamauchi, *Kobunshi Ronbunshu* **1989**, *46*, 661-665.
(20) S. Kadłubowski, J. Grobelny, W. Olejniczak, M. Cichomski, P. Ulański, *Macromolecules* **2003**, *36*, 2484-2492.
(21) A. Sáfrány, S. Kano, M. Yoshida, H. Omichi, R. Katakai, M. Suzuki, *Rad. Phys. Chem.* **1995**, *46*, 203-206.
(22) M. Graselli, E. Smolko, P. Harigittai, A. Sáfrány, *Nucl. Instrum. Meth. Phys. Res. B* **2001**, *185*, 254-261.
(23) K.-F. Arndt, T. Schmidt, R. Reichelt, *Polymer* **2001**, *42*, 6785-6791.
(24) T.G. Park, H.K. Choi, *Macromol. Rapid. Commun.* **1998**, *19*, 167-172.
(25) M. Kozicki, P. Kujawa, L. Pajewski, M. Kolodziejczik, J. Narebski, J.M. Rosiak, *Eng. Biomat.* **1999**, *2*, 11-17.
(26) S. Sakai, T. Ono, H. Ijima, K. Kawakami, *Biomaterials* **2002**, *23*, 4177-4183.
(27) R. Barbuchi, M. Consumi, A. Magnani, *Macromol. Chem. Phys.* **2002**, *203*, 1192-1300.
(28) T.I. Klokk, J.E. Melvik, *J. Microencapsulation* **2002**, *19*, 415-424.
(29) L.W. Chan, H.Y. Lee, P.W.S. Heng, *Int. J. Pharmaceutics* **2002**, *242*, 259-262.
(30) N. Nagasawa, H. Mitomo, F. Yoshii, T. Kume, *Polym. Degrad. Stab.* **2000**, *69*, 279-285.
(31) N.Q. Hien, N. Nagasawa, L.X. Tham, F. Yoshii, V.H. Dang, H. Mitomo, K. Makuuchi, T. Kume, *Rad. Phys. Chem.* **2000**, *59*, 97-101.
(32) Z.I Purwanto, L.A.M. van der Broek, H.A. Schols, W. Pilnik, A.G.J. Voragen, *Acta Alimentaria* **1998**, *27*, 29-42.
(33) A. Charlesby, S.H. Pinner, *Proc. Royal Soc. A* **1959**, *249*, 367-386.

Poly(ethylene oxide) Macromonomer Based Hydrogels as a Template for the Culture of Hepatocytes

Eliane Alexandre,[1] *Jacques Cinqualbre,*[1] *Daniel Jaeck,*[1] *Lysiane Richert,*[1,2] *François Isel,*[3] *Pierre J. Lutz**[3]

[1] Fondation Transplantation, Laboratoire de Chirurgie Expérimentale, 5, Avenue Molière, F-67200 Strasbourg Cedex, France
[2] Faculté de Médecine et de Pharmacie, Laboratoire de Biologie Cellulaire, 4, Place St Jacques, F-25030 Besançon Cedex, France
[3] Institut C. Sadron, CNRS, UPR22 F-67083 Strasbourg Cedex, France
Email: lutz@ics.u-strasbg.fr

Summary: Macromonomer based poly(ethylene oxide) (PEO) hydrogels were tested with respect to their ability to serve as a template for the survival and the growth of hepatocytes. Two systems were considered : either the surface of pre-existing hydrogels, with controlled structural parameters, were seeded with isolated rat hepatocytes or the hepatocytes were dispersed in physiological medium containing the macromonomer/initiator and heated to 37°C. In the first case, cells were examined at given times after spreading over two days. The results were compared to those observed for the dispersion of fibroblasts onto a surface of the same type of hydrogels. The effects of the structure of the hydrogels and its chemical nature on the extent of hepatocyte attachment (or encapsulation) and the morphology were investigated.

Keywords: biomaterials; crosslinking; hepatocyte encapsulation; hydrogels; polyethylene oxide

Introduction

Poly(ethylene oxide), (PEO) is an hydrophilic polymer which exhibits specific solution and solid state properties. Furthermore, the remarkable biocompatible properties of this polymer have already led to a wide number of biomedical applications[1-3]. Well-structured PEO hydrogels can be obtained directly in water or in physiological medium upon free radical homopolymerization of water-soluble bifunctional PEO macromonomers[4-7]. That approach combining polymerization in water and control in advance of the structural parameters of the resulting hydrogels presents several decisive advantages with respect to classical end-linking

 DOI: 10.1002/masy.200450653

process or irradiation techniques. Such macromonomer based hydrogels served as semi-permeable biocompatible membranes for an artificial pancreas[4,6]. They have also been tested regarding their capacity to serve as a template for the growth of nervous cells[5].They may provide an interesting scaffold for cell adhesion and a three dimensional space for cell proliferation. After some considerations on the synthesis of the hydrogels, the major part of the work will be devoted to the ability of such PEO hydrogels to serve as a template for the survival and the growth of hepatocytes. Two systems will be considered : either the hepatocytes will be seeded on the surface of pre-existing hydrogels or they will be dispersed in physiological medium containing the macromonomer/initiator and heated to 37°C whereupon crosslinking is to be expected.

Results and discussion

Synthesis of the hydrogels

PEO macromonomers with number average molar masses equal to 6500, 11500 or 15000 g.mol^{-1} and of controlled functionality were synthesized as described previously[4] and directly polymerized in water to hydrogels. Potassium peroxodisulfate or redox initiators were used as initiators (1 molar % versus double bond content) and the different reactions carried out at 37°C or 60°C. The gel point was reached between one and four hours dependent upon the temperature, the chain length of the macromonomer and /or the type of initiator. After preparation, the gels were placed in water for swelling. Once swollen to equilibrium, and free of linear non-connected chains, they were characterized in terms of their swelling behavior and uniaxial compression moduli according to procedures described in the literature. In most cases they were kept in water with 0.3 wt % sodium azide to avoid micro-organism proliferation. The physico-chemical characteristics of these PEO hydrogels have been given in several previous publications[4,6,7]. From these results, it can be concluded that PEO hydrogels over a large range of properties can be obtained by a rather simple procedure: direct polymerization of bifunctional macromonomers, easily accessible, in water solution. These properties can be controlled by the molar mass of the macromonomer, the concentration of the macromonomer and even the type of free radical polymerization initiator.

For identical preparation conditions and molar masses, we observed that with redox initiators, gel point is reached within one hour even at a temperature of 37°C. On the contrary, no crosslinking is noted when the reaction is conducted in the presence of potassium persulfate at 37°C. In that case a temperature around 60°C is a prerequisite to achieve crosslinking. The mechanical properties, for a given molar mass of macromonomer precursor are lower for hydrogels obtained in presence of the redox initiators.

Table 1 . Experimental conditions for the synthesis and physico-chemical characteristics of some hydrogels obtained by homopolymerization of bifunctional PEO macromonomers.

Reference	Precursor Molar Mass a)	Macromonomer b)	ε(%) c)	Q_V (water)	E_g (water)	E_g (THF)	Q_v (THF)
A	6500	20%	3.6%	8.77	91300	-	-
B	11500	20%	4%	11.14	58300	62800	7.42
C	15000	20%	6.5%	16.54	22500	-	-

a)Number average molar mass of the PEO precusor chain expressed in $g.mol^{-1}$
b)Wt-% of macromonomer to be crosslinked
c)Amount of extractable polymer in wt. %
d)Q_{VTHF} and Q_{Vwater} are the volume equilibrium swelling degrees in THF and in water, respectively
e)$E_{G\ THF}$ and $E_{G\ water}$ are the uniaxial compression moduli in THF and in water, respectively, expressed in Pa

Preliminary studies made on hydrogels confirmed that dense networks i.e. materials characterized by high values of uniaxial compression moduli are not well suited for cell-culture. Therefore, among the different types of available hydrogels, we selected those obtained via redox initiators. They are less dense that networks resulting from polymerization of PEO macromonomers at 60°C with potassium persulfate.

Growth of hepatocytes on pre-existing PEO hydrogels

Series of recent studies showed the importance of the matrice nature for hepatocyte attachment and increasing longevity for rat hepatocytes cultured respectively in sandwich configuration and on Matrigel^R compared to collagen singles layers. These observations have been confirmed recently by L.Richert et al.[8]. Hepatocytes can also attach to surfaces on which sugar derivates have been immobilized. L.G. Griffth et al. [9] coupled sugar derivates to different types of radiation cross-linked poly(ethylene oxide) hydrogels.

In the present work unmodified PEO hydrogels, whose synthesis has been discussed above, were tested regarding their potentiality to serve as a template for the growth and survival of hepatocytes. Hepatocytes were isolated from rat liver by a modification of the collagenase digestion method[8]. Cell preparation with a viability higher than 85 % was used for the further experiments. Surfaces of pre-existing hydrogels were seeded with isolated hepatocytes. Cells were examined at given times after spreading over two days. The effects of the structure of the surface of the hydrogels and its chemical nature on the extent of hepatocyte attachment and the morphology were investigated. The results were compared to those observed for the dispersion of fibroblasts onto a surface of the same type of hydrogels. Polystyrene (PS) was used as a reference surface. Hepatocytes attached rapidly onto the PS surface and form a monolayer. On the contrary, attachment was low on PEO hydrogel surfaces. The longer the precursor chains, the lower the attachment of the hepatocytes was. These results are coherent with observations made on PEO hydrogels seeded with fibroblasts[6]. The macromonomer based PEO hydrogels are constituted of PEO chains chemically crosslinked via small domains of hydrophobic polymethylmethacrylate units. Therefore some cells are also suspected to adsorb preferentially on these hydrophobic domains. This situation has yet to be clarified.

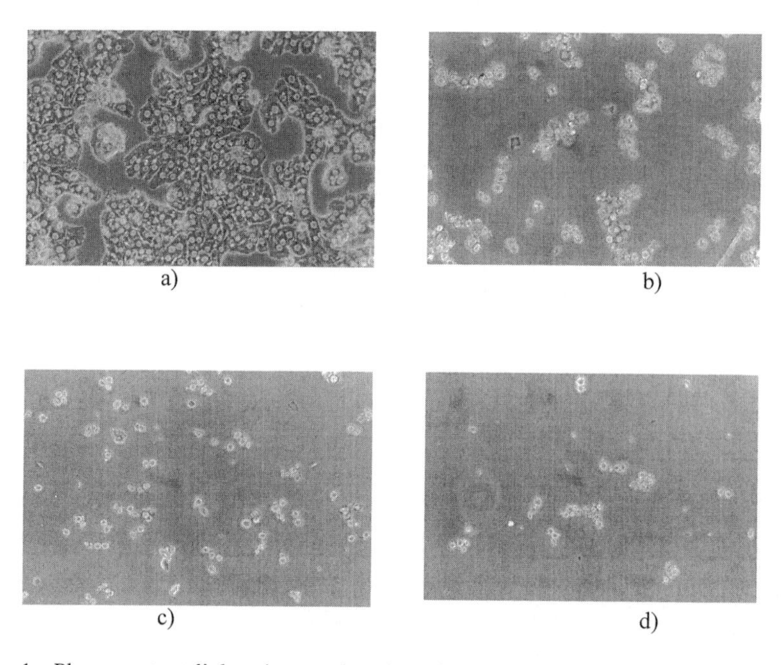

Figure 1 : Phase-contrast light micrographs of rats hepatocytes 24 h after seeding in various configurations (a) on a PS surface (b,c,d) on a PEO hydrogel of precursor molar mass (6500, 11500, 15000g.mol^{-1}) original modification X 50.

Encapsulation of cells during crosslinking

Surfaces of macromonomer based PEO hydrogels have just been shown to be well adapted for the growth of hepatocytes. The density of hepatocytes can be directly related to the nature of the surface and the crosslinking density of the hydrogel. In a previous publication[6], we showed that such macromonomer based hydrogels are also well suited as semi-permeable biocompatible membranes. Glucose and insulin diffuse through the material. These hydrogels may also be well designed as template for a tridimensional growth of hepatocytes into the material. To achieve that growth, hepatocytes have to be dispersed homogeneously in the gel. This is almost impossible for gels in the swollen state. One way to do it, could be to dry the gel and put it in a physiological medium containing the hepatocytes. Upon re-swelling hepatocytes should enter progressively the tridimensional structure. Incorporation of materials during re-swelling processes is a rather slow process and even over longer periods far from yielding homogeneous distribution of the hepatocytes. Earlier studies confirmed that bifunctional PEO macromonomers can be homopolymerized to hydrogels in less than one hour at 37°C. In addition, preliminary experiments confirmed the survival of the hepatocytes even in the presence of high concentrations of free radical polymerization initiators. This prompted us to proceed to crosslinking in the presence of isolated rat hepatocytes whereupon far more homogeneous dispersion of the hepatocytes should be reached. Fresh hepatocytes were dispersed in physiological medium containing the macromonomer precursor, the redox initiator and the solution was heated to 37°C. Two molar masses of precursor were selected 11 500 and 15 000 $g.mol^{-1}$ and the macromonomer weight concentration was by 20-wt %.

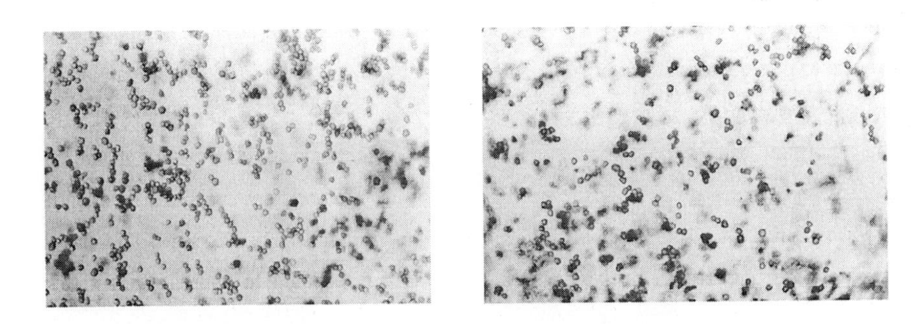

Precursor 11500$g.mol^{-1}$ Precursor 15 000 $g.mol^{-1}$

Figure 2 : Phase contrast light micrograph of rat hepatocytes encapsulated during free radical polymerization of bifunctional PEO macromonomers at 37°C after 1 hour in culture (magnification X50).

For purpose of comparison crosslinking was also performed on cell-free macromonomer solutions. In all cases crosslinking occurs generally after 30 to 60 min. Hydrogels free of encapsulated materials were purified and characterized according to usual procedures. Their properties are almost identical to hydrogels prepared earlier under similar conditions. No physico-chemical determinations could be made on the hydrogels containing the hepatocytes. In figure 2, we have presented the phase contrast light micrograph of hydrogels in which rat hepatocytes have been encapsulated. As it can be seen from these experiments, hepatocytes are dispersed in the hydrogels and are present at different levels in the gel. This is clearly revealed by the changes in coloration. These results could be further confirmed by figure 3, which shows the hepatocytes encapsulated in a PEO hydrogel (11500 $g.mol^{-1}$). The hepatocytes are dispersed in the hydrogel, but are much more concentrated at the bottom of the gel, due to the sedimentation of the cells during polymerization.

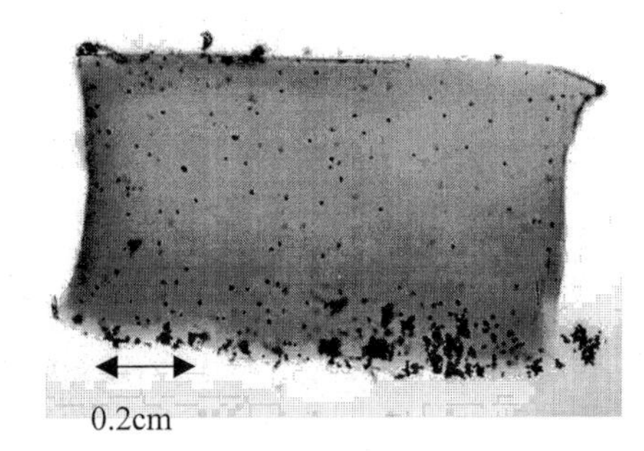

Figure 3: Phase contrast light micrograph (after fixation in osmium tetroxide) of rat hepatocytes encapsulated during free radical polymerization of bifunctional PEO macromonomers at 37°C after 3 hours in culture (magnification X50).

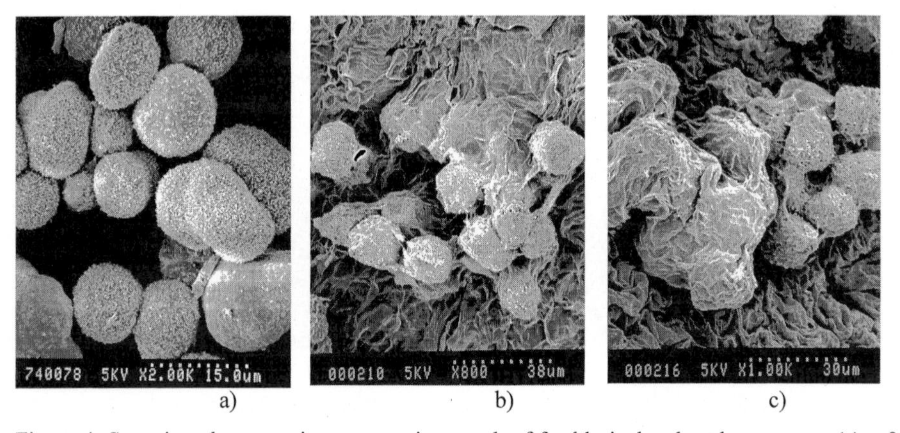

Figure 4: Scanning electron microscopy micrograph of freshly isolated rat hepatocytes (a), of rat hepatocytes encapsulated during free radical polymerization of bifunctional PEO macromonomers at 37°C after 3 hours (b) and 24 hours (c) in culture.

Figure 4 depicts the rat hepatocytes encapsulated in PEO hydrogel (11500 g.mol-1) after 3 hours in culture (4b) and after 24 hours in culture (4c). Figure 4a shows the freshly isolated rat hepatocytes and serves as control. After 3 hours in culture the cells appeared shrinked and in some cases presented holes in their membrane. This phenomenon was increased after 24 hours in culture. Almost identical conclusions could be drawn for hydrogels prepared from PEO macromonomer precursors of molar mass 15000g.mol^{-1}.

Conclusion

In the present work, PEO macromonomer based hydrogels were tested with respect to their potentiality to serve as a template for the growth of hepatocytes on the surfaces or in the hydrogels. The efficiency of the growth on surfaces is directly related to the physico-chemical characteristics of the hydrogels. Hepatocytes could also be incorporated directly in hydrogels during the crosslinking process. Survival of cells during few hours was observed. For longer periods, the survival shows some problems. Further work is going on along that line. In addition, the possibility to homopolymerize bifunctional PEO macromonomers to hydrogels directly in physiological medium opens new perspectives for the homogeneous incorporation of drugs or cells into biocompatible water swollen matrices.

Acknowledgements

The authors wish to express their acknowledgments to all their colleagues and coworkers who have contributed to various topics dealing with hydrogels and biomedical applications. Special thanks are addressed to Mrs C. Royer for the microscopy experiments. The authors thank also the French Education Ministry for financial support (ACI Technologie pour la Santé).

[1] M.J. Harris,"*Poly(ethylene Glycol) Biotechnical and Biomedical Applications*", Plenum Press,1992
[2] J.A. Hubbell, *Current Opinion in Biotechnology*, **1999**,*10*, 123
[3] P.J. Lutz, *Macromol. Symp.* **2001**, *164*, 277.
[4] B. Schmitt, E. Alexandre, K. Boudjema; P.J. Lutz, *Macromol. Symp.* **1995**, *93*, 117.
[5] K. Naraghi ; J. Soussand, J.M. Félix, S. Schimchowitsch, P.J. Lutz; *Polym. Prep., Am. Chem. Soc. Div. Polym. Chem. Boston,* (USA). **1998,** *39(2),* 196.
[6] B. Schmitt, E. Alexandre, K. Boudjema; P.J. Lutz, *Macromol. Biosci.* **2002**, *2,* 341.
[7] E. Alexandre, K. Boudjema, B. Schmitt, J. Cinqualbre, D. Jaeck, C.Lux, F. Isel and P.J. Lutz, *Polymeric Materials: Science & Engineering*, **2003,** *89,* 240
[8] L. Richert; D. Binda; G. Hamilton, C. Viollon-Abadie, E. Alexandre, D. Bigot-Lasserre, L. Bars, P. Coassolo, E. LeCluyse, *Toxicol. In Vitro* **2002**, *16,* 89.
[9] S. T. Lopina, G. Wu, E. W. Merrill, L. G. Cima, *Biomaterials*, **1996**,*17*, 559

The Sol-Gel Approach towards Thermo-Responsive Poly(N-isopropyl acrylamide) Hydrogels with Improved Mechanical Properties

*Wouter Loos, Filip Du Prez**

Ghent University, Department of Organic Chemistry, Polymer Chemistry Division, Krijgslaan 281 (S4), B-9000 Ghent, Belgium
Fax: +32(0)92644972
E-mail: filip.duprez@ugent.be

Summary: A new type of thermo-responsive hydrogels based on the polymer poly(N-isopropyl acrylamide) (PNIPAA) has been synthesized with the sol-gel technology. For the preparation of this type of nano-structured hydrogels, the inorganic silica phase was synthesized by the sol-gel process in the presence of an aqueous solution of high molecular weight PNIPAA. This combination of the organic and inorganic phases forms hybrid hydrogels with a semi-IPN morphology. The unique structure of these hydrogels improves the mechanical stability to a great extent as compared to conventional PNIPAA-hydrogels. This was shown by stress-strain experiments and the capability to absorb and desorb large amounts of water. The silica only slightly influences the transition temperature of the hydrogels but allows us to vary the thermo-responsive properties of the materials to a great extent.

Keywords: hydrogels; lower critical solution temperature; nanocomposites; poly(N-isopropyl acrylamide); silicas

Introduction

The hybrid hydrogels, produced by a sol-gel process, are a new kind of hydrogels, where nano-sized inorganic particles act as physical or chemical cross-links of the networks based on water-borne synthetic polymers. Although several silica composites based on water-soluble polymers have been reported in the literature, they mostly consist of an inorganic matrix and little attention is paid to their hydrogel properties.[1,2]

In general, these hybrid hydrogels represent a new class of materials where the advanced features of both hydrogels and an inorganic nano-particulate structure are combined.[3,4] Just like ordinary polymer hydrogels, they easily swell in water, exhibit good elasticity and high optical transparency. At the same time, they possess good mechanical properties, in particular high

 DOI: 10.1002/masy.200450654

strength characteristics.

For the production of the hybrid hydrogels, the well known sol-gel technology is applied.[5] Generally, this method can be represented as a two step network-forming process based on two fundamental chemical reactions - hydrolysis and condensation (Figure 1). Usually, tetrafunctional silicates such as tetramethoxysilane (TMOS) are used, but also other metal alkoxides (Al, Ti or Zr alkoxides). These two reactions are concurrent and interdependent and proceed under mild conditions. A number of variables influence the structure of the SiO_2-network: pH, solvent, water-to-alkoxide ratio and the type of the alkoxide used. Recently the sol-gel process has received much attention with respect to the design and preparation of polymer-inorganic hybrid materials. So far the most promising applications are based on novel characteristics and enhanced properties such as improved toughness, strength, modulus, impact and scratch resistance, optical transparency, thermal stability and electrical conductivity.[6-9] The structure of the inorganic phase can be controlled by changing the parameters of the sol-gel reactions.

$$Si(OR)_4 + 4\,H_2O \xrightarrow{\text{hydrolysis}} Si(OH)_4 + 4\,ROH$$

$$Si(OH)_4 \xrightarrow{\text{condensation}} SiO_2 + 2\,H_2O$$

Figure 1. Reaction scheme of the sol-gel process: hydrolysis and condensation of alkoxy derivatives of Si.

The goal of the present work is to apply the sol-gel process for the preparation of thermo-responsive hybrid hydrogels. In these "intelligent" gels, the thermo-responsive properties are introduced by the incorporation of a polymer that shows a "lower critical solution temperature" (LCST) in aqueous medium. For this research the well-known thermo-responsive polymer, poly(N-isopropyl acrylamide) (PNIPAA) has been chosen. PNIPAA possesses an LCST that is located at temperatures of 32-33°C.[10] These physiological temperatures open perspectives for numerous biomedical applications and drug delivery systems.[11-13]

The first PNIPAA-SiO_2 hybrid hydrogels have been reported by Kurihara et al.. In their system, PNIPAA and silica domains were linked through covalent bonds.[14] More recently Chujo et al. reported on conventional cross-linked PNIPAA networks that are combined with silica particles.[15]

In our work on the other hand, the principle of the hydrogel formation was to perform the sol-gel process in the presence of high molar mass polymer, without covalent bonds between the organic and inorganic phase. In this way, the silica particles act as physical cross-links for the polymer molecules and can be represented as a semi-interpenetrating polymer network (semi-IPN). In a previous report we already demonstrated that hybrid hydrogels as semi-IPNs based on another polymer with LCST-behaviour, poly(N-vinyl caprolactam) (PVCL), can better withstand the mechanical stress created during the swell and shrinkage processes in response to temperature changes in comparison to conventional hydrogels.[3] Compared to PVCL, PNIPAA shows a much different phase behaviour in water (Type II), leading to a discontinuous and more extensive shrinking of the corresponding hydrogels.[10] In this report, the reaction conditions and thermo-responsive properties of PNIPAA-containing hybrid hydrogels will be reported.

Experimental Part

Materials

Tetramethoxysilane (TMOS) (Acros, 99%), 2,2'-azobisisobutyronitrile (AIBN) (Merck-Schuchardt, >98%), ammonium persulfate ($(NH_4)_2S_2O_8$) (Aldrich, 98%) and N,N,N',N' tetramethylethylenediamine (TEMED) (Aldrich, 99%) were used as received. Benzene (Aldrich, 99+%) was refluxed over a sodium/benzophenon solution. N-isopropyl acrylamide (NIPAA) (Acros, 99%) was purified by recrystallization on hexane, followed by drying under vacuum and stored at 4°C.

Synthesis of Linear Poly(N-isopropyl acrylamide) (PNIPAA)

The high molecular weight PNIPAA (PNIPAA-1,2) were obtained by performing the radical polymerization in benzene with AIBN as initiator at an elevated temperature of 60°C. Lower molar mass PNIPAA (PNIPAA-3) was synthesized in water with the redox system $(NH_4)_2S_2O_8$ / TEMED.[16]

Synthesis of PNIPAA Organic/Inorganic Hybrid Materials

The synthesis of PNIPAA hybrid hydrogels was carried out in a test tube with a magnetic stirring rod. An aqueous solution of PNIPAA (distilled water) at pH equal to 12 is added together with a certain amount of TMOS. The reaction mixture is stirred for 5 minutes before being poured between two silylated glass plates, separated by a 3 mm thick spacer. By performing the gelation in the glass mould, the evaporation of water is prevented. The hydrogels have been investigated at least 24 hours after their preparation.

In order to explain the nomenclature in the discussion part, one synthetic procedure is described in detail: 0.4g TMOS (d = 1.02 g/ml) is added to a solution of 4 ml water containing 0.4g PNIPAA (9.1 wt.-% PNIPAA solution), thus V_{PNIPAA}/V_{TMOS} = 10/1 and the initial TMOS to water ratio in wt.-% is 9.1. The resulting ratio PNIPAA/SiO_2 is 50/50 (wt.-%) and the total composition, indicated as the ratio water/PNIPAA/silica (wt.-%), is equal to 84/8/8.

Methods of Analysis

Soluble fractions (SF) of the PNIPAA hybrid hydrogels are determined gravimetrically and are defined as $SF = 100.(W_0 - W_e)/W_0$. W_e and W_0 respectively denote the weight of extracted and dry hybrid material. The extraction proceeded in acetone during 6h.

The swelling degrees of the PNIPAA hybrid hydrogels in distilled water are determined gravimetically as a function of time and temperature. The equilibrium weight of the swollen samples is determined after a weight change of less than 1 wt.-%. The degree of swelling was defined as $S = 100.(W_{sw} - W_0) / W_0$, where W_{sw} and W_0 respectively denote the weight of the swollen and dried sample (vacuum, 60°C, 24 hours).

Stress-strain curves are determined at room temperature, at constant elongation speed (15mm/min) on 3 mm thick samples (width : 4mm).

Results and Discussion

Synthesis of Linear Poly(N-isopropyl acrylamide) (PNIPAA)

In the introduction part, the necessity for high molecular weight polymers to obtain stable hydrogels was already mentioned. The molecular weights and nomenclature of PNIPAA are shown in Table 1. The synthetic details are described in the experimental part.

Table 1. Molecular weights of PNIPAA.

Sample	M_v PNIPAA[1)] g/mol
PNIPAA-1	1.650.000
PNIPAA-2	1.050.000
PNIPAA-3	560.000

[1)] Molecular weights were determined by viscosimetry in water at 25 °C, using the Mark Houwink equation : $[\eta] = K \cdot M_v^a$ with $K = 14.5 \cdot 10^{-2}$ ml/g and a = 0.5.[17]

Synthesis and Characterization of PNIPAA-Silica Hybrid Materials

Regardless the fact that there are no covalent bonds present in the hydrogels, there are strong interactions between the two phases, i.e. the inorganic silica phase and the organic polymer phase. This is a consequence of both hydrogen bonds and physical entanglements. The existence of hydrogen bonds between the carbonyl groups of PNIPAA and the remaining silanol groups of the inorganic domains could be demonstrated by FTIR-spectroscopy. For example a shift of the carbonyl peak of PNIPAA from 1665 cm^{-1} for pure PNIPAA to 1639 cm^{-1} for PNIPAA present in a specific hybrid hydrogel is detected. This is due to the presence of silanol groups and thus the formation of hydrogen bonds.

The model to visualise the morphology has been described extensively in the previous report and is based on the work of Wilkes et al.[18] Instead of covalent bonds in the Wilkes model, the hybrid hydrogel can be represented as a semi-IPN consisting of cross-linked silica particles with entanglements and hydrogen bonds in between the particles and the high molar mass PNIPAA chains.

Another way to determine the stability of the physical cross-linked hydrogel is by the measurement of the soluble fraction and the equilibrium swelling degree (Table 2). From this table, one can see that the soluble fractions are quite low in the case of hybrid hydrogels with high molecular weight PNIPAA (PNIPAA-1 and -2) and TMOS weight fractions exceeding 4%. Under these conditions the number of physical entanglements and hydrogen bonds between both phases prevents the extraction of polymer chains and provides highly stable hydrogel materials in spite of the equilibrium swelling degrees up to 3000%. On the other hand, for lower molecular weight

PNIPAA (PNIPAA-3), much of the organic material is extracted from the gels and equilibrium swelling degrees could not be determined due to their instability.

Table 2. Soluble fractions and compositions of the PNIPAA hybrid hydrogels.

	Weight fraction TMOS to H_2O in reaction mixture	PNIPAA / SiO_2	Soluble fraction	Equilibrium swelling degree in water (20°C)
	wt.-%	wt.-%	%	%
PNIPAA-1	8	70/30	1.3	1800
	6	70/30	2.1	2400
PNIPAA-2	9	50/50	0	2200
	9	60/40	3.7	1800
	8	70/30	0.3	2200
	6	70/30	3.1	2900
	6	60/40	2.5	2500
	4	70/30	11.3	5400
	4	80/20	17.7	5100
PNIPAA-3	8	70/30	55	/
	6	70/30	58	/

As reported earlier for PVCL hybrid hydrogels, the swelling degrees can be tuned by the variation of the reaction conditions. The equilibrium swelling degree increases with the PNIPAA/SiO_2 ratio (up to 5000%), whereas it lowers with a higher TMOS concentration in the reaction mixture. The TMOS weight fraction determines the cross-linking density of the inorganic silica phase. Similar to conventional hydrogels, the equilibrium swelling degrees lower with increasing cross-linking density.

Mechanical Properties of the Hybrid Hydrogels

In order to demonstrate the enhancement of the mechanical stability of the hydrogels, the most obvious technique was to perform stress-strain tests (Table 3). As expected, the elasticity modulus increases with increasing silica fraction or decreasing water fraction. The good elasticity properties are demonstrated by the values of the elongation at break that vary from 160 up to 465%. Former investigations showed that stress-strain tests were impossible to perform on conventional PNIPAA-hydrogels with equal swelling degrees, indicating the improvement of mechanical stability of these hybrid hydrogels.

Table 3. Stress-strain data of PNIPAA hybrid hydrogels (PNIPAA-2).

Composition hybrid hydrogels : H_2O/PNIPAA/ SiO_2	(PNIPAA/ SiO_2)	Young modulus	Elongation at break
wt.-%		kPa	%
78/15.5/6.5	(70/30)	28	465
80/12/8	(60/40)	24	170
86/8.5/5.5	(60/40)	18	160
84/8/8	(50/50)	15	300
80/16/4	(80/20)	15	450
83/12/5	(70/30)	7.5	290

Thermo-responsive Properties of the Hybrid Hydrogels

The hybrid hydrogels owe their thermo-responsiveness to the nature of the organic phase, i.e. the LCST-polymer PNIPAA. In an aqueous solution at temperatures above the cloud point (T_{cp}) this polymer releases water molecules and undergoes a transition from a hydrophilic to hydrophobic state. This process generally results in two visually observable phenomena: a volume decrease and the appearance of turbidity.

The phase diagram of some PNIPAA hybrid hydrogels is shown in figure 2. It clearly shows the expulsion of water above the T_{cp} of PNIPAA. When the amount of silica is higher in comparison to PNIPAA, the expulsion is larger and, analogously to conventional cross-linked systems, it can be derived that the shrinkage effect is lower at higher cross-linking densities, i.e. higher TMOS weight percentage.

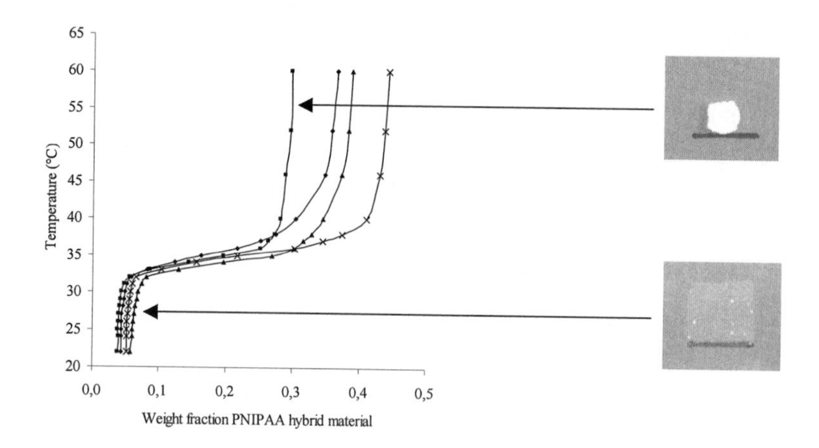

Figure 2. Phase diagram of PNIPAA hybrid hydrogels (PNIPAA-2) with composition: (a) PNIPAA/SiO_2 = 70/30; TMOS = 6 wt.-%, (b) 60/40; TMOS = 9 wt.-%, (c) 60/40; TMOS = 6 wt.-% en (d) 70/30; TMOS = 8 wt.-%. The shrinkage effect is illustrated for a gel with composition (a).

Although it is difficult to observe in figure 2, light transmission measurements showed that, in comparison to pure PNIPAA solutions, the presence of silica domains in the hybrid materials slightly lowers the LCST with about 2°C. Thus, the thermo-responsive behaviour still expresses itself above room temperature. The explanation of the decrease of the phase transition temperatures in the presence of the inorganic phase, is the competition between water and silanol groups to form hydrogen bonds with the carbonyl groups of PNIPAA. If these carbonyl groups no longer can freely interact with water molecules below the T_{cp}, the phase transition temperature of PNIPAA decreases.

Conclusion

Mechanically improved thermo-responsive hybrid hydrogels based on PNIPAA were developed by making use of the sol-gel technology. The synthesis was achieved by the in situ formation of an inorganic silica phase in the presence of high molecular weight PNIPAA. This methodology leads to micro-heterogeneous systems in which silica particles of nanometer dimensions act as physical cross-links for the PNIPAA molecules. Hydrogen bonds between the silica particles and

PNIPAA, together with physical entanglements are responsible for the strong interactions between the inorganic and organic phase. Stress-strain tests on highly swollen materials demonstrated that the unique structure of these hybrid hydrogels improves the mechanical stability to a great extent as compared to conventional hydrogels. The properties of the hydrogels can be easily tuned by changing the reaction conditions, whereas the influence of silica on the phase transition temperatures of PNIPAA is negligible.
As a general conclusion, it can be stated that the synthesis of organic/inorganic hybrid hydrogels is a successful concept to develop thermo-responsive hydrogels with improved mechanical properties.

Acknowledgement

W. Loos thanks the IWT (Institute for the Promotion of Innovation by Science and Technology in Flanders) for a doctoral fellowship. The Belgian Programme on Interuniversity Attraction Poles and the ESF programme SUPERNET are acknowledged for financial support.

[1] P. Hajji, L. David, J. Gerard, J. Pascault, G. Vigier, *J. Polym. Sci. Pol. Phys.* **1999**, 37, 3172
[2] C. Landry, B. Coltrain, J. Wesson, N. Zumbulyadis, J. Lippet, *Polymer* **1992**, 33, 1496
[3] W. Loos, S. Verbrugghe, E. Goethals, F. Du Prez, I. Bakeeva, V. Zubov, *Macromol. Chem. Phys.* **2003**, 204, 98
[4] T. Uragami, K. Okazaki, H. Matsugi, T. Miyata, *Macromolecules* **2002**, 35, 9156
[5] J. Brinker, G. Scherer, *"Sol-Gel Science"*, Academic Press, New York 1990
[6] J. Habsuda, G. Simon, Y. Cheng, D. Hewitt, D. Lewis, H. Toh, *Polymer* **2002**, 43, 4123
[7] B.M. Novak, *Adv. Mater.* **1993**, 5, 422
[8] D. McCarthy, J. Mark, D. Schaefer, *J. Polym. Sci. Pol. Phys.* **1998**, 36, 1167
[9] V. Bershtein, L. Egorova, P. Yakushev, P. Sissis, P. Sysel, L. Brozova, *J. Polym. Sci. Pol. Phys.* **2002**, 40, 1056
[10] F. Afroze, E. Nies, H. Berghmans, *J. Mol. Struct.* **2000**, 554, 55
[11] R. X. Zhou, W. Li, *J. Polym. Sci. Pol. Chem.* **2003**, 41, 152
[12] R. Langer, *Nature* **1998**, 392, 5
[13] K. Soppimath, T. Aminabhavi, A. Dave, S. Kumbar, W. Rudzinski, *Drug. Dev. Ind. Pharm.* **2002**, 28, 957
[14] S. Kurihara, A. Minagoshi, T. Nonaka, *J. Appl. Polym. Sci.* **1996**, 62, 153
[15] Y. Imai, N. Yoshida, K. Naka, Y. Chujo, *Polym. J.* **1999**, 31, 258
[16] F. Zheng, Z. Tong, X. Yang, *Eur. Polym. J.* **1997**, 33, 1553
[17] C. Boutris, E. G. Chatzi, C. Kiparissides, *Polymer* **1997**, 38, 2567
[18] H. Huang, G. Wilkes, *Polym. Bull.* **1987**, 18, 455

Macromol. Symp. **2004**, *210*, 493-499

Encapsulation of Living Cells with Polymeric Systems

Christian Schwinger,[1] *Albrecht Klemenz,*[2] *Karsten Busse,**[1] *Jörg Kressler*[1]

[1] Institute of Bioengineering, Department of Engineering Science, Martin-Luther-University Halle-Wittenberg, D-06099 Halle, Germany
[2] Institute of Medical Physics and Biophysics, Department of Medicine, Martin-Luther-University Halle-Wittenberg, D-06097 Halle, Germany
E-mail: karsten.busse@iw.uni-halle.de

Summary: Microencapsulation of cells producing recombinant proteins or hormones leads to immunoprotection and immobilization in culture or *in vivo*. We are investigating three different strategies for the production of calcium cross-linked alginate beads of a small size with immobilized and immunoprotected mammalian cells: a) the AirJet technology (coaxial gas flow extrusion), b) the vibrating nozzle technology, and c) the JetCutter technology. A alginate/poly-L-lysine/alginate complexation was used as the polymeric system. All three methods may be used for production of homogeneous beads with a diameter of approximately 350 μm. While the vibrating nozzle technique was limited to an alginate viscosity of 0.2 Pa·s or less, the AirJet and JetCutter technology were less sensitive to higher viscosities. High frequency Scanning Acoustic Microscopy is used for mechanical characterization of the microspheres as well as for investigation of surface properties.

Keywords: alginate; biocompatibility; microencapsulation; scanning acoustic microscopy

Introduction

Microencapsulation is a procedure where materials, such as enzymes, bacteria, yeast, or eucaryotic cells are enclosed within microscopic, semipermeable containers. Microencapsulation of mammalian cells is a novel and versatile tool of delivering therapeutically important natural or recombinant molecules *in vivo*. It can also have numerous applications as a platform for gene therapy of metabolic or neurological disorders and cancer.[1,2] Most of the previous work done on encapsulation of transgenic mammalian cells was concentrated on protection against the immune response of the host and on minimizing local inflammatory reactions generated by the microcapsules.[3,4] However, encapsulation techniques additionally are limited by physical factors, e. g. viscosity of the biopolymer solution, size distribution, or mechanical strength of the produced microcapsules. With the prevalent methods for bead formation, such as vibrating nozzles, gas jet droplet generators or laminar jet break-up, rather low concentrated alginate solutions can be employed.[5] In addition, beads with a small diameter (approximately 300 μm) are more difficult to produce than those

DOI: 10.1002/masy.200450655

with a larger diameter (> 500 μm) although smaller microcapsules may have several advantages, such as better oxygenation of encapsulated cells, smaller implant volume, and easier application to organs *in vivo*.[6,7]

A serious disadvantage of low viscosity alginate solutions is the lack of mechanical stability of the alginate hydrogel which is formed by cross-linking of the alginate molecules with polyvalent cations.[8] Therefore, covering solid beads with outer alginate layers and protective polycationic shells resulting in alginate-poly(L-lysine)-alginate (APA) microcapsules, or the use of Ba^{2+}- instead of Ca^{2+}- alginate has been suggested for improving the mechanical stability of hollow core microcapsules.[9,10]

The purpose of this study is to evaluate the physical properties of alginate microcapsules produced by three different methods, laminar gas flow (AirJet), vibrating nozzle, and JetCutter with the aim of optimizing the production of small size (approximately 350 μm) microcapsules suitable for biomedical applications in humans, and to investigate the proliferation of normal, neoplastic, or transgenic cells in these granules.

High-frequency Scanning Acoustic Microscopy (SAM) was used for investigation of mechanical properties in terms of acoustic impedance and 3D-surface topography of full alginate microspheres. Mean surface impedance was measured with SAM at 900 MHz with a spatial resolution of 1.5 μm. The sensitivity and reproducibility of SAM had to be increased considerably to receive and quantify signals in the very low impedance region. A multilayer analysis method was developed to get quantitative data with SAM at a microscopic level. The mechanical stiffness c_{11} was obtained from mass density and longitudinal ultrasound velocity, measured with a pulse echo method at 6 MHz.

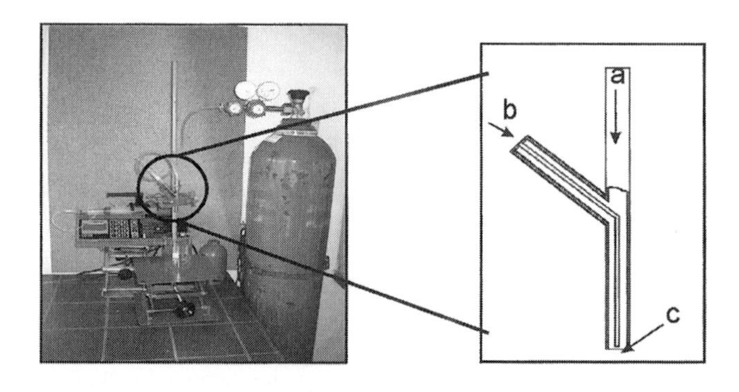

Figure 1. AirJet device used in our laboratory

Materials and methods

Sodium alginate powder was obtained from Inotech AG (Dottikon, Switzerland) and from Fluka (Buchs, Switzerland). All alginates as well as a 0.1 % (w/v) poly-L-lysine solution (PLL; M_w 25.700 g/mol; Sigma) were sterilized by filtration through a 0.22 μm filter (Merck), and stored at 4°C. The Ca solution consisted of 100 mM Calcium chloride, 10 mM MOPS (3-[N-Morpholino]propanesulfonic acid, ICN Biomedicals, Eschwege, Germany) and 0.85 % (w/v) NaCl. The solutions were adjusted to pH 7.4, sterilized by autoclaving, and stored at room temperature.

The murine fibroblast cell line GLI 328 [11,12] (from Dr. E. Otto, GTI Inc., Gaithersburg, MD) was maintained in DMEM with 1 g/L glucose (Biochrom KG, Berlin, Germany) with addition of 10 % (w/v) donor calf serum (CS; Gibco BRL Life Technologies, Karlsruhe, Germany) and 1 % penicillin/streptomycin (Gibco BRL) at 37°C in humid atmosphere containing 5 % (v/v) CO_2.

For encapsulation the AirJet apparatus[13-15] (self-made, Figure 1), the vibrating nozzle apparatus[16] from Inotech AG (Dottikon, Switzerland) and the JetCutter system[17] from geniaLab GmbH (Braunschweig, Germany) were used.

Microspheres from alginate type Fluka 71238 were produced by the AirJet and the JetCutter method with a concentration of 1.5% (w/v). For the vibrating nozzle method 1.5 % (w/v) sodium alginate from Inotech was used. 8 ml of a suspension with 2.0×10^6 GLI 328 cells/ml were added to all alginate solutions. The JetCutter and the AirJet systems were placed in a class two clean bench to allow bead production under semi sterile conditions. The vibrating nozzle system was operated under sterile GMP conditions. The droplets were collected in 200 ml Ca solution, formed beads were separated from the bath by sieving. Subsequent formation of outer layers was carried out according to a simplified alginate-poly-L-lysine-alginate (APA) protocol.[18]

The acoustic microscope SAM 2000 (Kraemer Scientific Instruments, Herborn, Germany) with a broadband lens (0.8 - 1.3 GHz, 100° aperture angle) was operated at 900 MHz. It works in burst mode.[19] All measurements were performed in a temperature controlled water tank at 25.0 ± 0.1°C.

Results

The size distribution of the produced microcapsules by AirJet is shown in Figure 2. The particle diameter and the arising size distribution depend on the volume flow of alginate and gas, the viscosity of the alginate, and the outer diameter of the capillary.[20] A monodisperse distribution can only be obtained if the capsule size is smaller then the diameter of the capillary.

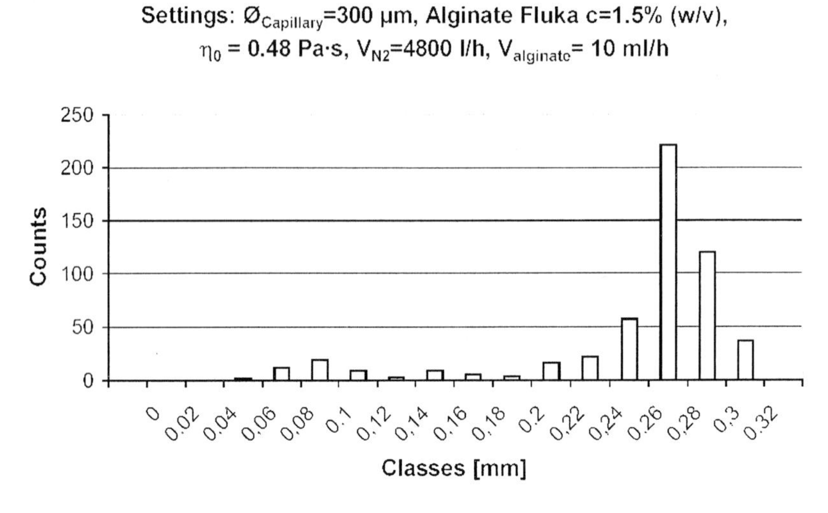

Figure 2. Size distribution with optimal settings of the AirJet apparatus.

After encapsulation of the cells by the different strategies, the cell growth shows a similar behavior. After one day, the cells formed a compact aggregate in the middle of the capsule. After approximately one week a significant cell growth can be seen and after four weeks the cells fill the whole capsule, as shown in Figure 3.

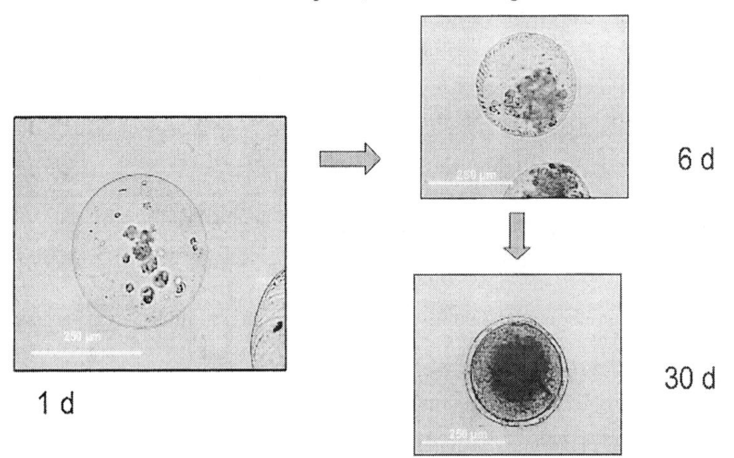

Figure 3. Cell growth (GLI 328) in a period of 30 days after encapsulation.

The results demonstrate that a custom made encapsulation AirJet device is well suitable for alginate encapsulation of mammalian cells and that it is comparable with a commercially available vibrating nozzle device (IEM-40, Inotech AG). The corresponding size distribution of the produced microcapsules is shown in Figure 4. Both encapsulation methods show the feasibility of generating uniform alginate microbeads for APA microcapsules.

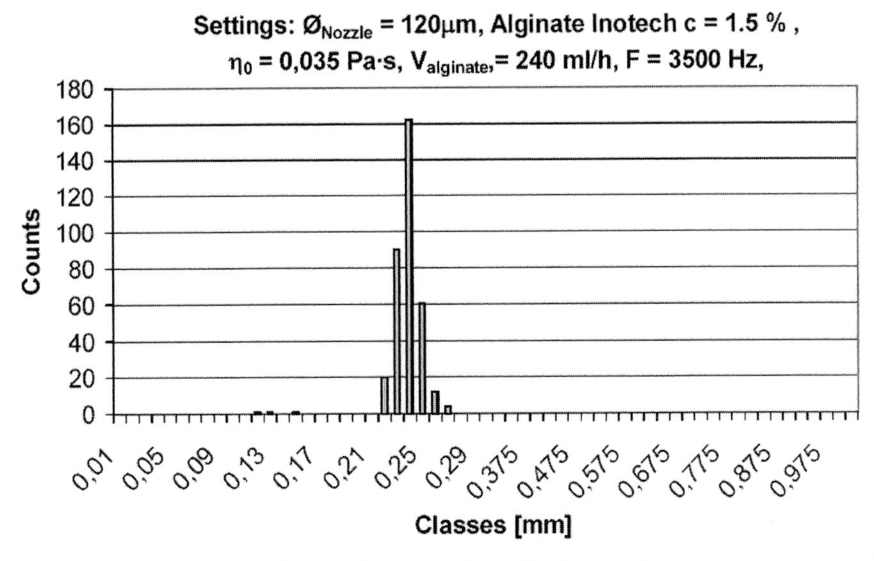

Figure 4. Size distribution with optimal settings of the vibrating nozzle apparatus.

The vibrating nozzle method is one of the most frequently used methods for large scale production of microbeads. It requires low viscosity of the biopolymer, however. Extrusion by coaxial gas flow (AirJet) is less sensitive against high viscosity polymers, but has a low throughput. It is therefore suitable for experimental purposes, but not for large scale production of microcapsules.

The JetCutter technology also seems to be appropriate for alginate encapsulation of living mammalian cells.[21] Small alginate beads (320 µm) containing viable cells could be produced at a very high throughput. The encapsulated murine fibroblasts formed colonies and proliferated at a considerable rate, which indicates that the mechanical stress during the encapsulation procedure is well tolerated and does not irreversibly damage the cells.

Sterilization or autoclaving of the parts of the JetCutter being in contact with the alginate-cell suspension, the use of sterile working solutions, and placement of the whole set-up in a clean bench may prevent contamination of the cultured beads and allow for microcapsule mass production under GMP conditions. Another important aspect is the mechanical stability of the beads. One possible and rather straightforward approach is to increase the polymer content of the hydrogel, e. g. to use higher concentrated alginate solutions (2 to 5 %). At present, the JetCutter is the only technology that is able to process alginate at such concentrations.

Scanning Acoustic Microscopy is a suitable and sensitive tool for measuring elasto-mechanical properties of alginate specimens in terms of quantitative acoustic impedance.[22] The alginate spheres are investigated in their presumed environment, i.e. water. There is no further preparation required that could possibly change mechanical properties, as for electron microscopic methods. The topography of the surface as well as detailed structural information with a resolution of 1.5 μm can be obtained (see Figure 5).

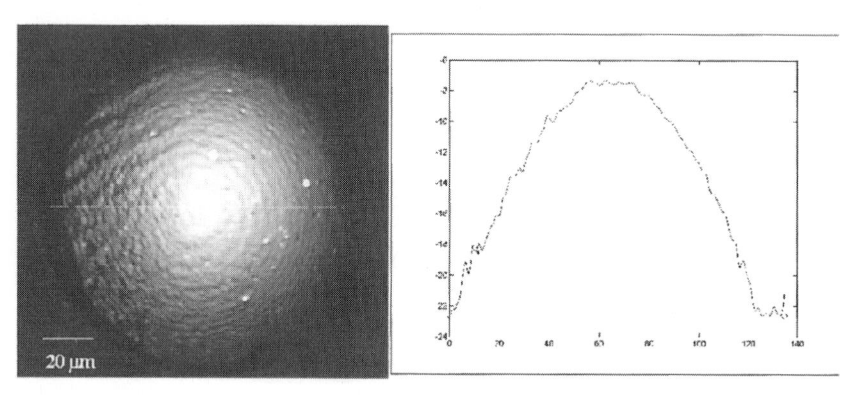

Figure 5. 2D-picture of an alginate sphere and the topography profile along the dotted hoizontal line.

The latter can be helpful for detailed studies of the surface inhomogeneities caused by material and by topography as well as for the optimization of the production process of the spheres. Bulk mechanical stiffness is estimated with low frequency ultrasound.

However, SAM is not only able to provide information on surface impedance of full microspheres but it also has the potential of investigating the elastic properties of capsule membranes. Because the SAM can be applied with both low- and high-frequency ultrasound, there are measurements at different stages of spatial resolution possible, beginning from bulk up to high resolution surface measurements.

Conclusion

All three methods under investigation may be used for production of homogeneous beads with a diameter of approximately 350 µm. The vibrating nozzle technique was limited to an alginate viscosity of 0.2 Pa·s or less. With the JetCutter technology alginate beads could be produced at a very high throughput. High frequency Scanning Acoustic Microscopy is used for mechanical characterization of the microspheres as well as for investigation of surface properties.

[1] J.J. Vallbacka, J.N. Nobrega, M.V. Sefton, *J Control Release* **2001**, *72(1-3)*, 93.
[2] M. Lohr, A. Hoffmeyer, J. Kroger, M. Freund, J. Hain, A. Holle, P. Karle, W.T. Knofel, S. Liebe, P. Muller, H. Nizze, M. Renner, R.M. Saller, T. Wagner, K. Hauenstein, W.H. Gunzburg, B. Salmons, *Lancet* **2001**, *357(9268)*, 1591.
[3] B. Kulseng, G. Skjak-Braek, L. Ryan, A. Andersson, A. King, A. Faxvaag, T. Espevik, *Transplantation* **1999**, *67(7)*, 978.
[4] S. Sakai, T. Ono, H. Ijima, K. Kawakami, *J Microencapsulation* **2000**, *17(6)*, 691.
[5] G. Klock, A. Pfeffermann, C. Ryser, P. Grohn, B. Kuttler, H.J. Hahn, U. Zimmermann *Biomaterials* **1997**, *18(10)*, 707.
[6] P. De Vos, B. De Haan, J. Pater, R. Van Schilfgaarde, *Transplantation* **1996**, *62(7)*, 893.
[7] R. Robitaille, J.F. Pariseau, F.A. Leblond, M. Lamoureux, Y. Lepage, J.P. Halle, *J Biomed Mater Res* **1999**, *44(1)*, 116.
[8] M. Peirone, C.J. Ross, G. Hortelano, J.L. Brash, P.L. Chang, *J Biomed Mater Res* **1998**, *42(4)*, 587.
[9] B. Thu, P. Bruheim, T. Espevik, O. Smidsrod, P. Soon-Shiong, G. Skjak-Braek, *Biomaterials* **1996**, 17(10), 1031 and 17(11), 1069.
[10] A. Gaumann, M. Laudes, B. Jacob, R. Pommersheim, C. Laue, W. Vogt, J. Schrezenmeir, *Exp Toxicol Pathol* **2001**, *53(1)*, 35.
[11] N.G. Rainov, *Hum Gene Ther* **2000**, *11*, 2389.
[12] Z. Ram, K.W. Culver, E.M. Oshiro, J.J. Viola, H.L. DeVroom, E. Otto, Z. Long, Y. Chiang, G.J. McGarrity, L.M. Muul, D. Katz, R.M. Blaese, E.H. Oldfield, *Nat Med* **1997**, *3*, 1354.
[13] A. Prokop, D. Hunkeler, S. DiMari, M.A. Haralson, T.G. Wang, *Adv in Pol Science* **1998**, *136*, 1.
[14] G.H.U. Wolters, W.M. Fritschy, D. Gerrits, R.J. van Schilfgaarde, *J Appl Biomat* **1992**, *3*, 281.
[15] C. Schwinger, J. Kressler, S. Koch, N.G. Rainov, A. Klemenz, *Proc ACS, PMSE* **2001**, *84*, 894.
[16] Inotech *www.inotech.ch/products.htm,* **2001**.
[17] U. Prüsse, J. Dalluhn, J. Breford, K.D. Vorlop, *Chem Eng Technol* **2000**, *23*, 1105.
[18] F. Lim, A.M. Sun, *Science* **1980**, *210*, 908.
[19] R.A. Lemons, C.F. Quate, *Appl Phys Lett* **1974**, *24*, 163.
[20] S. Koch, C. Schwinger, J. Kressler, C. Heinzen, N.G. Rainov, *J Microencapsulation* **2003**, *20(3)*, 303. C. Schwinger, A. Klemenz, K. Raum, J. Kressler, *Landbauforschung Völkenrode* **2002**, *SH241*, 51. C. Schwinger, PhD Thesis *„Vergleich verschiedener Verkapselungsmethoden zur Immobilisierung von Zellen"*, Martin-Luther-Universität Halle, 2003 (to appear).
[21] C. Schwinger, S. Koch, U. Jahnz, P. Wittlich, N.G. Rainov, J. Kressler, *J Microencapsulation* **2002**, *19(3)*, 273.
[22] A. Klemenz, C. Schwinger, J. Brandt, S. Koch, J. Kressler, *Acoustical Imaging* **2002**, *26*, 223. A. Klemenz, C. Schwinger, J. Brandt, J. Kressler, *J of Biomedical Materials Research* **2003**, *65A(2)*, 237.

Structural Characterization of Temperature-Sensitive Hydrogels by Field Emission Scanning Electron Microscopy (FESEM)

Rudolf Reichelt,[1] *Thomas Schmidt,*[2] *Dirk Kuckling,*[3] *Karl-Friedrich Arndt,**[2]

[1] Institut für Medizinische Physik und Biophysik, Westfälische Wilhelms-Universität Münster, Robert-Koch-Str. 31, D-48149 Münster, Germany
[2] Institut für Physikalische Chemie und Elektrochemie, Technische Universität Dresden, Mommsenstr. 13, D-01062 Dresden, Germany
Fax: +49-351-46332013; E-mail: karl-friedrich.arndt@chemie.tu-dresden.de
[3] Institut für Makromolekulare Chemie und Textilchemie, Technische Universität Dresden, Mommsenstr. 13, D-01062 Dresden, Germany

Summary: The submicrometer structure of the temperature-sensitive hydrogels was observed by field emission scanning electron microscopy (FESEM), using synthesized hydrogels of different outer size and shape. The hydrogel structure strongly depends on the homogeneity of the polymer chains during the crosslinking process. A porous structure of the poly(vinyl-methyl-ether) (PVME) bulkgel, synthesized by electron beam irradiation of a concentrated polymer solution, was observed in the swollen state because the phase transitions temperature is acquired through the crosslinking process. Photo-crosslinking reaction of the poly(*N*-isopropylacrylamide) (PNIPAAm) copolymer in the dry state to form PNIPAAm thin films leads to a rather homogeneous structure. In the shrunk state both gels possess structure being more compact than in the swollen state. We also synthesized PVME and PNIPAAm gels with small outer dimensions in the range of some 100 *nm*. Heating of the thermo-sensitive polymer in diluted solutions collapses the polymer chains or aggregates. The crosslinking reaction (initiated by electron beam or UV irradiation) of these phase separated structures produces thermo-sensitive microgels. These microgel particles of PVME and PNIPAAm are spherical shape having diameters in the range of 30 – 500 *nm*.

Keywords: electron microscopy; field emission scanning electron microscopy; microgels; PNIPAAM; poly(*N*-isopropylacrylamide); poly(vinyl-methyl-ether); porous structure; PVME; temperature-sensitive hydrogels

1. Introduction

Hydrogels are three-dimensional networks of crosslinked hydrophilic polymers swollen in water. Polymeric hydrogels are soft materials with properties that resemble those of materials from the human body (high water content). These materials are highly interesting for various medical

 DOI: 10.1002/masy.200450656

applications (see e.g. [1-5]). Especially, sensitive hydrogels with swelling/deswelling characteristics possess a high potential of biomedical applications. Thermo-responsive gels show discontinuous volume phase transition behavior. They are in a highly swollen state at temperatures below a critical temperature (T_c) and in a shrunk state above T_c. The volume phase transition temperature is near the lower critical solution temperature (LCST) of the corresponding non-crosslinked polymers. The effect of temperature induced phase transitions of hydrogels has been intensively studied on substituted acrylamides, i.e. on poly(*N*-isopropylacrylamid) (PNIPAAm) gels (e.g. [6-8]) (LCST about 33 °*C*).

For many applications it is necessary that the swelling/deswelling kinetics of the gel is fast. These kinetics depend on the gel size. Decreasing of this dimension decreases the time, in which the polymer gel will respond to an external stimulus. One possibility of reducing the dimension consists in the preparation of thin hydrogel layers. Kuckling et al. [9] synthesized thin layers of PNIPAAm by photo-crosslinking of its copolymer with 2-(dimethyl maleinimido)-*N*-ethyl-acrylamide (DMIAAm) as chromophore. The swelling behavior of the layers was characterized by surface plasmon resonance. Another possibility in preparing small gels is the synthesis of microgel particles. Pelton [10] reported on temperature-sensitive PNIPAAm microgels (and other acrylamides) produced by emulsion polymerization of the monomer. These microgel particles are almost spherical and show fast swelling/deswelling kinetics. Vo et al. [11] synthesized PNIPAAm microgels by photo-crosslinking of the PNIPAAm-DMIAAm copolymer.

Another well-known sensitive polymer is poly(vinyl-methyl-ether) (PVME) with a LCST of about 34 °*C* [12]. A highly concentrated solution of PVME can easily be transformed into a hydrogel by high-energy radiation, e.g. electron beam or γ-ray irradiation. These so-called "clean" methods lead to an additive-free crosslinking process (no need of initiator, crosslinker etc.). Irradiation of aqueous polymer solutions fixes the structure of the polymer chains in solution. Suzuki and Hirasa [13] found a strong dependence of the gel structure on the polymer concentration and the temperature of solution during irradiation with γ-rays. With increasing temperature the polymer chains aggregate and the crosslinking reaction forms a heterogeneous gel with a sponge-like structure. The porous structure strongly affects the swelling kinetics of PVME gels [14]. High-energy irradiation of diluted PVME solutions above LCST leads to temperature-sensitive microgel particles [15]. These microgels can be used as a stabilizer of the dispersion polymerization of pyrrole [16]. At high pyrrole concentrations polypyrrole particles with a needle-like structure are formed.

Crosslinking by γ-rays needs several days (depending on the dose rate of the γ-ray source) to transfer the polymer solution into a swollen hydrogel. A very efficient method in radiation chemistry is the initiation of reactions by electron beam. Due to the high dose rate and the orders of magnitude stronger electron-matter interaction the reaction time is several minutes only to form a gel. During this short reaction time the dissipated thermal energy caused an increase of the solution temperature. We found at our experimental conditions an increase of the temperature of about 2.5 *K* per 10 *kGy* [17].

Scanning Electron Microscopy (SEM) is a very powerful tool to investigate the structure of swollen hydrogels (see e.g. [17-20]). In former studies [20] the structure of PNIPAAm hydrogels synthesized by a free radical polymerization and crosslinking was investigated. It was shown that at room temperature the formed hydrogels have a sponge-like structure. It is of particular interest to compare the results of photo-crosslinked PNIPAAm hydrogel layers with their results.

In this paper, we present the results of the structural study of temperature-sensitive hydrogels of different outer dimensions in the swollen as well as shrunk state. The investigations were performed with high-resolution field emission scanning electron microscopy (FESEM) using cryo-prepared samples of the hydrogels.

2. Materials and Methods

2.1 Synthesis of hydrogels

PVME bulkgels were prepared by irradiation of a concentrated aqueous solution with electron beam (for further details see [17]). Oxygen-free PVME aqueous solution (20 *wt.%*) was dropped onto a small piece of thoroughly cleaned aluminum and covered with mica platelets. The Al-support was irradiated with a 0.5 *MeV* electron beam (radiation dose of 50 *kGy*) with a linear accelerator ELV-2 (Budker, Novosibirsk). After irradiation the gel-coated aluminum was immersed in distilled water for 15 *min* to remove non-crosslinked PVME.

Thin layers of PNIPAAm gel were prepared by photo-crosslinking of the copolymer of *N*-isopropylacrylamide and 2-(dimethyl-maleinimido)-*N*-ethyl-acrylamide (synthesis of the copolymer, see [21]). The solution of the copolymer was spin coated on a silicon wafer dried first at air-atmosphere and then in vacuum. The films were crosslinked by irradiation with UV light (360 – 430 *nm*). Further details of preparing were described earlier [9].

2.2 Synthesis of microgels

In diluted solutions (c_p < 0.5 *wt.%*) PVME molecules are not monomolecularly dissolved. By light scattering measurement we found aggregates with a radius of gyration of 100 – 200 *nm*. Above the phase transition temperature these aggregates collapse to spherical particles with a radius of 90 – 100 *nm* [15]. However, these particles do not precipitate because of the low density-differences between the polymer and the solvent. This phase separated structure can be fixed by irradiation with an electron beam at 50 *°C*. The PVME microgels were used without any further purification.

Microgel particles were prepared by photo-crosslinking of the PNIPAAm copolymer (see preparation of PNIPAAm films) in diluted aqueous solution [11]. To produce PNIPAAm microgels it was necessary to heat the solution above the transition temperature (45 *°C*). The precipitation of the copolymer by heating could only be prevented by using sodium dodecyl sulfate – SDS – as stabilizer. The crosslinking reaction furthermore was initiated by UV irradiation.

2.3 Field emission scanning electron microscopy

Synthesized thermo-sensitive hydrogels in different states were investigated with an "in-lens" field emission scanning electron microscope S-5000 (Hitachi Ltd., Japan) at low acceleration voltage. The secondary electron (SE) micrographs were taken from samples in swollen states at room temperature and in the shrunk state significantly above T_c (the used temperatures depend on the polymer). The structure formed during swelling of PVME and PNIPAAm bulkgels was fixed by rapid cooling with liquid ethane (cooled to 77 *K*). Under the experimental conditions used, the ice was in an amorphous state thus the network structure of the gels was not affected. The frozen water was removed by freeze-drying at 190 *K* for 6 *h* at about $5 \cdot 10^{-6}$ *Torr*. After drying the samples were rotary shadowed with about 2 *nm* platinum/carbon (Pt/C) at an elevation angle of 65°. The evaporation was performed in the high vacuum chamber of a freeze-etch devise (BAF 300 with turbo molecular pump, Balzers/Liechtenstein) at room temperature. The film thickness was measured with a quartz crystal film thickness monitor (Balzers QSG 201D, Balzers/Liechtenstein).

For preparation of the crosslinked microgels (PNIPAAm and PVME) in different swelling states a small droplet of a solution at 25 *°C* and 40 *°C* was placed onto a small fragment of silicon (Si)-wafer with the corresponding temperature. The microgels were allowed for approximately 90 *sec* to adsorbed physically onto the Si-wafer, before the excess solution was removed with filter paper [15]. The following rapid cooling and freeze-drying was as described above.

3. Results and Discussion

3.1 Poly(vinyl-methyl-ether) hydrogel

SE micrographs recorded at medium and high magnifications revealed that the swollen state of the PVME hydrogel at 25 °C (Fig. 1a, b) is characterized by three-dimensional sponge-like structure. The network consists of many small cavities in the range of several 10 – 100 nm. They are separated permeably from each other by thin membrane-like layers and represent the nano-reservoirs of the water. The thickness of these membranes is larger than of single network chains. It seems that some polymeric chains are aggregated forming this structure. The fine structure of the aggregated chains can be observed at high magnifications. The membrane-like layers are full of tiny holes with a typical hole-size in the order of 10 nm. This holey structure can be assigned to the structure of phase separated concentrated solution because the phase transition temperature was exceeded during the electron beam irradiation process. Above the transition temperature the polymer chains typically start to aggregate and to form polymeric bundles. These structures then were fixed by the high-energy electrons.

After heating the crosslinked PVME gel to 40 *°C*, the gel collapsed and shrunk. In this state the formed structure must be rather compact. The micrographs (Fig. 1c, d) show that the gel is still porous. The typical diameters of these cavities were in the order of up to a few 100 *nm* indicating a significant shrinkage of the mean pore size in comparison to the swollen state. Like in the swollen state these very small cavities represent the nano-reservoirs for the water. The remaining water in the gel after the collapse is the reason for the still high water content even at temperatures above LCST (degree of swelling of about 2 - 3 *g* water/*g* polymer). The membrane-like layers, which separate the cavities, appeared compact, i.e. no tiny holes in the layers were observed.

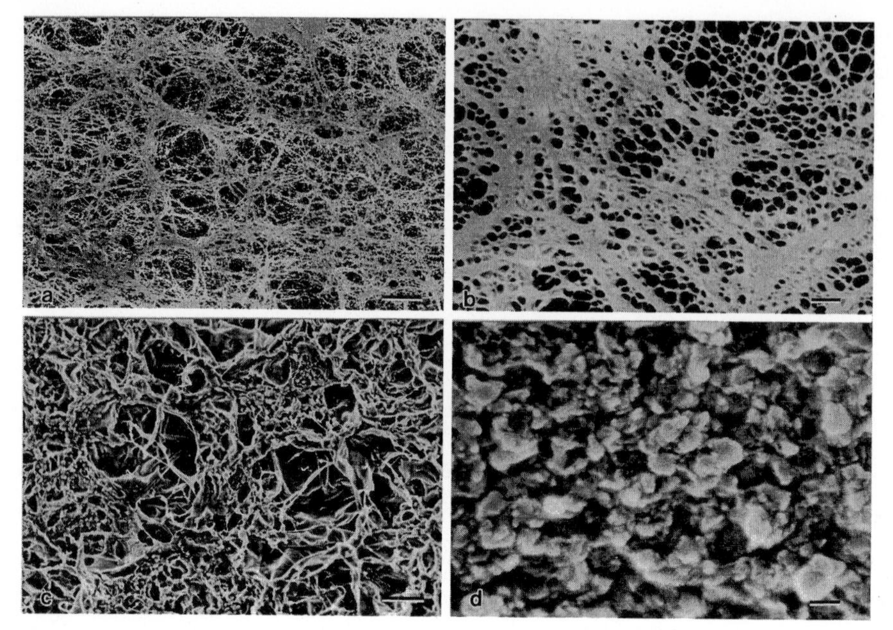

Figure 1. SE micrographs of freeze-dried PVME hydrogel synthesized by irradiation with an electron beam (a, b) in the swollen state (25 °C) and (c, d) in the shrunk state (40 °C) at different magnifications. The porous structure (due to the crosslinking of the phase separated polymer solution) of the swollen hydrogel collapsed above LCST to a more compact gel. The scale bars correspond to 1 μm (a, c) and 100 nm (b, d), respectively.

3.2 Poly(N-isopropylacrylamide) hydrogel

Fig. 2a, b display the structure of the PNIPAAm hydrogel in the swollen state at 25 *°C*. The low-magnification micrograph (Fig. 2a) displays the marginal zones of adjacent hydrogel dots synthesized by photo-patterning of the thin PNIPAAm film. The hydrogel was firmly attached to the support by an adhesion promoter [22]. A high-magnification micrograph of the PNIPAAm hydrogel (Fig. 2b) recorded in the middle region of the sample showed a gel structure with pores sizes between approximately 50 *nm* and 100 *nm*. The spatial network consisted of tiny polymeric ropes and sheet-like strands which typically have a thickness smaller than 10 *nm*.

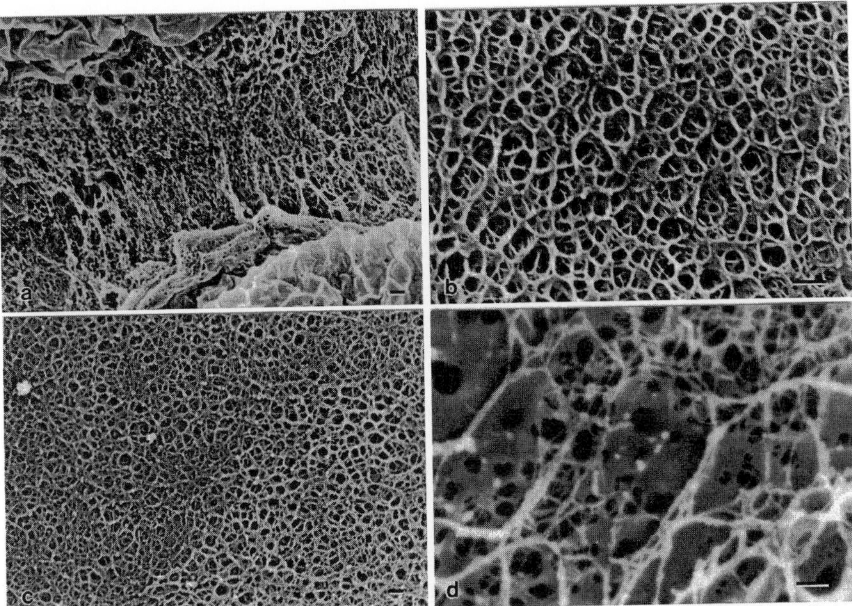

Figure 2. SE micrographs of freeze-dried PNIPAAm hydrogel synthesized by photo-crosslinking (a, b) in the swollen state (25 *°C*) and (c, d) in the shrunk state (35 *°C*) at different magnifications. The swollen PNIPAAm gel shows a homogeneously structure and generate polymer bundles due to the phase separation. The scale bars correspond to 100 *nm*.

In the shrunk state at 35 *°C* (Fig. 2c) the hydrogel surface appeared rather flat, i.e. grooves like in the swollen state were not found. A fine porous structure is visible. At high magnification a spatial network (Fig. 2d) of polymeric bundles with pores having diameters in the range of typically 10 *nm* could be observed.

3.3 Poly(vinyl-methyl-ether) microgel

Fig. 3 displays the structure of thermo-sensitive microgel particles synthesized by irradiation with a dose of 80 *kGy* at temperatures below (Fig. 3a) and above (Fig. 3b) the LCST.

In the swollen state at 25 *°C* the microgel particles are almost spherical with a creased surface. The particles seemed to have also a sponge-like internal structure. Small holes in the range of 10 *nm* in the polymeric material were observed sporadically at the particles surface. The outer diameter of these particles typically was in the range of 250 – 450 *nm*. Beside the particles we found at some locations of the sample an irregular polymeric net that can be assigned to the non-crosslinked polymers in the solution.

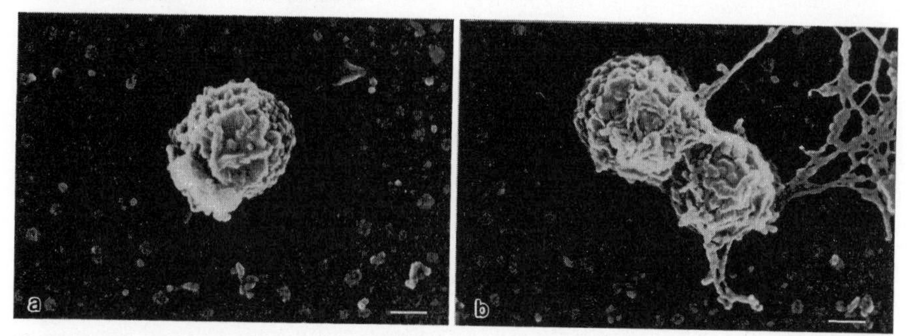

Figure 3. SE micrographs of freeze-dried PVME microgels synthesized by crosslinking the phase separated structure of diluted polymer solutions with electron beam (a) at 25 *°C* and (b) at 40 *°C*. Isolated particles in the swollen state have typically a diameter of approximately 300 *nm* and are porous. In the shrunk state the particles are compact. The scale bars correspond to 100 *nm*.

Above the phase transition temperature (40 *°C*) the microgels collapse to more compact particles having still a creased surface like in the swollen state. The shrinkage of particles above T_c is difficult to demonstrate with only a few imaged individual particles in single micrographs because of (i) one and the same particle cannot be imaged and compared in both states, and (ii) different particles usually differ significantly in diameter caused by the wide diameter-range in the swollen state.

3.4 Poly(N-isopropylacrylamide) microgel

Fig. 4 shows typical FESEM micrographs of PNIPAAm microgels in the swollen (Fig. 3a) and in the shrunk state (Fig. 3b). The outer dimension of the almost spherically formed PNIPAAm microgels is smaller than the one of the PVME microgels. In Fig. 4a "aggregates" consisting of several microgels having different diameters were found. The particles were also synthesized above the critical temperature. However, the surface of these microgels in the swollen state was rather smooth. There are two possible reasons for this effect: Firstly, for the stabilization of the suspensions SDS was used. The surfactant may be still at the surface of the particles smoothing their possibly creased surface. Secondly, a rather smooth surface could be caused by a higher crosslinking density. In the shrunk state (Fig. 4b) the microgel particles are collapsed and possess a creased surface which is similar to the one of the collapsed PVME microgels.

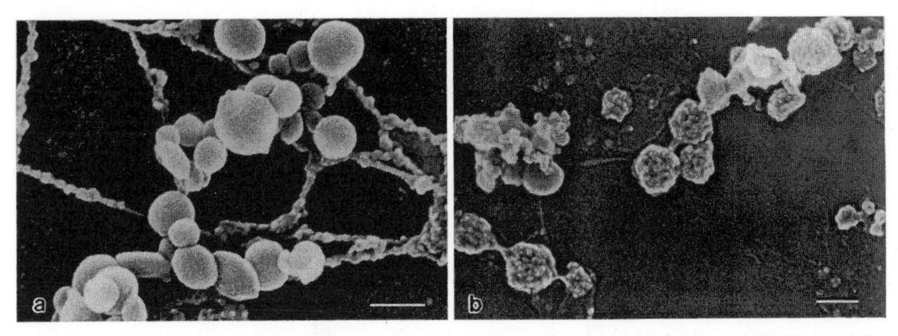

Figure 4. SE micrographs of PNIPAAm microgels synthesized by photo-crosslinking of the diluted polymer solution above LCST (a) in the swollen state (25 *°C*) and (b) in the shrunk state (40 *°C*). The swollen microgels have a smooth surface. In the shrunk state a compact rough surface is visible. The scale bars correspond to 100 *nm*.

4. Conclusions

The field emission scanning electron microscopy at low acceleration voltage proved to be a very valuable tool to observe the spatial structure of freeze-dried temperature-sensitive hydrogels in the swollen and in the shrunk state down to the dimensions of several nanometers. The cryo-preparation and subsequent Pt/C coating of these fragile structures with high water content (similar to biological specimens like soft issues) ensured the preservation of their structure also under high vacuum conditions required for electron microscopy. It is obvious that the observed structures would hardly be accessible by other high-resolution imaging methods like scanning force microscopy or conventional transmission electron microscopy thus FESEM represents an unique imaging method for hydrogels. The structure of thermo-sensitive hydrogels strongly depends on the synthesis conditions (Fig. 5). The network formation at lower temperatures or in the air-dried state leads to homogeneous structures (Fig. 5a). Crosslinking at temperatures above the phase transition leads to porous structures (Fig. 5b). In the case of diluted aqueous solution of thermo-sensitive polymers we use the phase separated structure (globular particles) to synthesize temperature-sensitive microgels. These particles are cross-linkable by UV irradiation (in case photo-sensitive PNIPAAm copolymers) or by irradiation with high-energy electrons (in case PVME). The structure of these microgel both in the swollen and in the shrunk state particles was almost globular with diameters ranging typically from about 100 *nm* to 500 *nm*.

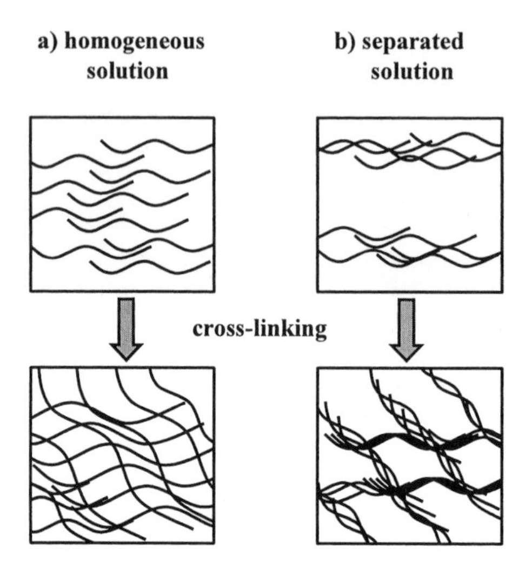

Figure 5. Schemes of crosslinking reaction of (a) homogeneous solution and (b) phase separated polymer solutions.

Acknowledgement

The authors are very grateful to Mrs. U. Keller (Institut für Medizinische Physik und Biophysik) for the expert sample preparation and the careful FESEM studies and to Mrs. G. Kiefermann (Institut für Medizinische Physik und Biophysik) for excellent photographic work. We are greatly indebted to Mr. H. Dorschner and Mr. G. Neubert (Institut für Polymerforschung Dresden e.V.) for the electron beam irradiation and Mr. C.D. Vo (Institut für Makromolekulare Chemie und Textilchemie) for the preparation of the PNIPAAm microgels. The financial support of this work by the Deutsche Forschungsgemeinschaft (DFG, Sonderforschungsbereich 287 "Reaktive Polymere") is gratefully acknowledged.

[1] Peppas, N.A. Hydrogels in Medicine and Pharmacy. CRC Press 1987, Boca Raton, Florida.
[2] Rosiak, J.M.; Ulanski, P.; Pajewski, L.A.; Yoshii, F.; Makuuchi, K. *Rad. Phys. Chem.* **1995**, *46*, 161-168.
[3] Hoffman, A.S. *Macromol. Symp.* **1995**, *98*, 645-664.
[4] Rosiak, J.M.; Yoshii, F. *Nucl. Instrum. Methods Phys. Res. B* **1999**, *151*, 56-64.
[5] Lee, K.Y.; Mooney, D.L. *Chem. Rev.* **2001**, *101*, 1869-1879.
[6] Hirowaka, Y.; Tanaka, T. *J. Chem. Phys.* **1984**, *81*, 6379-6380.
[7] Kishi, R.; Hirasa, O.; Ichijo, H. *Polym. Gels Networks* **1997**, *5*, 145-151.
[8] Nagaoka, N.; Safranj, A.; Yoshida, M.; Omichi, H.; Kubota, H.; Katakai, R. *Macromolecules* **1993**, *26*, 7386-7388.
[9] Kuckling, D.; Harmon, M.E.; Frank, C.W. *Macromolecules* **2002**, *35*, 6377-6383.
[10] Pelton, R. *Adv. Colloid Interface Sci.* **2000**, *85*, 1-33.
[11] Vo, C.D.; Kuckling, D.; Adler, H.-J.P.; Schönhoff, M. *Colloid. Polym. Sci.* **2001**, *280*, 400-409.
[12] Schäfer-Soenen, M.; Moerkerke, R.; Koningsveld, R.; Berghmans, H.; Dušek, K.; Šolc, K. *Macromolecules* **1997**, *30*, 410-416.
[13] Suzuki, M.; Hirasa, O. *Adv. Polym. Sci.* **1993**, *110*, 241-261.
[14] Kabra, B.K.; Akhetar, M.K.; Gehrke, S.H. *Polymer* **1992**, *33*, 990-995.
[15] Arndt, K.-F.; Schmidt, T.; Reichelt, R. *Polymer* **2001**, *42*, 6785-6791.
[16] Pich, A.; Lu, Y.; Adler, H.-J.; Schmidt, T.; Arndt, K.-F. *Polymer* **2002**, *43*, 5723-5729.
[17] Arndt, K.-F.; Schmidt, T.; Menge, H. *Macromol. Symp.* **2001**, *164*, 313-322.
[18] Kim, S.-H.; Chu, C.-C. *J. Biomed. Mat. Res.* **2000**, *49*, 517-527.
[19] Gabrelii, I.; Gatenholm, P. *J. Appl. Polym. Sci.* **1998**, *69*, 1661-1667.
[20] Matzelle, T.R.; Ivanov, D.A, Landwehr D.; Heinrich, L.A.; Herkt-Bruns, C.; Reichelt, R.; Kruse, N. *J. Phys. Chem. B* **2002**, *106*, 2861-2866.
[21] Kuckling, D.; Adler, H.-J., Ling, L.; Habicher, W.D., Arndt, K.-F. *Polym. Bull.* **2000**, *44*, 269-276.
[22] Kuckling, D.; Adler, H.J.; Arndt, K.-F.; Hoffmann, J.; Plötner, M.; Ling, L.; Wolff, T. *Polym. Adv. Technol.* **1999**, *19*, 345-352.

Electrical Properties of σ-Conjugated Polymers with Additives and Their Applications in Sensors

Stanislav Nešpůrek,[*1,2] Geng Wang,*[1] *Stanislav Böhm,*[3] *Marek Kořínek,*[3] *Hans-J. Adler*[4]

[1] Institute of Macromolecular Chemistry, Academy of Sciences of the Czech Republic, Heyrovský Sq. 2, 162 06 Prague 6, Czech Republic
E-mail: nespurek@imc.cas.cz
[2] Chemical Faculty of Technical University, Purkyňova 118, 612 00 Brno, Czech Republic
[3] Department of Organic Chemistry, Institute of Chemical Technology, Technická 5, 166 28 Prague 6, Czech Republic
[4] Institute of Macromolecular Chemistry and Textile Chemistry, Dresden University of Technology, Mommsenstr. 13, 01062 Dresden, Germany

Summary: The charge carrier transport in poly[methyl(phenyl)silylene] (PMPSi) proceeds predominantly along the σ-delocalized Si backbone with participation of interchain hopping and polaron formation. The charge carrier mobility increases with increasing electron affinity of acceptor dopands having zero dipole moments. On the other band, the hole drift mobility is influenced by the dipole moment of the dopand. The electrostatic charge-dipole interactions cause a broadening of the energy distribution of transport states, which results in a decrease in the charge carrier mobility. An addition of organic salts leads, under the conditions of increased humidity, to an increase in electrical conductivity and capacitance. This is demonstrated on the layers PMPSi/1,5-dimorpholino-1,5-diphenylpentamethinium perchlorate.

Keywords: polysilylenes, charge transport, doping, humidity sensors

Introduction

Polymers offer lots of advantages for sensor technologies: they are relatively low-cost materials, their fabrication techniques are quite simple and they can be deposited on various types of substrates. Sensors with electrical responses are usually fabricated in the form of vacuum-evaporated, cast or spin-coated thin films, Langmuir-Blodgett films, printed films of inks consisting of polymer matrix and particles of sensing additives, or spin-coated films prepared from a conjugated polymer, in which the sensing material is dispersed on molecular level. In the

 DOI: 10.1002/masy.200450657

last case usually π-conjugated polymers are used as charge-transporting media. In this contribution, we will show that also σ-conjugated polymers (e.g. polysilylenes) are suitable for the preparation of polymeric sensitive inks.

Experimental

Materials

Synthesis of poly[methyl(phenyl)silylene] (PMPSi; see Scheme 1): Oxide-free sodium was dispersed in boiling toluene under argon and a small amount of diglyme was added. Freshly distilled dichloro(methyl)phenylsilane in toluene solution was dropped into the dispersion. After the 2 h reaction at 140 °C, the reaction was quenched at room temperature by addition of a butyllithium solution. The residual sodium was reacted, under ice cooling, first with ethanol and then with H_2O. Crosslinked and insoluble portions were separated by centrifugation. The polymer was then precipitated by addition of isopropyl alcohol. The low-molecular-weight portion was separated by 1 h extraction with boiling ether. The polymer was then dried at 13 mbar for 48 h. The polymer, obtained in *ca.* 17% yield, possessed a unimodal but broad molar mass distribution ($M_w = 4 \times 10^4$ g mol^{-1}, $M_w / M_n = 2.7$; SEC averages, polystyrene standards).

Synthesis of 1,5-dimorpholino-1,5-diphenylpentamethinium perchlorate (MDPPMClO; see Scheme 1): Commercially available (Aldrich) 4-(trimethylsilyl)morpholine was after purification added to a solution of 2,6-diphenylpyrylium perchlorate in dry acetonitrile under nitrogen atmosphere. The reaction mixture was stirred for 4 h at room temperature and complete consumption of the starting material was checked by TLC. After evaporation to dryness in vacuum the solid residue was crystallized from ethanol.

Scheme 1. Chemical structures of materials used for humidity sensor fabrication.

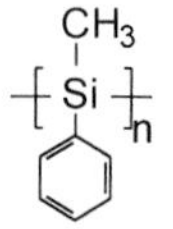

CH_3 Si n ClO_4^- O N O N +

poly[methyl(phenyl)silylene] (PMPSi)

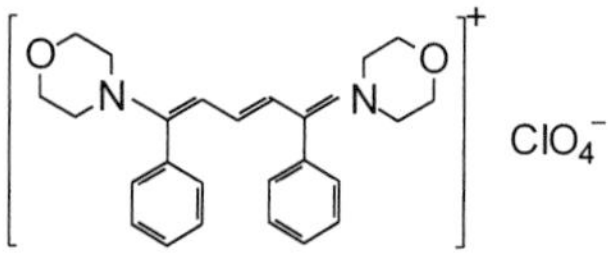

1,5-dimorpholino-1,5-diphenylpentamethinium perchlorate (MDPPMClO)

Thin film formation

Before deposition of films, the polymer was reprecipitated three times from a toluene solution with methanol and centrifuged at 12000 rpm for 15 min. After deposition (spin coating, 3000 rpm) on glass substrates from a toluene solution, the films were dried in vacuum (10^{-3} Pa) at 330 K for at least 4 h. Then, for the sensor element (the polymer system was deposited on a ceramic substrate), the system of interdigital gold electrodes (distance 1 mm) was vacuum evaporated on the top of the film. Samples for the time-of-flight (TOF) measurements were prepared in the form of sandwich structures ITO/sample/Al; Al electrode 60 nm thick was prepared by vacuum evaporation. The doped samples were prepared in the same way by spin coating of the corresponding mixed toluene solutions.

Measurements

DC and AC conductivities were measured using a Keithley 6517A electrometer and Hioki LCR Hitester 3532-50, respectively. The charge carrier mobility was measured by the TOF method in an electrical circuit consisting of a voltage source, sample and osciloscope (HP 54510A, 50 Ω input impedance) connected in series. The samples were illuminated by the 347 nm laser pulses (duration 20 ns) generated by a ruby laser (Korad model K1QS2) in conjunction with a frequency doubler through the transparent ITO electrode. The samples were kept in a vacuum cryostat (10^{-4} Pa) or under argon during the measurement.

Results and discussion

Charge carrier transport

Polysilylene consists of a chain of silicon atoms with three interacting sp^3 hybrid orbitals. Electron delocalization along the silicon backbone results from the interaction of sp^3 orbitals of adjacent silicon atoms[1]. The resonance integral between two sp^3 orbitals located on adjacent silicon atoms and pointing to each other, β_{vic}, is responsible for the formation of a Si-Si σ-bond. The degree of electron delocalization in the backbone is a function of the $\beta_{vic} / \beta_{gem}$ ratio, where β_{gem} is the resonance integral between two sp^3 orbitals localized on the same silicon atom. Electron delocalization is perfect if the ratio equals unity. Thus, the linear Si backbone behaves as a molecular wire due to the σ-conjugation. The mobilities of charges (holes) depend on the chemical nature and size of the side groups[2]. The "on-chain" hole mobility in PMPSi was found to amount to μ ~ 2×10^{-6} $m^2V^{-1}s^{-1}$ using time-resolved microwave photoconductivity[3,4].

A question arises why the value of "on-chain" mobility is so low. A possible reason is the formation of polarons. The strong electron-phonon coupling causes carrier self-trapping and creates a quasiparticle, a polaron, which can move only by carrying along the associated molecular deformation. The motion of such a charge carrier, dressed into a cloud of local deformation of the nuclear subsystem, can be phenomenologically described by introducing a temperature-dependent effective mass which is higher than the electron mass. A significant distortion of the PMPSi chain was recently found by Kim at al.[5] and by quantum-chemical calculations[6].

The shapes of photocurrent transients obtained by the TOF method were the following: after an initial drop, which is RC-limited, a plateau was reached followed by a tail. Transit time, t_0, was determined from the intersection of the asymptotes to the plateau and tail of the transient signal. Mobilities were determined from the conventional expression, $\mu = L^2/t_0U$, where L is the sample thickness and U is the applied voltage.

Figure 1 illustrates the dependences of the charge carrier mobility, μ, of a PMPSi sandwich sample (thickness ca. 2 μm) on $F^{1/2}$, the square root of the electric field strength, at different temperatures. In all cases, the mobility can be described by an $\exp(\beta F^{1/2})$ dependence for $F > 10^7$ Vm^{-1}. At lower field strengths, $\mu(F)$ becomes constant or increases slightly upon reducing F. These types of dependences are usually treated in the framework of the hopping disorder concept. The essential difference between the polaron and disorder models is that the latter, at variance with the former, implies a sufficiently weak electron-phonon coupling and the activation energy of charge transport reflects the static energy disorder of the hopping sites. In contrast, the polaron model suggests a strong electron-phonon coupling and a negligible contribution of energy disorder to the activation energy of the carrier mobility[7,8]. It was suggested[8-10] that the zero-field activation energy of the mobility, $E_a(F \rightarrow 0)$, obtained from the temperature dependence of the extrapolated charge mobility values to zero field $\mu(F \rightarrow 0)$, can be approximated by the sum of the disorder and polaron contribution as[9,11]

$$E_a(F \rightarrow 0) = E_a^{pol} + E_a^{dis} = \frac{E_p}{2} + \frac{4}{9}\frac{\sigma^2}{(kT)^2} \qquad (1)$$

where E_a^{pol} and E_a^{dis} are the polaronic and disorder contributions, σ is the energy width of the density-of-states (DOS) distribution, k is the Boltzmann constant, and T is temperature. Because it is experimentally difficult to determine energy E_p, the activation energy E_a ($F \rightarrow 0$) is often related to the effective energy width of the DOS distribution σ^*. Usually it is also difficult to distinguish experimental data between $\mu(1/T^2)$ and $\mu(1/T)$ dependences. Then, the right side of Eq. (1) is treated as an apparent (effective) activation energy which can be expressed as $E_a^{eff}(F \rightarrow 0) = (8/9)\,\sigma^{*2}/kT$.

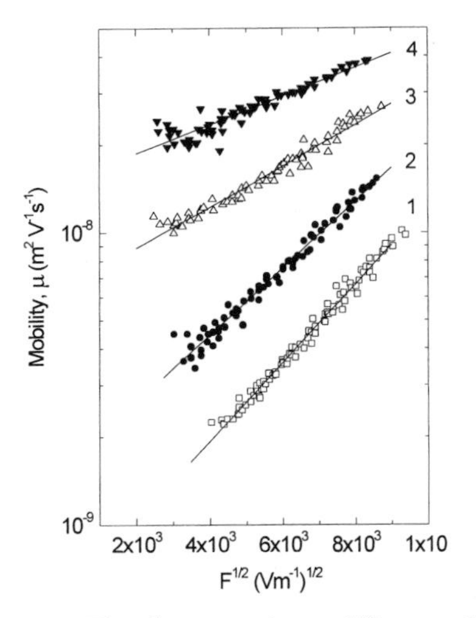

Figure 1. The charge carrier mobility vs. the square root of the electric field strength (F) for PMPSi, parametric in temperature; T = 295 K (curve 1), 325 K (2), 355 K (3), 358 K (4).

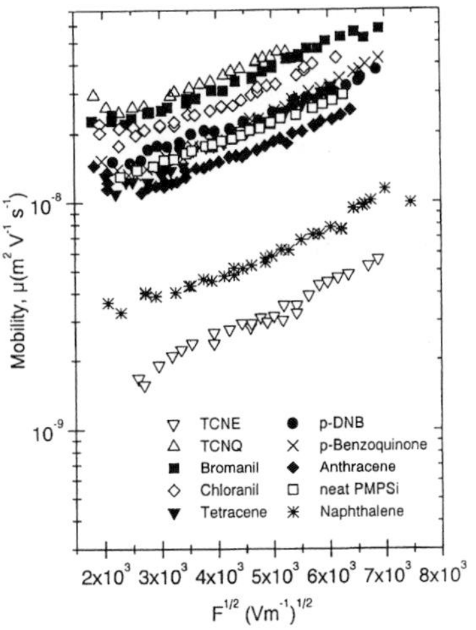

Figure 2. The charge carrier mobility vs. the square root of the electric field strength (F) for neat PMPSi (squares) and PMPSi doped with acceptors (3 mol %).

The polaron binding energy was determined for PMPSi[10] as E_p = 0.16 eV. Recently, Pan et al.[12] reported a smaller value of E_p = 0.08 eV. The E_p values are still an open question and need more measurements and discussions. The zero-field value of the activation energy of the mobility was measured, $E_a^{eff}(F \rightarrow 0)$ = 0.29 eV. The effective value of the half-width of the Gaussian distribution of the hopping states σ^* was determined from the above equation to be 0.093 eV (at room temperature). From thermostimulated luminescence studies[13], the value σ^* = 0.096 eV was obtained. The real value of the width of the distribution of hopping sites σ would be lowered by the polaron contribution ($E_a^{eff} - E_p/2 = 8\,\sigma^2/9\,kT$); this yields σ = 0.078 eV.

It is interesting to try to change the transport parameters by doping[11]. The dopands can be divided into three groups:

(i) Materials of acceptor type with high electron affinity and zero dipole moment, like 7,7,8,8-tetracyano-1,4-quinodimethane (TCNQ), 2,3,5,6-tetrabromo-1,4-benzoquinone (bromanil), 2,3,5,6-tetrachloro-1,4-benzoquinone (chloranil), tetracene, *p*-dinitrobenzene (*p*-DNB), *p*-benzoquinone and anthracene increase the charge carrier mobility and decrease the σ^* parameter. The mobility increases with increasing electron affinity of the acceptor (TCNE is an exception – the behaviour is not fully understood). At the moment it is not clear what is responsible for this effect, a decrease in the polaron binding energy or disorder parameter.

(ii) Dopands with high dipole moments, like *m*- and *o*-dinitrobenzene (*m*- and *o*-DNB), decrease the charge carrier mobility, while the effective σ^* parameter increases (see Table 1). The detail analysis shows that in this case charge-dipole interactions are important. Two effects must be mentioned in this context: (1) A broadening of the energy distribution of hopping states[11,14]; the mobility depends on the dipole moment (decreases with increasing dipole moment), molecular dimensions and concentration of the additive. (2) New local states are formed in the vicinity of dipolar species[15] even though these molecules do not necessarily have to act as trapping sites. However, contrary to the situation encountered in the case of both chemical traps and the "conventional" structural traps, the electron and hole dipolar traps are formed on different molecules: a molecule close to the negative pole of the dipole should be a hole trap, whereas a molecule close to the positive pole should trap electrons. The results of doping experiments are summarized in Table 1 and presented in Fig. 2 (*m*-DNB and *o*-DNB are not included because of the distinctness of the figure).

(iii) Naphthane does not act as an acceptor. Its presence in PMPSi increases the disorder effect.

Humidity sensor

Thin films for humidity sensors were deposited by spin coating on glass substrates. Both PMPSi and MDPPMClO were dissolved in toluene and mixed. After deposition, the sensor was kept in vacuum for 4 h at 330 K.

The humidity sensors are usually fabricated using hydrophilic polymers with some ionic groups[16], or polymeric matrices containing inorganic salts[17] like $FeCl_3$. Here, our sensing material, MDPPMClO, was prepared as an organic salt. Under the influence of humidity one can expect dissociation of the molecule and formation of two types of species: ion-pairs and free

Table 1. Values of hole mobilities, μ, their effective activation energies at zero electric field, E_a^{eff} $(F \to 0)$ and effective values of the half-widths of DOS, σ*. PMPSi films doped with 3 mol % additive.

Dopant	$10^8\mu$ #	E_a^{eff} $(F \to 0)$	A†	σ*	m‡
	(m^2/Vs)	(eV)	(eV)	(eV)	(D)
TCNQ[1]	5.71	0.241	1.7 ~ 1.8	0.084	
Bromanil	4.80	0.242	1.4	0.084	
Chloranil	4.12	0.247	1.3 ~ 1.4	0.085	
Tetracene	3.06	0.275	0.9 ~ 1.0	0.089	
p-DNB[2]	3.10	0.283	0.7	0.091	0.0
p-Benzoquinone[3]	2.74	0.286	0.7 ~ 1.8	0.091	
Anthracene	2.57	0.285	0.5	0.091	
None (neat PMPSi)	2.28	0.298	–	0.093	
m-DNB[4]	1.20	0.340	0.3	0.099	3.8
Naphthalene	0.77	0.375	0 ~ 0.15	0.104	
o-DNB[5]	0.50	0.460	0	0.116	6.0
TCNE[6]	0.43	0.463	2.2 ~ 2.9	0.116	

[1]7,7,8,8-tetracyano-1,4-quinodimethane, [2]1,4-dinitrobenzene, [3]1,4-benzoquinone, [4]1,3-dinitrobenzene, [5]1,2-dinitrobenzene [6]tetracyanoethylene, #measured at T = 295 K and F = 36 MV/m, †electron affinity and ‡dipole moment of the dopands.

charge carriers after full dissociation in external electric field. The formation of dipolar species results in the broadening of the transport hopping state distribution, as it follows from Table 1. Thus, the charge carrier mobility decreases and electric permittivity increases. On the other hand, full dissociation results in the formation of free charge carriers, which can move in the transport polymer matrix. Because electrical conductivity is the product of the unit charge, charge carrier mobility, μ, and free charge carrier concentration, n, the final change of the conductivity under the humidity exposure depends on the equilibrium of the changes $\Delta\mu$ and Δn. In our case the contribution to the conductivity of the increase in free charge carrier concentration is higher than the decrease in charge mobility; thus, both AC and DC conductivity increase. The changes of the steady-state current and capacitance for relative humidity changes from 45 to 96 % (fast change during about 2 s realized by the switching between dry and wet flowing air) are given in Fig. 3 The change of the current was about two orders of magnitude (thickness of the film was ca. 500 nm); for thinner layers even larger changes could be detected. Similar changes were also observed in AC conductivity. Figure 3(b) shows the capacitance changes. It could be pointed out that both current and capacitance changes were stable and fully reproducible.

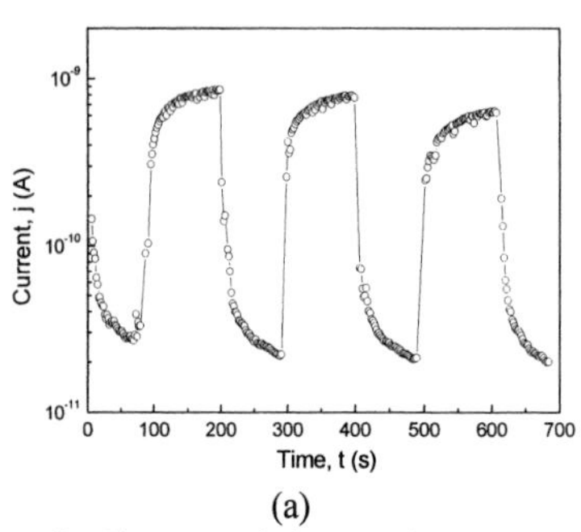

(a)

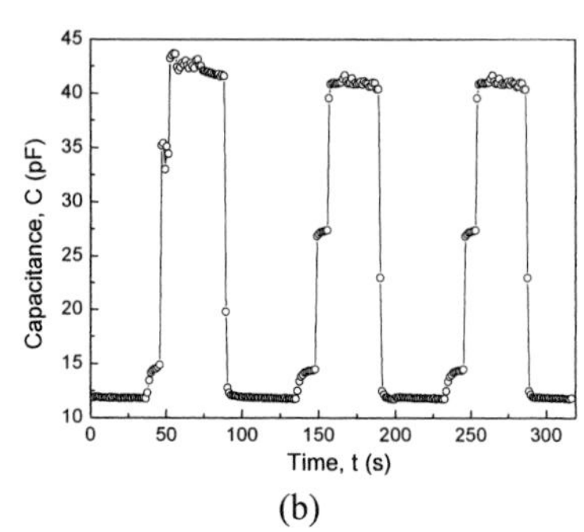

(b)

Figure 3. Changes of the steady-state current (a) and capacitance (b) of the system PMPSi : MDPPMClO (4 : 1 weight) for relative humidity changes from 45 to 96 %.

Acknowledgements

The research was supported by the Ministry of Education, Youth and Sports of the Czech Republic (grant No. OC D14.30 and No. ME 440). The financial support from European Graduate College "Advanced Polymer Materials" is gratefully appreciated.

[1] R. D. Miller, J. Michl, *Chem. Rev.* **1989**, *89*, 1359.
[2] F. C. Grozena, L. D. A. Siebbeles, J. M. Warman, S. Seki, S. Tagawa, and U. Schrf, *Adv. Mater.* **2002**, *14*, 228.
[3] S. Nešpůrek, V. Herden, M. Kunst, and W. Schnabel, *Synth. Met.* **2000**, *109*, 309.
[4] S. Nešpůrek, P. Toman, and J. Sworakowski, *Thin Solid Films* **2003**, *438-439*, 268.
[5] Y. R. Kim, M. Lee, J. R. G. Thorne, R. M. Hochstrasser, and J. M. Zeigler, *Chem. Phys. Lett.* **1988**, *145*, 75.
[6] P. Toman, S. Nešpůrek, J. W. Jang, and C. E. Lee, *Current Appl. Phys.* **2002**, *2*, 327.
[7] V. I. Arkhipov, E. V. Emelianova, A. Kadashchuk, I. Blonsky, S. Nešpůrek, D. S. Weiss, and H. Bässler, *Phys. Rev. B* **2002**, *65*, 165218.
[8] I. I. Fishchuk, A. Kadashchuk, H. Bässler, and S. Nešpůrek, *Phys. Stat. Solidi*, in press.
[9] S. Nešpůrek, A. Eckhardt, *Polym. Adv. Technol.* **2001**, *12*, 427.
[10] H. Bässler, P. M. Borsenberger, R. J. Perry, *J. Polym. Sci. B* **1994**, *32*, 1677.
[11] S. Nešpůrek, H. Valerián, A. Eckhardt, V. Herden, and W. Schnabel, *Polym. Adv. Technol.* **2001**, *12*, 306.
[12] I. Pan, M. Zhang, and Y. Nakayama, *J. Chem. Phys.* **1999**, *110*, 10509.
[13] A. Kadashchuk, N. Ostapenko, V. Zaika, and S. Nešpůrek, *Chem. Phys.* **1998**, *234*, 285.
[14] H. Valerián, E. Brynda, S. Nešpůrek, and W. Schnabel, *J. Appl. Phys.* **1995**, *78*, 6071.
[15] J. Sworakowski and S. Nešpůrek, in: A. Graja, B. R. Bulka, F. Kajzar (Eds.), *Molecular Low Dimensional and Nanostructured Materials for Advanced Applications,* Kluwer Academic Publishers, Dordrecht, **2002**.
[16] M. Šorm and S. Nešpůrek, *European Polym. J.* **1978**, *14*, 977.
[17] Y. Li and M. J. Yang, *Sensors and Actuators B* **2002**, *85*, 73.